Mathematics

Division

Mathematics for the Life Sciences

Mathematics for the Life Sciences

Erin N. Bodine
Suzanne Lenhart
Louis J. Gross

PRINCETON UNIVERSITY PRESS   Princeton and Oxford

Published by Princeton University Press, 41 William Street, Princeton,
New Jersey 08540
In the United Kingdom: Princeton University Press, 6 Oxford Street,
Woodstock, Oxfordshire, OX20 1TW
press.princeton.edu

Library of Congress Cataloging-in-Publication Data

Bodine, Erin N.
 Mathematics for the life sciences / Erin N. Bodine, Suzanne Lenhart, Louis J. Gross.
  pages cm

Summary: "The life sciences deal with a vast array of problems at different spatial, temporal, and
organizational scales. The mathematics necessary to describe, model, and analyze these problems is
similarly diverse, incorporating quantitative techniques that are rarely taught in standard undergraduate
courses. This textbook provides an accessible introduction to these critical mathematical concepts,
linking them to biological observation and theory while also presenting the computational tools needed
to address problems not readily investigated using mathematics alone. Proven in the classroom and
requiring only a background in high school math, Mathematics for the Life Sciences doesn't just focus on
calculus as do most other textbooks on the subject. It covers deterministic methods and those that
incorporate uncertainty, problems in discrete and continuous time, probability, graphing and data
analysis, matrix modeling, difference equations, differential equations, and much more. The book uses
Matlab throughout, explaining how to use it, write code, and connect models to data in examples chosen
from across the life sciences. The text provides undergraduate life science students with a succinct
overview of major mathematical concepts that are essential for modern biology, covers all the major
quantitative concepts that national reports have identified as the ideal components of an entry-level
course for life science students, and provides good background for the MCAT, which now includes
data-based and statistical reasoning. The book explicitly links data and math modeling, includes
end-of-chapter homework problems, end-of-unit student projects, select answers to homework problems,
and provides an online supplement with Matlab m-files and an R supplement. It prepares students to read
with comprehension the growing quantitative literature across the life sciences, gives an online answer
key, solution guide, and illustration package (available to professors)"—Provided by publisher.

Includes bibliographical references and index.
ISBN 978-0-691-15072-7 (hardback)
1. Biology—Mathematical models. 2. Mathematics. I. Lenhart, Suzanne. II. Gross, Louis J. III. Title.
QH323.5.B63 2014
570.1'51—dc23
                                                                              2014006493

British Library Cataloging-in-Publication Data is available

This book has been composed in LATEX
Printed on acid-free paper. ∞
Printed in the United States of America

10 9 8 7 6 5 4 3 2 1

Dedication

This book is dedicated to Bryan and Vellie Barnes, Peter and Phillip Andreae, and Marilyn Kallet and Heather Gross.

Contents

Preface   xiii

Acknowledgments   xix

UNIT 1   Descriptive Statistics   1

CHAPTER 1   Basic Descriptive Statistics   3

1.1   Types of Biological Data   3
1.2   Summary of Descriptive Statistics of Data Sets   4
1.3   Matlab Skills   9
1.4   Exercises   11

CHAPTER 2   Visual Display of Data   14

2.1   Introduction   14
2.2   Frequency Distributions   15
2.3   Bar Charts and Histograms   16
2.4   Scatter Plots   23
2.5   Matlab Skills   24
2.6   Exercises   27

CHAPTER 3   Bivariate Data and Linear Regression   30

3.1   Introduction to Linear Regression   30
3.2   Bivariate Data   31
3.3   Linear Analysis of Data   32
3.4   Correlation   37
3.5   Matlab Skills   41
3.6   Exercises   43

CHAPTER 4   Exponential and Logarithmic Functions   46

4.1   Exponential and Logarithmic Functions in Biology   46
4.2   Review of Exponential and Logarithm Properties   47

4.3   Allometry   54
4.4   Rescaling Data: Log-Log and Semilog Graphs   55
4.5   Matlab Skills   62
4.6   Exercises   67
UNIT 1   Student Projects   71

UNIT 2   Discrete Time Modeling   79

CHAPTER 5   Sequences and Discrete Difference Equations   84
5.1   Sequences   85
5.2   Limit of a Sequence   87
5.3   Discrete Difference Equations   90
5.4   Geometric and Arithmetic Sequences   92
5.5   Linear Difference Equation with Constant Coefficients   93
5.6   Introduction to Pharmacokinetics   97
5.7   Matlab Skills   100
5.8   Exercises   102

CHAPTER 6   Vectors and Matrices   107
6.1   Vector Structure: Order Matters!   108
6.2   Vector Algebra   110
6.3   Dynamics: Vectors Changing over Time   112
6.4   Matlab Skills   120
6.5   Exercises   120

CHAPTER 7   Matrix Algebra   123
7.1   Matrix Arithmetic   123
7.2   Applications   129
7.3   Matlab Skills   133
7.4   Exercises   138

CHAPTER 8   Long-Term Dynamics or Equilibrium   141
8.1   Notion of an Equilibrium   142
8.2   Eigenvectors   142
8.3   Stability   147
8.4   Matlab Skills   149
8.5   Exercises   149

CHAPTER 9   Leslie Matrix Models and Eigenvalues   152
9.1   Leslie Matrix Models   153
9.2   Long-Term Growth Rate (Eigenvalues)   156
9.3   Long-Term Population Structure (Corresponding Eigenvectors) 163
9.4   Matlab Skills   165
9.5   Exercises   168
UNIT 2   Student Projects   171

UNIT 3   Probability   175

CHAPTER 10   Probability of Events   177
10.1   Sample Spaces and Events   178
10.2   Probability of an Event   181

10.3   Combinations and Permutations   186
10.4   Binomial Experiments   188
10.5   Matlab Skills   189
10.6   Exercises   198

CHAPTER 11   Probability of Compound Events   201

11.1   Compound Events   201
11.2   Finding the Probability of a Compound Event   204
11.3   Probability Viewed as Darts Tossed at a Dart Board   209
11.4   Matlab Skills   210
11.5   Exercises   213

CHAPTER 12   Conditional Probability   216

12.1   Conditional Probability   217
12.2   Independence   220
12.3   Matlab Skills   225
12.4   Exercises   230

CHAPTER 13   Sequential Events   233

13.1   Partition Theorem   233
13.2   Bayes' Theorem   238
13.3   Exercises   242

CHAPTER 14   Population Genetics Models   246

14.1   Hardy-Weinberg Equilibrium   247
14.2   Hardy-Weinberg Selection Model   250
14.3   Exercises   253
UNIT 3   Student Projects   255

UNIT 4   Limits and Continuity   259

CHAPTER 15   Limits of Functions   261

15.1   Limit of a Function   262
15.2   Limit Properties   266
15.3   Matlab Skills   274
15.4   Exercises   277

CHAPTER 16   Limits of Continuous Functions   282

16.1   Right and Left Limits   283
16.2   Continuity   284
16.3   Intermediate Value Theorem   290
16.4   Matlab Skills   292
16.5   Exercises   295
UNIT 4   Student Projects   299

UNIT 5   Derivatives   303

CHAPTER 17   Rates of Change   305

17.1   Average Rate of Change   306
17.2   Estimating Rates of Change for Data   308
17.3   Velocity   309

17.4   Photosynthesis   311
17.5   Other Examples of Rates of Change   315
17.6   Definition of a Derivative at a Point   316
17.7   Matlab Skills   316
17.8   Exercises   320

CHAPTER 18   Derivatives of Functions   324

18.1   Concept of a Derivative   324
18.2   Limit Definition of a Derivative of a Function   326
18.3   Derivatives of Exponential Functions   330
18.4   Derivatives of Trigonometric Functions   334
18.5   Derivatives and Continuity   336
18.6   Derivatives of Logarithmic Functions   341
18.7   Matlab Skills   345
18.8   Exercises   349

CHAPTER 19   Computing Derivatives   352

19.1   Derivatives of Frequently Used Functions   353
19.2   The Chain Rule for the Composition of Functions   354
19.3   Quotient and Reciprocal Rules   359
19.4   Exponential Models   362
19.5   Higher Derivatives   369
19.6   Exercises   372

CHAPTER 20   Using Derivatives to Find Maxima and Minima   376

20.1   Maxima and Minima   377
20.2   First Derivative Test   377
20.3   Mean Value Theorem   382
20.4   Concavity   385
20.5   Optimization Problems   394
20.6   Matlab Skills   402
20.7   Exercises   404
UNIT 5   Student Projects   410

UNIT 6   Integration   413

CHAPTER 21   Estimating the Area under a Curve   414

21.1   The Area under a Curve   415
21.2   Increasing the Accuracy of the Area Estimation   426
21.3   Area below the Horizontal Axis   430
21.4   Matlab Skills   433
21.5   Exercises   436

CHAPTER 22   Antiderivatives and the Fundamental Theorem
              of Calculus   440

22.1   Definition of an Integral   441
22.2   Antiderivatives   442
22.3   Fundamental Theorem of Calculus   444
22.4   Antiderivatives and Integrals   446

22.5   Average Values   450
22.6   Matlab Skills   453
22.7   Exercises   456

CHAPTER 23   Methods of Integration   459

23.1   Substitution Method   459
23.2   Integration by Parts   465
23.3   Exercises   469

CHAPTER 24   Applications of Integrals to Area and Volume   471

24.1   The Area between Two Curves   472
24.2   The Volume of a Solid of Revolution   477
24.3   Density Functions   482
24.4   Exercises   485

CHAPTER 25   Probability in a Continuous Context   489

25.1   Expected Value and Median Value   493
25.2   Normal Distribution   495
25.3   Waiting Times   498
25.4   Matlab Skills   500
25.5   Exercises   507
UNIT 6   Student Projects   510

UNIT 7   Introduction to Differential Equations   513

CHAPTER 26   Separation of Variables   515

26.1   Separation of Variables Method   518
26.2   Matlab Skills   522
26.3   Exercises   527

CHAPTER 27   Equilibria and Limited Population Growth   529

27.1   Models of Limited Population Growth   531
27.2   Equilibria and Stability   535
27.3   Homeostasis   539
27.4   Exercises   541

CHAPTER 28   Implicit Differentiation and Related Rates   543

28.1   Explicitly and Implicitly Defined Functions   544
28.2   Implicit Differentiation   544
28.3   Related Rates   549
28.4   Exercises   551
UNIT 7   Student Projects   555

Bibliography   557
Appendix A   561
Appendix B   571
Answers to Selected Problems   579
Index   597

Preface

"Science is thought to be a process of pure reductionism, taking the meaning out of mystery, explaining everything away, concentrating all our attention on measuring things and counting them up. It is not like this at all. The scientific method is guesswork, the making up of stories. The difference between this and other imaginative works of the human mind is that science is then obliged to find out whether the guesses are correct, the stories true. Curiosity drives the enterprise, and the open acknowledgement of ignorance."

(Lewis Thomas, Sierra Club Bulletin, March/April 1982, p. 52)

The above quote from Lewis Thomas is a wonderfully succinct summary of the scientific process. It includes two main components that directly relate to the objectives of this text: the "making up of stories" and how we "find out whether the guesses are correct, the stories true." The "stories" we tell in science are called theories and models. Theories may be thought of as a general summary of how the world works, developed from many observations (in biology these would be in natural systems and in laboratory settings) and allowing us to generate hypotheses about what we expect to occur under different circumstances. Models are elaborations of particular aspects of theories, developed to elucidate specific issues that the theory can address.

A major theory in biology is that an organism's characteristics (morphology, size, physiology, behavior, etc.) are shaped by the processes of natural selection. Organisms with characteristics that enable them to better survive and reproduce are selected for. This means that, over time, all else being held fixed, organisms with those characteristics will increase in frequency, *if* such characteristics are heritable. Here, heritability means that the tendency to display such characteristics is possible to pass on to the organism's offspring. This is the essence of evolution by natural selection, and this general theory has produced an enormous number of hypotheses that have been found to be accurate from observations and experiments on a vast array of organisms and biological systems.

A large number of models rely on the theory of evolution to make predictions about the genetics, behavior, morphology, and relationships between organisms. These models allow us to predict the genetic response of bacteria populations in a human or livestock host to antibiotic use, the changes in strains of the HIV virus in an infected individual, the circumstances under which we expect to find organisms that reproduce once during their lifetime as compared to those that reproduce several times, and what shapes and sizes of leaves we expect to find on plants in desert versus those in temperate systems.

The second component of the above quote deals with determining whether our scientific stories are "correct." This is done by using data of all types to determine whether the model hypotheses and predictions are consistent with the data. We call this process "testing" or "evaluating" models, and the first part of this text deals with some of the basic methods for using data to test models. This is part of the field of statistics, and an objective in this text is to help you appreciate how we can use the methods of descriptive statistics to summarize and display data. It is then possible to use these data summaries to evaluate model hypotheses. Throughout this text, we will use data to point out the utility of mathematics in investigating biological questions. We have chosen examples with data from across the biological sciences, but we make no attempt to cover the huge breadth of examples that could be used to illustrate each mathematical topic.

Models in biology can be of several different types. They can be expressed verbally, mathematically, or in computer language. Models can also be "real" in the sense that biomedical science often uses animal models as substitutes for humans or uses cultures of cells to investigate responses that would be more difficult to determine from experiments on whole organisms. Similarly, rather than trying to carry out experiments on whole ecosystems, ecologists use microcosms and mesocosms (think of these as aquaria and terraria with different collections of species within them) to determine how the ecosystem might respond to changes, such as harvesting.

In using any of these models, we should ask the same basic questions: what is the objective for the model (e.g., what do you want to use the model to learn about), and how you would assess whether the model is useful for the purposes for which you intend to use it? In biomedical science, for example, there is a large literature on the use of animal models that provide some guidance as to whether a particular animal (or cell) is an appropriate model to use to address a question for which the ultimate goal is to relate the results to humans. Unfortunately, there is little agreement on the best ways to determine whether a particular model is "correct," and so mice may be used as a surrogate for humans, for example, in evaluating the efficacy of a new drug, but the response of the mice may be quite different from what a human response would be to the drug.

Our objective in this text is to help you learn about the variety of mathematical methods used to create and evaluate models in biology. The general aim of the text is to show how mathematics and associated computational tools may be used to explore and explain a wide variety of biological phenomena that are not easily understood with verbal reasoning alone or from simply analyzing the data that come from experiments. We provide an introduction to mathematical topics that have been found to be of use in analyzing problems arising across the biological sciences. So we expect that it will be useful if you are a student in biology, agriculture, forestry, wildlife, veterinary science, pre-medicine, or another pre-health profession. Our goal is to provide an overview rather than a detailed introduction to any particular topic. We assume that you have had prior exposure to high school algebra, geometry, and trigonometry.

Another objective of the text is to point out the utility of computational methods and to encourage you to develop a basic understanding of the underlying ideas of computer programming. While you have likely been using computers throughout your education and in many facets of your everyday life, we suspect that a large fraction of the readers of this text will have had little if any prior experience with the underlying programming (often called coding) that determines what a computer does. Computational methods have become more and more prevalent throughout biology, and we believe it is important for modern biologists to be aware of the basic ideas of computer algorithms. For that reason, we have chosen to illustrate examples throughout the text using the computer tool Matlab. This provides a means to address more complicated biological questions than we could otherwise do with only hand calculations or using a standard numerical calculator. We also use Matlab to help you learn about some of the basic coding constructs that underlie all digital computers.

Modeling is the creative process by which a model is developed. This includes determining the objectives for which the model is being developed, the choice and construction of the model, the evaluation criteria to be used to determine if the model is useful for the purposes for which it is being constructed, and an iterative procedure for model modification (or elimination if the approach is deemed unsuitable) to meet the objectives. Modeling is often inherent in the scientific process of observation, identifying patterns, hypothesis formulation, setting criteria to evaluate the hypotheses, abstracting the key features of the system under consideration, carrying out further observation or experiment, and evaluating the hypotheses based on the chosen criteria. The abstraction carried out in the scientific process typically involves a model or set of models that are applied to suggest appropriate experiments or observations and infer the implications of the assumptions inherent in the abstraction. The modeling process can point out the need for more data in order to create a useful model, and, as is true of science, models are modified regularly to incorporate new features and account for new data.

While this text includes many models, it is not our objective to teach you how to be effective in doing the modeling necessary to develop new models. Rather, we expect that the concepts and skills we incorporate in the text will help you appreciate the variety of components of the modeling process, including assessing hypotheses based on data, formulating a mathematical description of a system based on assumptions, and determining the implication of the assumptions by analyzing the model. The models we use throughout the text are designed to help you develop intuition for the concepts and techniques considered. So we use models mainly to motivate the utility of the mathematics being discussed and to illustrate how the mathematics can provide insight in the life sciences and be related back to observations. Given the proliferation of mathematical models in every area of the life sciences, our intent is also to help you read critically the literature in your chosen area. One of the benefits of learning mathematics is that the same concepts and methods often apply in diverse areas so that you will see the same mathematical tools used in very different areas of biology. Many fine texts that present models in particular subdisciplines of the life sciences are available. Our expectation is that after you complete this text, you will be well prepared to expand your expertise by reading through other, more specialized texts on modeling in biology in areas as diverse as pharmacokinetics, population ecology, cell signaling, and genomics.

Summary of Goals for This Text

- Develop a reader's ability to quantitatively analyze problems arising in the biological areas of interest to them.
- Illustrate the great utility of mathematical models to provide answers to key biological problems.
- Develop an appreciation of the diversity of mathematical approaches that are useful in the life sciences.
- Provide experience using computer software to analyze data, investigate mathematical models, and provide some exposure to programming.

The methods utilized in this text to meet these goals include the following:

1. Encouraging hypothesis formulation and testing for both the biological and the mathematical topics covered.
2. Encouraging investigation of real-world biological problems through the use of data.

3. Reducing rote memorization of mathematical formulae and rules through the use of Matlab.

4. Providing biological motivation for each main mathematical component introduced through a set of examples that are returned to regularly.

Note to Instructors

Although long considered the language of science, mathematics has slowly become more integral to biological fields as it has been historically in chemistry and physics. Although there is a long history of mathematical approaches in certain areas of biology, it has been only within the past two decades that an appreciation for the ubiquitous utility of mathematics has arisen across the life sciences. Several national reports (e.g., *BIO2010: Transforming Undergraduate Education for Future Research Biologists* from the National Research Council in 2003, the 2009 HHMI/AAMC report on *Scientific Foundations for Future Physicians*, and the NSF/AAAS 2011 *Vision and Change in Undergraduate Biology Education* report) have pointed out the need for quantitative and computational components to be closely linked to all areas of biology education.

Happily, a large number of new texts are devoted to the interface of mathematics and biology, with several of these focused on entry-level courses for life scientists. In most cases, these texts focus first on calculus topics. The pedagogy of the courses for which this text has been developed over the past two decades is that an entry-level course should provide an introduction to the variety of mathematical topics of importance in many areas of biology and that this goes well beyond the calculus that is typically part of an entry-level course sequence for undergraduates in science. The course was fostered by discussions at two workshops at the University of Tennessee, Knoxville, in the early 1990s at which gatherings of leading quantitative biologists encouraged a broad-based collection of material be included, with ample motivation by biological data. Additional information gathered by LJG noted that one of the most critical skills that biology faculty felt their undergraduates needed that was not being adequately addressed was interpreting and making graphs. Thus, this text starts with biological data and its analysis, developing basic concepts in descriptive statistics that are rarely included in entry-level math courses for life science students. Additional discussions with numerous faculty led LJG to compose a set of quantitative concepts that ideally all undergraduate life science students would be exposed to during their undergraduate careers. This set of concepts was included in the BIO 2010 report and includes rates of change, scale, equilibria, stability, structure, interactions, stochasticity, visualizing, and algorithms. This text provides introductory material for each of these concepts in a relatively simple mathematical framework using questions motivated by biological data.

One of the major themes of the undergraduate mathematics education reform movement over the past several decades has been the "Rule of Four." This acknowledges that students have diverse learning styles and that it is appropriate to incorporate alternative methods to enhance quantitative concept and skill comprehension through four complementary approaches: symbolic, numerical, graphical, and verbal. From our experience guiding life science students through quantitative courses over many decades, we have found that an additional approach can be beneficial: providing motivation for the concept or skill through observations and data on biological topics about which the students have some direct experience (e.g., heart rate changes during exercise, drug dosing, and growth of pets or gardens) or intuition from our activities as living beings (e.g., population changes and seasonal and diurnal changes of organism behavior). We thus posit that for interdisciplinary courses and readers, for which this text is designed, it is appropriate to follow a "Rule of Five" for conceptual development that includes biological data and examples that enhance intuition regarding the concepts.

In the units of this text, we try to incorporate each of these five components of learning development in the expectation that this will not only cover all potential learning styles of readers but additionally reinforce appreciation for the utility of mathematics by making explicit connections to biology. In this process, we realize that many readers may well have limited formal exposure to the science of biology (as opposed to our own direct experiences as organisms), so we have tried to limit inclusion of biological areas that typical students would not be exposed to until beyond the first year of their curriculum and about which they likely would have little intuition. This necessarily has focused our examples at levels of the biological hierarchy above the molecule and cell, with little emphasis on biochemistry, systems biology and biophysics of the cell, developmental biology, and neurobiology. We encourage instructors to enhance the biological topic coverage with examples from their own experience that may assist in student comprehension and would definitely appreciate suggestions based on your experiences in introducing these.

Using This Text

The topic coverage and approach we follow is concordant with suggestions in the reports on U.S. undergraduate life science education mentioned above. The text can be utilized in several ways to help institutions meet learning goals for life science students. This text has been used in draft form over the past 5 years for a two-semester, 3-credit-hour sequence taken by many life science and agriculture students at the University of Tennessee, Knoxville. In this form, it has been taught by a broad range of mathematics and life science faculty, instructors, and graduate students.

The format has been both one with small sections of approximately 30 students meeting three times weekly and one with a large-lecture format for up to 200 students with two large-class 50-minute meetings per week and a single 85-minute lab/practice session. The coverage has been essentially the entire text, with the first course covering Units 1 through 3 and the second covering Units 4 through 7. A summary of the course structure and discussion of the challenges in teaching a course covering the diverse topics included is in Gross [31].

While many national reports on life science education encourage incorporation of a diversity of quantitative concepts and skills as undergraduate learning objectives, a large fraction of formal course requirements for life science majors include only calculus. However, the MCAT exam structure has changed to include data-based and statistical reasoning while noting that an understanding of calculus is not required. We recognize that as a foundational quantitative approach in much of modern science and engineering, many institutions wish to maintain a calculus emphasis in their life science curricula. Thus, we have tried to make the text portion that focuses on continuous methods, Units 4 through 7, self-contained so that it can be utilized for a single-semester course if desired. Drafts of the text have been used in this manner for several years at Unity College, for which the course has begun with Sections 5.1 and 5.2 on sequences and their limits and then moved to Unit 4. In this case, if Matlab is being used, the instructor may wish to incorporate a initial lab period that covers the basic Matlab procedures and constructs included in earlier units.

If more class time is available in a single introductory course for which a calculus emphasis is desired, we suggest starting with Unit 1, continuing with Sections 5.1 and 5.2, and then moving to Units 4 through 7. This will provide background on many of the data analysis skills required for the MCAT (e.g., using, analyzing, and interpreting data in figures; using measures of central tendency and dispersion to describe data; and recognizing and interpreting linear, semilog, and log-log scales and calculating slopes from data found in figures). Linking this with a biostatistics course that includes hypothesis testing would then provide sufficient background for all quantitative components of the MCAT while also providing an overview of calculus. An alternative, for a single course that empasizes noncalculus topics, is to include Units 1 through 3.

This may then be followed with a biostatistics course for those students who wish to prepare only for the MCAT, a course based on Units 4 through 7 of this text, or a calculus course or sequence developed for a broader audience of students.

We have incorporated Matlab throughout this text both to introduce basic constructs used in computer coding and to introduce the concepts of algorithms that are applied throughout computational methods used in every area of science. However, the text components related to Matlab are included at the end of each chapter and may be skipped. Because of the ongoing growth of the use of the R language in many areas of biology, an online adjunct to this text is available that uses R rather than Matlab for all the sections at the end of chapters.

Acknowledgments

This text has its roots in a long-standing course sequence (Math 151-2) at the University of Tennessee, Knoxville, that was started by Tom Hallam in the late 1970s. For many years, the course sequence used a fine text by the late Michael Cullen and has been taught by over 50 instructors to thousands of students. An unpublished text by Simon Levin from which LJG taught at Cornell University had a similar emphasis on breadth of mathematical topics as the text by Michael, and we thank all three of these individuals for the inspiration that led to the series of workshops that LJG organized in the early 1990s and to this book.

Given the length of gestation of this text, there are many people with whom we have had conversations, participated with in workshops on math biology education, and have given presentations to about our efforts. We make no attempt here to acknowledge everyone involved over the years, but we particularly want to thank the following individuals who have contributed directly to the text, noting that we may not have always followed their advice and that whatever errors occur in judgment about topic and example inclusion are ours alone. Marco Martinez, Carrie Eaton, Jason Bintz, Evan Lancaster, Jillian Trask, Ashley Rand, Folashade Agusto, Heather Finotti, Vitaly Ganusov, Ben Levy, Steve Fassino, Jeremy Auerbach, Donna Stein, Michael Lawton, Ellie Abernethy, Jack Kilislian, and Rachael Miller Neilan provided many comments on the drafts of the text. Marco Martinez and Jason Bintz completed almost all of the problem solutions, and Tyler Massaro helped with R software.

The National Institute for Mathematical and Biological Synthesis (NIMBioS) has provided significant support for this text, and we appreciate the support of the National Science Foundation through Awards EF-0832858 and DBI-1300426 to the University of Tennessee, Knoxville, that provided us with an opportunity to develop this text but that also supported many researchers in mathematical biology over the past 5 years with whom we have discussed the text. We have received very helpful reviews of the text from several anonymous individuals who led us to rethink portions of the text, and we hope they will concur that it is improved because of their efforts. Several other individuals provided detailed comments on the text at various stages, and we appreciate their kindness in taking the time to assist us.

As all text authors know, the successful completion of a text depends on the kindness of family members. Our families have generously allowed us to spend long hours well beyond the "normal" workweek to develop this text, and we greatly appreciate their love and thoughtful allowance for our efforts away from family responsibilities. Thanks to Bryan and Vellie Barnes, Peter Andreae, and Marilyn Kallet.

Finally, the editorial guidance provided by several people at Princeton University Press has provided not only much-needed advice but also the time and flexibility to allow us to create a text that we hope will inspire many readers as it has inspired us during its development. Alison Kalett has been consistent and generous with her efforts to obtain excellent reviews, and her insight has greatly assisted us. Vickie Kearn gave us her support and encouragement and Mark Bellis and Quinn Fusting helped assure that the marketing and production processes went smoothly. Diane Kohnen and the staff of S4Carlisle Publishing dealt meticulously and patiently with many modifications of the initial draft.

Erin N. Bodine
Suzanne Lenhart
Louis J. Gross

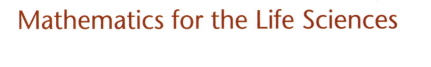

Mathematics for the Life Sciences

Descriptive Statistics

We begin our study of the mathematics used in the life sciences with a unit that explores how we understand data mathematically. Understanding data is a process, the steps of which are the following:

1. Collecting the data
2. Summarizing the data
3. Analyzing the data
4. Interpreting the results and reporting them

Note that before carrying out any of the above, you have presumably formulated some underlying question or hypothesis that you wish to use the data to address. There are a few key approaches through which we can address scientific questions:

(i) observation (natural history: see what occurs where and when and interpret the results based on differences in the locations or history),

(ii) experiment (vary aspects of the environment in order to tease apart how the biological components respond), and

(iii) theory (make assumptions about the natural world and analyze the implications of those assumptions using verbal, graphical, and mathematical arguments).

Each of these approaches involves quantitative methods, and an objective of this text is to provide you with an understanding of some of these methods.

Step 1 above involves the area of "design of experiments" in which the process of data collection is determined based on the objectives of the study and the limitations imposed (e.g., cost, time, available personnel, accessibility of the study area, etc.). Design implies that the scientist considers alternative methods to collect the data as well as the manner in which the factors deemed to affect the data collection are manipulated. Examples would be determining

- where and when to put out traps to collect animals in the field,
- how many replicates of an evaluation test to use in estimating the efficacy of a new drug,
- how many different levels of growth medium with what nutrient constituents to use in evaluating the impact of a new antibiotic on bacterial population growth in the lab,
- the response of an organism's respiration rate to temperature,
- how many different temperature treatments are applied, in what order, and for how long.

Step 2 in the process is typically called "descriptive statistics," in which the objective is to abstract out certain properties of the data in order to better interpret them. The assumption here is that the data are too complex for us to understand well by simply looking at them as lists or tables. The simplest example of this is the computation of an "average" value of the data. Many of us obtain a better grasp of a data set by having some summary of the data available, particularly in graphical form, rather than simply a tabular elaboration of the data. Note that whatever methods are utilized here, there is a loss of information associated with the description provided: the description (e.g., the average value of the data) does not include the full amount of information in the complete data set. An objective in descriptive statistics is to choose the appropriate level of description between complete enumeration of the data and a coarse simple summary (such as the average value) so as to be able to address the questions you posed in the first place. As an example, consider the height of all students in a course. Having these displayed as a long list would not be readily useful, whereas if we state that the average height of students is 165 cm, you have a simple means of comparing the students in the course to the students in another course. More information would be provided by a histogram (bar chart) of the heights of the students in the course, but even then there would be some loss of information since we could not develop from the histogram the full list of heights of all students in the course.

Step 3 in the process typically involves the area of inferential statistics, which consists of parameter estimation and hypothesis testing. Parameter estimation refers to using the data to determine estimates of values of particular interest (respiration rate, photosynthetic rate, hemoglobin level, etc.) from the observations. One might then use the data to evaluate hypotheses (respiration rate increases with temperature, the hemoglobin content of two species differs) in which one compares a "null hypothesis" (respiration rate is independent of temperature) to an "alternative hypothesis" (respiration rate increases with temperature).

Step 4 uses the results of the inferential statistics developed to evaluate the results of the observations and provide an interpretation of the results (there is a significant effect of temperature on respiration, and this implies that the species has limited latitudinal range due to the effects of temperature; two species differ significantly in their photosynthetic rate, and you expect one species to outcompete the other under certain environmental conditions).

This first unit of the text will focus on the descriptive statistics aspects of the above process. All life scientists are well served by being exposed to a formal statistics course that includes aspects of experimental design and hypothesis testing, so you are encouraged to enhance your training in this area beyond the limited coverage included here. The emphasis on descriptive statistics arises because of regular comments by life science practitioners that an extremely important aspect of quantitative training for their colleagues is the ability to interpret graphs and to utilize diverse graphical approaches to explain and interpret experiments.

Basic Descriptive Statistics

1.1 Types of Biological Data

Any observation or experiment in biology involves the collection of information, and this may be of several general types:

Data on a Ratio Scale

Consider measuring heights of plants. The difference in height between a 20-cm-tall plant and a 24-cm-tall plant is the same as that between a 26-cm-tall plant and a 30-cm-tall plant. These data have a "constant interval size." They also have a true zero point on the measurement scale, so that ratios of measurements make sense (e.g., it makes sense to state that one plant is three times as tall as another). A measurement scale that has constant interval size and a true zero point is called a "ratio scale." For example, this applies to measurements of weights (mg, kg), lengths (cm, m), volumes (cc, cu m), and lengths of time (s, min).

Data on an Interval Scale

Measurements with an interval scale but having no true zero point are of this type. Examples are temperatures measured in Celsius or Fahrenheit: it makes no sense to say that 40 degrees is twice as hot as 20 degrees. Absolute temperatures, however, are measured on a ratio scale.

Data on an Ordinal Scale

Data that can be ordered according to some measurements are on an ordinal scale. Examples would be rankings based on size of objects, the speed of an individual relative to another individual, the depth of the orange hue of a shirt, and so on. In some cases (e.g., size), there may be an underlying ratio scale, but if all that is provided is a ranking of individuals (e.g., you are told only that tomato genotype A is larger than tomato genotype B, not how much larger), there is a loss of information if we are given only the ranking on an ordinal scale. Quantitative comparisons are not possible on an ordinal scale (how can one say that one shirt is half as orange as another?).

Data on a Nominal Scale

When a measurement is classified by an attribute rather than by a quantitative, numerical measurement, then it is on a nominal scale (male or female; genotype AA, Aa or aa; in the taxa *Pinus* or in the taxa *Abies*; etc.). Often, these are called categorical data because you classify the data elements according to their category.

Continuous vs. Discrete Data

When a measurement can take on any conceivable value along a continuum, it is called continuous. Weight and height are continuous variables. When a measurement can take on only one of a discrete list of values, it is discrete. The number of arms on a starfish, the number of leaves on a plant, and the number of eggs in a nest are all discrete measurements.

1.2 Summary of Descriptive Statistics of Data Sets

Any time a data set is summarized by its statistical information, there is a loss of information. That is, given the summary statistics, there is no way to recover the original data. Basic summary statistics may be grouped as

(i) measures of central tendency (giving in some sense the central value of a data set) and
(ii) measures of dispersion (giving a measure of how spread out that data set is).

Measures of Central Tendency

Arithmetic Mean (the average)

If the data collected as a sample from some set of observations have values $x_1, x_2, \ldots, x_n$, then the mean of this sample (denoted by $\bar{x}$) is

$$\bar{x} = \frac{1}{n} \sum_{i=1}^{n} x_i = \frac{x_1 + x_2 + \cdots + x_n}{n}.$$

Note the use of the $\sum$ notation in the above expression, that is,

$$\sum_{i=1}^{n} x_i = x_1 + x_2 + \cdots + x_n.$$

Median

The median is the middle value: half the data fall above this and half below. In some sense, this supplies less information than the mean since it considers only the ranking of the data, not how much larger or smaller the data values are. But the median is less affected than the mean by "outlier" points (e.g., a really large measurement or data value that skews the sample). The LD 50 is an example of a median: the median lethal dose of a substance (half the individuals die after being given this dose, and half survive). For a list of data $x_1, x_2, \ldots, x_n$, to find the median, list these in order from smallest to largest. This is known as "ranking" the data. If n is odd, the median is the number in the $1 + \frac{n-1}{2}$ place on this list. If n is even, the median is the average of the numbers in the $\frac{n}{2}$ and $1 + \frac{n}{2}$ positions on this list.

Quartiles arise when the sample is broken into four equal parts (the right end point of the 2nd quartile is the median), quintiles when five equal parts are used, and so on.

Mode
The mode is the most frequently occurring value (or values; there may be more than one) in a data set.

Midrange
The midrange is the value halfway between the largest and smallest values in the data set. So, if x_{min} and x_{max} are the smallest and largest values in the data set, then the midrange is

$$\bar{x}_{mid} = \frac{x_{min} + x_{max}}{2}.$$

Geometric Mean
The geometric mean of a set of n data is the nth root of the product of the n data values,

$$\bar{x}_{geom} = \left(\prod_{i=1}^{n} x_i \right)^{1/n} = \sqrt[n]{x_1 \cdot x_2 \cdots x_n}.$$

The geometric mean arises as an appropriate estimate of growth rates of a population when the growth rates vary through time or space. It is always less than the arithmetic mean. (The arithmetic mean and the geometric mean are equal if all the data have the same value.)

Harmonic Mean
The harmonic mean is the reciprocal of the arithmetic mean of the reciprocals of the data,

$$\bar{x}_{harm} = \frac{n}{\sum_{i=1}^{n} \frac{1}{x_i}} = \frac{n}{\frac{1}{x_1} + \frac{1}{x_2} + \cdots + \frac{1}{x_n}}.$$

It also arises in some circumstances as the appropriate overall growth rate when rates vary.

Example 1.1 (Describing a Data Set Using Measures of Central Tendency)

After developing some heart troubles, John was told to monitor his heart rate. He was advised to measure his heart rate six times a day for 3 days. His heart rate was measured in beats per minute (bpm).

$$
\begin{array}{cccccc}
65 & 70 & 90 & 95 & 82 & 84 \\
61 & 83 & 120 & 83 & 72 & 70 \\
72 & 71 & 92 & 85 & 102 & 69
\end{array}
$$

(Continued)

(a) What was John's mean heart rate over the 3 days? Calculate the three different means (arithmetic, geometric, and harmonic).

(b) What was John's median heart rate?

(c) What were the modes of John's heart rate?

(d) What was the midrange of John's heart rate?

Solution:

(a) Arithmetic mean:

$$\bar{x} = \frac{65 + 70 + 90 + \cdots + 85 + 102 + 69}{18} = 81.4$$

Geometric mean:

$$\bar{x}_{\text{geom}} = (65 \times 70 \times 90 \times \cdots \times 85 \times 102 \times 69)^{1=18} = 80.3$$

Harmonic mean:

$$\bar{x}_{\text{harm}} = \frac{18}{\frac{1}{65} + \frac{1}{70} + \frac{1}{90} + \cdots + \frac{1}{85} + \frac{1}{102} + \frac{1}{69}} = 79.2$$

Notice that the three means do not yield equal values.

(b) Arranging the numbers from smallest to largest, we get

$$61 \quad 65 \quad 69 \quad 70 \quad 70 \quad 71 \quad 72 \quad 72 \quad 82$$
$$83 \quad 83 \quad 84 \quad 85 \quad 90 \quad 92 \quad 95 \quad 102 \quad 120$$

Since there are 18 data points, we take the average of the middle two numbers: 82 and 83. Thus, the median is 82.5.

(c) There are three modes in this data set: 70, 72, and 83.

(d) Midrange: $\bar{x}_{\text{mid}} = \dfrac{61 + 120}{2} = 90.5$. Notice that this is different from the median.

Measures of Dispersion

Range

The range is the largest minus the smallest value in the data set: $x_{\text{max}} - x_{\text{min}}$. This does not account in any way for the manner in which data are distributed across the range.

Variance

The variance is the mean sum of the squares of the deviations of the data from the arithmetic mean of the data. The *best* estimate of this (take a good statistics class to find out how *best* is defined) is the sample variance, obtained by taking the sum of the squares of the differences of the data values from the sample mean and dividing this by the number of data points minus one,

$$s^2 = \frac{1}{n-1} \sum_{i=1}^{n} (x_i - \bar{x})^2,$$

where n is the number of data points in the data set, x_i is the ith data point in the data set x, and $\bar{x}$ is the arithmetic mean of the data set x.

Standard Deviation

The variance has square units, so it is usual to take its square root to obtain the standard deviation,

$$s = \sqrt{\text{variance}} = \sqrt{\frac{1}{n-1} \sum_{i=1}^{n} (x_i - \bar{x})^2},$$

which has the same units as the original measurements. The higher the standard deviation s, the more dispersed the data are around the mean.

Both the variance and the standard deviation have values that depend on the measurement scale used. So measuring body weights of newborns in grams will produce much higher variances than if the same newborns were measured in kilograms. To account for the measurement scale, it is typical to use the coefficient of variability (sometimes called the coefficient of variance): the standard deviation divided by the arithmetic mean, which is dimensionless and has no units. This coefficient of variability is thus independent of the measurement scale used.

Example 1.2 (Describing a Data Set Using Measure of Dispersion)

In a summer ecology research program, Jane is asked to count the number of trees per hectare in five different sampling locations in King's Canyon National Park in California. Each sampling location is referred to as a plot, and each plot is a different size. Here are the data she collected:

Plot Size (hectares)	No. of Trees in Plot
1.50	20
2.30	31
1.75	43
3.10	58
2.65	29

Given the data Jane collected, (a) construct the data set that represents the number of trees per hectare for each of the five plots and then calculate the (b) range, (c) variance, and (d) standard deviation of the data set you constructed.

(*Continued*)

Solution:

(a) For each plot, the number of trees per hectare is

$$\frac{\text{\# trees in plot}}{\text{plot size}}.$$

For example, the first plot has $20/1.5 = 13.3$ trees/hectare. Thus, the data set that represents the number of trees per hectare for each of the five plots is

$$x = \{13.3, 13.5, 24.6, 18.7, 10.9\}.$$

(b) To calculate the range, we need to know x_{max} and x_{min} (the maximum and minimum values of the data set x). Looking at the data set constructed in (a), $x_{min} = 10.9$ and $x_{max} = 24.6$. Thus,

$$\text{range } = 24.6 - 10.9 = 13.7.$$

(c) Recall that to calculate the variance of a data set, you must first know the arithmetic mean of that data set. For the data set constructed in (a),

$$\bar{x} = \frac{13.3 + 13.5 + 24.6 + 18.7 + 10.9}{5} = 16.2.$$

Then, the variance is

$$s^2 = \frac{1}{5-1} \left[(13.3 - 16.2)^2 + (13.5 - 16.2)^2 + (24.6 - 16.2)^2 \right.$$

$$\left. + (18.7 - 16.2)^2 + (10.9 - 16.2)^2 \right]$$

$$= \frac{1}{4} \left[(-2.9)^2 + (-2.7)^2 + (8.4)^2 + (2.5)^2 + (-5.3)^2 \right]$$

$$= \frac{1}{4} [8.41 + 7.29 + 70.56 + 6.25 + 28.09]$$

$$= \frac{1}{4} [120.6]$$

$$= 30.15.$$

(d) Recall that the standard deviation of a data set is the square root of the variance of that data set. Thus, the standard deviation is

$$s = \sqrt{30.15} = 5.491.$$

Dispersion over Nominal Scale Data and the Simpson Index

All the above measures of dispersion apply to ratio scale data. For nominal scale data, there is no mean or variance that makes sense, but there certainly can be a measure of how spread out the data are among the various categories, a concept called diversity. In ecology, the two main factors taken into account when measuring diversity are richness and evenness. Species richness is the number of different species present, while evenness is a measure of the relative abundance of the different species making up the richness of an area. The area has uneven diversity if virtually all the individuals found are of one species with only rare individuals of the other species. The area has even diversity if all species have the same abundances. Simpson's index of diversity (SID) is one of several diversity indices. The SID represents the probability that two individuals randomly selected from a sample will belong to different species. In a certain area or sample, let

$$D = \sum_{i=1}^{S} \frac{n_i(n_i - 1)}{N(N - 1)},$$

where n_i is the number of individuals in species i, N is the total number of individuals, and S is the number of species. Then, the SID is

$$SID = 1 - D.$$

When SID is close to 1, the sample is considered to be highly diverse.

1.3 Matlab Skills

If you are not familiar with the software Matlab, review "Getting Started with Matlab" in Appendix A.

Entering Data Sets in Matlab

In Matlab, data sets are entered as arrays, and arrays are denoted with square brackets: [ ]. If we wanted to enter the trees per hectare data from Example 1.2, we would type

```
[13.3 13.5 24.6 18.7 10.9]
```

into Matlab. Notice that the data points in the set are separated by spaces. If we want to refer back to this data set using Matlab, we need to name the data set. In Example 1.2, we called the data set x. To call the data set x in Matlab, we type

```
x = [13.3 13.5 24.6 18.7 10.9]
```

into Matlab. Now, whenever we want to refer back to our data set, we can just use x instead of typing the entire data set again.

Table 1.1. Matlab commands for a variety of descriptive statistics. In each case, *x* refers to the data set.

Command	Description
`mean(x)`	Returns arithmetic mean of data set *x*
`prod(x)^(1/length(x))`	Returns geometric mean of data set *x*
`geomean(x)`	Returns geometric mean of data set *x* (using the Statistics Toolbox is available)
`length(x)/sum(1./x)`	Returns harmonic mean of data set *x*
`harmmean(x)`	Returns harmonic mean of data set *x* (using the Statistics Toolbox is available)
`median(x)`	Returns median of data set *x*
`mode(x)`	Returns mode of data set *x*
	(when there are multiple values occurring equally frequently,
	`mode(x)` Returns the smallest of those values)
`min(x)`	Returns minimum value of data set *x*
`max(x)`	Returns maximum value of data set *x*
`var(x)`	Returns the variance of data set *x*
`std(x)`	Returns the standard deviation of data set *x*

Calculating Descriptive Statistics in Matlab

Now that we know how to enter our data sets into Matlab, we can use Matlab to quickly compute basic descriptive statistics. Table 1.1 shows the commands for the descriptive statistics described earlier in this chapter.

Each of the commands in Table 1.1 returns its corresponding answer and names the answer ans. If we wish to save the answer for future use, we must name the output of the command. For example, if we wish to save the arithmetic mean, we can type

```
xbar = mean(x)
```

into Matlab. If you are typing this into the command window, you will see that the value that is returned is named `xbar`.

Notice there are no commands for calculating the range or the midrange. We can calculate these, however, by using the min and max commands. To calculate the midrange, we use

```
(min(x)+max(x))/2
```

and to calculate the range, we use

```
max(x)-min(x)
```

As an example, suppose we wanted to calculate the mean, median, mode, midrange, geometric mean, harmonic mean, range, variance, and standard deviation for the data set in Example 1.1.

The following shows the input typed into the command window (always proceeded by ») and its corresponding output:

```
                        ─── Command Window ───
>> y = [65 70 90 95 82 84 61 83 120 83 72 70 72 71 92 85 102 69]
y =
  Columns 1 through 11
     65     70     90     95     82     84     61     83    120     83     72

  Columns 12 through 18
     70     72     71     92     85    102     69

>> ybar = mean(y)
ybar =
   81.4444

>> ymed = median(y)
ymed =
   82.5000

>> ymode = mode(y)
ymode =
     70

>> ymidrange = (min(y)+max(y))/2
ymidrange =
   90.5000

>> ygeo = geomean(y)
ygeo =
   80.2747

>> yharm = harmmean(y)
yharm =
   79.1871

>> yrange = max(y)-min(y)
yrange =
     59

>> yvar = var(y)
yvar =
  217.3203

>> ystd = std(y)
ystd =
   14.7418
```

1.4 Exercises

1.1   The capacity for physical exercise (in seconds) was determined for each of 11 patients who were being treated for chronic heart failure.

    906   1320   711   1170   684   1200   837   1056   897   882   1008

(a) Determine the mean and the median of the data.
(b) Determine the geometric and harmonic means of the data.
(c) How do the three different measures of the mean differ?

1.2   Daily crude oil output (in millions of barrels) for the U.S. is shown below for the years 1971 to 1990.

$$9.45 \quad 9.40 \quad 9.25 \quad 8.75 \quad 8.30 \quad 8.10 \quad 8.25 \quad 8.70 \quad 8.55 \quad 8.60$$
$$8.55 \quad 8.65 \quad 8.70 \quad 8.70 \quad 8.91 \quad 8.60 \quad 8.20 \quad 7.70 \quad 7.20 \quad 6.75$$

Compute the mean, median, and mode for the data.

1.3   Suppose the scale of a data set is changed by multiplying each measurement by a positive constant. How would this affect the mean, median, mode, and range?

1.4   Ten hospital employees on a standard American diet agreed to adopt a vegetarian diet for 1 month. Below is the change in the serum cholesterol level (before − after).

$$49 \quad -10 \quad 27 \quad 13 \quad 36$$
$$19 \quad 48 \quad 21 \quad 8 \quad 16$$

(a) Compute the median and mean change in cholesterol.
(b) Compute the range, variance, and standard deviation of the data. Are the data fairly spread out or close together?

1.5   Twelve sheep were fed pingue (a toxin-producing weed of the southwestern United States) as a part of an experiment and died as a result. The time of death in hours after the ingestion of pingue for each sheep follows:

$$44 \quad 27 \quad 24 \quad 24 \quad 36 \quad 36$$
$$44 \quad 120 \quad 29 \quad 36 \quad 36 \quad 36$$

Compute the range, variance, and standard deviation of the sample.

1.6   The National Weather Service reports data on the number of hurricanes to strike the United States in decades in the last century (using the Saffir-Simpson category). Calculate the mean of the number of hurricanes per decade.

Decade	No. of Hurricanes
1901–1910	18
1911–1920	21
1921–1930	13
1931–1940	19
1941–1950	24
1951–1960	17
1961–1970	14
1971–1980	12
1981–1990	15
1990–2000	14

1.7   Consider these two sets of data [70]:

$$A = \{0, 5, 10, 15, 25, 30, 35, 40, 45, 50, 71, 72, 73, 74, 75, 76, 77, 78, 100\}$$
$$B = \{0, 22, 23, 24, 25, 26, 27, 28, 29, 50, 55, 60, 65, 70, 75, 85, 90, 95, 100\}$$

For both sets of data, calculate the range, median, the first quartile, and the third quartile. Do these values adequately represent the distribution in each data set?

1.8   Suppose the mean score on a national test is 400 with a standard deviation of 50. If each score is increased by 25, what are the new mean and standard deviation?

1.9   Suppose the mean score on a national test is 400 with a standard deviation of 50. If each score is increased by 25%, what are the new mean and standard deviation?

1.10   Use the following simple data set to calculate the SID for these trees in a particular plot [18]. Interpret your results as a probability.

Species of Trees	No. of Trees in Plot
Eastern rosebud	3
Black oak	4
Post oak	5
White pine	3
Honey locust	1

1.11   Below are some data from the Citizen Science program in the Great Smoky Mountains National Park that record the species of salamanders observed in a particular area in 2000 [18]. Calculate the SID for salamanders in this area using these data.

Species	No. of Salamanders
Desmog	3
Spotted dusky	7
Black bellied	22
Seal	16
Blue ridged Two lined	8
Imitator	2
Southern redback	1
Black chinned	1

Visual Display of Data

2.1 Introduction

Data arise in biology from many sources, including observations of biological objects (e.g., cells and individuals) in natural settings, observations of biological objects that are involved in some manipulations (e.g., experiments that modify the growing conditions of cells or individuals), and outputs of computer models designed to analyze biological systems. In all these situations, one of the first and primary tools is simply to look at the data. Human visual systems are excellent at rapidly discerning patterns in data; however, as you know from optical illusions, our visual systems can be "tricked" on occasion. Sometimes we seem to discern patterns when a more rigorous analysis would show that no such pattern is present. Our objective in this chapter is to introduce you to some of the most basic methods that are used to display data. There is an extensive literature on more complex methods than we describe here, and we encourage you to look at the books and Web sites of Edward Tufte, who has been called the "Leonardo da Vinci of data" by the *New York Times*, for many other examples.

Visual analysis of data has several objectives. These include providing a quick summary of the major patterns in the data (if any), pointing out whether the data might be better viewed in a different format or using a different scaling, and suggesting new possible patterns that might be discerned by modifying the display in some manner. A poorly displayed data set can be misleading to a viewer. An example is the case in which a population might "look" as if it is declining rapidly, but the scales are chosen to cover only a small range of sizes. Figure 2.1 shows census data from Philadelphia scaled so that the population appears to have dropped by a much higher fraction from 1950 to 2000 because the scale of the vertical axis starts at 1,500,000 rather than zero: the axis choice makes the population appear to have a much greater drop than it actually does as a fraction of the total population. Scales matter: both measurement scales (e.g., whether you are measuring an individual's height in millimeters or meters) and the scales used in graphing the data. The objective of graphs is often to convey major patterns of information to different viewers: a well-designed graph does this effectively.

For more complicated data sets, it is likely that multiple approaches and types of graphs might be needed. The height observations collected as described in Do We Grow in Our Sleep Student Projects for this Unit is an example of this: multiple aspects of each individual were involved.

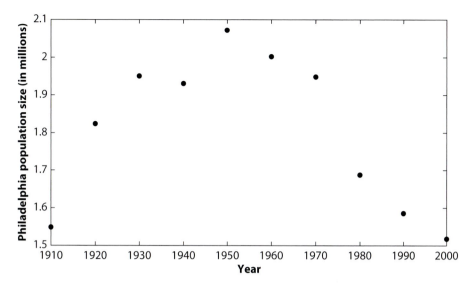

Figure 2.1 Philadelphia population census data from 1910 to 2000. The vertical axis is scaled to make the population increase from 1910 to 1920 and the population decrease from 1970 to 1980 look more dramatic than if the vertical axis showed population size 0 to 2,100,000.

Such data are called "multivariate" data because there are several different factors measured for each outcome (in this case, height change is the outcome, and the factors included hours of sleep, gender, and initial height). Multivariate data are by far the norm in biology. Drug effectiveness depends on many factors, including genetics, physiological status, and prior medical history. The spread of a disease depends on the specific populations of individuals initially infected, their movement patterns and interactions with others, any genetic predisposition to the disease, immunization history, and environmental factors. Keep the complications of multivariate factors in mind before becoming too enamored with the results that you obtain from any simple analysis of data.

2.2 Frequency Distributions

When making repeated measurements from an experiment (e.g., measuring how many white blood cells are in a 1-ml sample of whole blood), one might summarize the data pictorially by making a bar chart to display how frequently each count arises. Such graphs can be made for nominal data (the percentages of males versus females feeding female pups in meerkats; see fig. 1E in the paper by Clutton-Brock et al. [15]), ordinal data (the percentages of juveniles, subadults, yearlings, and adults contributing to babysitting in meerkats; see fig. 1A in [15]), discrete data (e.g., the frequency of meerkat litters of sizes 1, 2, 3, etc. pups), grouped discrete data (e.g., the numbers of white blood cells in a 1-ml whole blood sample, grouped by 0 to 100 cells, 101 to 200 cells, 201 to 300 cells, etc.), or continuous data, in which case we call the bar graph a histogram (e.g., the weights of individuals in the room, grouped by 5-kg increments).

In histograms, one either shows the range of the continuous values associated with each bar or shows a single number, that is, the midpoint of the interval covered by that bar. Generally, equal size intervals are used in histograms.

When viewing a bar graph or a histogram, one can see the distribution of the data. A commonly occurring distribution is the normal distribution. If the data are normally distributed, then about

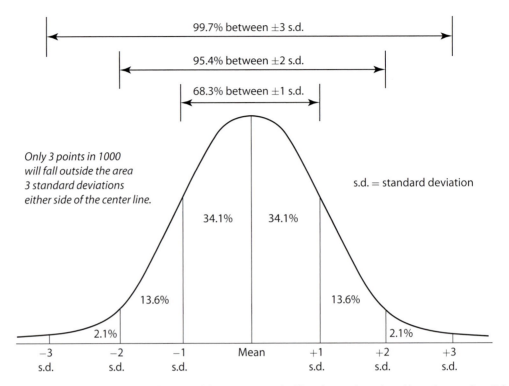

Figure 2.2 Curve showing normally distributed data. Center vertical line shows where the arithmetic mean lies. Other vertical lines indicate standard deviations from the mean.

68% of the values are within one standard deviation of the mean, about 95% of the values are within two standard deviations, and about 99.7% lie within three standard deviations. In any distribution, smaller s implies that data are closer together, and larger s implies that data are spread farther apart. See the nornal distribution curve in Figure 2.2.

Figure 2.3 from [11] shows the distribution of seed weights for the Wyoming big sagebrush plant (*Artemisia tridentata*) from four locations in northern Nevada. The figure shows the frequency at which seeds of different weights occurred when collected from plants at each of these locations. Note that each distribution has a single peak and appears to be similar in form to the "bell curve," or normal distribution. In fact, though, Busso and Perryman show that each of these seed weight distributions is not normal and that the distributions vary among the locations and from year to year, arguing that precipitation level affects the distribution of seed weights. Thus, although the normal distribution arises in many situations, even a distribution that looks as if it could be "normal" may not be. Formal statistical methods that are used to evaluate whether a distribution is "normal" are discussed in many statistics texts.

2.3 Bar Charts and Histograms

Bar Charts

Bar charts are easy to construct if the underlying variable is nominal (e.g., the number of female or male meerkats that are babysitting at some time) since you simply let the height of each bar be the number of observations associated with each nominal value. There is no issue in this case

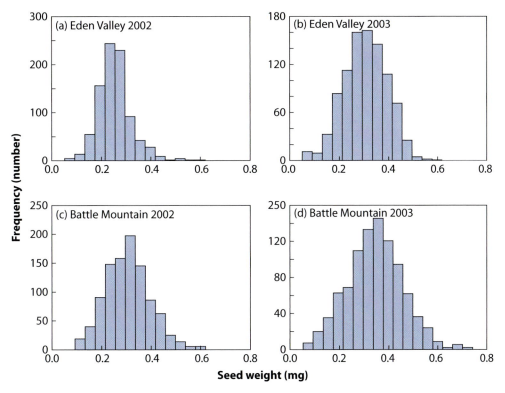

Figure 2.3 Frequency distribution of seed weight of *Artemisia tridentata* spp. *wyomingensis* from population samples in (a) Eden Valley, 2002 ($n = 880$); (b) Eden Valley, 2003 ($n = 938$); (c) Battle Mountain, 2002 ($n = 976$); and (d) Battle Mountain, 2003 ($n = 912$), in northern Nevada. Note the change of scale on the vertical axis for each of the four populations.

with defining the width of the range covered by each bar. If the underlying variable is ordinal but discrete (e.g., the number of tomatoes produced per plant must be an integer), then the only decision to be made is how to lump together the possible outcomes in selecting the range to be covered by each bar. In this case, generally, you should assign equal ranges to each bar and use 10 to 20 bars, assuming that the underlying data set has sufficient points to allow for more than 1 to 2 points per bar. For example, if the number of tomatoes per plant varied from 1 to 50, you could use 10 bars each of width 5 ($1 - 5, 6 - 10, \ldots, 46 - 50$). Difficulties arise here if the number of bars does not divide the range of the data evenly: there is no one solution to this because either one bar would have to be wider or narrower than the others or you must extend the range of the data beyond the observations. The usual rule is to have one of the bars cover a wider range of values than the other bars.

Example 2.1 (Poinsettia [60])

Poinsettias can be red, pink, or white. In one investigation of the hereditary mechanism controlling color, 182 progeny of a certain parental cross were categorized by color.

(Continued)

There are 108 red poinsettias, 34 pink poinsettias, and 40 white poinsettias. Draw a bar chart representing these data.

Solution:

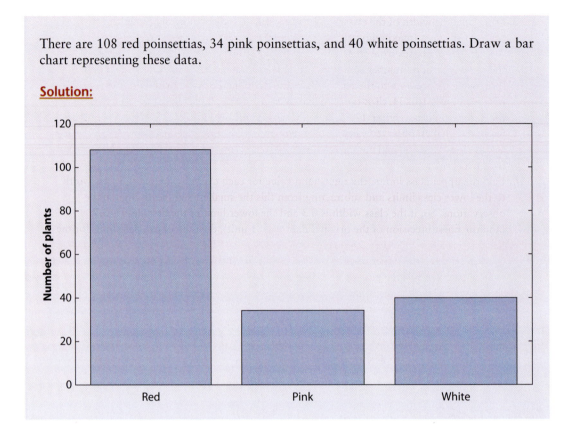

Histograms

Bar charts are used when the underlying data take on discrete values (e.g., the number of flowers on a single plant or the number of students in a class). In many situations, we are concerned with data sets that take on a continuous range of possible values, including individual height (measured in meters) and weight (measured in kilograms), heart rate (measured in beats per minute), and a leaf's rate of photosynthesis (measured in milligrams of CO_2 taken up per second per square centimeter of leaf surface). For data in which the underlying measurement can take on any possible value in a range, we use histograms rather than bar charts. Although these may look similar, the construction of a histogram explicitly acknowledges that the underlying data take on a range of values. In many cases, it may not be possible to measure these values to a fine degree. The ruler that you use to measure height may be readable only at the millimeter level, the balance you use to measure weight may be readable only down to 0.1 mg, and heart rate instruments typically display beats per minute as integer values. However, the underlying measurements are made on data that potentially take on any value within a range, and histograms are constructed to account for this.

CONSTRUCTING A HISTOGRAM

1. Decide on the number of classes (bars) for the histogram. Typically, choose this to be 10 to 20, but if there is a small data set, choose the number of classes to be 4 to 6. A rule of thumb might be that the number of observations divided by the number of classes be at least 4.

2. Determine the width of the classes (all widths will be the same) by dividing the range of the data (maximum value observed minus minimum value observed) by the number of classes from step 1. Round this value up so that the class width has the same precision (number of significant digits) as the measurements. For example, if the data are accurate to the tenths place and dividing the range by the number of classes gives 10.234, then the class width would be 10.3. If dividing the range by the number of classes gives 12.52, then the class width would be 12.6. Always round up.

3. Select the smallest observed value as the lower limit of the first class (e.g., the left-hand value for the first class) and add multiples of the class width from step 2 to obtain the lower limits for each class.

4. Find the upper class limits (the right-hand value for each class) by adding the class width to the lower class limits and subtracting from this the smallest significant digit in the observations. So, if the class width is 0.3 and the lower limit of the first class is 3.7 (and the finest precision of the original data was .1 unit), then the upper class limit for the first class is $3.7 + .3 - .1 = 3.9$, and the next class would have lower class limit 4.0. There is a gap between the upper class limit of one class and the lower class limit of the next class, and this gap is the precision of the original data. Note that the last class goes up through the highest value in the data set.

5. For the boundaries between classes, use the value that is halfway between the upper class limit of one class and the lower class limit of the next class. So in the above example, the class boundary would be 3.95, and the lowest class would be $3.65 - 3.95$, with the next class being $3.95 - 4.25$ and so on.

Example 2.2 (Constructing a Histogram Using Black Bear Data)

The black bear (*Ursus americanus*) population of the southern Appalachians has been observed and studied in detail since 1969 after bear managers grew concerned over declining bear populations in the region. During mark–recapture studies of the black bear population, many measurements of the bears were taken. Although a bear's weight might not be measured directly, the weight can be estimated from other measurements (total length, height at shoulder, neck circumference, and chest circumference). In 2002, Katie Settlage and her field research team collected the following weights for female black bears in the southern Appalachians (see [85] for further description of data and data collection procedures):

$$\begin{array}{ccccc}
60 & 85 & 95 & 85 & 115 \\
75 & 140 & 145 & 120 & 110 \\
90 & 115 & 75 & 125 & 80 \\
80 & 80 & 80 & 110 & 75 \\
120 & 150 & 38 & 118 &
\end{array}$$

where the weight is given in pounds. Let us go through the steps of constructing a histogram for this set of data.

(Continued)

1. Notice there are a total of 24 data points. Using the rule of thumb

$$\frac{\text{no. of data points}}{4},$$

we get $24/4 = 6$. We will use 6 classes.
2. The smallest value in the data set is 38; the largest value is 150. Thus, the range of the data is

$$\text{range} = 150 - 38 = 112.$$

To get the width of each class, we divide the range by the number of classes and round up to the precision of the data:

$$\text{class width} = 112/6 = 18.667 \approx 19.$$

3. The lower limit for each class is as follows:

Class	Calculation		Lower Limit
1	minimum data value		38
2	$38 + 19$		57
3	$38 + 2(19)$ or	$57 + 19$	76
4	$38 + 3(19)$ or	$76 + 19$	95
5	$38 + 4(19)$ or	$95 + 19$	114
6	$38 + 5(19)$ or	$114 + 19$	133

4. The precision of the data is to the ones place, so the smallest significant digit is 1. Thus, the upper limit for each class is as follows:

Class	Calculation	Upper Limit
1	$38 + 19 - 1$	56
2	$57 + 19 - 1$	75
3	$76 + 19 - 1$	94
4	$95 + 19 - 1$	113
5	$114 + 19 - 1$	132
6	$133 + 19 - 1$	151

5. Now we can find out class boundaries and determine the number of data points in each class:

Class	Lower Boundary	Upper Boundary
1	37.5	56.5
2	56.5	75.5
3	75.5	94.5
4	94.5	113.5
5	113.5	132.5
6	132.5	151.5

(*Continued*)

Notice that because of the precision of the boundaries, no data point can ever fall exactly on a boundary; it must be in one class or another. With this, we can make our histogram:

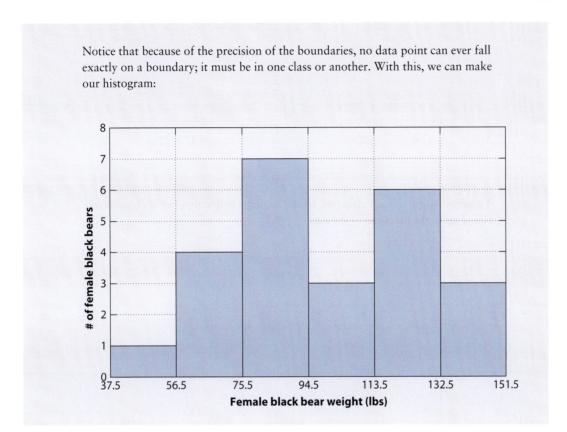

Table 2.1. A table that combines all the information needed to construct the histogram generated in Example 2.2.

Class	Lower Limit	Upper Limit	Lower Boundary	Upper Boundary	Frequency	Relative Frequency
1	38	56	37.5	56.5	1	1/24
2	57	75	56.5	75.5	4	1/6
3	76	94	75.5	94.5	7	7/24
4	95	113	94.5	113.5	3	1/8
5	114	132	113.5	132.5	6	1/4
6	133	151	132.5	151.5	3	1/8

Note that in Example 2.2, while showing each step carefully, we generated three different tables. Once you understand the process of generating a histogram, you can easily combine these steps into the production of one table (see Table 2.1).

Table 2.1 contains a column labeled "Relative Frequency." The ***relative frequency*** of a class is the frequency of that class divided by the total number of data points. We can make a histogram using the frequency or the relative frequency. Figure 2.4 shows the histogram from Example 2.2 displayed as a relative frequency histogram.

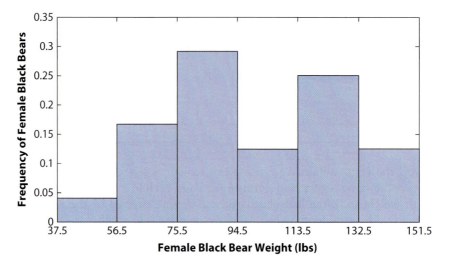

Figure 2.4 Histogram from Example 2.2 displayed as a relative frequency histogram.

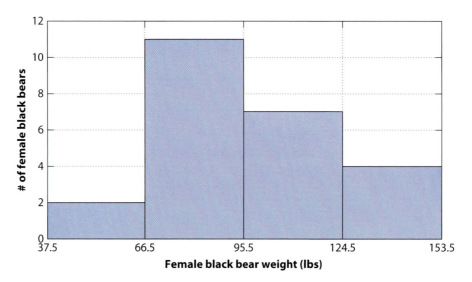

Figure 2.5 Histogram of female black bear weight using four classes.

Describing Data Distributions via Histograms

In looking at a histogram, we would like to have some vocabulary to describe features of the data that become obvious when displayed visually. For example, in Example 2.2, we might notice that the histogram seems to have two peaks. When we are presented with histogram data that have two peaks, we describe the data set as *bimodal*. Thus, the female black bear weight data could be described as bimodal. If a histogram has only a single peak, we describe the data as being *unimodal*. If a histogram has more than two peaks, we describe the data as being *multimodal*.

However, when classifying data as unimodal, bimodal, or multimodal, we must be careful. Suppose that in Example 2.2 we had chosen only four classes (instead of six). In that case, we would have produced the histogram shown in Figure 2.5. Notice that in this histogram, the data appear to be unimodal.

2.4 Scatter Plots

Bar charts and histograms are used to visually display single sets of measurements. In many cases, however, the value obtained from a measurement is related to other characteristics or factors affecting the same object being measured. Thus, a child's birth weight is related to the mother's birth weight [63], the leaf photosynthesis rate is affected by the light level, and the number of eggs laid by a spider may be related to the size of the mother [64].

The objective of a scatter plot is to determine visually whether there appears to be any relationship between two different observations or characteristics of a single biological object. Creating a scatter plot is easy once one decides which characteristics are to be included. Each axis is chosen to cover the range of measurements for that observation set. One measurement is chosen for the horizontal axis and the other for the vertical. Which is chosen for which axis does not matter unless you have some reason to believe that one measurement causes changes in the other, in which case it is normal to use the horizontal axis for the measurement that is viewed as the causative factor. Then, for each object measured, a single point is plotted on the graph at the appropriate coordinates. So, if we measure the number of eggs laid by 38 different spiders, the scatter plot will have 38 different points corresponding to the number of eggs laid and the size (perhaps measured as length in millimeters) of each spider (see Figure 2.6).

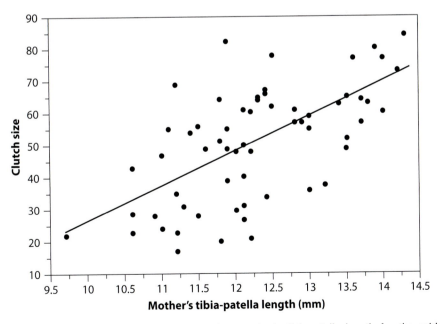

Figure 2.6 Clutch size (number of eggs) plotted against mother's tibia-patella length for the spider species *Holocnemus pluchei*. Image from [64].

A scatter plot provides only a visual display of the relationship between two measurements of some objects. It is not possible from a scatter plot alone to decide whether one factor "determines" the value of the other (e.g., one factor causes the other to have a particular value). For example, leaf length and leaf width may be closely linked in many deciduous plants in that scatter plots of these measurements on leaves show a close relationship between width and length (indeed, this is used to measure leaf area; see [51]), but this is not meant to imply that a certain width leaf "causes" the leaf to have a certain length. We say that width and length are "correlated," which we will define more carefully in the next chapter, but being "correlated" does not imply causation.

Scatter plots, though easy to construct, may not be very useful in the situation in which multiple factors affect the outcome of a particular experiment. The yield of a plant (in terms of fruit produced, weight of grain produced, or number of seeds produced) depends on multiple environmental and biological factors (e.g., disease level) that may be interrelated. Although it is possible to produce scatter plots in three (or more) dimensions, interpreting these is much more difficult than interpreting simple scatter plots with two axes. The field of multivariate statistics and the tools of data mining have been developed to tease apart the kinds of information we obtain from scatter plots in situations that have many interacting factors. Although the methods are much more complex than those we discuss here, the objectives are the same: to discern patterns from data that are not easily seen from looking at the raw data simply as lists of numbers.

2.5 Matlab Skills

Histograms

The command in Matlab used to make histograms is

```
hist(x)
```

where x is the data set. This command will automatically use 10 classes. If a different number of classes, say, five, is desired, then we would use the command

```
hist(x,5)
```

where the second entry in the function `hist` is the number of desired classes. The graph that is produced when using the function `hist` will have no axis labels and that the values along the horizontal axis label the midrange point of each class. To add axis labels to any graph in Matlab, use the commands

```
xlabel('horizontal label name here')
ylabel('vertical label name here')
```

after you have used the `hist` command. To change the values shown on the horizontal axis, use the function `set`. First, set the range of values for the horizontal axis using

```
set(gca,'XLim',[xmin,xmax])
```

where `gca` refers to the current graph that Matlab is manipulating, `'XLim'` indicates that you want to set the limits of the horizontal axis of the current graph, `xmin` is the minimum value you want for the horizontal axis, and `xmax` is the maximum value you want for the horizontal axis. Next, we want to change the values that appear on the horizontal axis to correspond with class boundaries. The values shown on the axes are known as ticks. The values on the horizontal axis are known in Matlab as XTicks. To change the values of the XTicks so that they correspond to the class boundaries, we use

```
set(gca,'XTick',TickArray)
```

where `'XTick'` indicates that you want to set the tick marks of the horizontal axis on the current graph and `TickArray` is the array of values you want to show up as ticks on your horizontal axis.

The m-file (see Appendix A for a description of m-files) that is used to create the histogram shown in Example 2.2 is shown below.

```
                          ─── BlackBearHist.m ───
1    % M-file to create Female Black Bear Histogram
2
3    % Enter data
4    w = [60 85 95 85 115 75 140 145 120 110 90 115 75 125 80 80 80 ...
5    80 110 75 120 150 38 118];
6
7    % Calculate range
8    range = max(w)-min(w);
9
10   % Calculate class width
11   cw = ceil(range/6);
12
13   % Make Histogram with labels
14   hist(w,6)
15   xlabel('Female Black Bear Weight (lbs)')
16   ylabel('# of Female Black Bears')
17   set(gca,'XLim',[min(w)-0.5,min(w)-0.5+6*cw])
18   set(gca,'XTick',[min(w)-0.5 : cw : min(w)-0.5+6*cw])
```

Refer to Appendix A for descriptions of the function `ceil` and how to form an array using the structure `[a:b:c]`. Note the lines 4 and 5 in this which illustrates the use of . . . to split a long line into two pieces. The . . . tells Matlab to concatenate (join) the two lines you type into a single line.

Scatter Plots

If you have a set of n points $\{(x_1, y_1), (x_2, y_2), \ldots, (x_n, y_n)\}$, the command in Matlab to make a scatter plot of the x values versus the y values is

```
plot(x,y,'.')
```

where x is the array for the set $x = \{x_1, x_2, \ldots, x_n\}$, y is the array for the set $y = \{y_1, y_2, \ldots, y_n\}$, and `'.'` indicates that you want to plot only the points (no lines connecting the points).

Suppose you have the following data:

x	2	5	2	4	6
y	4	7	5	8	11

To make a scatter plot, you could type the following into the command window of Matlab:

```
──────── Command Window ────────
>> x = [2 5 2 4 6];
>> y = [4 7 5 8 11];
>> plot(x,y,'.')
```

When you press "Enter" after the last command, a new window will pop up in Matlab containing an image of the scatter plot. Notice that it is hard to see all of the data points since the horizontal axis starts at the minimum x data point and ends at the maximum x data point and the vertical axis starts at the minimum y data point and ends at the maximum y data point. To fix this, use the commands

```
xlim([xmin xmax])
ylim([ymin ymax])
```

where the function `xlim` sets the horizontal axis to start at the value `xmin` and end at the value `xmax` and `ylim` sets the vertical axis to start at the value `ymin` and end at the value `ymax`.

We can add axis labels using the same `xlabel` and `ylabel` commands we used with `hist`. The following sequence of commands typed into the command window of Matlab will produce a scatter plot with axis labels where all the data are easily seen:

```
──────── Command Window ────────
>> x = [2 5 2 4 6];
>> y = [4 7 5 8 11];
>> plot(x,y,'.')
>> xlim([0 8])
>> ylim([2 13])
>> xlabel('x')
>> ylabel('y')
```

The `plot` function also allows us to make plots of lines and curves. Suppose we want to plot the line given by the equation $f(t) = 1.25t + 2.25$ and the curve given by the equation $g(t) = t^3 - 5t^2 + 8t + 3$. Since the `plot` function plots sets of points, it would seem that we must first create the $(t, f(t))$ and $(t, g(t))$ sets of points to plot. However, the `plot` function is clever, and if we define the set of t points to use, it can compute the corresponding $f(t)$ and $g(t)$ value within the `plot` function. Suppose we want to graph both functions from $t = 0$ to $t = 8$. Then we could create the t points using the following command:

```
──────── Command Window ────────
>> t = [0 : 0.1 : 8]
```

This command creates a vector with 81 entries. The first entry is 0, the second entry is 0.1, and the entries continue to increase by increments of 0.1 until they reach the value of 8.0. Now, we can plot our two functions $f(t)$ and $g(t)$:

```
──────────── Command Window ────────────
>> t = [0 : 0.1 : 8]
>> plot(t, 1.25*t + 2.25, '-')
>> figure
>> plot(t, t.^3 - 5*t.^2 + 8*t + 3, '-')
```

By entering the command `figure` between the two plots, we create another figure window so that the second plot does not erase the first plot by plotting over it. Notice also that we use `'-'` to indicate that we want to plot a line. If we want to plot the two functions on the same window, we would type the following into the command window:

```
──────────── Command Window ────────────
>> t = [0 : 0.1 : 8]
>> plot(t, 1.25*t + 2.25, 'r-', t, t.^3 - 5*t.^2 + 8*t + 3, 'g-')
>> legend('f(t)','g(t)')
>> xlabel('t')
```

The command `legend` creates a legend for the graph where the first entry (in single quotes) names the first set of points that are plotted and the second entry names the second set of points that are plotted. Notice that instead of just `'-'`, we use `'r-'` to indicate that we want a red line and `'g-'` to indicate that we want a green line. To find all the possible colors that can be used, type

```
──────────── Command Window ────────────
>> help plot
```

into the command window.

2.6 Exercises

2.1   Make a bar chart for the number of eastern bluebirds counted in the Breeding Bird Surveys in Tennessee from 1970 to 2010. Comment on any changes noted over time (adapted from USGS BBS).

Year	No. of Eastern Bluebirds
1970	200
1975	300
1980	125
1985	250
1990	425
1995	450
2000	575
2005	675
2010	500

2.2 These data were generated by a normal distribution with a mean of 10 and a standard deviation of 2. Make a histogram using 6 classes. The data were rounded to be integers.

$$4, 5, 12, 13, 7, 8, 11, 8, 11, 10, 8, 10, 10, 8, 10, 10, 11, 12, 8, 7, 8, 9, 10, 3, 12$$

2.3 The length of time an outpatient must wait for treatment is a variable that plays an important part in the design of outpatient clinics. The waiting times (in minutes) for 50 patients at a pediatric clinic are as follows:

35	22	63	6	49	19	15	83	46	19
16	31	24	29	26	68	42	57	64	8
23	47	21	51	7	40	19	46	16	32
108	33	55	32	22	36	25	27	37	58
39	10	42	72	13	51	45	77	16	28

Construct the following histograms:

(a) by hand using 5 classes
(b) with Matlab using 6 classes
(c) with Matlab using 10 classes
(d) with Matlab using 12 classes

After constructing the histograms, answer the following questions. Be clear and concise in your responses.

(e) What was the general shape of the distribution of the data in (a)? Unimodal? Bimodal?
(f) Did the shape of the distribution change as you changed the number of classes used?
(g) Given your response to (f), what might you conclude about the data?

2.4 The regulations of the Board of Health specify that the fluoride level in drinking water should not exceed 1.5 parts per million (ppm). Each of the 11 measurements below represent average fluoride levels over 15 days in ppm.

$$0.75 \quad 0.86 \quad 0.84 \quad 0.97 \quad 0.94 \quad 0.89 \quad 0.88 \quad 0.78 \quad 0.77 \quad 0.76 \quad 0.82$$

(a) Find the range of the data.
(b) Construct a relative frequency histogram for these data using an appropriate number of classes (by hand).

2.5 (From [47]) The Arithmetic Skills Test (A.S.T.) is a standardized basic mathematics skills test administered to general mathematics students at Western Kentucky University. The test has 30 multiple-choice questions with a minimum passing score of 21. The scores from a recent exam are given.

19	25	22	23	24	19	22	19	24	24
27	18	20	23	25	28	10	18	23	19
21	17	23	26	24	26	21	14	27	18
19	24	23	21	23	20	22	25	25	16
24	19	22	15	22	26	19	8	22	24

Construct a histogram for the data using the appropriate number of classes.

2.6   (From [47]) In a medical study, 70 guinea pigs were infected by a virus. The survival times (in days) are given.

40	45	53	56	56	57	58	62	64	73
74	79	78	80	81	81	81	82	83	83
84	85	89	90	91	92	92	97	99	99
99	99	101	102	102	102	103	104	107	108
109	113	114	118	121	123	126	128	137	138
134	144	145	147	156	162	174	174	179	184
191	198	208	214	247	249	328	383	403	511

Construct a relative frequency histogram for the guinea pig survival times.

2.7   (From [38]) The following observations are from data on lead concentration (in micrograms per cubic meter) recorded at an air monitoring station.

6.7	5.4	5.2	6.0	8.7	6.0	6.4	8.3	5.3	5.9
7.6	5.0	6.9	6.8	4.9	6.3	5.0	6.0	7.2	8.0
8.1	7.2	10.9	9.2	8.6	6.2	6.1	6.5	7.8	6.2
8.5	6.4	8.1	2.1	6.1	6.5	7.9	15.1	9.5	10.6
8.4	8.3	5.9	6.0	6.4	3.9	9.9	7.6	6.8	7.3
8.5	11.5	7.0	7.1	6.0	9.0	10.1	8.0	6.8	8.6
9.7	9.3	3.2	6.4						

Construct a frequency histogram for these data.

2.8   The following data are scores from an introductory math course for biology majors. Construct a histogram for these data.

66	50	92	60	87	60	64	83	95	99
76	50	69	68	97	63	50	60	72	80
81	72	93	92	86	62	61	65	78	82
85	64	81	88	66	65	79	96	95	88
84	83	59	60	64	76	99	76	68	73
85	90	70	71	60	90	84	80	68	86

Construct a histogram for these data. Consider that the data consist of the median grade on this test.

2.9   Suppose you have the following data representing a population over time with t in years. Make a scatter plot of these data and describe in words how the population is changing. Label the axes of your plot.

t	1	2	3	4	5	6
y	10	40	45	47	48	49

2.10   The following data represent two interacting populations as time t progresses. Make a scatter plot of t, x data and a scatter plot of t, y data. Then describe in words how the populations are changing. Label the axes of your plots.

t	1	2	3	4	5	6
x	10	5	2	10	5	2
y	2	4	3	2	4	3

Bivariate Data and Linear Regression

3.1 Introduction to Linear Regression

While visual analysis of data is useful, often it is necessary to use more exact methods to determine whether indeed there is a sufficiently strong relationship between two measurements to possibly eliminate the necessity of making both measurements. In many fish, there is a strong relationship between length and body mass, so that measuring the length alone can provide a very good estimate of the individual's weight. Organisms with determinant growth patterns may also have a very strong relationship between their size and age, so that measurements of individual size may be useful in estimating an individual's age. In these cases, an objective is to determine an equation that allows you to estimate one measurement on the basis of another measurement that might be easier to make.

A linear relationship between two measurements is the simplest description of a strong relationship since if this were to hold, knowing one measurement allows you to find the other simply by looking at the associated point on the line (unless, of course, the line is horizontal or vertical, in which case you learn nothing about one measurement by knowing the other measurement). Simple, linear relationships arise in many biological situations, particularly if the data are rescaled to account for factors that interact in a nonlinear manner.

Linear analysis also forms the basis for determining whether an hypothesis about experimental results should be accepted. This arises by doing a scatter plot in which the horizontal axis is the observed of some measurement on a biological object and the vertical axis is the predicted value of an associated measurement according to the hypothesis. An example of this is whether the observed frequency of codons (sequences of three nucleotides that code for amino acids to produce proteins) in amino acids is that expected from chance based on the number of different codons that give the same amino acid. The attached figure 3.1 illustrates that at least for this data set on codon frequencies in vertebrates, the slope of the line is close to one, providing evidence

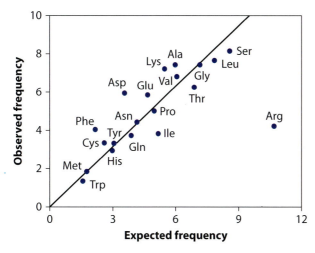

Figure 3.1 Plot of observed versus expected frequency of amino acids in vertebrate mRNA determined from the number of codons for an amino acid and the observed frequency of nucleotide bases [86].

that codon use is well predicted by the assumption of random occurrence (but there is evidence for a bias in use of codons that is not evident in this data set; see [25].)

3.2 Bivariate Data

One objective in using descriptive statistics is to help one understand whether there is a relationship between two measurements associated with an observation (e.g., whether leaf length and leaf width are related, whether the weights of bats are related to their age since birth, or whether the respiration rate of an individual is related to the temperature of the environment around that individual). Such data are called bivariate data and refer to measurements made on two variables per observation.

Bivariate data are often displayed as ordered pairs $(x_1, y_1), (x_2, y_2), \ldots, (x_n, y_n)$ or in a table:

x	x_1	x_2	x_3	$\cdots$	x_n
y	y_1	y_2	y_3	$\cdots$	y_n

The simplest type of relationship that would exist between two measurements x and y is that they are *proportional*:

$$y = kx.$$

where k is the proportionality constant.

A slight extension of this is that they are ***linearly related***:

$$y = mx + b.$$

where m and b are constants. A *regression* is a formula that describes the relationship between variables. A *linear regression* is one in which the relationship can be expressed in the equation of a line. In this case, an objective is to find the equation for a line that best describes the relationship.

3.3 Linear Analysis of Data

The first step is to see if indeed this process makes any sense at all. That is, do the data appear to be linearly related? This step is usually accomplished by constructing a scatter plot of the data. This means plotting them as x, y points on a graph. The choice of ranges for the axes matters a great deal here. A poor choice for the range of values on one of the axes could lead to all the points occupying a very small portion of the graph, making it impossible to tell whether there is any potential relationship in the data set. Thus, choose your axes so that their range is approximately the same as that in your data. So, if you have a set of body weights ranging from 5 to 160 grams, it would be reasonable to choose the axis for body weight to run from 0 to 160 grams or so.

Which is the x (horizontal) axis, and which is the y (vertical)? If you have some reason to expect that one of the measurements is the dependent one (e.g., body weight depends on age rather than the reverse), then choose the one that is dependent (weight) as the y-axis and the independent one (age) as the x-axis. If you do not have any reason to suspect that one of the measurements is caused by the other, then the choice of which measurement is plotted on which axis does not matter. Just choose one and go with it.

The next step is to eyeball the data and see if there appears to be any relationship. If the scatter plot looks like the points might be described at least approximately by a line (don't worry if there is a lot of scatter about any line you might draw on the graph), then it is reasonable to proceed with fitting a line. Eyeballing the data and making a guess at a line without doing any calculations or using any program will not give you an exact answer, but it will be useful later when checking to see that the line that is obtained from using a computer or another method makes sense. To eyeball, quickly draw a line through the data, eyeball a rough slope, and then write down the formula for the line using the point-slope form, that is,

$$y - y_1 = m \cdot (x - x_1),$$

where m is the slope and the point you have chosen is (x_1, y_1), or use the point-intercept form if one can easily estimate where the line crosses the y-axis, that is,

$$y = m \cdot x + b,$$

where b is the y-intercept. In doing this, be sure that you keep your units straight so that you know the units of each measurement and thus the units of the slope you have calculated. Note that given two points on the line, (x_1, y_1) and (x_2, y_2), the slope is

$$m = \frac{y_2 - y_1}{x_2 - x_1}.$$

SUMMARY OF FREEHAND LINEAR FIT STEPS

1. See if the data appear to be linearly related. Make a scatter plot of the data. This means plotting them as (x, y) points on a graph. Choose your axes so that the range is approximately the same as that in your data.
   Which is the x- (horizontal) axis and which is the y- (vertical) axis?
   - If one is the dependent (e.g., body weight depends on age rather than the reverse), then choose the one that is dependent (weight) as the y-axis and the independent one (age) as the horizontal x-axis.
   - If you do not have any reason to expect that one of the measurements is dependent on the other, then choose one and go with it.

   Eyeball the data and see if there appears to be any relationship. If the scatter plot looks like the points might be described approximately by a line, then it is reasonable to proceed with fitting a line.
2. Draw a freehand line that estimates the line of best fit. Draw a line through the data, trying to minimize the vertical distance between the line and the points.
3. Using two points ON THE LINE (not necessarily data points), find the equation of the line.

Example 3.1 (Freehand Linear Fit)

Given the set of bivariate data

x	2	5	2	4	6
y	4	7	5	8	11

find the equation of a line that fits the data.

Solution: Let us use the Freehand Linear Fit Steps given in the box above

1. If we make a scatter plot of the data, we see that there seems to be a linear relationship between the data points. Since we do not know what x and y stand for, we will plot x on the x-axis and y on the y-axis.

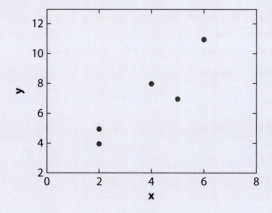

(Continued)

2. We can freehand draw a line through the data that attempts to minimize the distance between the line and the points.

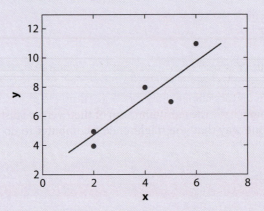

3. Next we pick two points ON THE LINE to determine the equation of the line. Notice that none of the data points fall on the line, so we DO NOT use the data points to estimate the equation of the line.

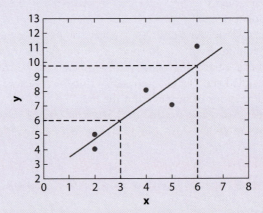

We choose the two points $(3, 6)$ and $(6, 9.75)$. Notice that both of these points are ON THE LINE. First, we calculate the slope of the line using the two points,

$$m = \frac{y_1 - y_2}{x_1 - x_2} = \frac{6 - 9.75}{3 - 6} = \frac{-3.75}{-3} = 1.25.$$

So far, we have

$$y = 1.25x + b.$$

To find b, we use the equation and one point on the line $(3, 6)$:

$$6 = 1.25(3) + b$$

$$6 = 3.75 + b$$

$$2.25 = b.$$

Thus, the equation for our freehand line is

$$y = 1.25x + 2.25.$$

What if the data do not appear to be linearly related? In this case, don't try to fit a straight line. Later we'll discuss ways to transform the data to see if a different relationship might be a better choice.

Least-Squares Fit

If it looks like a linear relationship is a reasonable assumption, the next step is to find the "least-squares fit." This can be done automatically in Matlab (see Matlab Skills at the end of the chapter), and many calculators can also do this. The idea is to choose a line that "best fits" the data in that the data points have the minimum sum of their vertical distances from this particular line. The following is one way that you might code a computer to do this:

    (i) Guess at the equation for the line $y = mx + b$.
    (ii) Measure the vertical distance from each data point to the line you have chosen.
    (iii) Sum up the distances chosen in (ii).
    (iv) Change the slope m and the intercept b to see if you can reduce the sum obtained in (iii).
    (v) Continue this until you get tired or you have done the best you can.

One of the potential problems with the above is what "distance" to use in (ii). If you allow some distances to be positive and some negative (e.g., for a point above the line and a point below the line), then the distances can cancel, which is not appropriate. So we want all distances to be positive, and the standard way to do this is to square each vertical distance found in (ii) and produce the sum of the square distances to get (iii). The line we would get by going through step (v) would then be called the "least-squares fit" since it chooses the line so as to minimize the sum of the square deviations of points from the line.

It turns out that it is not necessary to go through steps (i) to (v) above at all. It can be proven that the "best" values of the slope m and the intercept b can be obtained from a relatively simple formula that just uses the x- and y-values for all the data points.

Suppose that there is a set of n data points (x_i, y_i) where you have plotted these using a scatter plot and it appears that a linear relationship between them is reasonable. We now use $\hat{y}, \hat{m}$, and $\hat{b}$ for the y-value, slope, and y-intercept of the least-squares line. The least-squares line (regression line) that best fits these data,

$$\hat{y}_i = \hat{m}x_i + \hat{b},$$

has the regression coefficients $\hat{m}$ and $\hat{b}$ chosen so as to minimize the sum of the errors squared:

$$\sum_{i=1}^{n} (y_i - \hat{y}_i)^2 = \sum_{i=1}^{n} \left[y_i - \left(\hat{m}x_i + \hat{b} \right) \right]^2.$$

This says that the regression line that "best fits" the data is the line chosen to minimize the sum of the differences (squared) between the data points y_i and the y-values predicted by the regression line $\hat{y}_i$.

The values of the regression coefficients are calculated as

$$\hat{m} = \frac{S_{xy}}{S_{xx}},$$

where

$$S_{xx} = \sum_{i=1}^{n} x_i^2 - \frac{1}{n}\left(\sum_{i=1}^{n} x_i\right)^2 = \sum_{i=1}^{n}(x_i - \bar{x})^2 \tag{3.1}$$

and

$$S_{xy} = \sum_{i=1}^{n} x_i y_i - \frac{1}{n}\left(\sum_{i=1}^{n} x_i \sum_{i=1}^{n} y_i\right) = \sum_{i=1}^{n}(x_i - \bar{x})(y_i - \bar{y}) \tag{3.2}$$

and

$$\hat{b} = \bar{y} - \hat{m}\bar{x},$$

where $\bar{x}$ and $\bar{y}$ are the arithmetic means (or averages) of the x and y data, respectively. Note that $y_i - \hat{y}_i$ is called the ith residual or error. We denote

$$\text{RES} = \sum_{i=1}^{n}(y_i - \hat{y}_i)^2,$$

which is the sum of the residuals squared.

Note that the point $(\bar{x}, \bar{y})$ is on the least-squares regression line.

Example 3.2 (Least-Squares Linear Regression)

Using the data in Example 3.1, let us find the best-fit line using a least-squares linear regression. First, note that $\bar{x} = 3.8$ and $\bar{y} = 7$. Next, we calculate S_{xx} and S_{xy}.

$$S_{xx} = (2 - 3.8)^2 + (5 - 3.8)^2 + (2 - 3.8)^2 + (4 - 3.8)^2 + (6 - 3.8)^2$$
$$= 12.8$$
$$S_{xy} = (2 - 3.8)(4 - 7) + (5 - 3.8)(7 - 7) + (2 - 3.8)(5 - 7)$$
$$+ (4 - 3.8)(8 - 7) + (6 - 3.8)(11 - 7)$$
$$= 18$$

Now, we can calculate the slope:

$$\hat{m} = \frac{S_{xy}}{S_{xx}} = \frac{18}{12.8} = 1.4.$$

Next, we calculate the y-intercept:

$$\hat{b} = 7 - (1.4)(3.8) = 1.7.$$

Now, we write the equation for our best-fit line using the least-squares linear regression method:

$$\hat{y} = 1.4\hat{x} + 1.7.$$

Notice that the equation for the line we freehand drew in Example 3.1 was not the best-fit line by the least-squares method.

Once you have a linear least squares fitted line, you can proceed to use it to do the following:

- *interpolate*: find the y-value predicted by the linear fit for an x-value that falls in the range of the x-values in your data set
- *extrapolate*: find the y-value predicted by the linear fit for an x-value that falls outside the range of the x-values in your data set

---

Example 3.3 (Interpolation and Extrapolation)

Using the data and best-fit line found in Example 3.2, estimate the y-values that correspond to the following x-values: (a) $x = 5$ and (b) $x = 10$. Which corresponds with performing an interpolation and which with an extrapolation?

Solution: The equation of the best-fit line found in Example 3.2 was

$$y = 1.4x + 1.7.$$

(a) $y = 1.4(5) + 1.7 = 8.7$
   Since $x = 5$ falls in the range of x-values in the data set, this is an interpolation.
(b) $y = 1.4(10) + 1.7 = 15.7$
   Since $x = 10$ falls outside of the range of x-values in the data set, this is an extrapolation.

3.4 Correlation

How can we tell if a linear fit is any good?

This is where we make use of the notion of a correlation. In common parlance, we say that two measurements are "correlated" if there appears to be some relationship between them, though this relationship need not be causal. Thus, leaf length and width might be related to each other, but neither is caused by the other. They might simply be related because of the age of the leaf or the environmental conditions under which the leaf developed (e.g., better nutrients and water could produce a larger leaf).

We will use a formal definition of correlation, which essentially tells how close two measurements are to being linearly related. Note that this definition is restricted to describing a measure for linear relationships. If two measurements are related but not linearly, then the correlation may not imply that the measurements are closely related when they actually are. For example, human body weight is certainly related to age as an individual grows, but growth is not linear at all, so a correlation may not be the best way to that these two variables are related.

It is important to note that correlation does NOT imply causation. Rather, it is appropriate to say that the complex dependence between two variables that are very correlated may be adequately summarized as a linear one.

Correlation is measured by the ***correlation coefficient***, for which the small Greek letter ρ (spelled "rho," pronounced like "row") is typically used. The calculation of ρ follows easily

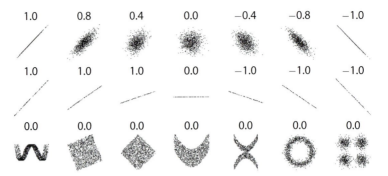

Figure 3.2 Various sets of data with their corresponding correlation coefficients shown above each data set. Notice in the bottom row that there can be a correlation in the data, but the value for the correlation coefficient ρ is zero because there is not a linear correlation. Figure taken from [87].

from the x- and y-values of the data set:

$$\rho = \frac{S_{xy}}{\sqrt{S_{xx}S_{yy}}}, \tag{3.3}$$

where

$$S_{yy} = \sum_{i=1}^{n} y_i^2 - \frac{1}{n}\left(\sum_{i=1}^{n} y_i\right)^2 = \sum_{i=1}^{n}(y_i - \bar{y})^2 \tag{3.4}$$

and S_{xx} and S_{xy} are defined by Equations 3.1 and 3.2, respectively. Note that $-1 \leq \rho \leq 1$. Figure 3.2 shows correlation coefficients for various types of data sets.

Next, we develop another way to interpret ρ and to measure how well the regression line fits a data set. Recall that to find the regression line, we minimized the sum of the errors squared (RES):

$$\text{RES} = \sum_{i=1}^{n}\left(y_i - \hat{y}_i\right)^2.$$

Define the total sum of squares (TSS) of the data set as

$$\text{TSS} = \sum_{i=1}^{n}(y_i - \bar{y})^2$$

(note that $\text{TSS} = S_{yy}$) and the sum of squares of the regression (SSR) as

$$\text{SSR} = \sum_{i=1}^{n}\left(\hat{y}_i - \bar{y}\right)^2,$$

where $\hat{y}_i$ is the point on the regression line that corresponds with x_i. It can be shown that

$$\text{TSS} = \text{RES} + \text{SSR}.$$

Note that if not all of the y_i are exactly on the regression line, RES $\neq 0$ and TSS $>$ SSR. If all the y_i points are exactly on the line, that is, $y_i = \hat{y}_i$ for all i, then RES $= 0$ and TSS $=$ SSR. Thus, the closer the points are to the regression line, the closer TSS is to SSR.

The *coefficient of determination* is defined to be

$$R^2 = \frac{\text{SSR}}{\text{TSS}}.$$

So, as the data points get close to being exactly on a line, SSR gets close to TSS, and thus R^2 gets close to 1. When R^2 is close to 1, the points are said to be highly correlated, meaning that a very large proportion of the TSS is accounted for by the regression (SSR).

It is possible to show that

$$\text{SSR} = \frac{S_{xy}^2}{S_{xx}}$$

and thus

$$R^2 = \frac{\text{SSR}}{\text{TSS}} = \frac{S_{xy}^2}{S_{xx}S_{yy}} = \rho^2.$$

Therefore, the correlation coefficient, ρ, can be thought of as measuring how well a regression line fits a data set. Thus, the correlation coefficient is also is a measure of the strength of the straight-line relationship between the two measurements, scaled in such a way that if two measurements fall exactly on a straight line with positive slope, then $\rho = 1$, while if they fall exactly on a line with negative slope, $\rho = -1$. In these cases, we say that the data are *perfectly positively correlated* or *perfectly negatively correlated*. If $\rho = 0$, we say that the data are *uncorrelated*, but again this does not mean that the data are not related. For example, if the data fell on a parabola $y = (x - 1)^2$ for values of x between 0 and 2, the correlation would be near zero, but the data are certainly closely related (see Figure 3.2 for other examples). If $R^2 = 0.95$ for a certain data set, then we say that the regression line accounts for 95% of the variance in the data.

Example 3.4 (Correlation)

How correlated are the data presented in Example 3.1?

Solution: We need to calculate ρ. We have already calculated $S_{xx} = 12.8$ and $S_{xy} = 18$ in Example 3.2. However, we still need to calculate S_{yy}.

$$S_{yy} = (4 - 7)^2 + (7 - 7)^2 + (5 - 7)^2 + (8 - 7)^2 + (11 - 7)^2$$
$$= 30$$

Thus,

$$\rho = \frac{S_{xy}}{\sqrt{S_{xx}S_{yy}}} = \frac{18}{\sqrt{12.8 \cdot 30}} = 0.92.$$

Since this is a value very close to 1, we would say that the x and y data are highly correlated.

Example 3.5 (Negative Correlation)

The size of a tree is typically denoted by measuring the diameter of the tree at breast height (diameter at breast height, or DBH). One would expect that the larger trees are, the less they could grow in a particular area. Let us test this hypothesis by looking at a set of data for trees. In each area, the number of trees per acre is recorded along with the average DBH.

DBH	22.9	30.0	30.3	27.8	24.1	28.2	26.4	12.8	39.7	38.0
Trees/acre	31.5	18.3	18.0	21.3	28.4	20.7	23.7	100.7	10.5	11.4

Solution: We would like to see if there is a negative correlation between the DBH and the trees/acre. To calculate the correlation coefficient, we must calculate S_{xy}, S_{xx}, and S_{yy}. Let DBH be the horizontal variable,

$$x = \{22.9, 30.0, 30.3, 27.8, 24.1, 28.2, 26.4, 12.8, 39.7, 38.0\},$$

and the trees/acre be the vertical variable,

$$y = \{31.5, 18.3, 18.0, 21.3, 28.4, 20.7, 23.7, 100.7, 10.5, 11.4\}.$$

Next, to calculate S_{xx}, S_{yy}, and S_{xy}, we need to know the means of x and y. You should check that $\bar{x} = 28.02$ and $\bar{y} = 28.45$. Now,

$$S_{xx} = (22.9 - 28.02)^2 + (30.0 - 28.02)^2 + \cdots + (39.7 - 28.02)^2 + (38.0 - 28.02)^2$$
$$= 521.1$$
$$S_{yy} = (31.5 - 28.45)^2 + (18.3 - 28.45)^2 + \cdots + (10.5 - 28.45)^2 + (11.4 - 28.45)^2$$
$$= 6188.2$$
$$S_{xy} = (22.9 - 28.02)(31.5 - 28.45) + \cdots + (38.0 - 28.02)(11.4 - 28.45)$$
$$= -1530.9$$

$$\rho = \frac{S_{xy}}{\sqrt{S_{xx}S_{yy}}} = \frac{-1530.9}{\sqrt{521.1 \cdot 6188.2}} = -0.85.$$

Yes, there is a negative correlation between the average DBH and the trees/acre. Notice that to check for negative versus positive correlation, we must check ρ since the R^2 value will always be positive:

$$R^2 = 0.7225.$$

Thus, the regression line accounts for 72.25% of the variance in the data. Although R^2 is a good measure of the correlation of two sets of data, it will not reveal positive or negative correlation.

3.5 Matlab Skills

Linear Regression

Matlab can easily determine the slope and *y*-intercept for the best line through a set of data. Use the command

```
C=polyfit(X,Y,1)
```

which produces the vector C, in which the first value is the best fit for the slope and the second is the best fit for the intercept for the least-squares fit of the vector of data Y (on the vertical axis) to the vector of data X (on the horizontal axis).

For example, if we wanted to find the least-squares regression line for the data in Example 3.1, we could type into the command window:

```
                              Command Window
>> x = [2 5 2 4 6];
>> y = [4 7 5 8 11];
>> C = polyfit(x,y,1)
C =
    1.4062    1.6562
```

Thus, the equation for the least-squares regression line would be

$$\hat{y} = 1.4062x + 1.6562.$$

Interpolation and Extrapolation

To interpolate and extrapolate in Matlab, use the command

```
yhat=polyval(C,xhat)
```

where C is the vector resulting from applying the polyfit function to the data set, xhat is the horizontal axis value at which you would like to estimate the corresponding vertical axis value, and yhat is the corresponding vertical axis value.

Continuing from our example above, to find estimated *y*-values for $x = 3$ and $x = -1$ using the linear regression for the data in Example 3.1, we could type the following into the command window:

```
                              Command Window
>> yhat1 = polyval(C,3)
yhat1 =
    5.8750
>> yhat2 = polyval(C,-1)
yhat2 =
    0.2500
```

Correlation Coefficients

Again, Matlab makes it easy to calculate the correlation coefficient of two vectors using

```
rho=corrcoef(X,Y)
```

where X is the vector of data on the horizontal axis, Y is the vector of data on the vertical axis, and rho is an array (in our case a 2 × 2 array) in which the correlation coefficient is shown on the off diagonal (i.e., the first-row, second-column entry and the second-row, first-column entry). Many calculators will compute this as well. Note that the correlation coefficient is a dimensionless number: the dimensions of the measurements cancel out when this number is calculated.

Continuing from our example above, if we want to find the correlation coefficient for the data in Example 3.1, we could type the following in the command window:

─────────────────── Command Window ───────────────────

```
>> rho=corrcoef(x,y)
rho =
      1.0000      0.9186
      0.9186      1.0000
```

Thus, for our data set, $\rho = 0.9186$.

Although in the examples above we typed all of our commands into the command window, we could also write an m-file that executes all of these commands. Below is an example of such an m-file.

─────────────────── LSR.m ───────────────────

```
1    % Filename: LSR.m
2    % M-file to
3    %     - compute equation for least squares regression line
4    %     - plot least squares regression line on a graph with the data
5    %     - compute the correlation coefficient
6    %     - compute the coefficient of determination
7    % LSR = least square regression
8
9    % Enter the data
10   x = [2 5 2 4 6];
11   y = [4 7 5 8 11];
12
13   % Find the equation for the LSR line
14   C = polyfit(x,y,1);
15   % Display the equation
16   fprintf('Eqn for LSR: yhat = %f x + %f\n',C(1),C(2))
17
18   % Find the yhat value for each x value
19   yhat = polyval(C,x);
20
21   % Plot the data and the LSR line
22   plot(x,y,'k.',x,yhat,'k-')
23   xlabel('x')
24   ylabel('y')
25   xlim([min(x)-1 max(x)+1]) %Set x-axis a little wider than data
26
27   % Find the correlation coefficient
28   rho = corrcoef(x,y);
29   % Display the correlation coefficient
30   fprintf('rho = %f\n',rho(1,2))
31   % Display the coefficient of determination
32   fprintf('The regression line accounts for %5.2f%% ',rho(1,2)^2*100)
33   fprintf(' of the variance in the data. \n')
```

See Appendix A for details on creating and running an m-file. When this m-file is run in the command window, the output looks like the following:

```
──────────────────── Command Window ────────────────────
>> LSR
Eqn for LSR: yhat = 1.406250 x + 1.656250
rho = 0.918559
The regression line accounts for 91.86% of the variance in the data
```

The graph is shown in Figure 3.3.

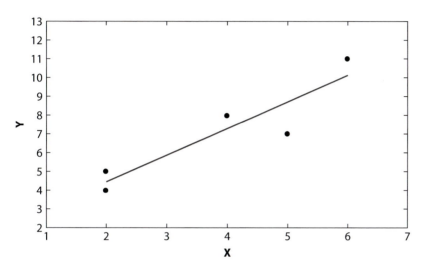

Figure 3.3 Graph produced by the m-file LSR.m.

3.6 Exercises

3.1   How correlated are the x- and y-values in Exercise 2.10?

3.2   Suppose that the regression line for a set of data is $y = 2.1x + b$ and that the line passes through the point $(1, 6)$. What is the relationship between the mean of the x-values and the mean of the y-values?

3.3   For the eastern bluebird data from Exercise 2.1, make a scatter plot of years and number of bluebirds, draw a freehand line, and find the equation of your freehand line.

For each of the problems 3.4 to 3.9, do the following:

(a) Make a scatter plot.
(b) Draw a freehand line.
(c) Find the equation of your freehand line.
(d) Calculate the least-squares regression line for the data. Do this by hand for the first problem and use Matlab for the remainder of the problems.
(e) Compute the correlation coefficient. Again, do this by hand for the first problem and use Matlab for the remainder of the problems. Using the correlation coefficient, state

how much confidence you place in a prediction based on the regression line. State whether you think the data are strongly or weakly related.

(f) Choose a value that falls in the range of the data that you plotted on the horizontal axis but that is not one of the data points. Interpolate the corresponding vertical axis value using the equation for the least-squares regression line.

(g) Choose a value that falls outside the range of the data you plotted on the horizontal axis. Extrapolate the corresponding vertical axis value using the equation for the least-squares regression line.

3.4   The average length and width of various bird eggs are given in the following table.

Bird Name	Width (cm)	Length (cm)
Canada goose	5.8	8.6
Robin	1.5	1.9
Turtledove	2.3	3.1
Hummingbird	1.0	1.0
Raven	3.3	5.0

3.5   An investigator was interested in examining the effect of different doses of a new drug on pulse rate in humans. Four doses were used in the experiment. Three people were randomly assigned each of the four doses. After a pre study pulse rate was recorded for each individual, subjects were injected with the appropriate drug dose. Pulse rates were again recorded an hour later. The changes in pulse rates in beats per minute (bpm) are listed below.

Dose (mL/kg of body weight)	1.5	1.5	1.5	2.0	2.0	2.0	2.5	2.5	2.5	3.0	3.0	3.0
Change in pulse rate (bpm)	20	21	19	16	17	17	15	13	14	8	10	8

3.6   Suppose that we are interested in the relation between carbon monoxide (CO) concentrations and the density of cars in some geographical area. The hundreds of cars per hour to the nearest 500 cars and the concentration of CO in parts per million (ppm) at a particular street corner are measured. The results are as follows:

Cars/hour	1.0	1.0	1.0	1.5	1.5	1.5	2.0	2.0	3.0	3.0	3.0	3.0
[CO]	9.0	6.8	7.7	9.6	6.8	11.3	12.3	11.8	20.7	19.2	21.6	20.6

3.7   (From [60]) In a study of a free-living population of the snake *Vipera bertis*, researchers caught and measured nine adult females. Their body lengths x and weights y are shown in the table below.

Length (cm), x	60	69	66	64	54	67	59	65	63
Weight (g), y	136	198	194	140	93	172	116	174	145

3.8   The data below give the infant mortality rate (MR) per 1000 live births in the United States for the period of 1960–1979.

Year	1960	1965	1970	1971	1972	1973	1974	1975	1976	1977	1978	1979
MR	26.0	24.7	20.0	19.1	18.5	17.7	16.7	16.1	15.2	14.1	13.8	13.0

3.9   Researchers in a laboratory experiment measured fuel consumption rates in miles per gallon (mi/gal) of a car at different speeds in miles per hour (mph) in an attempt to study the fuel economy.

mph	15	23	30	35	42	45	50	54	60	65
mi/gal	14	17	20	24	26	23	18	15	11	10

3.10   (a) Complete a least-squares regression for the data in Example 3.1 to find the equation of a best-fit line of the y data in terms of the x data.

(b) Complete a least-squares regression for the data in Example 3.1 to find the equation of a best-fit line of the x data in terms of the y data.

(c) Solve the equation in (b) for y in terms of x. Is this the same equation that you obtained in (a)? Explain why you think the answers were the same or different.

(d) Compute the R^2 value in each case and interpret the results.

CHAPTER

Exponential and Logarithmic Functions

4.1 Exponential and Logarithmic Functions in Biology

If you were to plot data points obtained from biological measurements (e.g., mean brain weight as compared to mean total body weight for mammals of various sizes), you would often find that the data do not fall on a straight line. There are a variety of reasons for this. One example is illustrated by the human auditory system through which we hear sounds. Imagine yourself in a classroom with one other person in it and that person drops a pencil (assume that the floor does not have a rug). You would likely hear the pencil drop. Now imagine the same person dropping the pencil with the classroom full of people talking before class starts. You likely would not hear the pencil fall at all, but it certainly is making the same "sound" (e.g., the physics of the situation has not changed). Why does this happen? It is because our hearing is better at detecting relative rather than absolute differences between sound levels. Of course, this is very simplified since it also depends on the frequency of the sound, the person's age, and how much exposure he or she has had to previous loud sounds (e.g., rock concerts with too much amplification or having one's iPod turned up too loud), but in general much of our perception does not occur in a linear manner. Our perceptions are tuned to detect not "additive differences" but rather "multiplicative differences."

Another example is population growth. Imagine algae growing in a Petri dish, starting from a single cell. Through time, the cell will split (or else die), and then each new cell will split again and continue to do so; thus, the total number of cells in the Petri dish does not increase linearly (in an additive manner) through time but rather multiplicatively (by doubling). If you were to plot the number of cells through time, it would not increase linearly.

The above are two examples why exponentials and logarithms are used so much in biology. Exponentials are used to describe something that increases (or decreases) in a multiplicative manner. Logarithms are a way to rescale something that is increasing (or decreasing) in a multiplicative manner so as to measure its increase in a new way that does increase (or decrease) linearly. This arises, as you will recall, from the fact that the logarithm of a product is the sum of the logarithms of the components of the product, that is, $\log(ab) = \log(a) + \log(b)$. This would be a good time to refresh your memory of logarithms and exponentials by reading the following section.

4.2 Review of Exponential and Logarithm Properties

Exponential Functions

An **exponential function** is a function of the form

$$f(x) = a^x,$$

where $a > 0$ is a fixed real number called the **base** of the exponent. Note: This is different from the **power function** x^k, where the base is the variable and the exponent k is a fixed constant.

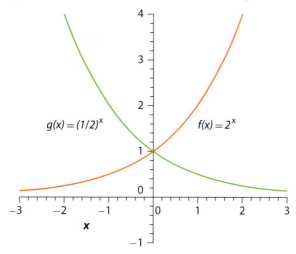

Consider the functions $f(x) = 2^x$ and $g(x) = \left(\frac{1}{2}\right)^x$. Note that $f(x)$ is an increasing function and that $g(x)$ is a decreasing function. Also, as x grows larger, $f(x)$ grows larger and larger and $g(x)$ decreases to be closer and closer to 0. So, we call $f(x)$ an **exponential growth function** ($a > 1$) and $g(x)$ an **exponential decay function** ($0 < a < 1$).

What happens if $a = 1$? Then we have a the function $f(x) = 1^x = 1$, which is a constant function.

Laws of Exponents

1. $a^{(x+y)} = a^x \cdot a^y$
2. $a^{(x-y)} = a^x \cdot a^{-y} = \frac{a^x}{a^y}$
3. $a^{xy} = \left(a^x\right)^y$
4. $a^x b^x = (ab)^x$
5. $a^0 = 1$
6. $a^{-x} = \frac{1}{a^x}$

Logarithmic Functions

A **logarithmic function** is the inverse of an exponential function and has the form $f(x) = \log_b(x)$, where $b > 0$ is a fixed real number referred to as the **base** of the logarithm. The function

$f(x) = \log_b(x)$ is the inverse of $g(x) = b^x$, that is,

$$\log_b\left(b^x\right) = b^{\log_b x} = x.$$

Recall that $\log_2 8 = 3$ since 3 is the exponent used to raise the base 2 to get 8, that is, $2^3 = 8$. Notation:

- log usually refers to "log base 10" or $\log_{10}$
- ln always refers to "log base e" or $\log_e$, where $e \approx 2.71828\ldots$

Laws of Logarithms

1. $\log_a(xy) = \log_a x + \log_a y$
2. $\log_a(\frac{x}{y}) = \log_a x - \log_a y$
3. $\log_a x^k = k\log_a x$
4. $\log_a a = 1$ and $\log_a 1 = 0$
5. $\log_a x = (\log_{10} x)/(\log_{10} a)$
6. $\log_a x = (\ln x)/(\ln a)$.

Now we will formally find the inverse of $f(x) = a^x$. Let $y = f(x)$. Then, taking the log of both sides gives

$$\begin{aligned}
y &= a^x \\
\log y &= \log(a^x) \\
\log y &= x \log a \\
x &= \frac{\log y}{\log a} \\
x &= \log_a y.
\end{aligned}$$

Since $f^{-1}(y) = x$, we can write

$$f^{-1}(y) = \log_a y.$$

So, we see that the exponential and logarithm functions are indeed inverses of each other.

Note that the range of the exponential function consists of positive real numbers, implying that the domain of the logarithmic function consists of positive real numbers.

Example 4.1 (Solving for an Exponent)

Solve $5^x = 0.23$ for x.

Solution: Taking ln of both sides,

$$5^x = 0.23$$

(Continued)

$$\ln 5^x = \ln 0.23$$
$$x \ln 5 = \ln 0.23$$
$$x = \frac{\ln 0.23}{\ln 5}$$

Now, use your calculator to find that $x = -0.91$.

Notice that in solving for an exponent, we need to use a log function.

Example 4.2 (Working with Logs)

If $\log_a x = 2.1$ and $\log_a y = 0.45$, compute $\log_a(x^3 y)$.

Solution: Using properties of logs,

$$
\begin{aligned}
\log_a(x^3 y) &= \log_a x^3 + \log_a y \\
&= 3 \log_a x + \log_a y \\
&= 3(2.1) + 0.45 \\
&= 6.3 + 0.45 \\
&= 6.75.
\end{aligned}
$$

Example 4.3 (More Working with Logs)

Solve $\log_4(3x) = 1.4$ for x.

Solution: Using the meaning of logarithm base 4,

$$
\begin{aligned}
\log_4(3x) &= 1.4 \\
3x &= 4^{1.4} \\
x &= \frac{1}{3} 4^{1.4}.
\end{aligned}
$$

Now, use your calculator to find that $x = 2.3$.

Logarithms and exponentials are frequently used to calculate the half-life of radioactive substances or the time required for a population to double.

The *half-life* of a radioactive substance is how long it takes for N amount of that substance to decay to $\frac{1}{2}N$ amount.

The *doubling time* of a population is how long it takes N individuals in that population to reproduce and become $2N$ individuals.

Example 4.4 (Half-Life of a Drug in the Body)

A usual assumption in pharmacokinetics, the field which studies the changes and effects of drugs in an individual, is that the amount of drug in a person's body decays exponentially. We assume that the drug is administered intravenously so that the amount of the drug in the bloodstream jumps almost immediately to its highest level. Then, as time passes, the amount of the drug in the bloodsteam decays exponentially.

Let $C(t)$ denote the amount of the drug measured in milligrams in the bloodstream at time t (in days), and let C_0 be the initial amount of drug in the bloodstream (i.e., the amount of drug administered in milligrams). Then we can write

$$C(t) = C_0 e^{-kt},$$

where $k > 0$ is known as a *decay constant*. The larger the value of k, the more quickly the drug decays in the bloodstream.

(a) If the drug has a half-life of 10 days, what is the value of k?
(b) What percent of the administered amount of drug remains in the bloodstream after 4 hours?

Solution:

(a) A half-life of 10 days means that when $t = 10$, we should have $\frac{1}{2}$ of the initial amount of drug, C_0. We can write this mathematically as

$$C(10) = \frac{1}{2}C_0.$$

Using the above equation, we know that

$$C(10) = C_0 e^{-10k}.$$

Properties of logarithms and exponentials give

$$\frac{1}{2}C_0 = C(10) = C_0 e^{-10k}$$

$$\frac{1}{2}C_0 = C_0 e^{-10k}$$

$$\frac{1}{2} = e^{-10k}$$

$$\ln \frac{1}{2} = \ln\left(e^{-10k}\right)$$

$$\ln \frac{1}{2} = -10k$$

$$k = \frac{\ln\left(\frac{1}{2}\right)}{-10} \approx 0.069.$$

(Continued)

(b) Using the results from (a), our equation for the decay of the drug in the bloodsteam is

$$C(t) = C_0 e^{-0.069t}.$$

Recall that t is measured in days; thus, if we want to know how much is left after 4 hours, we need to convert this to units of days:

$$4 \text{ hours} \times \frac{1 \text{ day}}{24 \text{ hours}} = \frac{1}{6} \text{ days}.$$

Thus, we want to know how much of C_0 is left after $\frac{1}{6}$ days:

$$
\begin{aligned}
C\left(\frac{1}{6}\right) &= C_0 e^{-0.069 \times \frac{1}{6}} \\
&= 0.989 C_0.
\end{aligned}
$$

Thus, 98.9% of the administered dose is left after 4 hours.

Example 4.5 (Oxygen Consumption in Salmon [29])

Biologists studying salmon have found that the oxygen consumption of yearling salmon (in appropriate units such as mg of oxygen per kg of body mass per hour) increases exponentially with the speed of swimming according to the function

$$f(x) = 100e^{0.6x},$$

where x is the speed in feet per second. Find each of the following:

(a) The oxygen consumption when the fish are not moving.
(b) The oxygen consumption at a speed of 2 ft/s.
(c) If a salmon is swimming at 2 ft/s, how much faster does it need to swim to double its oxygen consumption?

Solutions:

(a) When the fish are not moving, their speed is 0 ft/s. Thus, when $x = 0$,

$$f(0) = 100e^{0.6 \times 0} = 100e^0 = 100 \times 1 = 100.$$

Thus, when the fish are not moving, their oxygen consumption is 100 units.
(b) Here $x = 2$ and

$$f(2) = 100e^{0.6 \times 2} = 100e^{1.2} \approx 332.$$

Thus, at a speed of 2 ft/s, oxygen consumption is 332 units.

(Continued)

(c) We want to solve for x when oxygen consumption is $2 \times f(2) = 2 \times 332 = 664$ units:

$$
\begin{aligned}
664 &= 100e^{0.6x} \\
6.64 &= e^{0.6x} \\
\ln 6.64 &= \ln e^{0.6x} \\
\ln 6.64 &= 0.6x \\
x &= \frac{\ln 6.64}{0.6} \approx 3.16 \text{ ft/s.}
\end{aligned}
$$

Note that 3.16 ft/s is $3.16 - 2 = 1.16$ ft/s faster than 2 ft/s. Thus, to get double the oxygen consumption than when swimming 2 ft/s, the salmon would have to swim 1.16 ft/s faster.

Another example using exponential functions involves the amount of light hitting a surface. Light level affects many biological processes and is typically measured as the radiant flux density arriving at a surface, in units such as photons per square mm of surface area per minute. Let $I(x)$ denote the radiant flux density of light in a medium (e.g., water or glass) at distance x from the surface at which the light is entering the medium.

The **Beer-Lambert Law** states that if I_0 is the radiant flux density of light at the surface, that is, $I_0 = I(0)$, then

$$I(x) = I_0 a^x,$$

where $0 < a < 1$ depends on the medium. The value of a determines how effective the medium is at absorbing or reducing the transmission of light. The closer the value of a to 1, the larger the value of I will be for larger x (i.e., deeper water). If a is close to 1, then the water is very clear. If a is very close to 0, then the water is very murky.

Example 4.6 (Beer-Lambert Law [16])

Suppose sunlight is shining on a lake of murky water where the radiant flux density of light 1 meter below the surface is 1/4 of that at the surface.

(a) Calculate the constant a in Beer-Lambert's Law for this murky water.
(b) Find the radiant flux density of photosynthetically active radiation light intensity at 1.5 meters below the surface if the radiant flux density at the surface is 2000 μ mol m^{-2} s^{-1}.

(*Continued*)

Solution:

(a) We know that $I(1) = \frac{1}{4}I_0$. Substitute this information into the Beer-Lambert formula:

$$I(x) = I_0 a^x$$
$$I(1) = I_0 a^1$$
$$(1/4)I_0 = I_0 a$$
$$a = 1/4.$$

(b) We now have the equation $I(x) = 2000(1/4)^x$ since $I_0 = 2000$. Then
$I(1.5) = 2000(1/4)^{1.5} = 250 \, \mu \, \text{mol m}^{-2} \, \text{s}^{-1}$.

Logarithmic functions are often used to represent functions with extremely large ranges or domains. We will see more of this in Section 4.3. Here we consider the example of the Richter scale.

Example 4.7 (Richter Scale)

The Richter scale assigns a single number to quantify the magnitude of an earthquake. The scale is a base-10 logarithmic scale that is obtained by calculating the logarithm of the combined horizontal amplitude of the largest displacement from zero on a seismometer's output. Measurements have no limits and can be either positive or negative.

We use the following equation to calculate the Richter magnitude of an earthquake:

$$R = \log_{10} \frac{A}{A_0},$$

where A_0 is the smallest amplitude of ground movement that a seismometer can detect (this depends on the epicentral distance of the location) and A is amplitude of ground movement during an earthquake.

The Northridge, California, earthquake (January 17, 1994) had a magnitude of 6.7: $R_N = 6.7 = \log_{10} \frac{A_N}{A_0}$. We can translate this to $A_N = A_0 10^{6.7}$. The Kobe, Japan, earthquake occurred exactly one year later (January 17, 1995) and had a magnitude of 7.2: $R_K = 7.2 = \log_{10} \frac{A_K}{A_0}$. We can translate this to $A_K = A_0 10^{7.2}$. Now, if we want to find the amplitude of ground movement of the Kobe earthquake compared to the Northridge earthquake, we find the ratio A_K/A_N:

$$\frac{A_K}{A_N} = \frac{A_0 10^{7.2}}{A_0 10^{6.7}}$$

(Continued)

$$= \frac{10^{7.2}}{10^{6.7}}$$

$$= 10^{7.2-6.7}$$

$$= 10^{0.5}$$

$$= 3.1623.$$

So, the amplitude of the Kobe earthquake was over 3 times larger than that of the Northridge earthquake.

The largest recorded earthquake occurred in Japan in 1933 and had a Richter magnitude of 8.9. Using the same process as above, we find that the 1933 quake was 158.5 times more intense than the Northridge quake and 50 times more intense than the Kobe quake.

4.3 Allometry

A common relationship that arises over and over again in biology is that of allometry between two measurements. Two variables x and y are said to be **allometrically related** if $y = ax^b$, where a and b are real constants. Notice that the variable x is now the base in the allometric function $y = ax^b$, while x is the exponent in exponential functions.

Allometric relationships describe different aspects of a single organism:

- Length versus volume
- Surface area versus volume
- Body weight versus brain weight
- Body weight versus blood volume

Note that typically $x > 0$ since negative quantities do not usually have any biological meaning (would negative measurement make sense in the examples above?).

Example 4.8 (Elephants)

It has been determined that for any elephant, the surface area of the body can be estimated as an allometric function of trunk length. For African elephants, the allometric exponent is 0.74. If a particular elephant has a surface area of 200 ft^2 and a trunk length of 6 ft, what is the expected surface area of an elephant with a trunk length of 7 ft?

Solution: First, recall that an allometric function looks like $y = ax^b$, where x and y are variables and a and b are constants. It is given that surface area is an allometric

(*Continued*)

function of trunk length. Let $x =$ trunk length and $y =$ surface area. Additionally, the exponent in our function is 0.74. Thus, our function looks like

$$y = ax^{0.74}.$$

To find the constant a, use data from a single elephant. Since an elephant with a surface area of $200 \, \text{ft}^2$ has a trunk length of 6 ft, for that particular elephant,

$$x = 6 \text{ and } y = 200.$$

Substituting this into the equation, we get

$$200 = a \cdot (6)^{0.74}$$
$$= a \cdot 3.77.$$

We find $a = 53.05$, and the allometric relation is

$$y = 53.05x^{0.74}.$$

To find the expected surface area of an elephant with a trunk length of 7 ft substitute 7 ft for the trunk length in the allometric equation:

$$y = 53.05(7)^{0.74} = 223.9 \, \text{ft}^2.$$

Note that this is the expected surface area of an elephant with trunk length 7 ft; it might not be the actual area. The equations describing allometric relationships are determined by fitting curves to data, as will be discussed in the next section. The data point for any particular individual elephant may not fall on the curve, but the points on the curve are a good approximation of the elephant trunk length and surface area data. See the next section on rescaling data for more details, and see [68] for other ways to estimate the surface area for an elephant, in particular for Indian elephants.

4.4 Rescaling Data: Log-Log and Semilog Graphs

Data can be displayed with different types of scales on the horinzontal and vertical axes. Frequently in plots of biological data, the horizontal axis may be labeled

$$\ln(x)$$

and the vertical axis may be labeled

$$\ln(y),$$

where x and y are the biological variables under consideration. Such a graph is called a log-log graph. If the horizontal axis is labeled

$$x$$

and the vertical axis is labeled

$$\ln(y),$$

where x and y are the biological variables under consideration, then the graph is called semilog. The next two examples illustrate semilog graphs.

Log-log and semilog graphs are closely connected to allometric and expontial functions. If our data are close to an allometric or exponential function, the data can be rescaled and the transformed data will look like a line on a log-log or semilog graph.

To understand how to rescale data, we need to first understand how to rewrite exponential and allometric equations so that they look like equations for lines. First, recall that the equation for a line is

$$y = mx + b,$$

where x and y are the variables, m is the slope of the line, and b is the y-intercept (i.e., $y = b$ when $x = 0$).

Next, consider the exponential function $f(x) = ac^x$, where a and c are constants. Take the natural log of both sides of the equation to obtain

$$\ln\left(f(x)\right) = \ln\left(ac^x\right)$$
$$= \ln a + \ln c^x$$
$$= \ln a + x \ln c.$$

Does this look like the equation of a line? Look closer, using $y = \left(f(x)\right)$:

$$\underbrace{\ln y}_{\text{variable}} = \overbrace{\ln a}^{\text{constant}} + \underbrace{x}_{\text{variable}} \overbrace{\ln c}^{\text{constant}}.$$

This is a line with y-intercept $\ln a$ and slope $\ln c$ (if $c = e \approx 2.71828\ldots$, then $\ln c = 1$). Notice that the original vertical axis variable $f(x)$ has been rescaled to $\ln\left(f(x)\right)$. Notice also that we could have just as easily used log instead of ln.

Now, consider the allometric function $g(x) = ax^c$, where a and c are constants. Again, take the natural log of both sides of the equation to obtain

$$\ln\left(g(x)\right) = \ln\left(ax^c\right)$$
$$= \ln a + \ln\left(x^c\right)$$
$$= \ln a + c \ln x.$$

Again, if we take a closer look, this is the equation of a line using $y = (g(x))$:

$$\underbrace{\ln y}_{\text{variable}} = \overbrace{\ln a}^{\text{constant}} + \overbrace{c}^{\text{constant}} \underbrace{\ln x}_{\text{variable}},$$

where the y-intercept is $\ln a$ and the slope is c. Here, in the allometric case, the vertical axis variable $g(x)$ has been rescaled to $\ln(g(x))$, and the horizontal axis variable has been rescaled to $\ln x$. Notice that this is different from the exponential case, in which only the vertical axis was rescaled.

What is the big picture?

    ★ Given an exponential equation, if we rescale the y-axis to be logarithmic, we get an equation for a line.

    ★ Given an allometric equation, if we rescale both the x-axis and the y-axis to be logarithmic, we get an equation for a line.

Example 4.9 (Algae Growth)

Imagine algae growing in a Petri dish, starting from a single cell. After some time, the cell will split (or else die). Then each of the two cells will split and so on. Thus, the number of cells in the Petri disk increases not linearly (i.e., in an additive manner) but rather multiplicatively (by doubling each time). Suppose that the doubling time is 1 day. Thus, if we start off with one cell, after 1 day we will have 2 cells, after 2 days 4 cells, after 3 days 8 cells, and so on:

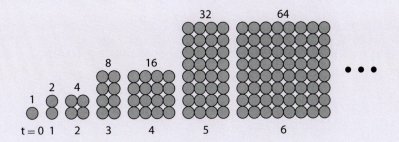

Let us see what happens when we rescale the y-axis data.

x-axis	y-axis	y-axis rescaled
Time t	No. of cells N	$\ln N$
0	1	0
1	2	0.693
2	4	1.386
3	8	2.079
4	16	2.773
5	32	3.466
6	64	4.159

(*Continued*)

When graphed, when the *y*-axis data are rescaled to be logarithmic, the data form a line:

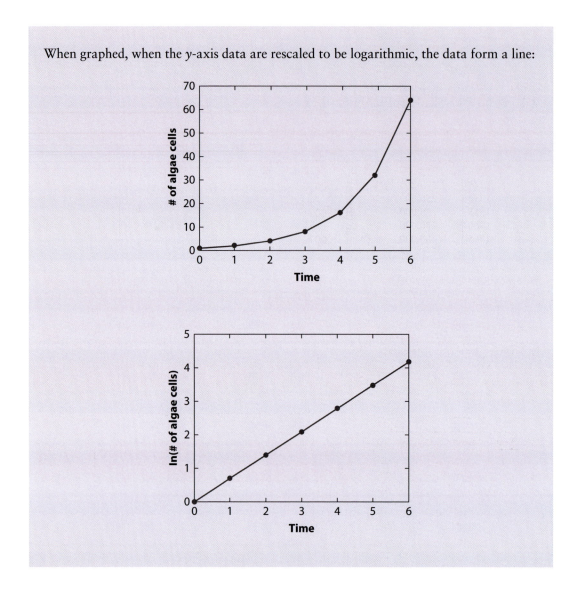

Linking Rescaling with Linear Regressions

Suppose that you were given a set of data that appeared to have an exponential or allometric trend. How would you determine the exponential or allometric function that best describes that set of data? If the data had a linear trend, then we could use linear regression to find the best fit line for the data. We have just seen how we can transform exponential and allometric equations to linear equations with rescaling. Using this same idea, we can rescale data so that they appear to be linear; we may then use the techniques of linear regression to find the equation that best describes the data.

Example 4.10 (Harvesting Bluefish [25])

For the first example, consider data on bluefish harvesting in the Chesapeake Bay, reported by the *Daily Press*, a newspaper in Virginia.

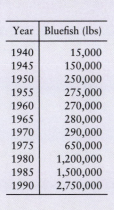

Year	Bluefish (lbs)
1940	15,000
1945	150,000
1950	250,000
1955	275,000
1960	270,000
1965	280,000
1970	290,000
1975	650,000
1980	1,200,000
1985	1,500,000
1990	2,750,000

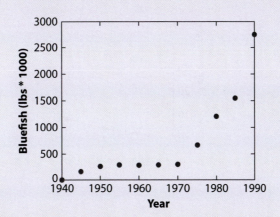

A quick glance at the graph shows that the data are not linearly related. Let us try rescaling the y-axis data on a logarithmic scale. Additionally, on the horizontal axis, let $x = \frac{1}{5}$ (year − 1940).

x	y	$\ln y$
Year	Bluefish (lbs)	Rescaled
0	15,000	9.616
1	150,000	11.918
2	250,000	12.429
3	275,000	12.525
4	270,000	12.506
5	280,000	12.543
6	290,000	12.578
7	650,000	13.385
8	1,200,000	13.998
9	1,500,000	14.221
10	2,750,000	14.827

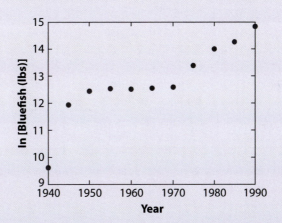

With the exception of the data point corresponding to the year 1940, the data appear to be more linearly related when the y-axis data are rescaled. Fitting a least-squares regression line to these data using the methods described in Section [3.3], we find that the equation for the best fit line is

$$\ln y = 0.3797x + 10.8784,$$

(*Continued*)

where x is the year and y is the pounds of bluefish harvested. We can convert this equation to an exponential equation:

$$\ln y = 0.3797x + 10.8784$$

$$e^{\ln y} = e^{0.3797x + 10.8784}$$

$$y = e^{0.3797x} \times \underbrace{e^{10.8784}}_{\text{constant}}$$

$$y = (53,019)\, e^{0.3797x}.$$

Using this equation, we could extrapolate how many fish were caught in 2000. First, we compute

$$
\begin{aligned}
x &= \frac{1}{5}(2000 - 1940) \\
&= 12.
\end{aligned}
$$

Then

$$
\begin{aligned}
y &= (53,019)\, e^{0.3797(12)} \\
&= 5,049,529.
\end{aligned}
$$

Thus, we estimate that in the year 2000, about 5.05 million pounds of bluefish were caught.

Example 4.11 (Mutation Rates [29, 36])

Researchers studying the relationship between the generation time of a species and the mutation rate for genes that cause deleterious effects gathered the following data:

Species	Generation Time (in years)	Genomic Mutation Rate (per generation)
D. melanogaster/D. pseudoobscura	0.1	0.070
D. melanogaster/D. simulans	0.1	0.058
D. picticornis/D. silvestris	0.2	0.071
Mouse/rat	0.5	0.50
Chicken/Old World quail	2	0.49
Dog/cat	4	1.6
Sheep/cow	6	0.90
Macaque/New World monkey	11	1.9
Human/chimpanzee	25	3.0

(Continued)

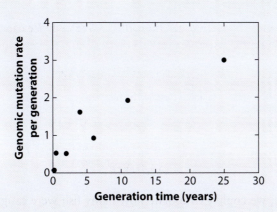

Let x = the generation time and y = the genomic mutation rate. Rescaling both the vertical and the horizontal axes, gives the following data:

$\ln x$	$\ln y$
−2.3	−2.659
−2.3	−2.847
−1.6	−2.645
−0.7	−0.693
0.7	−0.713
1.4	0.470
1.8	−0.105
2.4	0.642
3.2	1.099

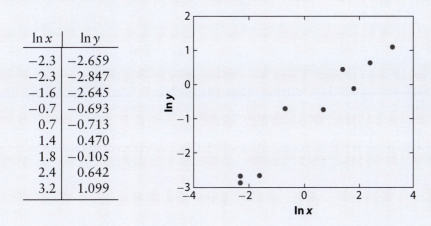

Fitting a least-squares regression line to these data using the methods described in Section [3.3], gives the equation for the best fit line

$$\ln y = 0.7097 \ln x - 1.0316.$$

We can convert this to an allometic equation using the properties of logarithms and exponentials:

$$\ln y = \ln x^{0.7097} - 1.0316$$

$$e^{\ln y} = e^{\ln x^{0.7097} - 1.0316}$$

$$e^{\ln y} = e^{\ln x^{0.7097}} e^{-1.0316}$$

$$y = x^{0.7097} e^{-1.0316}$$

$$y = 0.3564 x^{0.7097}.$$

(Continued)

Using this equation, if we knew that a certain species had a generation time of 5 years, we could interpolate to find the genomic mutation rate of this species:

$$y = 0.3564(10)^{0.7097} = 0.1827.$$

Thus, we would estimate that this particular species has a genomic mutation rate of 0.1827 mutations per generation.

4.5 Matlab Skills

Exponential and Logarithmic Functions in Matlab

In Matlab, to compute the value e^x, where x is some number, we use the function

```
exp(x)
```

This function can also be used on arrays of numbers. If x is an array, then `exp(x)` will return an array where each value is e raised to the corresponding value in x. Examples of using `exp` on both a number and an array are shown below:

```
———————————— Command Window ————————————
>> a = 5;

>> x = [1 2 3 4 5];

>> exp(a)
ans =
      148.4132

>> exp(x)
ans =
      2.7183    7.3891    20.0855    54.5982    148.4132
```

The function for computing the natural logarithm in Matlab, `log(x)`, works similar to `exp(x)`. Note that Matlab uses `log` for computing the natural logarithm. If you want to compute the logarithm of a different base, use the formula

$$\log_a x = \frac{\ln x}{\ln a}.$$

For example,

$$\log_{10} 5 = \frac{\ln 5}{\ln 10}.$$

Like the `exp` function, the `log` function can be applied to an array. So `log(y)`, where y is an array of numbers, will return an array where each value is the natural logarithm of the corresponding value in the array `y`. Some examples of using the function `log` are shown below:

```
                          ──── Command Window ────
>> a = 5;

>> b = 10;

>> x = [1 2 3 4 5];

>> log(a)
ans =
        1.6094

>> log(b)/log(10)
ans =
        1

>> y = log(x)
y =
        0      0.6931     1.0986     1.3863     1.6094

>> z = exp(y)
z =
        1.0000    2.0000    3.0000    4.0000    5.0000
```

Rescaling Data and Linear Regressions

In Section 4.4, we learned how to rescale data to use linear regression techniques to fit a least-squares regression line to the data. Matlab can be used to accomplish this task.

In Example 4.10, we looked at the total pounds of bluefish caught in the Chesapeake Bay every 5 years. In the example, we rescale the data by letting $x = \frac{1}{5}(\text{year} - 1940)$ (a linear rescaling) and $\ln y = \ln(\text{lbs of bluefish})$ (a logarithmic rescaling). Next, we plotted the data. Then we computed the equation for the least-squares regression line for the $(x, \ln y)$ data. Below is an m-file that does all these calculations in Matlab. Additionally, the m-file computes the correlation coefficient for the $(x, \ln y)$ data.

```
                          ──── Bluefish.m ────
1    % Filename: Bluefish.m
2    % M-file to
3    %      - enter bluefish data
4    %      - rescale bluefish data
5    %      - compute equation for least squares regression line
6    %      - plot least squares regression line on a graph with the data
7    %      - compute the correlation coefficient
8    % LSR = Least Squares Regression
9
10   % Enter year array
11   year = [1940 : 5 : 1990];
12   % Rescale year array to get x data
13   x = (year - 1940)/5;
14
15   % Enter pounds of bluefish caught
16   y = [15000
```

```
17           150000
18           250000
19           275000
20           270000
21           280000
22           290000
23           650000
24           1200000
25           1500000
26           2750000]';
27   % The ' after the array changes the column array to a row array
28
29   % Find the equation for the LSR line
30   C = polyfit(x,log(y),1);
31   % Display the equation
32   fprintf('Eqn for LSR: ln y = %f x + %f\n',C(1),C(2))
33
34   % Find the lnyhat value for each x value
35   lnyhat = polyval(C,x);
36
37   % Plot the data and the LSR line
38   plot(x,log(y),'k.',x,lnyhat,'k-')
39   xlabel('Year (rescaled)')
40   ylabel('ln(Pounds of bluefish caught)')
41   xlim([min(x)-1 max(x)+1]) %Set x-axis a little wider than data
42
43   % Find the correlation coefficient
44   rho = corrcoef(x,log(y));
45   % Display the correlation coefficient
46   fprintf('rho = %f\n', rho(1,2))
```

See Appendix A for details on creating and running an m-file. When this m-file is run in the command window, the output looks like the following:

```
──────────────── Command Window ────────────────
>> Bluefish
Eqn for LSR: ln y = 0.379678 x + 10.878423
rho = 0.908055
```

The graphical output is shown in Figure 4.1.

Sometimes, we are given a set of data and need to decide which type of function (linear, exponential, or allometric) best describes how the data are related (see Exercises 4.12, 4.13, and 4.14). To do this, we compare the correlation coefficients of the original data with (1) the data logarithmically rescaled only in the vertical axis variable and (2) the data logarithmically rescaled in both the horizontal and the vertical axis variables.

Suppose that we are given data and must decide what type of function best describes the relationship of the data. The data in Table 4.1 are for the body weights in grams (g) and pulse rate in beats per minute (bpm) for various mammals. The data are from [14, 25].

The following is an m-file that computes the correlation coefficient for the (x, y) data, for the $(x, \ln y)$ data, and for the $(\ln x, \ln y)$ data.

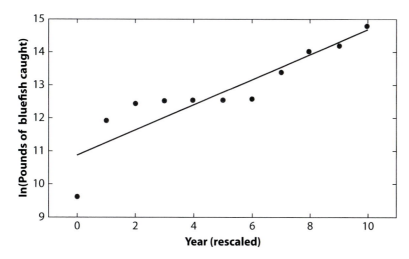

Figure 4.1 Graph produced by the m-file Bluefish.m.

Table 4.1. Data on mammal pulse rates relative to body weight [14, 25].

Mammal	Body Weight (g), x	Pulse Rate (bpm), y
Vesperugo pipistrellus	4	660
Mouse	25	670
Rat	200	420
Guinea pig	300	300
Rabbit	2000	205
Little dog	5000	120
Big dog	30,000	85
Sheep	50,000	70
Man	70,000	72
Horse	450,000	38
Ox	500,000	40
Elephant	3,000,000	48

```
                        ─── MammalPulseRates.m ───
1    % Filename: MammalPulseRates.m
2    % M-file to
3    %    - enter mammal pulse rates data
4    %    - calc. correlation coeff. for y vs x
5    %    - calc. correlation coeff. for ln y vs x
6    %    - calc. correlation coeff. for ln y vs ln x
7
8    % Enter the data
9    x = [ 4
10          25
11          200
12          300
13          2000
14          5000
```

```
15              30000
16              50000
17              70000
18              450000
19              500000
20              3000000 ]'; % body weight data
21    y = [ 660
22              670
23              420
24              300
25              205
26              120
27              85
28              70
29              72
30              38
31              40
32              48 ]'; % pulse rate data
33
34    % Calculate correlation coeff. for y vs x
35    rho = corrcoef(x,y);
36    % Display the correlation coefficient
37    fprintf('(x,y)        rho = %f\n', rho(1,2))
38
39    % Calculate correlation coeff. for ln y vs x
40    rho = corrcoef(x,log(y));
41    % Display the correlation coefficient
42    fprintf('(x,ln y)     rho = %f\n', rho(1,2))
43
44    % Calculate correlation coeff. for ln y vs ln x
45    rho = corrcoef(log(x),log(y));
46    % Display the correlation coefficient
47    fprintf('(ln x, ln y) rho = %f\n', rho(1,2))
```

When this m-file is run in the command window, the output looks like the following:

```
———————————————— Command Window ————————————————
>> MammalPulseRates
(x,y)        rho = -0.334064
(x,ln y)     rho = -0.442814
(ln x, ln y) rho = -0.976793
```

From the output, we see that when both the x and the y data are scaled logarithmically, the correlation coefficient has the largest absolute value. Thus, if choosing between a linear, exponential, and allometric function, an allometric function will best fit the data. To find the equation of the allometric function, use the `polyfit` function to obtain the equation of the line (where $\ln x$ is the horizontal variable and $\ln y$ is the vertical variable) and then transform the function into an equation of y in terms of x:

```
———————————————— Command Window ————————————————
>> polyfit(log(x),log(y),1)
ans =
    -0.2461    7.0732
```

Thus, the equation for the least-squares regression line is

$$\ln y = -0.2461 \ln x + 7.0732.$$

We can then solve this equation for y in terms of x:

$$\ln y = -0.2461 \ln x + 7.0732$$

$$e^{\ln y} = e^{-0.2461 \ln x + 7.0732}$$

$$y = e^{\ln x^{-0.2461}} \cdot e^{7.0732}$$

$$= \left(x^{-0.2461}\right)(1179.9)$$

$$= 1179.9 x^{-0.2461}.$$

4.6 Exercises

4.1   Solve each of the exponential equations for t.

(a) $5\left(2^t\right) = 40$
(b) $4\left(3^{2t}\right) = 12$
(c) $(e^{-t}) = 0.4$
(d) $6(0.6)^t = 10$
(e) $5 - 2^{3t} = 3$
(f) $4e^{0.05t} = 8$
(g) $7e^{-0.2t} = 28$
(h) $3^{t^2 - t} = 9$

4.2   Solve each of the following logarithmic equations for x.

(a) $\log_{10} x = 1000$
(b) $\ln(2x + 3) = 5$
(c) $\log_2(x + 4) = 6$
(d) $\log_2\left(\log_3 2x\right) = 4$

4.3   Salmonella bacteria grow rapidly at room temperature. If some bacteria are left on the cutting board when a chicken with salmonella is cut up and they get into a salad, the population of bacteria begins growing. Suppose that the number present in the salad after t hours is given by

$$f(t) = 300 \cdot 6^t.$$

(a) If the salad is left out at room temperature, how many bacteria are present 1 hour later?
(b) How many were present initially in the salad?
(c) How often do the bacteria double?

4.4   (From [16]) In marine biology, the most important zone in the sea is the *photic zone*, where photosynthesis can take place. For marine phytoplankton, the photic zone must end at the depth where about 1% of surface light penetrates.

(a) Near Cape Cod, Massachusetts, the depth of the photic zone is about 16 meters. Find the constant a in the Beer-Lambert Law $I = I_0 a^x$.

(b) In very clear waters in the Caribbean, 50% of the light at the surface reaches a depth of about 13 meters. Estimate the depth of the photic zone (using Beer-Lambert's Law).

(c) Which water location, Cape Cod or the Caribbean, had a higher value for a? Does this make sense? Explain.

4.5   A farmer wishes to have about 400 trout in a pond on his farm and stocks the pond with 100 trout initially. After 2 months, there are approximately 160 trout in the pond.

(a) Find an exponential function for the number of trout in the pond using the information above.

(b) At what time will the number of trout reach 400?

4.6   Suppose that the island area A and the number of species S (of a certain taxa) on that island are given by

$$S = \alpha A^{\beta},$$

where α and β are constants.

(a) What kind of function is this?

(b) Suppose that you are studying land and freshwater birds among a small cluster of islands. The first island has an area of 50 square miles, and you count 13 species of birds. The second island has an area of 125 square miles, and you count 17 species of birds. Given this information, find the values of α and β so that you have a description of how bird species count depends on island size.

(c) If you count 24 species of birds on a nearby island, estimate the area of this island.

4.7   The paper [20] shows that leaf biomass L (measured in kg dry weight) is an allometric function of root biomass R (in kg dry weight). Suppose that the log-log (base-10) graph approximately goes through the points $(-2, -2)$ and $(1, 0)$.

(a) Give an equation that expresses L as a function of R.

(b) Compare two plants with plant A having twice the root biomass of plant B. How do the leaf biomasses compare?

4.8   Suppose that for a certain data set, the semilog (base-10) graph goes through the points $(2, -2)$ and $(3, -4)$. Using only this information, give an equation for y as a function of x.

4.9   A seed germination study has followed the number of seeds germinating in flats in a greenhouse for several years. Starting with 10,000 seeds in a flat for a particular species and letting $G(t)$ be the number of ungerminated seeds at year t, it is found that when graphing t on the horizontal axis and $\log_{10} G(t)$ on the vertical axis, you get a line with slope -0.477. The data fall very close to this line.

(a) Give an equation for $G(t)$.

(b) In what year will the number of ungerminated seeds left in the flat be less than 100?

4.10   (From [29, 57]) The basal metabolic rate (in kcal/day) for large anteaters is given by

$$f(x) = 19.7x^{0.753},$$

where x is the anteater's weight in kilograms (kg).

(a) Find the basal metabolic rate for anteaters weighing (i) 5 kg and (ii) 25 kg.

(b) Suppose that the anteater's weight is given in pounds (lbs) rather than kg. Given that 1 lb = 0.454 kg, find a function $x = g(z)$ giving the anteater's weight in kg if z is the animal's weight in lbs.

(c) Write out the function $f(g(z))$, that is, the basal metabolic rate for anteaters given the anteaters weight in lbs.

4.11   Consider Example 4.10.

(a) Find an equation for y (bluefish in pounds) using x = year,

(b) Find an equation for y (bluefish in pounds) using x = year/1000.

(c) Compare how these equations differ from the equation derived in Example 4.10 by extrapolating how many fish were expected to be caught in the year 2000. Explain why the answers might differ.

4.12   (From [14, 26]) Warm-blooded animals use large quantities of energy to maintain body temperature due to heat loss through the body's surface. In fact, biologists believe that the primary energy drain on a resting warm-blooded animal is the maintenance of body temperature. The table below shows the body weights in grams (g) and pulse rates in beats per minute (bpm) for nine different bird species.

Bird	Body Weight (g), x	Pulse Rate (bpm), y
Canary	20	1000
Pigeon	300	185
Crow	341	378
Buzzard	658	300
Duck	1100	190
Hen	2000	312
Goose	2300	240
Turkey	8750	193
Ostrich	71,000	65

(a) Make a hypothesis about how you think body weight and pulse rate are related in birds? How might your hypothesis explain birds maintaining their body heat?

(b) Compute, by hand, the correlation coefficient for y versus x.

(c) Compute, by hand, the correlation coefficient for $\ln(y)$ versus x.

(d) Compute, by hand, the correlation coefficient for $\ln(y)$ versus $\ln(x)$.

(e) What type of function (linear, exponential, or allometric) best describes the relationship between the weight and pulse rate data? Explain. Find the equation for the the function that best describes the data as y in terms of x.

4.13   (From [13, 60]) Researchers measured the diameters of 20 trees in a central Amazon rain forest and used ^{14}C-dating to determine the ages of these trees. The data are given in the following table.

Diameter (cm), x	Age (yr), y	Diameter (cm), x	Age (yr), y
180	1372	115	512
120	1167	140	512
100	895	180	455
225	842	112	352
140	772	100	352
142	657	118	249
139	582	82	249
150	562	130	227
110	562	97	227
150	552	110	172

Consider the use of diameter, x, as a predictor of age, y.

(a) Make a scatter plot of age on the vertical axis and diameter on the horizontal axis and fit a regression line to the data (using the freehand method).

(b) Using Matlab, make a scatter plot of age on the vertical axis and diameter on the horizontal axis. Fit a regression line to the data using the least-squares method and compute the coefficient of determination.

(c) Using the information you have gathered so far, would you hypothesize that there is a linear relationship between the age data and the diameter data? Explain why or why not.

(d) Using Matlab, plot $\ln(y)$ versus x, fit a least-squares regression line to these transformed data, and compute the coefficient of determination.

(e) Using Matlab, plot $\ln(y)$ versus $\ln(x)$, fit a least-squares regression line to these transformed data, and compute the coefficient of determination.

(f) What type of function (linear, exponential, or allometric) best describes the relationship between the age data and the diameter data? Explain. Write the function for y in terms of x.

4.14 (From [7, 29]) In an attempt to measure how the pace of city life is related to the size of the city, two researchers measured the mean speed of pedestrians in 15 cities by measuring the mean time it took them to walk 50 feet.

City	Population (x)	Speed (ft/s) (y)
Brno, Czechoslovakia	341,948	4.81
Prague, Czechoslovakia	1,092,759	5.88
Corte, France	5491	3.31
Bastia, France	49,375	4.90
Munich, Germany	1,340,000	5.62
Psychro, Crete	365	2.67
Itea, Greece	2500	2.27
Iráklion, Greece	78,200	3.85
Athens, Greece	867,023	5.21
Safed, Israel	14,000	3.70
Dimona, Israel	23,700	3.27
Netanya, Israel	70,700	4.31
Jerusalem, Israel	304,500	4.42
New Haven, CT, USA	138,000	4.39
Brooklyn, NY, USA	2,602,000	5.05

Using Matlab, complete the following.

(a) Plot the original pairs of numbers (x, y). Is the pattern linear or nonlinear?

(b) Compute the coefficient of determination, R^2, for the data.

(c) Plot y against $\ln x$. Are the data more linear now than in part (a)?

(d) Compute R^2 for y against $\ln x$. Is R^2 closer to 1 than in part (b)? What does this say about how the data are related?

(e) Find the equation for the least-squares regression line of y against $\ln x$.

Unit 1 Student Projects

Height and Weight Data

This experiment is designed to motivate thinking about how data are collected, summarized, analyzed, and interpreted. Through this experiment, we would like to answer the question, "In a population, does weight increase with height?" Before the experiment even begins, write down what you think the answer to this question is. If you wrote yes, are there exceptions when that might not be true? Write down any exceptions you can think of. Ask a few other students around you what they wrote. In doing this, you are forming a *hypothesis* about what you expect to find from the results of your experiment. Through this experiment and the skills you will learn in this unit of the book, Descriptive Statistics, we will learn how to test whether your hypothesis is invalid or whether it may be true.

Instructor Tasks

1. Pass out small slips of paper to each student present in class on the day of this experiment. Each student should write down his or her height in inches (i.e., a student who is 5 feet 7 inches tall is $5(12) + 7 = 67$ inches tall) and weight in pounds.
2. Next, collect the height and weight data in a hat, bowl, or bag.
3. Lead the discussion outlined in Student Tasks.

Student Tasks

Height and weight data for the entire population of the class have been collected. However, in most biological experiments, data for all individuals in a population are difficult (and sometime impossible) to obtain. Thus, we would like to "sample" the class population. We will do this by reaching blindly into the collection data and pulling out little slips of paper. Notice that in this sampling method, we will not know from whom each piece of data came. Now, as a class, you must make the first decision in this experiment:

1. What is the appropriate sample size?

How many little slips of paper should be pulled out of the hat? How many data points are necessary to be representative of the entire population? How many data points would be needed before we started to see any trend that might appear in the data? Consider all this and decide as a class how many data points you will use in your sample of the class population.

Once you have decided how many data points will be in your sample,

2. Choose someone to pull the data from the hat and someone to record the data on the board where the entire class can see.
3. When recording the data, remember that each data point contains two pieces of information: height and weight. Consider using a table or ordered pairs where it is obvious which weights go with which heights.

Next, recall that we would like to answer the question, "Does weight increase with height?" How might we display the data so that we can answer this question? Think of how you have seen data displayed before.

4. As a class, decide on one or two methods of displaying the data that might assist in answering our question. Choose a student or two to write or draw these data displays on the board so that the entire class can see.
5. Using the various displays of data, are there any trends that agree or disagree with your hypothesis? Discuss any such trends you find with the class. Does everyone agree?
6. Write down any conclusions you have from this experiment, including whether the data supported your hypothesis. ■

Do We Grow in Our Sleep?

This assignment requires you to make measurements on yourself twice a day for 4 days and to submit the results by email to your instructor.

The measurements to be taken are your height (in millimeters) just before you go to sleep at night and just after you wake up in the morning as well as how many hours of sleep you had that evening. Make these measurements in bare feet, tape a paper ruler onto a wall that you can stand in front of and use a book or other flat, hard surface placed on top of your head to determine your height. A ruler that you can use is available at [88]. In doing this, tape the paper ruler on the wall so that the zero edge is below the top of your head and measure (using another ruler) the height to the zero edge in millimeters. We are most concerned with this set of observations with the difference in heights between evening and morning, not in the precise height (so don't worry if the overall height is not correct to the millimeter).

Since each student in the class will be collecting his or her own measurement, it is important that the formatting of results be uniform. In compiling your results, use the following format, with spaces between each measurement:

Table 4.2. An example of what your formatted results would look like.

Day #	Student ID#	Gender (1 = female, 2 = male)	Age (in years)	Height at night	Hours of sleep	Height in morning
1	000226633	1	20	1575	8	1577
2	000226633	1	20	1574	6	1575
3	000226633	1	20	1574	7	1577
4	000226633	1	20	1575	4	1574

See Table 4.2 for an example of what your results should look like.

Email your data set to your instructor as a text in the body of an email (do not attach a file).

Instructor Tasks

1. Combine the data sets provided by each student so that there is a single file containing all data. In doing this, ensure that there is no identifying information about any particular student so that the data remain private.
2. Post this data set in a location where all students can obtain it (e.g., a class BlackBoard site) and inform all students after it has been posted.
3. Assign the students the tasks outlined below.
4. In the class meeting for which this assignment is due, have the class break into small groups of three or four. Ask them to talk amongst themselves about the hypotheses they made, the conclusions they drew regarding whether the data indicated their hypothesis should be accepted, and what they learned about data quality from this.
5. After the small-group discussions, ask what hypotheses they generated, have them write these as a summary so that everyone can see them, and determine if there is concordance among the class about the hypotheses and the evaluation of these from the data.
6. Finally, show the Web site for your university's institutional review board. Explain what it does and explain why collecting the data as you have done is in concordance with the board's policies.

Student Tasks

The instructor will post both the data from all students in the class (with no personal identifying information) and a Matlab code with directions for running it and describing the output it will create. Your assignment is the following:

1. Make one or more hypotheses before analyzing the data about what you expect to observe from this set of height observations.
2. Using the Matlab skills you have learned in this unit, write an m-file (see Appendix A for a description of writing and running an m-file) that performs an analysis of the compiled data. Your analysis should answer the following questions:
   (a) Is a change in height during sleep affected by age?
   (b) Is a change in height during sleep affected by hours of sleep?

3. Write a two-page report (not including figures) that describes the analysis you performed and that summarizes the findings of your analysis. Your report should have four sections:
   (a) Introduction: Describe the problem put forth and any hypotheses you had about the results of the experiment.
   (b) Methods: Describe the mathematics you employed to solve the described problem.
   (c) Results: Describe in words and with graphs the results of your analysis.
   (d) Conclusion: Indicate any conclusions that you can draw from your results. Comment on any hypotheses you made in light of your findings. What are other possible factors affecting change in height during sleep that could not be tested with the given data set?
4. Turn in a copy of your report along with a copy of your m-file. ■

Dendrology Measurements

A dendrologist is a scientist who studies trees and other woody plants. In studying trees, there are a variety of measurements that a dendrologist can make to describe the size of a tree. One such measurement is the diameter at breast height (DBH). As the name implies, this is a measurement of the diameter of the tree at breast height, which is defined as 4.5 feet (1.37 m) above the forest floor on the uphill side of the tree. DBH is measured using either a diameter tape or a tree caliper. Another measurement is tree height. It is usually not practical to measure the height of a tree directly, but it can be estimated using a variety of methods (see [89] for some common methods). The next set of measurements involve the crowns of trees. The crown of a tree includes the branches, leaves, and reproductive structures extending from the trunk. One can measure the width of the crown at its widest point (crown width), the height from the forest floor to the crown (height to crown), and the height from the bottom of the crown to the top of the crown (crown height). A final type of measurement is the number of trees per acre. If the trees/acre measurement is associated with the previously mentioned tree measurements, then that tree is in the plot or stand of trees where the trees/acre measurement was taken.

The data in Table 4.3 consist of a set of 20 tree measurements taken by the Oregon Transect Ecosystem Research (OTTER) Project (for details on how the data were collected and the objectives of the project, visit [90]).

Student Tasks

This project is to be completed individually by each student using Matlab.

1. Make one or more hypotheses before analyzing the data about what you expect to observe from this set of tree measurements.
2. Using the Matlab skills you have learned in this unit, write an m-file (see Appendix A for a description of writing and running an m-file) that performs an analysis of the data in Table 4.3. Your analysis should do the following:
   (a) Find the range, mean, median, standard deviation, and variance of each of the six columns of data in Table 4.3.

Table 4.3. Data collected by the Oregon Transect Ecosystem Research (OTTER) Project ([90]).

DBH (in.)	Height (ft)	Crown Width (ft)	Height to Crown (ft)	Trees/Acre	Crown Height (ft)
22.9	142	15	102	31.5	40
30.0	146	18	96	18.3	50
30.3	145	19	64	18.0	81
27.8	139	19	75	21.3	64
24.1	157	22	81	28.4	76
28.2	153	22	72	20.7	81
26.4	155	17	78	23.7	77
12.8	92	13	65	100.7	27
39.7	179	24	113	10.5	66
38.0	171	19	84	11.4	87
25.5	153	16	94	25.4	59
36.6	200	28	115	12.3	85
43.3	181	28	110	8.8	71
21.4	145	18	51	36.0	94
29.2	149	14	76	19.3	73
54.2	172	28	103	5.6	69
31.9	161	17	117	16.2	44
31.5	168	14	107	16.6	61
35.7	176	22	123	12.9	53
21.8	142	26	85	34.7	57

(b) Produce histograms for each of the three sets of data that deal with the crowns of the trees.

(c) For the three sets of bivariate data, (1) DBH versus height, (2) height versus ln(trees/acre), and (3) ln(DBH) versus ln(trees/acre), do the following:

   (i) Create a scatter plot displaying the bivariate data with an appropriate range for both the horizontal and the vertical axes.

  (ii) Find and display the equation for the least-squares regression line for the bivariate data.

 (iii) Plot the least-squares regression line on the scatter plot with the bivariate data.

 (iv) Compute the correlation coefficient for the set of bivariate data.

3. Write a two-page report (not including figures) that describes the analysis you performed and that summarizes the findings of your analysis. Your report should have four sections:

(a) Introduction: Describe the problem put forth and any hypotheses you had about the results of the experiment.

(b) Methods: Describe the mathematics you employed to solve the described problem.

(c) Results: Describe in words and with graphs the results of your analysis. In this section, include

- a table that gives the statistics results for each of the six columns of data;
- a paragraph that discusses the distribution of the data on crown width, crown height, and height to crown (include appropriate histograms); and
- a few paragraphs that discuss the correlation of (1) DBH versus height, (2) height versus trees/acre, and (3) ln(DBH) versus trees/acre (include appropriate scatter plots with regression lines).

(d) Conclusion: What conclusions can you draw from your results? Comment on any hypotheses you made in light your findings. Why do you think the crown data are distributed the way they are? What might cause this distribution? From the regression analysis, why do certain data sets have a positive or a negative correlation? What might cause these correlations?

4. Write a paragraph that describes how comfortable you feel writing code in Matlab after this project and how much you feel this project has contributed to your understanding of the statistics and linear regression material covered in this unit. These comments will not be graded.

5. Turn in a copy of your report along with a copy of your m-file and your comments on writing Matlab code (from item 4).   ■

Why Study Life Science

Compose two coherent paragraphs (approximately one typed page) on the following topic: Describe why you have chosen to study an area of the life sciences. In writing this, describe one area of the life sciences that is of interest to you and explain why you have a particular interest in that area.

Clear, concise writing is critical to the scientific process. Your composition will be evaluated for content, clarity, and grammar.   ■

Tobacco Growth and Nitrosamine Levels Measurements

Given a data set from 20 burley tobacco barns in Tennessee during the 2008 and 2009 growing seasons, you will investigate how abiotic factors (relative humidity and temperature) in a barn affect the level of carcinogens in the tobacco plants. RH is the number of hours above the threshold of 90% relative humidity in each barn, while T is the number of hours above the threshold of 70 degrees Fahrenheit, in each barn. TSNA is the level of tobacco-specific nitrosamines (some of which are carcinogens) found in the tobacco plants at the end of curing.

You should perform some analysis on the abiotic factor data to see which factor (RH or T) has the stronger correlation with TSNA production.

See [21] for more details about how these data were collected and analyzed.

TSNA	RH	T
0.238	20	63
0.387	4	6
0.592	24	106
0.797	46	0
0.812	68	0
1.228	105	4
1.249	39	0
1.846	35	77
2.215	5	110
2.260	60	272
2.269	65	272
2.276	40	345
2.439	92	1
3.317	59	132
3.569	170	62
3.628	179	343
3.792	149	2
4.988	384	320
6.797	317	300
8.710	379	240

1. Make one or more hypotheses before analyzing the data about what you expect to observe from the set of abiotic factors and their effect on TSNA production.
2. Write an m-file that performs an analysis of the data in the table. Your analysis should do the following:
    (a) Find the range, mean, median, standard deviation, and variance of each of the three columns of data in the table.
    (b) For the two sets of bivariate data, (1) TSNA versus RH and (2) TSNA versus T, do the following:
        i. Create a scatter plot displaying the bivariate data with appropriate ranges for both the horizontal and the vertical axes.
        ii. Find and display the equation for the least-squares regression line for the bivariate data.
        iii. Plot the least-squares regression line on the scatter plot with the bivariate data.
        iv. Compute the correlation coefficient for the set of bivariate data.
3. Results and Discussion: Describe in words and with graphs the results of your analysis:
    (a) A table containing the statistical results for each of the six columns of data
    (b) A paragraph description of the correlation between (1) TSNA versus RH and (2) TSNA versus T.
4. Conclusion: What conclusions can you draw from your results? Which abiotic factor has a stronger correlation with TSNA production? Comment on the validity of your hypothesis. Does it make sense to expand the analysis of the effect of abiotic factors on TSNA production (interaction between RH and T)? ■

Extension: You could try to express TSNA production as a linear function of both RH and T. Look for a tool in Matlab to do this type of linear regression.

Discrete Time Modeling

This unit of the text focuses on how we can use mathematics to mimic biological situations in which it is reasonable to view the underlying biological variables as changing at discrete time intervals. Examples of this will include cases in which we wish to follow how a single variable changes over discrete time intervals, such as the number or density of an insect pest in a particular region measured every week, the blood concentration of a drug measured either just after or just prior to each administration of the drug, and the frequency of a particular genotype measured in subsequent generations of a population of experimental organisms, such as fruit flies. We will go on to expand this to situations in which we wish to follow a collection of linked variables measured at discrete time periods. Examples of this will include following the age structure of a population every year (e.g., every 10 years, the U.S. Census determines the number of U.S. residents of different discrete age groupings); the fraction of a landscape measured every year that is forest, grassland, agricultural land, and so on; and the amount of a drug found in tissues or organs following each administration of the drug.

Many of these models are used for populations that have discrete generations (e.g., many insects and plants). For these populations, the size of a generation typically depends upon the size of the previous generation. Thus, the state variable that describes the population will typically be x_n for the population size (or density) in generation n. We can then extend these models to cases in which we follow more than a single state variable at each time so that rather than following a total population size at each time interval, we follow how many individuals there are of each type, one example being the number of males and females and another being the number of individuals in each age grouping. We will start this unit by considering the case of following a single state variable to serve as the basis for many practical applications, including fisheries and insect pest management models.

The key assumption in discrete-time models is that there is a fixed time period, the time step (e.g., a year or a generation time), which is appropriately chosen so that the population size at the next time step is a simple function of the population size at a previous (sometimes several previous) time step. There are many situations in which this is clearly an approximation; for example, human populations breed and reproduce continuously, not at fixed time steps. The latter half of this text will show how to produce models for continuously breeding populations (this is part of calculus), but in some cases discrete-time models are an adequate representation of continuously breeding populations. Thus, much of human demography (the study of the structure of human populations) utilizes discrete time models (typically using 5 years as the time step, so a human population is broken into age classes of 0 to 4 years old, 5 to 9 years old, etc.), even though humans breed continuously.

We will start with the simplest possible population model in discrete time, in which we view individuals in the population as replacing themselves with b individuals in the next generation. What do we expect to happen in this case as time passes? Obviously, if $b > 1$, we expect that the population will increase, and if $b < 1$, the population will decrease since each individual is not completely replaced. You may wonder what it means if we let $b = 1.5$ since it is not clear that 0.5 individuals makes sense. What does half an individual mean? At least two interpretations are reasonable. One is that the state variable for the model is density (individuals per unit area). Then, $b = 1.5$ means that for each unit of density currently present, the population has 1.5 units of density at the next time step. So if there were 100 individuals per km^2 present, then in the following time period there would be 150 individuals per km^2. Another interpretation is that $b = 1.5$ represents the average number of new individuals produced per individual present in the previous generation, where this average is taken over the entire population. Thus, $b = 1.5$ would arise if half the individuals produce one offspring and die and the other half produce two offspring and then die. The average reproduction rate is thus 1.5 individuals per time step.

This simple model is described by letting x_n be the population size at generation n so that

$$x_{n+1} = bx_n.$$

If the initial population size is x_0, then it is easy to show that

$$x_n = x_0 b^n.$$

What happens here as we look at the population after many generations? If $b > 1$, then x_n continues growing without bound as n increases, and we say that the limit of the sequence of population sizes does not exist, while if $b < 1$, then x_n gets closer and closer to zero. In fact, x_n can get as close to zero as you might want by choosing a large enough n sufficiently large. This model is called geometric growth, or Malthusian growth (after Thomas Malthus, who first described the implications of geometric growth on human populations as eventually outstripping the potential agricultural production of food). You have already seen geometric growth from plotting x_n versus n using semilog axes, in which you get points on a straight line. Geometric growth is the same as exponential growth, but it occurs in discrete time.

If we write down the list $x_0, x_1, x_2, x_3, \ldots$, then we get the list of population sizes indexed by the generation. Mathematically, this list is called a sequence: it is determined by a rule (a function) that assigns some number (here it is population size) to each nonnegative integer (the generation). As we argued above, if the sequence is geometric and $b < 1$, then the sequence gets closer and closer to zero (but never quite reaches it) as n gets larger. In this case, a population following this model goes to extinction, where we might define extinction as arising when a population has less than one individual present. So extinction occurs at the first generation n at which $x_n < 1$. We say then that the sequence has a limit, which is zero. If $b > 1$, then x_n continues growing as n increases, never approaching any particular number or population size, so no limit exists. In the real world, no population can continue growing forever, so the case of $b > 1$ can happen for some finite time period, such as at the beginning of a flu epidemic if x_n counts the number of individuals with the flu, but eventually the value of b must change so that $b < 1$.

The idea of a "limit" therefore means that a sequence of numbers x_n approaches a particular value as n gets larger and larger. If the sequence represents population size, then the population size approaches some "limiting value," a long-term equilibrium value, as time goes on. Just think of the limit as the long-term population size, if it exists. Clearly, in the real world, there are numerous factors that cause populations to vary in time, so an equilibrium population size may not be observed.

Now consider situations in which there are multiple state variables or measurements of interest at any discrete time interval. A central theme in biology is that biological objects of interest have structure that may affect the way those objects should be analyzed. Moving up the hierarchical biological scale, examples include the various components of cells, the cellular composition of tissues, the tissue composition of individual organisms, the composition of different individuals that make up a population, the species that make up a community, and the types of habitat across a landscape. Each of these biological entities has a substructure. Think of different types of individuals within a population: the males and females of different ages in a meerkat population, the demographic (age, location, economic status, etc.) structure of a human population, the chemical composition of cells, and so on. We will present mathematical concepts that provide a way to describe the composition of these biological objects and a mechanism to analyze how the composition might change through time or across space.

A motivating example that we will use regularly concerns patterns of species across a landscape. Think of taking aerial photographs of a plot of land once each decade, following the changes in pattern across the landscape. On a real landscape, this could be looking at species, communities of plants (deciduous hardwoods, herbaceous, grassland, etc.), or land use (agricultural, urban, suburban, etc.). For each decade, you have a photo in which you might be able to classify each area as containing mostly one of a few species. Then one summary you might use for the landscape is to describe it by the fraction of the total area that is covered by each of the species. Imagine generating a map of the total area in which different colors on the map represent which type of organism (grass, blackberry bushes, hickory trees, etc.) occupies each area on the map.

A description of this landscape at a particular time could use a list of numbers giving the fraction of the landscape of each species, in other words, a vector of numbers that sum to one. Pictorially, this description corresponds to bar charts that show the number of squares in each landscape covered by each type of organism. There is a loss of information in going from a picture of the landscape to a simple list of numbers representing the fraction of the landscape of each type. All information about the spatial arrangement is lost; that is, we cannot tell from simply a list of numbers whether all the grass is clustered on one side of the landscape or whether it is dispersed throughout the landscape. The ordered list of what fraction of landscape is in each of the different species is called a vector. The ordering of the items in the list matters since switching the order without respectively changing the fractions would give a misrepresentation of the landscape.

As time goes on (e.g., we take a picture of the landscape every decade), the vector describing the fraction covered by each type of organism will change. If we could determine the rules by which this vector changes, we would have a basis for a model of the landscape's dynamics (called succession in ecology). From this model, we might then be able to determine, from only a few decades of pictures, what the landscape might look like many decades from now (the process that we called extrapolation when we were talking about regressions).

The field of study concerning landscapes, how they can be described, and how they change is part of the field of geography. The main tool used to analyze changes is called Geographical

Figure 4.2 Aerial photos of Amazon Park in Eugene, Oregon (1936–2009). The photos taken from 1936 to 1995 were posted on [91], and the 2009 photo was created using Google Maps [92]. © 2014 TerraMetrics, map data © 2014 Google.

Information Systems (GIS), which is software that allows you to view, build, and manipulate spatial maps.

One objective in this part of the text is to construct the mathematical tools needed to analyze temporal changes in vectors describing landscapes as they are observed from one time period (e.g., a year or a decade) to the next. However, this does not apply only to landscapes. It applies to any biological entity that can be broken down into discrete classes. So we will be able to use this to describe changes in the age breakdown of the human population of a city or a region. This is the basis for the field of demography: the study of changes in the structure of human populations through time and space. But this is just as easily applied to any population, be

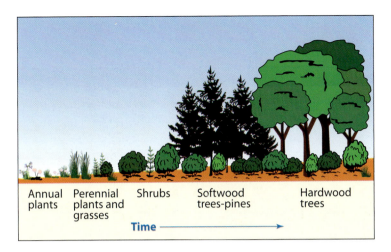

Annual   Perennial   Shrubs   Softwood   Hardwood
plants   plants and   trees-pines   trees
         grasses
Time ───────────────────────▶

Figure 4.3 Simple sketch of ecological succession. Image source [93]. Image courtesy Michael Pidwirny.

that black bears in the Great Smoky Mountains National Park, killer whales in the Pacific, or meerkats in South Africa. More than that, however, it applies to an immense number of biological problems, including those of cell population dynamics (think of bacterial populations for which you track the fraction of bacteria that are resistant to a set of different antibiotics), behavior (think of the fraction of a group of meerkats carrying out a set of different behaviors), and pharmacokinetics (think of the fraction of a drug that is infused and distributed across a set of different tissues and how this changes through time; there would be a class here called "removed" in which the drug has been excreted).

5

CHAPTER

Sequences and Discrete Difference Equations

One of the main uses of mathematics in biology is to have a standard mechanism to study a biologically interesting measured variable (body size, drug concentration in the blood, respiration rate, etc.) as it changes in time or space or because of changes in a factor that affects the variable. In many cases, we take measurements only at particular times or for particular values of the factors that affect the measurement. So the measurements are made only for discrete values. We might then like to use these measurements to make inferences about what the variable would be eventually or at other values of the factor affecting the measurement.

You have all experienced this as your body weight and height changed from birth through the early stages of your life. Indeed, you may well have (or a parent may have) a chart showing your height plotted at the times you were measured during health checkups. These measurements may have been made at regular or irregular times, but when plotted on the standard growth curves for humans, it is entirely possible to predict quite accurately your adult height, assuming that proper nutrition is continued. The set of your heights measured at different times form what we call a sequence, meaning a list of numbers indexed by some integer. Thus, for a sequence of height measurements $\{H_n\}$, the value H_n would be that your height in centimeters at the time that your height was measured the nth time, where n is an integer.

This chapter concerns how we can use sequences from data or from theory to determine how biologically relevant variables change. One of the main objectives is to learn how to use mathematical formulae to describe a sequence and what the long-term implications are for the variable being measured. One example is using your sequence of heights measured from birth to age 10 to predict what your adult height would be. Note that we are using height rather than weight as an example because adult height is much more constrained and does not change much over adulthood compared to body weight (you might consider why this is so).

5.1 Sequences

In this section, we will consider properties of sequences of real numbers that represent biological quantities changing as the time steps change. However, the underlying measurements may be made not only across time but also as some other factor affecting the biological measurement varies. For example, drug concentration in the blood a day after taking a pill may vary with the magnitude of a dose so that we could consider a sequence of different blood concentrations depending not on time but on dose level. An objective is to derive a mathematical formula for the variable being measured (e.g., blood concentration) as some factor is varied (e.g., dose level). We'll start with an easily understood measurement in time.

   Every Christmas, the Audubon Society invites birders across the United States to participate in the Christmas Bird Count. Data for various birds have been collected for over 100 years. The data are posted on the Audubon Society's Web site at [93]. Suppose that we form a sequence using the data collected on red cardinals each year. The count in the first year would be a_1, the count in second year would be a_2, the count in the third year would be a_3, and so on.

Example 5.1 (Cardinals)

Birders in Tennessee have reported seeing the northern cardinal (*Cardinalis cardinalis*) since 1959. The numbers reported are given in the Table 5.1. We could make a sequence using these data. If a_n represents the count at year n, where $n = year - 1959$, then

$$a_0 = 2206, \; a_1 = 2297, \; a_3 = 2650, \ldots, a_{50} = 6896, \; a_{51} = 6190, \; a_{52} = 6739.$$

This particular sequence has 53 terms in the sequence.

Table 5.1. Data collected by Tennessee birders for the Audubon Christmas Bird Count. Data reflect the total count of northern cardinals sighted in Tennessee.

Year	Count	Year	Count	Year	Count	Year	Count
1959	2206	1972	3696	1985	5359	1998	5439
1960	2297	1973	4989	1986	4321	1999	4367
1961	2650	1974	3779	1987	5044	2000	6045
1962	2277	1975	4552	1988	3092	2001	4632
1963	2242	1976	3872	1989	5388	2002	6974
1964	2213	1977	4049	1990	4079	2003	4528
1965	2567	1978	4037	1991	4416	2004	6875
1966	3152	1979	3475	1992	4828	2005	5154
1967	2186	1980	4448	1993	4291	2006	6631
1968	2998	1981	3660	1994	4861	2007	7051
1969	2628	1982	5141	1995	4662	2008	4882
1970	3450	1983	4890	1996	4827	2009	6896
1971	2829	1984	3500	1997	4377	2010	6190
						2011	6739

A *real sequence* is a function $f : \mathbb{N} \to \mathbb{R}$ with domain being the natural numbers $\mathbb{N}$ and the range being contained in the real numbers $\mathbb{R}$. We write $a_n = f(n)$ and call a_n the nth term in the sequence.

Some real sequences can be expressed by an algebraic formulation.

Example 5.2 (Example Sequence)

Suppose we defined the sequence $a_n = f(n) = (-1)^n \frac{2n}{n+1}$. Find the first 5 terms of this sequence.

<u>Solution:</u> We can use the function definition of the sequence to find each term.

$$f(1) = (-1)\frac{2 \cdot 1}{1+1} = -1$$

$$f(2) = (-1)^2\frac{2 \cdot 2}{2+1} = \frac{4}{3}$$

$$f(3) = (-1)^3\frac{2 \cdot 3}{3+1} = -\frac{6}{4} = -\frac{3}{2}$$

$$f(4) = (-1)^4\frac{2 \cdot 4}{4+1} = \frac{8}{5}$$

$$f(5) = (-1)^5\frac{2 \cdot 5}{5+1} = -\frac{10}{6} = -\frac{5}{3}$$

As an example of a sequence that is not based on measurements made at different times, consider a standard method used to analyze one of the most important biological processes: photosynthesis. Photosynthesis is the major process by which green plants transform the energy in sunlight into carbon compounds that provide the base of the food resources for life on earth. A major factor that affects photosynthesis is light availability. A standard experiment is to determine how photosynthesis is affected by light level (or irradiance), and this is often done using leaves on plants exposed to different light levels. At each light level, the leaf's photosynthetic rate is measured after the leaf has had sufficient time (typically a few minutes) to acclimate to the light level. Photosynthesis is measured on a per-unit-leaf area basis to compare plants that have leaves of different sizes and shapes. Its units are typically μmoles of oxygen (O_2) per m^2 of leaf area per second, so it is a rate of oxygen released per unit leaf area per unit time. The data obtained from this experiment create what is called the "light response" for the individual plant of that species. Note that the photosynthetic rate of leaves can also be determined by measuring the carbon uptake rate (amount of CO_2 absorbed by the plant), which is closely related to the O_2 release rate.

Table 5.2 (derived from data presented in figure 3 in [22]) shows two light responses for leaves of soybean plants. A graphical representation of the data is shown in Figure 5.1. The data were collected while the plants were experiencing scarce rainfall and were thus experiencing drought stress. Each of the points is obtained by letting leaves sit at a particular light level for a period of time long enough to have a stable photosynthetic rate; the photosynthetic rate is measured, and these rates are averaged across several leaves. The two sets of data shown are for leaves from

Table 5.2. Data presented in [22] of the photosynthetic flux density (PPFD) and net photosynthetic oxygen evolution rate (A_{O_2}, rate of production of breathable oxygen) for soybean leaves located on the upper and lower portions of the plant.

Lower Leaves		*Upper Leaves*	
P_n **PPFD** μmol m^{-2} s^{-1}	I_n A_{O_2} μmol m^{-2} s^{-1}	P_n **PPFD** μmol m^{-2} s^{-1}	I_n A_{O_2} μmol m^{-2} s^{-1}
3.64335	−2.29730	3.77583	−2.56757
19.1441	−0.76577	16.7815	−0.94595
36.9634	0.40541	37.0076	0.49550
57.1895	1.66667	57.2779	1.84685
86.9988	2.47748	72.2929	2.47748
104.376	2.92793	89.8472	3.28829
126.612	3.28829	104.906	4.00901
153.661	3.46847	129.681	4.54955
178.259	3.64865	154.500	5.18018
210.166	3.73874	179.319	5.81081
271.661	4.18919	211.447	6.35135
369.745	4.27928	273.119	7.16216
485.029	4.45946	371.600	8.06306
649.289	4.54955	487.016	8.51351
828.343	4.81982	651.541	9.14414
997.328	4.54955	833.002	9.32432
		1002.08	9.23423

the upper part of the soybean plant and those from the lower part. The light level is measured as photosynthetic photon flux density (PPFD), giving the amount of photons in the spectral range usable by a plant for photosynthesis on a per-unit-area-per-unit-time basis. The units are μmoles of photons per m^2 of leaf area per second. You can think of this as the number of photons used by a green plant over a unit surface area of the leaf per unit time. The highest light levels are at about 2200 μmol m^{-2} s^{-1} (which is full sunlight on a very clear day), so the measurements illustrated have been made at light levels well below this. We can define I as the light level, and thus I_n is the nth light level at which a photosynthesis measurement was taken. Recall that the light level is measured as the PPFD in μmol m^{-2} s^{-1}. Additionally, we define P_n as the net photosynthetic oxygen evolution rate, measured as the μmoles of O_2 released per m^2 of leaf area per second at light level I_n. Note that the sequence for P_n is different for upper and lower leaves and that the sequence of data points arises from measuring P only at certain I values. In theory, measurements could be made at any I level, but in practice, this is not easily done, so the experiment is done at a sufficient number of light levels to provide a good understanding of leaf response to light.

5.2 Limit of a Sequence

Our objective in this section is to point out how we can use mathematical descriptions of sequences to analyze the behavior of biological processes of interest. We will later use this as

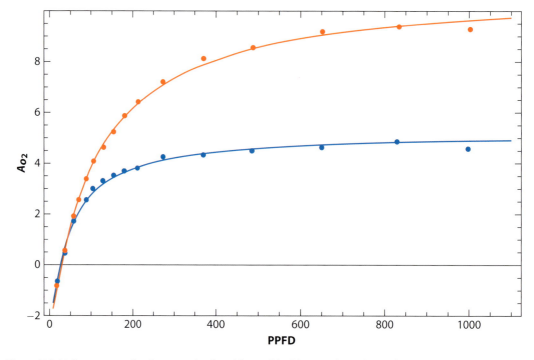

Figure 5.1 Light responses for the upper (red) and lower (blue) leaves of a soybean plant. Data derived from [22].

the basis for understanding how population numbers change through time. As an example here, consider the soybean leaf photosynthetic rates described above. From Figure 5.1, it appears that when the light level increases, the photosynthetic rates approach a particular maximum value. For the lower-level leaves, this is at approximately 5 μmol m^{-2} s^{-1}, and for upper-level leaves, it is at about 10 μmol m^{-2} s^{-1}. This value, typically called P_{max}, gives the light-saturated photosynthetic rate, which is the limit of the sequence of measurements of photosynthetic rate as the PPFD is increased in the experiments. In this section, we want to define this idea of a "limit" and show how it can be used in cases for which we have mathematical formulae for the sequence rather than only the data, as we illustrated with photosynthetic rate.

Suppose that we have a sequence $\{a_n\}$ whose terms are getting closer and closer to L as n gets larger and larger, where we denote these increasing values of n by $n \to \infty$ with the symbol ∞ called infinity. Then we say that the *limit* of the sequence $\{a_n\}$ is L. Formally, we write

$$\lim_{n \to \infty} a_n = L$$

if the sequence $\{a_n\}$ converges to the limit L (a finite number). If the sequence $\{a_n\}$ increases toward infinity that is, it continues to increase and is eventually larger than any number we choose, or decreases toward negative infinity (that is, gets eventually smaller than any negative number we might choose) as $n \to \infty$, then we say that the sequence $\{a_n\}$ does not have a limit. Thus, note that not all sequences have limits.

Example 5.3 (Photosynthetic Capacity of Leaves)

Consider the data shown in Table 5.2 and Figure 5.1. Looking at the photosynthetic rates for lower leaves as light level is increased, we can estimate that the limit of the photosynthetic rate for the lower leaves is roughly $5 \, \mu\text{mol m}^{-2} \text{ s}^{-1}$, while the limit of this for the upper leaves is roughly $10 \, \mu\text{mol m}^{-2} \text{ s}^{-1}$. As we defined P_n above, another way of stating this is that

$$\lim_{n \to \infty} P_n = 5$$

for lower leaves and

$$\lim_{n \to \infty} P_n = 10$$

for upper leaves. The differences between the upper and lower leaves arises because of physiological differences in the leaves based on their growing location in the plant. Upper leaves typically have a higher density of chloroplasts and are thicker than lower leaves. Note that I really cannot increase to infinity, as there is a maximum value for irradiance, and one needs to be careful in extrapolating beyond the available data. Although we cannot tell from the data what the P value would be for $I = 2200$, it is reasonable to assume that it is very close to the limiting value we estimated without more data to suggest otherwise.

Note that in Figure 5.1, there are curves representing the general trends of the data points for the upper (red data points) and lower (blue data points) leaves. Each of these curves is a type of *Michaelis-Menten curve*, which has the form

$$P = \frac{abI}{aI + b} - d, \tag{5.1}$$

where a, b, and d are positive constants; I is the light level; and P is the photosynthetic rate. Technically, a standard Michaelis-Menten curve has the form $y = \frac{abI}{aI+b}$. Given the standard Michaelis-Menten form, if a and b are positive constants, then y will be positive for all positive values of I. However, the data show that P is negative for some positive values of I, indicating that at low light levels, the soybean plant is actually losing CO_2 and absorbing O_2. Thus, we subtract the positive constant d from the standard Michaelis-Menten curve, where $-d$ is the *dark respiration rate*, that is, the respiration rate of the soybean plant when there is no light, or $P(0) = -d$.

Using a nonlinear least-squares regression, we can find the values of a, b, and d so that the curve minimizes the distance between the data and the curve. Recall our discussion of linear regressions in Chapter 3. Nonlinear regressions are similar to linear regression with the exception that the underlying model for the data is not a line (in this case, we use the Michaelis-Menten curve). The "best-fit" equations for the curves in Figure 5.1 are

$$P(I) = \frac{1.62I}{0.195I + 8.29} - 3.13 \tag{5.2}$$

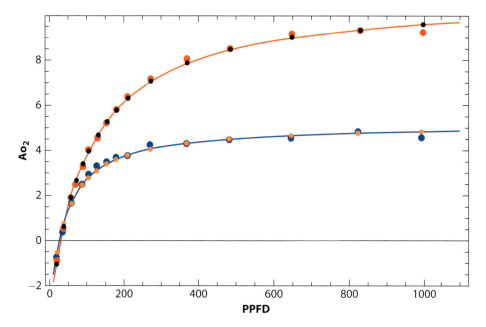

Figure 5.2 Light responses for the upper (red) and lower (blue) leaves of a soybean plant. Data derived from [22]. The red curve is given by Equation (5.3) and the blue curve is given by Equation (5.2). The black and orange points are the sequences $\{\hat{P}_n\}$ calculated using the I_n for the upper and lower leaves, respectively.

for the lower leaves (blue curve) and

$$P(I) = \frac{1.84I}{0.132I + 14.0} - 3.03 \tag{5.3}$$

for the upper leaves (red curve).

If we evaluate the function in Equation (5.1) only at the light level values in the sequence $\{I_n\}$, then we create the sequence

$$\hat{P}_n = P(I_n) = \frac{abI_n}{aI_n + b} - d. \tag{5.4}$$

In this way, we are able to define the sequence $\{\hat{P}_n\}$ as a function of the sequence I_n. Note that the sequences $\{P_n\}$ and $\{\hat{P}_n\}$ will not have exactly the same values. The sequence $\{P_n\}$ represents the actual data collected, while the sequence $\{\hat{P}_n\}$ represents estimations of the data points based on a curve that was fit to the data. Figure 5.2 shows the sequences for $\{P_n\}$ (red for upper leaves and blue for lower leaves) and $\{\hat{P}_n\}$ (black for upper leaves and orange for lower leaves). Notice the points representing $\{\hat{P}_n\}$ are always on the curve for $P(I)$ while the actual data points deviate from this curve.

5.3 Discrete Difference Equations

Suppose that we have a population that doubles each year. We could construct a sequence where x_n represents the size of the population at each time step. Since the population is doubling each year, $x_{n+1} = 2x_n$; that is, the population at time step $n + 1$ is twice what it was at time step n, using yearly time steps.

We will study equations showing biological quantities that change over time. For example, the Fibonacci sequence

$$1, 1, 2, 3, 5, 8 \ldots$$

(originally involving generations of rabbit populations) is given by

$$x_0 = 1, x_1 = 1$$

and then for $n = 1, 2, \ldots$

$$x_{n+1} = x_n + x_{n-1}.$$

Note that x_{n+1} is a function of the two previous terms in the sequence, meaning that this equation for x_{n+1} is called a second-order difference equation. Note that the equation above, $x_{n+1} = 2x_n$, is called a first order difference equation since each new term in the sequence depends only on the previous term.

Suppose that a population at time step $n + 1$ depends on the population values at all the previous time steps; then we can represent the population at time step $n + 1$ by

$$x_{n+1} = f(x_n, x_{n-1}, \ldots, x_0). \tag{5.5}$$

That is, x_{n+1} can be written as a function of terms that come before it in the sequence. We refer to Equation (5.5) as a **difference equation**. If the x_{n+1} term can be written solely as a function of the x_n term, that is,

$$x_{n+1} = f(x_n),$$

then we say that this sequence is built from a **first-order difference equation**. The term *first order* means that to find x_{n+1}, you need only use the previous value, x_n. Notice that if we know x_0 (the initial value in the sequence), then we can determine all the terms in the sequence, coming from a first-order difference equation.

Example 5.4 (First-Order Difference Equation)

A population of doves increases by 3% each year. Let x_n be the size of the population in year n. Then $x_{n+1} = x_n + .03x_n = 1.03x_n$. Thus, the first-order difference equation that describes the population is

$$x_{n+1} = 1.03x_n.$$

Notice that if we know x_0, then

$$x_1 = 1.03x_0$$

$$x_2 = 1.03x_1 = 1.03(1.03x_0) = (1.03)^2 x_0$$

$$\vdots$$

$$x_n = (1.03)^n x_0.$$

Note that as n increases, the terms of x_n continue to grow so that $\lim n \to \infty$, x_n doesn't exist.

5.4 Geometric and Arithmetic Sequences

The sequence in the dove example above is called a geometric sequence since its population changes by a multiplication factor each time step.

A *geometric sequence* is defined by

$$x_{n+1} = rx_n,$$

where r is a fixed real number. Notice that if we know x_0, then

$$x_1 = rx_0$$
$$x_2 = rx_1 = r(rx_0) = r^2x_0$$
$$x_3 = rx_2 = r(r^2x_0) = r^3x_0$$
$$\vdots$$
$$x_n = r^nx_0.$$

The general solution to the difference equation represents the x_n in terms of x_0, n, and other given constants. For a geometric sequence, the general solution is

$$x_n = r^nx_0.$$

When $0 < r < 1$, the sequence decays to zero. It has a limit of zero, meaning that as n gets large $(n \to \infty)$, $x_n \to 0$. For $r > 1$, the terms of the sequence increase exponentially (see Example 5.4).

Example 5.5 (Wild Hares)

A population of wild hares increases by 13% each year. Currently, there are 200 hares. If x_n is the number of hares in the population at the end of year n, find

(a) the difference equation relating x_{n+1} to x_n,
(b) the general solution to the difference equation found in (a), and
(c) the number of hares in the population at the end of 6 years from now.

Solution: Notice that $x_0 = 200$.

(a) Since the population increases by 13% each year,

$$x_{n+1} = \underbrace{x_n}_{\text{from year before}} + \underbrace{0.13x_n}_{\text{increase}} = 1.13x_n.$$

(Continued)

(b) Using the notation of the definition of a geometric sequence, here $r = 1.13$; thus, the general solution to the difference equation in (a) is

$$x_n = (1.13)^n x_0 = 200(1.13)^n.$$

(c) For $n = 6$, $x_6 = 200(1.13)^6 \approx 416$. Thus, at the end of year 6, there are approximately 416 hares.

If a population increases by a fixed number d each time period, we say that the sequence is an *arithmetic sequence*:

$$x_{n+1} = x_n + d.$$

Suppose we know the intial value x_0; then the general solution to an arithmetic sequence is

$$x_1 = x_0 + d$$
$$x_2 = x_1 + d = (x_0 + d) + d = x_0 + 2d$$
$$\vdots$$
$$x_n = x_0 + nd.$$

5.5 Linear Difference Equation with Constant Coefficients

A first-order difference equation is *linear* if it takes the form

$$x_{n+1} = a_n x_n + b_n,$$

where a_n and b_n are sequences of constants. In this section, we are interested in linear first-order difference equations where $a_n = a$ and $b_n = b$ for all n. That is, we can write the difference equation as

$$x_{n+1} = ax_n + b.$$

Let us find the general solution to this difference equation.

First, notice that if $a = 1$, then we are in the case of the arithmetic sequence, and we have already found the general solution:

$$x_n = x_0 + nb.$$

Second, if $b = 0$, we have a geometric sequence with the general solution $x_n = a^n x_0$.

If $a \neq 1$ and $b \neq 0$, then to find the general solutions, we use the following steps.

Finding the General Solution to $x_{n+1} = ax_n + b$ where $a \neq 1$

Step 1: We first solve the *homogeneous difference equation*, which is the difference equation without the constant b, that is, $x_{n+1} = ax_n$. We already know that the solution to this has the form $a^n c$, where c is some constant. We cannot say that $c = x_0$ here because a solution to the homogeneous equation is not necessarily a solution to the full equation.

Step 2: Next, we construct a *particular solution* p_n. We will assume that the particular solution is a constant, that is, $p_n = K$ for all n. We now need to determine K. For p_n to be a solution to the difference equation, it must satisfy

$$p_{n+1} = ap_n + b.$$

However, $p_n = K$; thus,

$$K = aK + b$$
$$K - aK = b$$
$$(1 - a)K = b$$
$$K = \frac{b}{1 - a}.$$

Thus, our particular solution is $p_n = \frac{b}{1-a}$.

Step 3: Finally, we form the general solution as the sum of the homogeneous and particular solutions, that is,

$$x_n = ca^n + \frac{b}{1 - a}.$$

Notice that $x_0 = c + \frac{b}{1-a}$; thus, $c = x_0 - \frac{b}{1-a}$.

To see why building a general solution in this way works, consider two solutions, s_n and w_n, to

$$x_{n+1} = ax_n + b.$$

Let $y_n = s_n - w_n$; then y_n satisfies this difference equation:

$$y_{n+1} = s_{n+1} - q_{n+1} = (as_n + b) - (aw_n + b) = a(s_n - w_n) = ay_n.$$

So it must be true that y_n is a solution of the homogeneous equation, giving $y_n = ca^n$ for a constant c. Any general solution can be written as $s_n = ca^n + w_n$.

Therefore, the general solution to the difference equation is

$$x_n = \left(x_0 - \frac{b}{1 - a} \right) a^n + \frac{b}{1 - a}.$$

Note that in the general solution of the difference equation, there are two different possibilities as $n \to \infty$. If $0 < a < 1$, then x_n gets closer and closer to $b/(1 - a)$ and so $\lim_{n\to\infty} x_n = \frac{b}{1-a}$. If $a > 1$, then the first term in the solution gets larger and larger so that $\lim_{n\to\infty} x_n$ doesn't exist. Let us consider an example of a population that could be modeled by a linear difference equation with constant coefficients. What would each term in the difference equation represent?

$$\underbrace{x_{n+1}}_{\text{pop. @ } t = n + 1} = \underbrace{ax_n}_{\text{pop growth or decline}} - \underbrace{b}_{\text{fixed decrease}} \tag{5.6}$$

Notice that in Equation (5.6), the population grows or declines first (with the factor a), and then a fixed amount b is being removed. Thus, this model could represent a population that is being harvested by a fixed amount at each time step after growth or decline. If the constant b was being added in the equation, then the difference equation might model a population (after growth or decline) that was being augmented by a fixed amount each time step (think of a fish population being restocked each season).

Furthermore, it may be advantageous to think of the constant a, representing population growth or decline, as the difference between the birthrates and the death rates of the population. That is, if β is the proportion by which the population increases because of births at each time step and δ is the proportion by which the population decreases because of deaths at each time step, then $a = 1 + \beta - \delta$. If $a > 1$, then the population is growing (in the absence of harvesting); if $a < 1$, then the population is decreasing (in the absence of harvesting); and if $a = 1$, then in the absence of harvesting, the population will remain constant.

Example 5.6 (Fisheries)

Consider a lake fish population whose yearly birthrate is 1.2 and whose yearly death rate is 0.7. Each year, fishing is allowed until 1200 fish are caught. Thereafter, fishing is banned. Currently, there are 12,230 fish in the lake.

(a) Write a difference equation for the lake fish population and find the general solution.

(b) How many fish are in the lake after 5 years?

(c) If the resource managers of the lake wanted the population to remain constant each year, what level of harvesting should they allow?

Solution: Let x_n be the size of the fish population at the end of year n. Then $x_0 = 12,230$.

(a) Since $\beta = 1.2$ and $\delta = 0.7$, then $a = 1 + \beta - \delta = 1 + 1.2 - 0.7 = 1.5$. Thus, $x_{n+1} = 1.5x_n - 1200$. The general solution is then

$$x_n = \left(x_0 - \frac{-1200}{1 - 1.5}\right)(1.5)^n + \frac{-1200}{1 - 1.5}$$

$$= (12,230 - 2400)(1.5)^n + 2400$$

$$= 9830(1.5)^n + 2400.$$

(b) For $n = 5$, $x_n = 9830(1.5)^5 + 2400 \approx 77,047$. Thus, there are approximately 77,047 fish in the lake after 5 years.

(c) To keep the population constant, we would want $x_{n+1} = x_n$ for all n. Specifically, $x_1 = x_0$. Let h be the harvesting level we wish to find. Thus, using $x_1 = x_0$ in

$$x_1 = 1.5x_0 - h,$$

(Continued)

we want to solve for h:

$$x_0 = 1.5x_0 - h$$
$$(1 - 1.5)x_0 = -h$$
$$-0.5x_0 = -h$$
$$-0.5(12,230) = -h$$
$$h = 6115.$$

Thus, if the lake resource managers allow 6115 fish each season to be caught, the fish population size will remain constant from year to year.

Example 5.7 (Stocking a Lake)

Consider a lake fish population whose yearly birthrate is 0.5 and whose yearly death rate is 0.7. Currently, there are 12,230 fish in the lake. How many fish are needed to stock the lake each year so that the population remains constant?

Solution: The birthrates and death rates correspond to $\beta = 0.5$ and $\delta = 0.7$, and thus $a = 1 + \beta - \delta = 1 + 0.5 - 0.7 = 0.8$. Thus, the population is naturally decreasing. We can construct the difference equation:

$$x_{n+1} = 0.8x_n + \alpha,$$

where x_n is the size of the fish population at the end of year n and α is the number of fish that are added to the lake each year. Notice that $x_0 = 12,230$. To find the number of fish needed to stock the lake each year to keep the population constant, we use $x_1 = x_0$ and solve $x_1 = 0.8x_0 + \alpha$ for α:

$$x_0 = 0.8x_0 + \alpha$$
$$(1 - 0.8)x_0 = \alpha$$
$$0.2x_0 = \alpha$$
$$\alpha = 0.2(12,230) = 2446.$$

Thus, each year, the lake must be stocked with an additional 2446 fish to maintain a constant population size.

5.6 Introduction to Pharmacokinetics

Population ecology models are not the only applications for linear difference equations with constant coefficients. Let us next consider how to utilize linear difference equations to construct a drug dosage scheme for a patient. Suppose that an initial amount of drug, b mg, is introduced into the body. If no additional doses are given, the amount of drug in the body will gradually diminish. For most drugs, it is assumed that the amount of drug remaining in the body after t hours is $x(t) = be^{-kt}$, where $k > 0$ is a drug-specific rate-of-decay constant. Recall that we considered a similar scenario in Example 4.4. Now, suppose an additional dose of size b is given to the patient every τ hours (see Figure 5.3). If we let x_n be the amount of drug left in the body when the nth dose is given, then when the $(n+1)$th dose is administered, there is $e^{-k\tau}x_n$ of the drug remaining from the nth dose, and then amount b of the drug is added.

Denoting $a = e^{-k\tau}$, our difference equation is.

$$x_{n+1} = \underbrace{ax_n}_{\text{drug remaining}} + \underbrace{b}_{\text{new dose}}$$

Note that $0 < a < 1$ and $x_0 = b$. Using the general solution to this difference equation, we obtain

$$
\begin{aligned}
x_n &= \left(x_0 - \frac{b}{1-a}\right)a^n + \frac{b}{1-a} \\
&= \left(b - \frac{b}{1-e^{-k\tau}}\right)(e^{-k\tau})^n + \frac{b}{1-e^{-k\tau}} \\
&= \left(\frac{b - be^{-k\tau} - b}{1-e^{-k\tau}}\right)e^{-k\tau n} + \frac{b}{1-e^{-k\tau}} \\
&= \frac{-be^{-k\tau}e^{-k\tau n}}{1-e^{-k\tau}} + \frac{b}{1-e^{-k\tau}} \\
&= \frac{-be^{-k\tau(1+n)} + b}{1-e^{-k\tau}}
\end{aligned}
$$

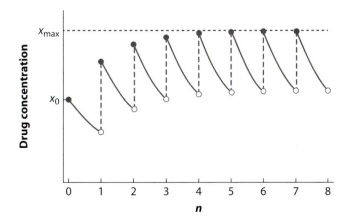

Figure 5.3 Drug dose of size b begin given every one unit of time.

$$x_n = \frac{b}{1 - e^{-k\tau}} \left(1 - e^{-k\tau(1+n)}\right).$$

Thus, we see that the general solution to the drug dose difference equation is

$$x_n = \frac{b}{1 - e^{-k\tau}} \left(1 - e^{-k\tau(1+n)}\right).$$

By considering what happens as n gets larger and larger we can find the limit of x_n. Because the exponential term in the formula gets smaller and smaller as n gets larger and larger, it can be replaced by 0, so that

$$\lim_{n \to \infty} \frac{b}{1 - e^{-k\tau}} \left(1 - e^{-k\tau(1+n)}\right) = \frac{b}{1 - e^{-k\tau}} (1 - 0) = \frac{b}{1 - e^{-k\tau}}.$$

Let

$$x_{max} = \frac{b}{1 - e^{-k\tau}}.$$

Notice that $x_n = x_{max} \left(1 - e^{-k\tau(1+n)}\right)$. Since $0 < 1 - e^{-k\tau} < 1$ for all n,

$$b < x_{max}$$

and

$$x_n < x_{max}$$

for all n. So, we see that if we give the same drug dose at regular intervals, the amount of drug in the body will always remain less than x_{max}. We can use this information to design a drug dosing scheme. Suppose that we knew that a certain drug was lethal or produced detrimental side effects if the amount of drug in the body ever exceeds x_{max} mg. We can solve the x_{max} equation for b to find what drug dose amount will ensure that the total amount of drug in the patient's body never exceeds x_{max} mg:

$$x_{max} = \frac{b}{1 - e^{-k\tau}}$$

$$b = x_{max} \left(1 - e^{-k\tau}\right).$$

Notice that the size of b depends on x_{max}, k, and τ. If it happens that the drug dose size b is set, we could instead solve the x_{max} equation for τ to find the dosing period that will allow the amount of drug in the patient to always remain below x_{max}:

$$x_{max} = \frac{b}{1 - e^{-k\tau}}$$

$$1 - e^{-k\tau} = \frac{b}{x_{max}}$$

$$e^{-k\tau} = 1 - \frac{b}{x_{max}}$$

$$-k\tau = \ln\left(1 - \frac{b}{x_{max}}\right)$$

$$\tau = -\frac{1}{k} \ln \left(1 - \frac{b}{x_{\max}} \right).$$

Note that we used $b < x_{\max}$ to get

$$0 < \frac{b}{x_{\max}} < 1,$$

which means that the natural logarithm,

$$\ln \left(1 - \frac{b}{x_{\max}} \right),$$

is negative and that $\tau > 0$.

Example 5.8 (Drug Dosing Period)

A certain cancer treatment drug must never exceed 1200 mg in the body. The decay rate of the drug is 0.05 mg per hour (i.e., $k = 0.05$). If the dose size is 250 mg, how many hours should pass between each dose?

<u>Solution:</u> Since $x_{\max} = 1200$, $k = 0.05$, and $b = 250$, when we solve $x_{\max} = \frac{b}{1-e^{-k\tau}}$ for τ, we get

$$\tau = -\frac{1}{k} \ln \left(1 - \frac{b}{x_{\max}} \right)$$

$$= -\frac{1}{0.05} \ln \left(1 - \frac{250}{1200} \right)$$

$$\approx 4.672.$$

Thus, the patient should take another dose roughly ever 4.7 hours, or every 4 hours and 42 minutes. Obviously, this timing may be hard to keep track of, so the doctor might tell the patient to take the drug every 5 hours. Notice, however, that it would be detrimental for the doctor to tell the patient to take the drug every 4.5 hours. If the patient took another dose every 4.5 hours, the drug would exceed the maximum amount that should be in the patient's body.

Example 5.9 (Tuberculosis Drug Dosing)

The standard short-course treatment for tuberculosis is a combination of drugs over a 6-month period. The drug ethambutol is given during the first 2 months of the treatment. A tuberculosis patient is prescribed a 1000-mg dose to be taken daily (every 24 hours)

(*Continued*)

for the first 2 months. It is known that ethambutol has a half-life in the human body of approximately 4 hours. Given the patient's weight, ethambutol is known to have severe side effects if the amount of the drug in the body exceeds 1500 mg. Will this patient experience these side effects? How would this change if the half-life of the drug were 10 hours (i.e., four times as long)?

Solution: First, we know that $b = 1000$, $\tau = 24$, and $x_{max} = 1500$. The piece of information we are missing is the decay constant k. We can use the information about the half-life of the ethambutol in the body to determine the value of k. Recall from Chapter 4 how to find the decay rate using information about the half-life. Using the equation $x(t) = x_0 e^{-kt}$, we know that when $t = 4$,

$$\frac{1}{2} x_0 = x_0 e^{-4k}$$

$$\frac{1}{2} = e^{-4k}$$

$$\ln\left(\frac{1}{2}\right) = -4k$$

$$-\frac{1}{4}\ln\left(\frac{1}{2}\right) = k.$$

Using $b = 1000$, $\tau = 24$, and $k = -\frac{1}{4}\ln\left(\frac{1}{2}\right)$, we find that

$$x_{max} = \frac{1000}{1 - e^{0.25\ln(0.5)(24)}} \approx 1016.$$

Thus, the amount of ethambutol in the body of the patient will never exceed 1016 mg, and therefore the patient will not experience severe side effects.

If, however, the half-life of ethambutol is 10 hours, then $k = -\frac{1}{16}\ln\left(\frac{1}{2}\right)$, and we will find that

$$x_{max} = \frac{1000}{1 - e^{0.0625\ln(0.5)(24)}} \approx 1547.$$

Thus, when the half-life is 16 hours, the amount of drug in the body will approach 1547 mg in the limit. Thus, we would expect after some time that the patient would begin to experience the severe side effects.

5.7 Matlab Skills

Given a difference equation $x_{n+1} = f(x_n)$, we would like to utilize Matlab to generate a plot of x_n for some finite set of n values. To simulate a discrete difference equation model, we first need to learn how to implement *for loops* in Matlab.

Loops

Often we would like to use Matlab to perform the same operation over and over with only a slight change. It is tedious to type the same commands (with a slight change) repeatedly. To remove the tediousness of this task, we use what are known in computer programming as *for loops*. The basic structure of a *for loop* is

```
for i = [ % some array of values ]
    % some commands that change only as i changes
end
```

where i is known as the ***index*** of the loop. Although it is typical to use the letter i for the index, you may use whatever label you like. Other common index labels are j, k, and count.

Suppose that we wanted to sum up the numbers from 1 to 100. We could do this with a *for loop*.

```
                              ─── Sum100.m ───
1   % Filename: Sum100.m
2   % M-file to
3   % - Sum the numbers 1 to 100
4
5   total = 0;              % inital total to 0
6   for i = 1:100           % loop through 1, 2, ..., 100
7     total = total + i;    % add next value to total
8   end
9   fprintf('total = %d\n', total)
```

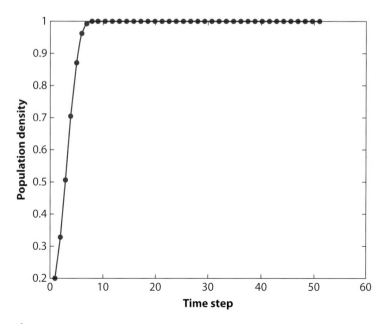

Figure 5.4 Output from `LogisticDifferenceEqn.m`.

When this m-file is run in the Command Window, the output looks like the following:

```
———————————— Command Window ————————————
>> Sum100
total = 5150
```

Suppose that we want to model a population that grows according to the difference equation

$$x_{n+1} = \underbrace{x_n}_{\text{Population density at time step } n} + \underbrace{rx_n(1 - x_n)}_{\text{Growth term}},$$

where the value r is referred to as the *intrinsic growth rate* of the population. This difference equation is known as the **logistic difference equation**, and the general solution of this equation cannot be found using the methods presented here. Thus, we will use Matlab to explore what happens to the population over some finite number of n values. As with Sum100.m above, we use a *for* loop to determine the values of x_n. We will use an intrinsic growth rate of 0.8 and an initial value of $x_0 = 0.2$:

```
———————————— LogisticDifferenceEqn.m ————————————
1   % Filename: LogisticDifferenceEqn
2   % M-file to simulate the logistic difference equation
3
4   % Set the values of r and x0
5   r = 0.8;
6   x0 = 0.2;
7
8   % Through the loop we will fill the values x_n into the array x
9   % In the first iteration of the loop, the value of x(1) needs to
10  %  be known, we set that value before starting the loop
11  x(1) = x0;
12  for n = 1:50
13      x(n+1) = x(n) + r*x(n)*(1-x(n));
14  end
15
16  % Plot the results
17  plot(x,'r.-')
18  xlabel('Time Step')
19  ylabel('Population Density')
```

When we run this file, we generate the graph shown in Figure 5.4.

5.8 Exercises

5.1   The table below (derived from data presented in figure 3 in [22]) shows light responses for the leaves of soybean plants. The data were collected during normal rainfall conditions after a period of 30 days during which there was scarce rainfall. Thus, the plants had recently experienced a short drought. Each of the points is obtained by letting leaves sit at a particular light level (measured as the PPFD in μmol of photons per m^2 of leaf area per second) for a period long enough to have a stable photosynthetic rate. The photosynthetic rate is measured as the net photosynthetic oxygen evolution rate (A_{O_2} in μmol of O$_2$ per m^2 of leaf area per second), and these rates are averaged across several leaves.

Lower Leaves		Upper Leaves	
PPFD $\mu\text{mol m}^{-2}\text{ s}^{-1}$	A_{O_2} $\mu\text{mol m}^{-2}\text{ s}^{-1}$	PPFD $\mu\text{mol m}^{-2}\text{ s}^{-1}$	A_{O_2} $\mu\text{mol m}^{-2}\text{ s}^{-1}$
21.9	−0.23	16.4	−1.58
42.3	1.22	29.1	−0.68
57.5	2.21	44.3	0.32
72.7	3.29	54.4	1.04
90.2	4.10	105.0	4.28
107.9	5.18	127.7	5.45
133.1	6.44	152.9	6.89
155.6	7.52	180.4	7.97
180.6	8.42	215.5	9.59
277.5	11.13	275.5	12.03
374.1	13.11	375.0	15.00
490.1	14.73	491.4	17.52
657.5	16.35	656.9	20.14
836.9	17.25	836.9	22.30
1003.7	17.61	1006.3	22.84

(a) For the lower leaves, create a sequence of the photosynthetic rates. How many elements are in this sequence?

(b) As the light level is increased, does the photosynthetic rate of the lower leaves approach a limit? If so, estimate the value of this limit.

(c) As the light level is increased, does the photosynthetic rate of the upper leaves approach a limit? If so, estimate the value of this limit.

5.2   Find the first five terms of each of the following sequences. If possible, find the $\lim_{n\to\infty} a_n$ by considering very large values of n.

(a) $a_n = \dfrac{n}{3n + 1}$

(b) $a_n = \dfrac{2n}{n^2 + 1}$

(c) $a_n = (-1)^n \left(\dfrac{1}{4}\right)^n$

(d) $a_n = 5 - n$

5.3   Find the general solution and $\lim_{n\to\infty} x_n$ (if it exists) for each of the following difference equations. Assume in each case that $x_0 = 10$.

(a) $x_{n+1} = 0.3x_n$

(b) $x_{n+1} - 3x_n = 5x_n$

(c) $x_{n+1} = x_n + 6$

(d) $x_n - x_{n-1} = 8$

(e) $x_{n+1} - 3x_n = 5$

(f) $x_{n+1} = 2x_n + 2$

(g) $x_{n+1} = x_n - 4$

(h) $x_{n+1} + 2x_n = 4 - x_n$

(i) $3x_{n+1} - 2x_n + 1 = 0$

5.4   Find the general solution for each of the following difference equations. Plot the x_n data versus time for 20 time steps. Assume in each case that $x_0 = 10$.

(a) $x_{n+1} = -1.3x_n$

(b) $x_{n+1} = 0.2x_n + 5$

5.5   A herd of newly introduced elk in a wildlife management area in the Cumberland Mountains in Tennessee increases its numbers by 10% each year. Let x_n be the number in the population at the end of year n.

(a) Find the difference equation relating x_{n+1} to x_n.
(b) Solve for x_n if $x_0 = 50$.

5.6   The probability that an adult elk survives the year is 90%. Suppose that a herd starts with 50 adult individuals and let x_n be the number of original adult elk still alive after n years. Assume that no individuals are added to the population.

(a) Write the difference equation relating x_{n+1} to x_n.
(b) Find the general solution to the difference equation.
(c) Find the number of original adult elk in the herd after 4 years.

5.7   The body eliminates 10% of the amount of pain reliever drug present each hour. Let x_n be the amount of drug (in milligrams) in the body n hours after the initial dose of 180 mg.

(a) Relate x_{n+1} to x_n.
(b) Find the general solution to the difference equation.
(c) How many milligrams of the drug remains in the body after 1 hour?

5.8   A population of buffalo can increase its numbers by about 10% each year. Let x_n be the population count after n years and assume that h buffalo are removed from the herd at the end of each year.

(a) Find x_n if $x_0 = 1000$ if $h = 20$.
(b) With $h = 20$, would this population ever go negative? (If so, the model would not make biological sense.)
(c) Again assuming that $x_0 = 1000$, find the largest h so that $x_{10} \geq 1500$.

5.9   Suppose that in a lake a population of trout increases its own numbers by 10% each year. After the births occur each year, 100 young trout are added each year to build up the population. Let x_n denote the size of the population after n years, starting with a population of 1000 trout.

(a) When does $x_n \geq 2000$ happen?
(b) After that time found in part (a), the lake is no longer stocked, and fishermen will catch 400 fish per year (not worrying about the size of the fish). What is the fate of the population?

5.10   (From [16]) The body eliminates 10% of a certain drug each hour. Suppose that doses of 200 milligrams are given every 6 hours.

(a) Find the maximum drug level in the body.
(b) Find the amount of drug in the body 24 hours after the first dose.
(c) How frequently should the drug be administered to build up a maximum drug level of 1000 mg?

5.11   (From [16]) A level of more than 1500 mg of a certain drug in the body is considered unsafe. Individual doses are 250 mg, and the drug is removed from the body according to the exponential decay equation

$$x(t) = x_0 e^{-0.1t},$$

where t is measured in hours. How frequently can the drug be safely administered?

5.12 (From [16]) The charge on a nerve cell (*neuron*) is increased by 1 millivolt every 2 milliseconds. Individual charges decay exponentially according to the formula

$$x(t) = x_0 e^{-0.05t},$$

where t is measured in milliseconds. Thus, if x_0 is the present charge on the cell, the charge remaining after 2 milliseconds is

$$x(2) = x_0 e^{-0.1} \text{ millivolts.}$$

Let x_n be the charge on the cell after $2n$ milliseconds.

(a) Show that $x_{n+1} = \left(e^{-0.1}\right) x_n + 1$ and solve for x_n if $x_0 = 0$.
(b) The "all or nothing" law asserts that the neuron will fire as soon as the total charge on the cell exceeds a certain threshold value. If the neuron fires as soon as the charge exceeds 4 millivolts, how frequently will the neuron fire?

5.13 Use `LogisticDifferenceEqn.m` to determine what happens to the population when $x_0 = 0.2$ and

(a) $r = 1.0$
(b) $r = 2.0$
(c) $r = 2.5$
(d) $r = 2.6$
(e) $r = 2.7$
(f) $r = 2.8$
(g) $r = 2.9$
(h) $r = 3.0$

Describe the changes you see in the population density over time as the value of r (sometimes known as the intrinsic growth rate of the population) increases.

5.14 When we construct difference equation models where two events are occurring, it is important to given careful consideration to which event occurs first in each time step. Consider the example where a population is harvested every time step by a fixed amount. During that time step, does this harvesting occur after natural population growth (i.e., after the breeding season) or before natural population growth? If we assume that natural growth can be modeled as logistic growth, then in the first case where harvesting occurs after population growth, the population would be modeled by the difference equation

$$x_{n+1} = x_n + ax_n(1 - x_n) - b,$$

where a is the growth rate and b is the fixed amount by which the population is harvested. In the second case where harvesting occurs before population growth, the population would be modeled by the difference equation

$$x_{n+1} = (x_n - b) + a(x_n - b)\left(1 - (x_n - b)\right),$$

where, again, a is the growth rate and b is the fixed amount by which the population is harvested.

Write an m-file similar to `LogisticDifferenceEqn.m` to model the population under each harvesting scenario. Use an initial population density of 0.2 and a fixed harvest of 0.1 (this is 10% of the population density). Then simulate both models for the following:

(a) An intrinsic growth rate of 1.4. Compare the results of the two different models.
(b) An intrinsic growth rate of 2.5. Compare the results of the two different models.
(c) An intrinsic growth rate of 3.0. Compare the results of the two different models.
(d) How do the results for each model change as the value of a is increased?

Vectors and Matrices

We pointed out in the introduction to this unit that one way to summarize an image of a landscape, such as an image of your hometown obtained using Google Earth, is by making a list of the fraction of the image that is of each "type." Here the type could be bare soil, grass, forest, roads, lakes, buildings, and so on. There is some loss of information when we summarize an image this way since the spatial arrangement of buildings, roads, forests and so on is not included if we have only a list of the fraction of area in each type. In geography, these "types" are typically called land classifications (or simply "classes"), and we illustrated how you might determine the changes in a landscape over time by following how the fraction of landscape in each class changes. This list of fractions can be organized as a mathematical object called a "vector," and remember that the order of numbers in this list matters since these numbers refer to the fractions in each class. If we were to switch the order of classes (e.g., if we initially had grass first and forest second on our list), then we would need to switch the order of the fractions on the list as well, or else we would not be correctly representing the landscape.

Our objective in this chapter is to point out the properties of these ordered lists of numbers, or vectors. We'll show how we can use these properties to help us compare one landscape to another and how to describe changes through time of a landscape by following how the vector describing the landscape changes. In ecology, this allows us to track succession across a landscape, but the same methods can be used to follow many biological quantities that can be modeled using vectors. Such quantities include the distribution of a drug in different body tissues and the fraction of a population that is in different stages of infection by a disease, such as being susceptible, exposed, infected, or immune.

A key mathematical tool that we'll need is an extension of a vector to an array of numbers arranged in a rectangle. A rectangular array of numbers is called a "matrix" (yes, the movie series used this term but dealt with much more complicated collections of information than we'll consider here). We'll see that a matrix is a useful way to summarize, in a specially ordered way, information about how one class transitions to another type. For example, we'll use a matrix to describe what fraction of the forest on a landscape changes (or transitions) to buildings in some time period, what fraction of the area that is bare soil changes to grass in the same period, and so on. Similarly, we can use a matrix to describe, for individuals in a population who are susceptible to a disease, what fraction of these individuals becomes exposed or infected with the disease over some time period. So the methods form the basis for understanding how to describe changes through time for many situations in which we can classify the state of the system in a discrete set of classes and for which it is reasonable to assume that a certain fraction of each class changes in a given time period. In this sense, the underlying changes are linear in that a proportion of each class changes each time period.

6.1 Vector Structure: Order Matters!

It turns out that the mathematics that allows us to describe changes in the structure of biological entities or landscapes is described simply by building an appropriate set of rules for manipulating numbers that are arranged in particular orders. So, if we had a landscape that was 20% grass, 50% shrubs, and 30% trees, we could represent this in the vector

$$\mathbf{v} = \begin{bmatrix} 0.2 \\ 0.5 \\ 0.3 \end{bmatrix},$$

where the first component represents the proportion of landscape that is grass, the second component the proportion of landscape that is shrub, and the third component the proportion that is trees. Notice that we use a boldface lowercase letter to denote a vector. The order matters because the way we describe these structures depends on the order in which you list the numbers. So a vector that describes the fraction of the landscape in (grass, shrub, trees) as

$$\mathbf{v} = \begin{bmatrix} 0.2 \\ 0.5 \\ 0.3 \end{bmatrix}$$

is different from the vector

$$\mathbf{w} = \begin{bmatrix} 0.5 \\ 0.2 \\ 0.3 \end{bmatrix}.$$

These represent very different landscapes: the order of the numbers matters. So we have to build up a way of manipulating numbers in which order matters. We essentially have to come up with a way to manipulate vectors in the same way you used the algebraic manipulations (addition, subtraction, multiplication, and division) for single numbers (single numbers are called "scalars" to differentiate them from vectors, which consist of lists of numbers in particular orders). Just as numbers can be represented in general by a letter, vectors can be represented by letters, and the elements of a vector (the numbers that make it up) can be letters (algebraic entities) as well.

Example 6.1 (Constructing a Vector to Represent Wetland Composition)

Suppose that we were asked to construct a vector that would represent the composition of a coastal wetlands landscape. What would that mean? First, we would have to decide what classes make up coastal wetlands. Coastal wetlands are areas near the coast where the soil is submerged or soaked with moisture either permanently or seasonally. Coastal wetlands are found between areas that are always wet (i.e., oceans, seas, bays, and estuaries) and areas that are always dry (i.e., grasslands and forests).

We could divide wetlands into land that is submerged underwater and land that is not submerged (that would make two classes). However, we might decide that having two classes is not sufficient to accurately describe the variation in landscape in the wetlands. Thus, it might be appropriate to introduce a third class: land that is not submerged but that is heavily saturated with moisture. After we decide what the classes are, we must decide how to order them. Here, let us use submerged, saturated but not submerged, and dry. If we let

- u = proportion of wetlands that are submerged or underwater,
- s = proportion of wetlands that are not submerged but are saturated with moisture, and
- d = proportion of wetlands that are dry,

then the vector $\mathbf{v}$ representing the coastal wetlands composition would be

$$\mathbf{v} = \begin{bmatrix} u \\ s \\ d \end{bmatrix}.$$

If at some particular time t 65% of the wetlands were submerged, 30% saturated, and 5% dry, then we could write

$$\mathbf{v}(t) = \begin{bmatrix} 0.65 \\ 0.30 \\ 0.05 \end{bmatrix}.$$

We could divide wetlands into the types of plants that grow there. As an example, the types of flora that grow in the Queensland coastal wetland in Australia (see [95] for more information) are mangroves, seagrasses, grasses, sedges/rushes, ferns, wet heaths, trees and shrubs, and salt-marsh plants. This would give eight classes. Just as above, we could construct an ordering for these classes and then construct a vector representing the composition of the Queensland coastal wetlands based on the proportion of the total area in which each of the above types of flora grow.

A vector can also be represented as a bar chart where each element of the vector is represented by one bar whose height is given by the value of the vector element. For example, the vector we constructed in Example 6.1 could be represented as a bar chart with three bars whose heights are 0.65, 0.30, and 0.05 (see Figure 6.1).

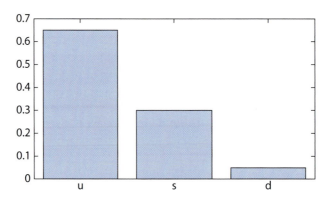

Figure 6.1 Bar chart of the vector we formed in Example 6.1.

6.2 Vector Algebra

In working with vectors, a few basic mathematical operations are key. Suppose that we have two vectors,

$$\mathbf{v} = \begin{bmatrix} v_1 \\ v_2 \\ \vdots \\ v_n \end{bmatrix} \quad \text{and} \quad \mathbf{w} = \begin{bmatrix} w_1 \\ w_2 \\ \vdots \\ w_n \end{bmatrix},$$

each having n elements, and a real number c. Then we have the following rules for vector algebra:

Rules for Vector Algebra

1. Addition:

$$\mathbf{v} + \mathbf{w} = \begin{bmatrix} v_1 \\ v_2 \\ \vdots \\ v_n \end{bmatrix} + \begin{bmatrix} w_1 \\ w_2 \\ \vdots \\ w_n \end{bmatrix} = \begin{bmatrix} v_1 + w_1 \\ v_2 + w_2 \\ \vdots \\ v_n + w_n \end{bmatrix}$$

It is important to note that only two vectors of the same length (i.e., having the same number of elements) can be added together. For example, if $\mathbf{v}$ had length 2 and $\mathbf{w}$ had length 5, the two vectors could not be added together.

2. Scalar Multiplication:

$$c\mathbf{v} = c \begin{bmatrix} v_1 \\ v_2 \\ \vdots \\ v_n \end{bmatrix} = \begin{bmatrix} cv_1 \\ cv_2 \\ \vdots \\ cv_n \end{bmatrix}$$

Let us consider some examples where these operations might be used.

Example 6.2 (Converting a Vector from Proportions to Hectares)

In Example 6.1, we formed a vector to describe the structure of a coastal wetlands at a particular time t. The elements of the vector were proportions of the total area that were submerged, saturated, and dry. Suppose that we knew the total area of the landscape to be 160 hectares. How would we determine how many hectares were in each of the three classes?

Solution: You might say, "Just multiply each entry by 160 hectares," which would be correct. However, this is exactly scalar multiplication with a vector!

$$160 \begin{bmatrix} 0.65 \\ 0.30 \\ 0.05 \end{bmatrix} = \begin{bmatrix} 104 \\ 48 \\ 8 \end{bmatrix}$$

Notice that the elements in the vector on the left-hand side of the equation are unitless, while the elements in the vector on the right-hand side of the equation have units of hectares. Also, notice that if we divide any single element in

$$\begin{bmatrix} 104 \\ 48 \\ 8 \end{bmatrix}$$

by the sum of elements in this vector, then we get back the corresponding proportion for that element.

Example 6.3 (Changing Landscape Composition via Land Acquisition)

Suppose that we were managing some national park. For the entire park, the flora composition is 8% annual plants, 23% perennial plants and grasses, 26% shrubs, 24% softwood trees and pines, and 19% hardwood trees. The total area of the park is 1250 hectares. The park is now considering purchasing 300 hectares of farmland located next to the park, of which 5% is annual plants, 90% is perennial plants and grasses, and 5% is softwood trees and pines. If the national park acquires this new land, what will be the new flora composition for the park?

Solution: First let us construct the composition vectors for the park ($\mathbf{p}$) and for the farmland ($\mathbf{f}$), where the class ordering is (1) annual plants, (2) perennial plants and grasses, (3) shrubs, (4) softwood trees and pines, and (5) hardwood trees:

$$\mathbf{p} = \begin{bmatrix} 0.08 \\ 0.23 \\ 0.26 \\ 0.24 \\ 0.19 \end{bmatrix} \quad \text{and} \quad \mathbf{f} = \begin{bmatrix} 0.05 \\ 0.90 \\ 0 \\ 0.05 \\ 0 \end{bmatrix}$$

(Continued)

Notice that we cannot directly add these two proportions together because the **p** vector represents proportions of a total of 1250 hectares, while the **f** vector represents proportions of a total of only 300 hectares. First, we must convert both vectors to having units of hectares, then we may add the vectors:

$$1{,}250\mathbf{p} + 300\mathbf{f} = \begin{bmatrix} 100.0 \\ 287.5 \\ 325.0 \\ 300.0 \\ 237.5 \end{bmatrix} + \begin{bmatrix} 15 \\ 270 \\ 0 \\ 15 \\ 0 \end{bmatrix} = \begin{bmatrix} 115.0 \\ 557.5 \\ 325.0 \\ 315.0 \\ 237.5 \end{bmatrix}$$

Currently, our answer is in units of hectares. If we would like to turn our answer back to proportions, we must divide each element in the vector by the sum of all the elements in the vector:

$$\text{total hectares} = 115.0 + 557.5 + 325.0 + 315.0 + 237.5 = 1550$$

So the vector describing the flora composition by proportions of the total 1550 hectares is

$$\frac{1}{1550} \begin{bmatrix} 115.0 \\ 557.5 \\ 325.0 \\ 315.0 \\ 237.5 \end{bmatrix} = \begin{bmatrix} 0.075 \\ 0.377 \\ 0.210 \\ 0.204 \\ 0.153 \end{bmatrix}.$$

6.3 Dynamics: Vectors Changing over Time

Consider again our coastal wetlands example (see Example 6.1). Suppose that we represent the composition of the coastal wetlands based on wetness; that is, we use the classes in the following order:

- u = proportion of wetlands that are submerged or underwater,
- s = proportion of wetlands that are not submerged but are saturated with moisture, and
- d = proportion of wetlands that are dry.

Because of the process of ecological succession (see Figures 4.3 and 6.2), over a large enough scale of time, we expect areas that are submerged to become only saturated and areas that are saturated to become dry. How could we represent this process?

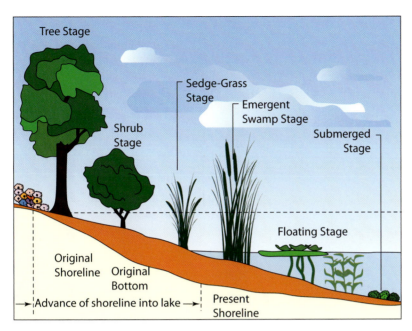

Figure 6.2 Depiction of wetlands succession. From Marlene Weigel. *UXL ENCYCLOPEDIA OF BIOMES 2 V3*, 2E. © 2010 Gale, a part of Cengage Learning, Inc. Reproduced by permission. www.cengage.com/permissions.

Example 6.4 (Ecological Succession: Simple Case)

Suppose that we were looking at a total area of 100 hectares, which is initially all in submerged wetlands. Every 10 years, 5% of submerged wetlands become saturated wetland, 12% of saturated wetlands become dry, and all hectares that are dry remain dry. After 10 years, how much of the land is submerged, how much is saturated, and how much is dry? After 20 years?

Solution: Let us start with our vector representing the fractional composition of the 100 hectares of wetland:

$$\mathbf{v}(0) = \begin{bmatrix} 1 \\ 0 \\ 0 \end{bmatrix}.$$

After 10 years, the proportion of submerged wetlands remaining, $u(10)$, is the amount that we started with, $u(0) = 1$ (100% of the area is submerged) minus the $5\% = 0.05$ that becomes saturated, that is,

$$\begin{aligned} u(10) &= u(0) - 0.05\, u(0) \\ &= (1 - 0.05)\, u(0) \\ &= 0.95\, u(0) \\ &= 0.95 \ \text{ since } u(0) = 1. \end{aligned}$$

(Continued)

Thus, after 10 years, 0.95, or 95%, of the wetlands are submerged. Since the total area is 100 hectares, after 10 years, 95 hectares are submerged.

After 10 years, the proportion of saturated wetlands remaining, $s(10)$, is the amount that we started with, $s(0) = 0$ minus the $12\% = 0.12$ that becomes dry plus the 5% of submerged that becomes saturated, that is,

$$s(10) = s(0) - 0.12\, s(0) + 0.05\, u(0)$$
$$= (1 - 0.12)\, s(0) + 0.05\, u(0)$$
$$= 0.88\, s(0) + 0.05\, u(0)$$
$$= 0.05 \text{ since } u(0) = 1 \text{ and } s(0) = 0.$$

Thus, after 10 years, 0.05, or 5%, of the wetlands are saturated. That is, 5 hectares are saturated.

After 10 years, the proportion of dry land, $d(10)$, is the amount that we started with, $d(0) = 0$ plus the 12% of saturated wetland that become dry. Recall that once an area is dry, it remains dry, so there is no transfer of dry land to another type. Mathematically, we write this as

$$d(10) = d(0) + 0.12\, s(0)$$
$$= 0 \text{ since } s(0) = 0 \text{ and } d(0) = 0.$$

Thus, after 10 years, there are still no dry hectares in our 100-hectare area. Thus, after 10 years, the vector representing the composition of the 100 hectares of wetland is

$$\mathbf{v}(10) = \begin{bmatrix} 0.95 \\ 0.05 \\ 0 \end{bmatrix}.$$

What about after 20 years? We know that the composition of the wetlands at 10 years, and we know how the area will change over the next 10 years, so we can compute $\mathbf{v}(20)$:

$$u(20) = u(10) - 0.05\, u(10)$$
$$= (1 - 0.05)\, u(10)$$
$$= 0.95\, u(10)$$
$$= 0.9025 \text{ since } u(10) = 0.95.$$
$$s(20) = s(10) - 0.12\, s(10) + 0.05\, u(10)$$
$$= (1 - 0.12)\, s(10) + 0.05\, u(10)$$
$$= 0.88\, s(10) + 0.05\, u(10)$$
$$= 0.0915 \text{ since } u(10) = 0.95 \text{ and } s(10) = 0.05.$$
$$d(20) = d(10) + 0.12\, s(0)$$
$$= 0.006 \text{ since } s(10) = 0.05 \text{ and } d(10) = 0.$$

(*Continued*)

Thus, after 20 years, the vector representing the composition of the 100 hectares of wetland is

$$\mathbf{v}(20) = \begin{bmatrix} 0.9025 \\ 0.0915 \\ 0.006 \end{bmatrix}.$$

Notice that the elements in the vector still sum to 1.

As we can see from Example 6.4, if we know the composition of a landscape at some time step t and we want to find the composition at the next time step $t+1$, deriving the equations each time can be tedious. Let us now develop some mathematical notation to make the process quicker.

First, let us write the equations we developed in Example 6.4 in a more general form:

$$u(t+1) = 0.95\, u(t)$$

$$s(t+1) = 0.88\, s(t) + 0.05\, u(t)$$

$$d(t+1) = d(t) + 0.12\, s(t),$$

where one time step is 10 years. Now, we will rewrite these equations in such a way that we can make use of the vector structure:

$$\mathbf{v}(t) = \begin{bmatrix} u(t) \\ s(t) \\ d(t) \end{bmatrix},$$

$$u(t+1) = 0.95\, u(t) + 0.00\, s(t) + 0.00\, d(t) \tag{6.1}$$

$$s(t+1) = 0.05\, u(t) + 0.88\, s(t) + 0.00\, d(t) \tag{6.2}$$

$$d(t+1) = 0.00\, u(t) + 0.12\, s(t) + 1.00\, d(t). \tag{6.3}$$

Let us form an array that has three rows and three columns. We will call this a 3×3 matrix. The elements of the matrix will be the coefficients of $u(t)$, $s(t)$, and $d(t)$ in Equations (6.1) to (6.3). Thus, our matrix looks like

$$\begin{bmatrix} 0.95 & 0 & 0 \\ 0.05 & 0.88 & 0 \\ 0 & 0.12 & 1 \end{bmatrix},$$

where each row corresponds to the coefficients in one equation and each column corresponds to the coefficients of one class; that is, the first row gives the coefficients of Equation (6.1), and the first column gives the coefficients of $u(t)$ in each equation. Notice that each column sums to 1. Furthermore, we see that the entries along the diagonal

$$\begin{bmatrix} 0.95 & 0 & 0 \\ 0.05 & 0.88 & 0 \\ 0 & 0.12 & 1 \end{bmatrix}$$

are the proportion of land in each class that remain the same during a single 10-year period. Thus, the second entry along the diagonal, 0.88, tells us that 88% of the second class (saturated)

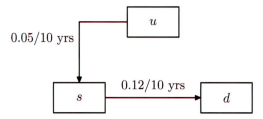

Figure 6.3 Flow diagram of the transfer matrix in Equation (6.4).

remains in the second class each time step of 10 years. The entries not along the diagonal tell us how land changes from one class to another:

	moving from		
	$u(t)$	$s(t)$	$d(t)$
$u(t+1)$	0.95	0	0
$s(t+1)$	0.05	0.88	0
$d(t+1)$	0	0.12	1

(rows labeled *moving into*)

For example, the entry in the second row, first column, indicates that 5% of the first column class, $u(t)$, will move into the second-row class, $s(t+1)$, at the next time step. Now, our mathematical shorthand for Equations (6.1) to (6.3) will be

$$\begin{bmatrix} u(t+1) \\ s(t+1) \\ d(t+1) \end{bmatrix} = \begin{bmatrix} 0.95 & 0 & 0 \\ 0.05 & 0.88 & 0 \\ 0 & 0.12 & 1 \end{bmatrix} \begin{bmatrix} u(t) \\ s(t) \\ d(t) \end{bmatrix}$$

or

$$\mathbf{v}(t+1) = \begin{bmatrix} 0.95 & 0 & 0 \\ 0.05 & 0.88 & 0 \\ 0 & 0.12 & 1 \end{bmatrix} \mathbf{v}(t). \tag{6.4}$$

The operation on the right-hand side of the equals sign is **matrix multiplication**, which will be described in more detail in Section 7.1. The 3×3 matrix in Equation (6.4) is sometimes referred to as a **transfer matrix** since it describes how the landscape is transfered from one state to another. Transfer matrices can be easily represented as a flow diagram, such as the one in Figure 6.3.

Modeling Structural Change in General

Now, let us write this more generally to answer future questions involving changing structures of biological systems. Suppose that we have n classes describing our landscape (or whatever structure we are modeling). Then we can write the vector describing the composition of the landscape at time step t as

$$\mathbf{v}(t) = \begin{bmatrix} v_1(t) \\ v_2(t) \\ \vdots \\ v_n(t) \end{bmatrix},$$

where the kth entry $v_k(t)$ is the proportion of the landscape that is of class k at time step t. Since this is a proportion, we must have

$$0 \leq v_k(t) \leq 1 \quad \text{for all } k.$$

Also, note that since each class is a proportion of the total landscape, the sum of the proportions in all the classes must equal 1, that is,

$$\sum_{k=1}^{n} v_k(t) = v_1(t) + v_2(t) + \cdots + v_n(t) = 1.$$

Next, we create an array with n rows and n columns, that is, an $n \times n$ matrix,

$$A = \begin{bmatrix} a_{11} & a_{12} & \cdots & a_{1n} \\ a_{21} & a_{22} & \cdots & a_{2n} \\ \vdots & \vdots & \ddots & \vdots \\ a_{n1} & a_{n2} & \cdots & a_{nn} \end{bmatrix}$$

where a_{ij} is the proportion of class j that becomes or moves into class i. Again, we must have each column of our matrix $\mathbf{A}$ sum to 1, that is,

$$\sum_{i=1}^{n} a_{ij} = 1 \quad \text{for each } j.$$

Now we can describe the composition of the landscape at time step $t + 1$ as

$$\begin{bmatrix} v_1(t+1) \\ v_2(t+1) \\ \vdots \\ v_n(t+1) \end{bmatrix} = \begin{bmatrix} a_{11} & a_{12} & \cdots & a_{1n} \\ a_{21} & a_{22} & \cdots & a_{2n} \\ \vdots & \vdots & \ddots & \vdots \\ a_{n1} & a_{n2} & \cdots & a_{nn} \end{bmatrix} \begin{bmatrix} v_1(t) \\ v_2(t) \\ \vdots \\ v_n(t) \end{bmatrix}$$

or

$$\mathbf{v}(t+1) = \mathbf{A}\mathbf{v}(t).$$

Now that we have developed a mathematical notation for describing models of structural change, let us look at some examples. The first example is a landscape change model similar to Example 6.4.

Example 6.5 (Ecological Succession: More Complex Case)

Suppose that we want to model how the composition of a coastal wetland changes over time. Again, we will structure the wetlands into submerged, saturated, and dry. After several decades of data collection, we know that, every 10 years,

(Continued)

- 5% of submerged wetlands become saturated,
- 1% of submerged wetlands become dry,
- 12% of saturated wetlands become dry,
- 2% of saturated wetlands become submerged again,
- 6% of dry wetlands become saturated again, and
- 1% of dry wetlands become submerged again.

This is summarized by the following flow diagram:

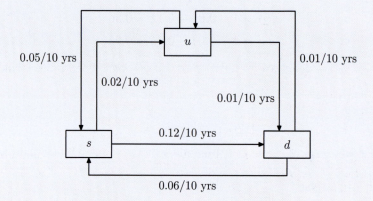

Find the matrix that describes how the composition of these coastal wetlands changes every 10 years.

<u>Solution:</u> First, notice that since we have three classes, we will need to construct a 3×3 matrix (three rows and three columns):

$$
\begin{array}{c|ccc}
 & \multicolumn{3}{c}{\textit{moving from}} \\
 & u(t) & s(t) & d(t) \\
\hline
u(t+1) & a_{11} & a_{12} & a_{13} \\
s(t+1) & a_{21} & a_{22} & a_{23} \\
d(t+1) & a_{31} & a_{32} & a_{33}
\end{array}
$$

moving into

Recall that a_{ij} is the proportion of class j that becomes or moves into class i. Here we will list submerged as class 1, saturated as class 2, and dry as class 3. Let us use the information given to find the a_{ij} values:

- 5% of submerged wetlands ($j = 1$) become saturated ($i = 2$), that is, $a_{21} = 0.05$.
- 1% of submerged wetlands ($j = 1$) become dry ($i = 3$), that is, $a_{31} = 0.01$.
- 12% of saturated wetlands ($j = 2$) become dry ($i = 3$), that is, $a_{32} = 0.12$.
- 2% of saturated wetlands ($j = 2$) become submerged ($i = 1$), that is, $a_{12} = 0.02$.
- 6% of dry wetlands ($j = 3$) become saturated ($i = 2$), that is, $a_{23} = 0.06$.
- 1% of dry wetlands ($j = 3$) become submerged ($i = 1$), that is, $a_{13} = 0.01$.

(*Continued*)

Thus, our matrix becomes

$$\begin{bmatrix} a_{11} & 0.02 & 0.01 \\ 0.05 & a_{22} & 0.06 \\ 0.01 & 0.12 & a_{33} \end{bmatrix}.$$

Notice that we still need to enter values for a_{11}, a_{22}, and a_{33}. Recall that each column must sum to 1. We can determine the value of the entries along the diagonal by utilizing this fact. Thus,

$$a_{11} = 1 - 0.05 - 0.01 = 0.94$$

$$a_{22} = 1 - 0.02 - 0.12 = 0.86$$

$$a_{33} = 1 - 0.01 - 0.06 = 0.93$$

Thus, the matrix that describes how the composition of these coastal wetlands changes every 10 years is

$$\begin{bmatrix} 0.94 & 0.02 & 0.01 \\ 0.05 & 0.86 & 0.06 \\ 0.01 & 0.12 & 0.93 \end{bmatrix}.$$

If the time step is 10 years (i.e., $t = 1$ indicates 10 years have passed, $t = 2$ indicates 20 years have passed, etc.), then the model describing how the composition of these coastal wetlands changes every 10 years is

$$\begin{bmatrix} u(t+1) \\ s(t+1) \\ d(t+1) \end{bmatrix} = \begin{bmatrix} 0.94 & 0.02 & 0.01 \\ 0.05 & 0.86 & 0.06 \\ 0.01 & 0.12 & 0.93 \end{bmatrix} \begin{bmatrix} u(t) \\ s(t) \\ d(t) \end{bmatrix}.$$

In this next example, we will construct a simple model to describe the spread of an infectious disease.

Example 6.6 (The Spread of the Common Cold)

For this model, we will divide a population into two classes of people: (1) susceptible and (2) infected. During the fall semester last year, a particular dormitory had a common cold epidemic. Each day, 40% of the dorm residents became infected with the cold, while only 20% of those infected recovered and became susceptible again.

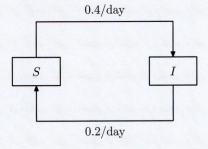

(*Continued*)

Construct a model to describe how this two-class structured dorm population changes each day.

Solution: Let the susceptibles be class 1 and the infecteds be class 2; thus, we need to build a 2×2 matrix,

$$\begin{bmatrix} a_{11} & a_{12} \\ a_{21} & a_{22} \end{bmatrix},$$

to describe how this population structure changes each day. Next, we calculated the entries in our matrix:

- 40% of susceptibles become infected each day (from class 1 to class 2), $a_{21} = 0.4$.
- 20% of the infected recover and become susceptible each day (from class 2 to class 1), $a_{12} = 0.2$.
- The proportion that remain susceptible is $a_{11} = 1 - 0.4 = 0.6$.
- The proportion that remain infected is $a_{22} = 1 - 0.2 = 0.8$.

Thus, the model describing how this dorm population changes from time step t to time step $t + 1$ is

$$\mathbf{v}(t+1) = \begin{bmatrix} 0.6 & 0.2 \\ 0.4 & 0.8 \end{bmatrix} \mathbf{v}(t),$$

where the time step is 1 day.

6.4 Matlab Skills

See Appendix A.3 for information on how to work with vectors and matrices in Matlab.

6.5 Exercises

6.1   If you have not already done so, look at the Student Project on Land Classification given at the end of this Unit. Describe the different classes you decided on to describe the landscape of your hometown. Describe the scheme you developed for deciding what proportion of the landscape was in each class. What is the vector that corresponds to the landscape classification you developed for the Google Earth image of your hometown?

6.2   Let

$$x = \begin{bmatrix} 3 \\ 1 \\ 0 \end{bmatrix}, \quad y = \begin{bmatrix} 2 \\ -1 \\ 3 \end{bmatrix}, \quad z = \begin{bmatrix} 0 \\ -1 \\ 4 \end{bmatrix}, \quad v = \begin{bmatrix} 1 \\ 0 \\ 0 \\ 4 \end{bmatrix}, \quad \text{and} \quad u = \begin{bmatrix} -1 \\ 2 \\ 3 \\ 0 \end{bmatrix}.$$

Compute (if possible) each of the following vectors:

(a) x + y
(b) x − y

(c) v + x
(d) 2v − 2u
(e) u − y
(f) 2 (y − z) + 4z
(g) 3v + 4u
(h) v − (2u − v)

6.3 A basic model for the spread of the herpes simplex virus divides a population into three categories: (1) susceptible (does not have the virus), (2) infected and shedding the virus (can infect others), and (3) quiescent (infected but not shedding the virus). After completing a survey of the population, researchers discover that 65% of the population is susceptible, 20% are currently shedding virus, and the remaining proportion of the population is in the quiescent phase. Construct a vector that represents what proportion of the population is in each category.

6.4 In an ant colony, there are three different classes of ants: (1) workers, (2) males, and (3) queens. Workers and queens are female. A particular ant colony contains 250 workers, 180 males, and 2 queens.

(a) Construct a vector that represents the number of ants in each class.
(b) Construct a vector that represents the proportion of ants that are males and females. Indicate which entry in your vector represents females.

6.5 Construct the transfer matrix that describes the following flow diagrams.

(a) This flow diagram represents the dynamics of a nonfatal disease with susceptible (S) and infected (I) classes.

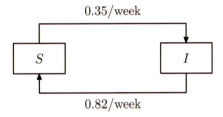

(b) This flow diagram represents the dynamics of ecological succession with classes for grass (G), shrubs (S), and trees (T).

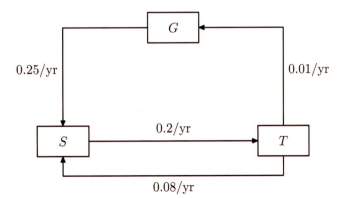

(c) This flow diagram represents the dynamics of strontium 90 cycling through an ecosystem. The ecosystem is divided into grasses (G), soil (S), streams (W), and dead organic matter (O).

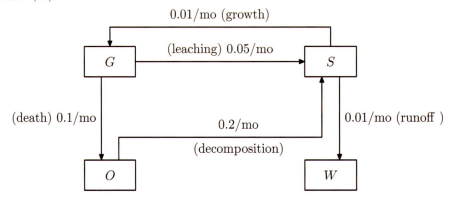

6.6 Suppose that each day, 30% of the dorm residents became infected with a cold, while only 20% of those infected recovered and became susceptible again. Draw the flow diagram for this situation and give the corresponding transfer matrix.

6.7 We will divide a population into three classes of people: (1) susceptible, (2) infected, and (3) recovered. Those individuals who recovered are immune temporarily, and then their immunity wanes. A particular dormitory had a flu epidemic. Each day, 40% of the dorm residents became infected, while only 20% of those infected recovered and became immune. Those recovered and temporarily immune students became susceptible at a rate of 1% per day. Draw the flow diagram for this situation and give the corresponding transfer matrix.

6.8 Herpes simplex virus (HSV) is the virus that causes the disease herpes. One of the simplest population models for HSV (without including treatment) is to divide the population into three categories: (1) susceptible, (2) infected and shedding the virus (people in this class can infect other individuals), and (3) quiescent (people in this class are infected but not shedding the virus). Suppose that each year, 2% of the susceptible population becomes infected and moves into the infected and shedding virus category. Additionally, each day, 28.6% of those infected and shedding the virus move into the quiescent stage, and 2.9% of those in the quiescent stage start to shed the virus again. Note that the units are not the same for each of the given transfer rates.

(a) Draw a flow diagram to represent the dynamics of HSV in the population. Be sure to label your diagram carefully.
(b) Construct a transfer matrix to represent how the population of susceptible, shedding, and quiescent individuals changes each day.

6.9 Describe in sentences what situation is described by the flow diagram in Exercise 6.5b.

6.10 A copier machine is always in one of two states: either working or not working (broken). If it is working today, then there is an 80% chance that it will be working tomorrow. If it is not working today, there is a 20% chance that it will not be working tomorrow. Draw the flow diagram for this situation and give the corresponding transfer matrix.

Matrix Algebra

The previous chapter illustrated how to use a matrix to characterize the changes of areas across landscapes by summarizing the fraction of area that transitions from one class to another. This can be calculated by using the procedure of matrix multiplication with a vector, describing the change of the landscape in one time period. We also saw how to use this approach to describe the changes in the disease state of a population from one time period to another, which is a critical need in epidemiology. It is natural to ask whether we can use the same approach to determine how a landscape changes and a disease spreads over longer periods of time. We'll see in this chapter that matrix multiplication allows us to make these calculations easily.

Matrix multiplication can also be used to follow how somewhat more complicated systems change, including those that are not able to be described by a single vector. So we'll see here how to use matrix multiplication to evaluate the costs over a time period for providing food (bird seed in our example) to different species that have different diet preferences. The idea is that, as long as the underlying process is linear so that costs are proportional to the amount of seed provided in the bird-feeding case, we can relatively easily calculate total costs for supplying multiple types of feed to different species. This underlies many practical applications in livestock-feeding operations in which different feeds may contain different nutrients and in which it is necessary to calculate the costs for the different feedstocks and sum them to obtain total costs.

7.1 Matrix Arithmetic

We saw in Section 6.3 the benefits in using matrix multiplication to describe changing structures in biological systems. Here we describe the mathematics of matrix multiplication and other matrix operations. After we have learned how to manipulate matrices algebraically, we will return to the examples we considered in Section 6.3 and carry out some of these matrix operations.

First, let us formally define a matrix. A *matrix* is a two-dimensional array of numbers. We will use boldface uppercase letters to denote matrices. Recall that we use boldface lowercase letters to denote vectors. If we say that $\mathbf{A}$ is an $n \times m$ matrix, we mean that is has n rows and m columns. To denote a particular entry of the matrix $\mathbf{A}$, we use a_{ij}, where i is the row number and j is the column number. Thus, in the matrix

$$\mathbf{A} = \begin{bmatrix} 1 & 2 \\ 3 & 4 \end{bmatrix}$$

we have

$$a_{11} = 1$$
$$a_{12} = 2$$
$$a_{21} = 3$$
$$a_{22} = 4$$

Next, we describe matrix addition and multiplication.

Matrix Addition

Let $\mathbf{A}$ and $\mathbf{B}$ be $n \times m$ matrices; then

$$\mathbf{A} + \mathbf{B} = \begin{bmatrix} a_{11} & \cdots & a_{1m} \\ \vdots & \ddots & \vdots \\ a_{n1} & \cdots & a_{nm} \end{bmatrix} + \begin{bmatrix} b_{11} & \cdots & b_{1m} \\ \vdots & \ddots & \vdots \\ b_{n1} & \cdots & b_{nm} \end{bmatrix} = \begin{bmatrix} a_{11}+b_{11} & \cdots & a_{1m}+b_{1m} \\ \vdots & \ddots & \vdots \\ a_{n1}+b_{n1} & \cdots & a_{nm}+b_{nm} \end{bmatrix}.$$

Notice that to add two matrices they must be of the same size (i.e., both matrices have the same number of rows, and both have the same number of columns).

Let us try a simple example.

Example 7.1 (Matrix Addition Example)

Find the sum of $\mathbf{A}$ and $\mathbf{M}$ if

$$\mathbf{A} = \begin{bmatrix} 2 & 4 & 8 \\ 6 & 3 & 7 \end{bmatrix} \text{ and } \mathbf{M} = \begin{bmatrix} 2 & 5 & 5 \\ 9 & 1 & 7 \end{bmatrix}.$$

Solution:

$$\begin{bmatrix} 2 & 4 & 8 \\ 6 & 3 & 7 \end{bmatrix} + \begin{bmatrix} 2 & 5 & 5 \\ 9 & 1 & 7 \end{bmatrix} = \begin{bmatrix} 4 & 9 & 13 \\ 15 & 4 & 14 \end{bmatrix}.$$

Matrix Scalar Multiplication

Let $\mathbf{A}$ be an $n \times m$ matrix and c be a real number. Then

$$c\mathbf{A} = c\begin{bmatrix} a_{11} & \cdots & a_{1m} \\ \vdots & \ddots & \vdots \\ a_{n1} & \cdots & a_{nm} \end{bmatrix} = \begin{bmatrix} ca_{11} & \cdots & ca_{1m} \\ \vdots & \ddots & \vdots \\ ca_{n1} & \cdots & ca_{nm} \end{bmatrix}.$$

Notice that using matrix addition and matrix scalar multiplication, we can define matrix subtraction as

$$\mathbf{A} - \mathbf{B} = \mathbf{A} + (-1)\mathbf{B},$$

where the matrix $\mathbf{B}$ is being multiplied by the real number -1.

If we have a matrix where all the entries are 0, then we denote this with a boldface zero, that is, $\mathbf{0}$. Next, we develop some matrix addition properties.

Properties of Matrix Addition

Let $\mathbf{A}$, $\mathbf{B}$, and $\mathbf{C}$ be $m \times n$ matrices and a and b be real numbers. Then

1. $\mathbf{A} + \mathbf{B} = \mathbf{B} + \mathbf{A}$
2. $\mathbf{A} + \mathbf{0} = \mathbf{0} + \mathbf{A} = \mathbf{A}$
3. $\mathbf{A} + (-\mathbf{A}) = -\mathbf{A} + \mathbf{A} = \mathbf{0}$
4. $\mathbf{A} + (\mathbf{B} + \mathbf{C}) = (\mathbf{A} + \mathbf{B}) + \mathbf{C}$
5. $(a + b)\mathbf{A} = a\mathbf{A} + b\mathbf{A}$
6. $a(\mathbf{A} + \mathbf{B}) = a\mathbf{A} + a\mathbf{B}$

Matrix multiplication is a bit more complex than matrix addition, but with a little practice, you will be multiplying matrices in your sleep in no time.

Matrix Multiplication

If $\mathbf{A}$ is an $m \times n$ matrix and $\mathbf{B}$ is an $n \times p$ matrix, then their product is an $m \times p$ matrix denoted by $\mathbf{AB}$. If $\mathbf{C} = \mathbf{AB}$ and c_{ij} denotes the entry in $\mathbf{C}$ at position (i, j), then

$$c_{ij} = \sum_{r=1}^{n} a_{ir}b_{rj} = a_{i1}b_{1j} + a_{i2}b_{2j} + \cdots + a_{in}b_{nj}$$

for each pair i and j with $1 \le i \le m$ and $1 \le j \le p$.

This is a type of "row by column" operation; to get the "ij" entry of $\mathbf{AB}$, multiply the ith row entries of $\mathbf{A}$ by the corresponding jth column entries of $\mathbf{B}$ and then sum the products.

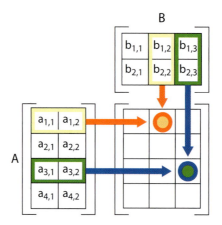

Figure 7.1 An illustration of matrix multiplication $\mathbf{AB}$, where $\mathbf{A}$ is a 4×2 matrix and $\mathbf{B}$ is a 2×3 matrix. Illustration taken from [96].

Notice that in order to multiply two matrices together, the number of columns of the matrix being multiplied on the left must equal the number of rows of the matrix being multiplied on the right. Additionally, the product of the two matrices will have the same number of rows as the matrix being multiplied on the left and the same number of columns as the matrix being multiplied on the right. Thus,

$$\overbrace{(m \times \underbrace{n) \cdot (n}_{\text{equal}} \times p)}^{\text{product is an } m \times p \text{ matrix}}.$$

A good tool for remembering which elements are multiplied in each sum is to think of matrix multiplication pictorially. For example, if A is a 4×2 matrix and B is a 2×3 matrix, to determine the entry in the first row, second column of the resulting 4×3 matrix, the elements across the first row of $\mathbf{A}$ would be multiplied by the elements down the second column of $\mathbf{B}$ (see Figure 7.1).

Example 7.2 (Matrix Multiplication)

Let

$$\mathbf{A} = \begin{bmatrix} 1 & 0 & 2 \\ -1 & 3 & 1 \end{bmatrix} \text{ and } \mathbf{B} = \begin{bmatrix} 3 & 1 \\ 2 & 1 \\ 1 & 0 \end{bmatrix}.$$

What are $\mathbf{AB}$ and $\mathbf{BA}$?

<u>Solution:</u> First, let us make sure matrix multiplication is valid for $\mathbf{AB}$:

$$\overbrace{(2 \times \underbrace{3) \cdot (3}_{\text{equal}} \times 2)}^{\text{product is an } 2 \times 2 \text{ matrix}}.$$

(Continued)

Thus, the product of **AB** will be a 2×2 matrix:

$$\begin{bmatrix} 1 & 0 & 2 \\ -1 & 3 & 1 \end{bmatrix} \cdot \begin{bmatrix} 3 & 1 \\ 2 & 1 \\ 1 & 0 \end{bmatrix} = \begin{bmatrix} (1 \cdot 3) + (0 \cdot 2) + (2 \cdot 1) & (1 \cdot 1) + (0 \cdot 1) + (2 \cdot 0) \\ (-1 \cdot 3) + (3 \cdot 2) + (1 \cdot 1) & (-1 \cdot 1) + (3 \cdot 1) + (1 \cdot 0) \end{bmatrix} = \begin{bmatrix} 5 & 1 \\ 4 & 2 \end{bmatrix}.$$

Next, let us check that the multiplication **BA** is valid:

$$\overbrace{(3 \times \underbrace{2) \cdot (2}_{\text{equal}} \times 3)}^{\text{product is an } 3 \times 3 \text{ matrix}}.$$

Thus, the product **BA** will be a 3×3 matrix:

$$\begin{bmatrix} 3 & 1 \\ 2 & 1 \\ 1 & 0 \end{bmatrix} \begin{bmatrix} 1 & 0 & 2 \\ -1 & 3 & 1 \end{bmatrix} = \begin{bmatrix} (3 \cdot 1) + (1 \cdot -1) & (3 \cdot 0) + (1 \cdot 3) & (3 \cdot 2) + (1 \cdot 1) \\ (2 \cdot 1) + (1 \cdot -1) & (2 \cdot 0) + (1 \cdot 3) & (2 \cdot 2) + (1 \cdot 1) \\ (1 \cdot 1) + (0 \cdot -1) & (1 \cdot 0) + (0 \cdot 3) & (1 \cdot 2) + (0 \cdot 1) \end{bmatrix}$$

$$= \begin{bmatrix} 2 & 3 & 7 \\ 1 & 3 & 5 \\ 1 & 0 & 2 \end{bmatrix}.$$

Note that **AB** $\neq$ **BA**. This demonstrates that matrix multiplication is not commutative; that is, changing the order of multiplication changes the product. There are some special matrices for which changing the order of multiplication does not change the product, but in general this is not the case.

Next, let us consider an example of a matrix times a vector.

Example 7.3 (More Matrix Multiplication)

Let

$$\mathbf{A} = \begin{bmatrix} 1 & 0 \\ 2 & 1 \end{bmatrix} \text{ and } \mathbf{x} = \begin{bmatrix} 3 \\ -1 \end{bmatrix}.$$

Does the product **Ax** exist? If so, what is it? Does the product **xA** exist? If so, what is it?

Solution: Notice that **x** is a vector. We can think of a vector as a matrix that has only one column. So, **x** is a 2×1 matrix. The process for multiplication remains the same.

First, we check if the product **Ax** is possible under matrix multiplication:

$$\overbrace{(2 \times \underbrace{2) \cdot (2}_{\text{equal}} \times 1)}^{\text{product is an } 2 \times 1 \text{ matrix}}.$$

(Continued)

The multiplication is possible and produces a 2×1 matrix:

$$\begin{bmatrix} 1 & 0 \\ 2 & 1 \end{bmatrix}\begin{bmatrix} 3 \\ -1 \end{bmatrix} = \begin{bmatrix} (1 \cdot 3) + (0 \cdot -1) \\ (2 \cdot 3) + (1 \cdot -1) \end{bmatrix} = \begin{bmatrix} 3 \\ 5 \end{bmatrix}.$$

Now, we check to see if the product $\mathbf{x}\mathbf{A}$ is possible under matrix multiplication.

$$(2 \times \underbrace{1) \cdot (2}_{\text{not equal}} \times 2).$$

Thus, the matrix multiplication $\mathbf{x}\mathbf{A}$ is not possible.

Next, we define an *identity matrix*. If $\mathbf{A}$ is an $n \times n$ matrix where $a_{ii} = 1$ for all i and $a_{ij} = 0$ for $i \neq j$, that is,

$$\mathbf{A} = \begin{bmatrix} 1 & 0 & \cdots & 0 \\ 0 & 1 & \cdots & 0 \\ \vdots & \vdots & \ddots & \vdots \\ 0 & 0 & \cdots & 1 \end{bmatrix},$$

then $\mathbf{A}$ is known as an identity matrix. An identity matrix is square (the number of columns equals the number of rows) and has 1s on its diagonal. Usually, the identity matrix is denoted $\mathbf{I}$.

Before continuing with more examples, let us take a look at some of the properties of matrix multiplication. Matrix multiplication has the associative and distributive propertues as long as the sizes of the matrices are such that the operations are defined. But, in general, matrix multiplication is not commutative.

Properties of Matrix Multiplication

Let $\mathbf{A}, \mathbf{B}$, and $\mathbf{C}$ be matrices and a and b be real numbers. Provided that the following matrix multiplications are defined, then

1. $\mathbf{A}(\mathbf{BC}) = (\mathbf{AB})\mathbf{C}$
2. $\mathbf{A}(\mathbf{B} + \mathbf{C}) = \mathbf{AB} + \mathbf{AC}$
3. $(\mathbf{B} + \mathbf{C})\mathbf{A} = \mathbf{BA} + \mathbf{CA}$
4. If $\mathbf{A}$ is an $m \times n$ matrix and $\mathbf{I}$ is the $n \times n$ identity matrix, then $\mathbf{AI} = \mathbf{A}$. Furthermore, if $m = n$, then $\mathbf{AI} = \mathbf{IA} = \mathbf{A}$.
5. $(a\mathbf{A})(b\mathbf{B}) = (ab)\mathbf{AB}$
6. Matrix multiplication is not necessarily commutative. That is, in general, $\mathbf{AB} \neq \mathbf{BA}$.

7.2 Applications

Now, let us return to the examples presented in Section 6.3 and carry out the matrix multiplications where needed.

Example 7.4 (Ecological Succession)

In Example 6.4 and Section 6.3, we developed a simple model for ecological succession of a coastal wetland. Given a vector describing the landscape composition at time step t, our model found the landscape composition at time step $t + 1$ (i.e., 10 years later):

$$\mathbf{v}(t + 1) = \begin{bmatrix} 0.95 & 0 & 0 \\ 0.05 & 0.88 & 0 \\ 0 & 0.12 & 1 \end{bmatrix} \mathbf{v}(t).$$

Now, if we start with the composition of our total area being 100% submerged, what is the composition after 20 years?

Solution: We are going to solve this problem in two different ways.

Method 1: First, we find the composition 10 years later ($t = 1$):

$$v(1) = \begin{bmatrix} 0.95 & 0 & 0 \\ 0.05 & 0.88 & 0 \\ 0 & 0.12 & 1 \end{bmatrix} \begin{bmatrix} 1 \\ 0 \\ 0 \end{bmatrix} = \begin{bmatrix} 0.95 \\ 0.05 \\ 0 \end{bmatrix}.$$

Next, we find the composition another 10 years later ($t = 2$):

$$v(2) = \begin{bmatrix} 0.95 & 0 & 0 \\ 0.05 & 0.88 & 0 \\ 0 & 0.12 & 1 \end{bmatrix} \begin{bmatrix} 0.95 \\ 0.05 \\ 0 \end{bmatrix} = \begin{bmatrix} 0.9025 \\ 0.0915 \\ 0.0060 \end{bmatrix}.$$

Thus, the composition after 20 years is 90.25% submerged, 9.15% saturated, and 0.6% dry.

Method 2: For this method, let

$$A = \begin{bmatrix} 0.95 & 0 & 0 \\ 0.05 & 0.88 & 0 \\ 0 & 0.12 & 1 \end{bmatrix}.$$

Notice that $\mathbf{v}(1) = \mathbf{A}\mathbf{v}(0)$ and $\mathbf{v}(2) = \mathbf{A}\mathbf{v}(1)$, and thus

$$\mathbf{v}(2) = \mathbf{A}\mathbf{v}(1) = \mathbf{A}\left(\mathbf{A}\mathbf{v}(0)\right) = (\mathbf{A}\mathbf{A})\mathbf{v}(0).$$

(Continued)

Is **AA** a valid matrix multiplication? Yes, because both are 3×3 matrices. We will denote **AA** as $\mathbf{A}^2$. Thus,

$$\mathbf{v}(2) = \mathbf{A}^2\mathbf{v}(0) = \begin{bmatrix} 0.95 & 0 & 0 \\ 0.05 & 0.88 & 0 \\ 0 & 0.12 & 1 \end{bmatrix} \begin{bmatrix} 0.95 & 0 & 0 \\ 0.05 & 0.88 & 0 \\ 0 & 0.12 & 1 \end{bmatrix} \begin{bmatrix} 1 \\ 0 \\ 0 \end{bmatrix}$$

$$= \begin{bmatrix} 0.9025 & 0 & 0 \\ 0.0915 & 0.7744 & 0 \\ 0.0060 & 0.2256 & 1 \end{bmatrix} \begin{bmatrix} 1 \\ 0 \\ 0 \end{bmatrix} = \begin{bmatrix} 0.9025 \\ 0.0915 \\ 0.0060 \end{bmatrix}.$$

In general, if we have a matrix model

$$\mathbf{v}(t+1) = \mathbf{A}\mathbf{v}(t)$$

and some initial composition vector $v(0)$, then we can derive the following:

$$\mathbf{v}(1) = \mathbf{A}\mathbf{v}(0)$$

$$\mathbf{v}(2) = \mathbf{A}\mathbf{v}(1) = \mathbf{A}\mathbf{A}\mathbf{v}(0) = \mathbf{A}^2\mathbf{v}(0)$$

$$\mathbf{v}(3) = \mathbf{A}\mathbf{v}(2) = \mathbf{A}\left(\mathbf{A}^2\mathbf{v}(0)\right) = \mathbf{A}^3\mathbf{v}(0)$$

$$\vdots$$

$$\mathbf{v}(t) = \mathbf{A}^t\mathbf{v}(0).$$

Example 7.5 (More Complex Ecological Succession)

In Example 6.5, we developed a more complex model for ecological succession of a coastal wetland. Given a vector describing the landscape composition at time step t, our model projected that the landscape composition at time step $t+1$ (i.e., 10 years later) would be:

$$\mathbf{v}(t+1) = \begin{bmatrix} 0.94 & 0.02 & 0.01 \\ 0.05 & 0.86 & 0.06 \\ 0.01 & 0.12 & 0.93 \end{bmatrix} \mathbf{v}(t).$$

Now, if we start with the composition of our total area being 100% submerged, what is the composition after 20 years?

Solution: After 20 years ($t = 2$),

$$\mathbf{v}(2) = \begin{bmatrix} 0.94 & 0.02 & 0.01 \\ 0.05 & 0.86 & 0.06 \\ 0.01 & 0.12 & 0.93 \end{bmatrix}^2 \begin{bmatrix} 1 \\ 0 \\ 0 \end{bmatrix}$$

$$= \begin{bmatrix} 0.8847 & 0.0372 & 0.0199 \\ 0.0906 & 0.7478 & 0.1079 \\ 0.0247 & 0.2150 & 0.8722 \end{bmatrix} \begin{bmatrix} 1 \\ 0 \\ 0 \end{bmatrix} = \begin{bmatrix} 0.8847 \\ 0.0906 \\ 0.0247 \end{bmatrix}.$$

Thus, after 20 years, of the total coastal wetland, 88.47% is submerged, 9.06% is saturated, and 2.47% is dry.

So far, in using matrices to construct models, we have used only *column vectors* (matrices with only one column) and *square matrices* (matrices with the same number of rows as columns). In our next example, we consider a matrix model that requires a nonsquare matrix.

Example 7.6 (Bird Feeding [29])

A wildlife center finds that the number of cardinals and finches that visit a bird feeder on a regular basis varies with the season, as shown in the table below.

Season	Cardinals	Finches
Spring	25	31
Summer	20	35
Fall	22	29
Winter	36	20

The wildlife center has also found that cardinals and finches eat different amounts of sunflower seed, corn, and millet (in grams) when they visit the feeder, as shown in the table below.

Feed Type	Cardinals	Finches
Sunflower	32	10
Corn	10	18
Millet	30	27

Each type of feed has a different cost: sunflower seeds are 3¢/gram, corn is 2¢/gram, and millet is 1¢/gram.

   (a) What is the total cost of seed for each season of the year in order to feed the cardinals and finches that regularly visit the feeder?
   (b) How many grams of each kind of seed are used over an entire year?
   (c) What is the total cost of seed in a year?
   (d) Suppose that the wildlife center puts bird feeders in four more locations, for a total of five feeders. The locations are very distant from each other but have similar habitats, so the number of birds and the amount they eat is expected to be the same as the original feeder. Calculate the total amount of each type of seed expected to be used in each season.

Solution: Each question can be answered using matrix multiplication. First, let us represent the information we have in matrices. We have three different sets of data: (1) numbers of each type of bird that visits the feeder determined by season, (2) amount of each type of

(Continued)

feed the birds eat determined by bird species, and (3) cost of each type of feed. Each set of information can be represented as a matrix. Let

$$
P = \begin{array}{c} \\ \text{Spring} \\ \text{Summer} \\ \text{Fall} \\ \text{Winter} \end{array}
\begin{array}{cc} \text{Cardinals} & \text{Finches} \\ \begin{bmatrix} 25 & 31 \\ 20 & 35 \\ 22 & 29 \\ 36 & 20 \end{bmatrix} \end{array}, \quad
Q = \begin{array}{c} \\ \text{Cardinals} \\ \text{Finches} \end{array}
\begin{array}{ccc} \text{Sunflower} & \text{Corn} & \text{Millet} \\ \begin{bmatrix} 32 & 10 & 30 \\ 10 & 18 & 27 \end{bmatrix} \end{array},
$$

$$
\text{and } R = \begin{array}{c} \\ \text{Sunflower} \\ \text{Corn} \\ \text{Millet} \end{array}
\begin{array}{c} \text{Cost} \\ \begin{bmatrix} 3 \\ 2 \\ 1 \end{bmatrix} \end{array}
$$

(a) To find the total cost for each season, first find **PQ**, which calculates how much of each type of seed is eaten each season of the year:

$$
PQ = \begin{bmatrix} 25 & 31 \\ 20 & 35 \\ 22 & 29 \\ 36 & 20 \end{bmatrix} \begin{bmatrix} 32 & 10 & 30 \\ 10 & 18 & 27 \end{bmatrix},
$$

$$
= \begin{array}{c} \\ \text{Spring} \\ \text{Summer} \\ \text{Fall} \\ \text{Winter} \end{array}
\begin{array}{ccc} \text{Sunflower} & \text{Corn} & \text{Millet} \\ \begin{bmatrix} 1{,}110 & 808 & 1{,}587 \\ 990 & 830 & 1{,}545 \\ 994 & 742 & 1{,}443 \\ 1{,}352 & 720 & 1{,}620 \end{bmatrix} \end{array}.
$$

Now, multiply **PQ** by **R**, the cost matrix, to get the total cost of seed for each season:

$$
(PQ)R = \begin{bmatrix} 1{,}110 & 808 & 1{,}587 \\ 990 & 830 & 1{,}545 \\ 994 & 742 & 1{,}443 \\ 1{,}352 & 720 & 1{,}620 \end{bmatrix} \begin{bmatrix} 3 \\ 2 \\ 1 \end{bmatrix} =
\begin{array}{c} \\ \text{Spring} \\ \text{Summer} \\ \text{Fall} \\ \text{Winter} \end{array}
\begin{array}{c} \text{Cost} \\ \begin{bmatrix} 6{,}533 \\ 6{,}175 \\ 5{,}909 \\ 7{,}116 \end{bmatrix} \end{array}
$$

These costs are in cents, so multiply by 0.01 to convert to dollars. The total cost of seed is \$65.33 in spring, \$61.75 in summer, \$59.09 in fall, and \$71.16 in winter.

(b) The totals of the columns of the matrix **PQ** give a matrix whose elements represent the total number of grams of each seed used over an entire year. Call this matrix **T** and write it as a row vector:

$$
T = \begin{array}{c} \\ \text{Entire Year} \end{array}
\begin{array}{ccc} \text{Sunflower} & \text{Corn} & \text{Millet} \\ \begin{bmatrix} 4{,}446 & 3{,}100 & 6{,}195 \end{bmatrix} \end{array}.
$$

(*Continued*)

Thus, 4,446 grams of sunflower seed, 3,100 grams of corn, and 6,195 grams of millet are used over an entire year.

   (c) For the total cost of seed in a year, find the product of the matrix **T**, the matrix showing the total amount of each seed for a year, and the matrix **R**, the cost matrix (by seed). (Notice that **T** is a 1×3 matrix and that **R** is a 3×1 matrix. Thus, if we multiply **TR**, we will get a 1×1 matrix, which is just a scalar value:

$$\mathbf{TR} = \begin{bmatrix} 4{,}446 & 3{,}100 & 6{,}195 \end{bmatrix} \begin{bmatrix} 3 \\ 2 \\ 1 \end{bmatrix} = \begin{bmatrix} 25{,}733 \end{bmatrix}.$$

Thus, the total cost is $257.33. Of course, we could have calculated this same answer by summing the elements of the column vector **PQR** that we found in (a).
   (d) Multiply **PQ** by the scalar 5, as follows:

$$5(\mathbf{PQ}) = 5 \begin{bmatrix} 1{,}110 & 808 & 1{,}587 \\ 990 & 830 & 1{,}545 \\ 994 & 742 & 1{,}443 \\ 1{,}352 & 720 & 1{,}620 \end{bmatrix} = \begin{bmatrix} 5{,}550 & 4{,}040 & 7{,}935 \\ 4{,}950 & 4{,}150 & 7{,}725 \\ 4{,}970 & 3{,}710 & 7{,}215 \\ 6{,}760 & 3{,}600 & 8{,}100 \end{bmatrix}.$$

The total amount of corn used in the fall, for example, is 3,710 grams.

Note that in all of the models presented in this chapter, nothing is created or destroyed; the elements of the matrix represent entities that are only changing states. In the landscape succession examples, we never created or destroyed land, but the land could change from one state to another. In the disease model, there were no births or deaths, but individuals could move between infected and susceptible states. This type of model is known as a Markov chain (named after Andrey Markov). Markov chain models have the Markov property that future states depend *only* on the present state and are independent of past states. In our matrix models case, "state" refers to the vector describing the distribution of the system at a particular time. This means that if we know the state of the system at time $t = 5$, we can predict the future, even though we may not know what the system did from times $t = 0$ to $t = 4$. The word "chain" in "Markov chain" refers to the fact that the model has a discrete number of possible states.

7.3 Matlab Skills

Matrix Operations in Matlab

Since Matlab is built to handle matrices easily, basic matrix operations in Matlab are calculated using algebraic manipulations that mimic standard algebra. If you have two matrices of the same size and want to add or subtract them, use the + operator or – operator, respectively. If you have two matrices that you want to multiply (and each is the appropriate size), then use the * operator.

If you try to add or subtract matrices that do not have the same dimension or try to multiply matrices that do not have compatible dimensions, Matlab will return an error message. Matlab also makes it easy to multiply a scalar number by a matrix. Again, we use the * operator:

```
─────────────────── Command Window ───────────────────
>> A = [5 1; 2 3]
A =
      5      1
      2      3

>> B = [3 -1 0; 1 0 2]
B =
      3     -1      0
      1      0      2

>> C = [-1 5; 7 0]
C =
     -1      5
      7      0

>> A+C
ans =
      4      6
      9      3

>> A-C
ans =
      6     -4
     -5      3

>> A*B
ans =
     16     -5      2
      9     -2      6

>> A*C
ans =
      2     25
     19     10

>> 5*A
ans =
     25      5
     10     15

>> (1/5)*A
ans =
    1.0000    0.2000
    0.4000    0.6000

>> A/5
ans =
    1.0000    0.2000
    0.4000    0.6000

>> B*A
??? Error using ==> mtimes
Inner matrix dimensions must agree.
```

```
>> A-B
??? Error using ==> minus
Matrix dimensions must agree.

>> B+A
??? Error using ==> plus
Matrix dimensions must agree.
```

Notice the last three commands produced error messages. Notice also that `(1/5)*A` produces the same matrix as `A/5`. Thus, we see that we can use the `/` operator as a shortcut for multiplying by $\frac{1}{5}$.

In addition to having these standard operations, Matlab has a few operations that make working with matrices easier. Suppose that we want to subtract a constant value from every entry in a matrix, such as subtracting 2 from every entry in

$$\mathbf{A} = \begin{bmatrix} 5 & 1 \\ 2 & 3 \end{bmatrix}.$$

To do this by hand, we would calculate

$$\begin{bmatrix} 5 & 1 \\ 2 & 3 \end{bmatrix} - 2 \begin{bmatrix} 1 & 1 \\ 1 & 1 \end{bmatrix}.$$

However, Matlab has a shorthand method to do this. In Matlab, we can type `A-2` to get the same result. When we subtract a scalar value from a matrix in Matlab, Matlab assumes that we want to multiply that scalar number by the appropriately size matrix filled with 1s and then subtract that matrix from the original one:

```
━━━━━━━━━━━━━━━━━━━━━ Command Window ━━━━━━━━━━━━━━━━━━━
>> A = [5 1; 2 3];

>> A - 2
ans =
         3      -1
         0       1

>> A + 2
ans =
         7       3
         4       5
```

Displaying Model Output

There are two different ways we might choose to display the output for a matrix model. We could display the size of each class or element of a matrix at each time step in a table, or we could create a graph of the size of each class over time.

CREATING TABLES OF OUTPUT

We can use a for loop to help display tables of data. For example, suppose that we want to display how the landscape structure changes over time given the ecological succession model using in Example 8.1. We can use an `fprintf` statement inside a loop to print out a table that displays this information. See Appendix A.5 for details on using the `fprintf` command:

```
———————————— EcoSuccessionTable.m ————————————
1    % Filename: EcoSuccessionTable.m
2    % M-file to
3    % - Print out a table showing landscape structure over time
4
5    % Enter the transfer matrix
6    T = [ 0.94 0.02 0.01
7          0.05 0.86 0.06
8          0.01 0.12 0.93 ];
9
10   % Enter the initial state
11   x0 = [ 1 ; 0 ; 0 ];
12
13   % Print header for table
14   fprintf(' t     u     s     d\n')
15   fprintf('------------------------\n')
16
17   % Fill in table using a for loop
18   for t = [1 2 3 4 5 10 20 30 40 50 100 200]  % array of times
19       x = T^t * x0;              % matrix multiplication
20       u = x(1,1);                % get u value for this time step
21       s = x(2,1);                % get s value for this time step
22       d = x(3,1);                % get d value for this time step
23       fprintf('%3d  %5.3f  %5.3f  %5.3f\n',t,u,s,d)  % print it all out
24   end
```

When this m-file is run in the command window, the output looks like the following:

```
———————————— Command Window ————————————
>> EcoSuccessionTable
     t     u     s     d
    ------------------------
     1   0.940   0.050   0.010
     2   0.885   0.091   0.025
     3   0.834   0.124   0.043
     4   0.787   0.151   0.063
     5   0.743   0.173   0.084
    10   0.569   0.236   0.194
    20   0.368   0.274   0.358
    30   0.273   0.284   0.444
    40   0.227   0.287   0.486
    50   0.205   0.289   0.506
   100   0.185   0.291   0.524
   200   0.184   0.291   0.525
```

PLOTTING TIME DYNAMICS

Now that we can produce tables of time-series data, it would be nice to transfer that information to a plot so that we can view the information graphically. We can do this by creating the time-series data that we want using a loop and then using the `plot` command. The following m-file plots time-series data for the ecological succession model from Example 8.1 for time steps $t = 0$ to $t = 200$:

_____ EcoSuccessionPlot.m _____

```
1   % Filename: EcoSuccessionPlot.m
2   % M-file to
3   % - Print out a table showing landscape structure over time
4
5   % Enter the transfer matrix
6   T = [ 0.94 0.02 0.01
7          0.05 0.86 0.06
8          0.01 0.12 0.93 ];
9
10  % Enter the initial state
11  x0 = [ 1 ; 0 ; 0 ];
12
13  % We will create a matrix x that has three columns.
14  % Each column will contain time series data for one class.
15  % Each row will correspond to a time step.
16  x = zeros(201,3);
17
18  x(1,:) = x0;         % Data for time step t = 0
19
20  % Use for loop to generate times series data
21  for t = 1:200
22      x(t+1,:) = T^t * x0;   % Data for time step t
23  end
24
25  % Time series information for proportion underwater is
26  % in the first column
27  u = x(:,1);
28
29  % Time series information for proportion saturated but
30  % not underwater is in the second column
31  s = x(:,2);
32
33  % Time series information for proportion dry is in the
34  % third column
35  d = x(:,3);
36
37  % Generate plot
38  time = [0:200];
39  plot(time,u,'r-',time,s,'g-',time,d,'b-')
40  legend('u', 's', 'd')
41  xlabel('Time step t')
42  ylabel('Proportion of Wetlands')
```

The plot resulting from this m-file is shown in Figure 7.2.

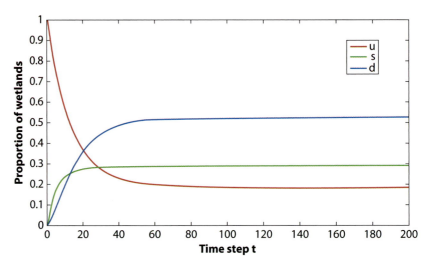

Figure 7.2 Plot generated by the m-file EcoSuccessionPlot.m.

7.4 Exercises

7.1 Find the values of x and y so that the matrix equalities hold.

(a) $\begin{bmatrix} 5 & -1 \\ 2x & y \end{bmatrix} = \begin{bmatrix} 5 & -1 \\ 9 & 4-y \end{bmatrix}$

(b) $\begin{bmatrix} 3 \\ x \\ -1 \end{bmatrix} = \begin{bmatrix} 3 \\ -4 \\ 0 \end{bmatrix}$

(c) $\begin{bmatrix} 2 & x-3 & -5 \\ 9 & 1 & 3-y \end{bmatrix} = \begin{bmatrix} 2 & 1 & -5 \\ 9 & 1 & 4 \end{bmatrix}$

7.2 Given the matrix

$$A = \begin{bmatrix} 4 & -3 & 2 & 0 \\ 5 & -2 & 0 & 6 \\ \frac{1}{2} & 3 & 7 & -\frac{5}{2} \end{bmatrix},$$

calculate $a_{31} + a_{13}$ and $a_{34} + a_{24}$.

7.3 Using the matrix A from the previous exercise, use Matlab to calculate A^2 and A^4.

7.4 Use the following matrices to do the following calculations (if possible).

$$\mathbf{A} = \begin{bmatrix} 2 & -1 \\ 5 & 3 \end{bmatrix} \quad \mathbf{B} = \begin{bmatrix} 4 & 0 & 2 \\ -3 & 1 & -2 \end{bmatrix} \quad \mathbf{C} = \begin{bmatrix} 0 & -1 & 1 \\ 1 & 2 & 0 \end{bmatrix}$$

$$\mathbf{D} = \begin{bmatrix} 4 & 2 \\ -1 & 3 \\ 0 & 0 \end{bmatrix} \quad \mathbf{E} = \begin{bmatrix} 3 \\ 4 \\ -1 \end{bmatrix} \quad \mathbf{F} = \begin{bmatrix} 0 & -1 & 6 \\ 2 & 0 & 3 \\ -1 & 0 & 0 \end{bmatrix}$$

(a) 1.2**A**

(b) **CD**

(c) **DC**
(d) **B + C**
(e) **AB + AC**
(f) **DA**
(g) **AB**
(h) **A + B**
(i) **DF**
(j) **(AC) F**
(k) **2A − BD**
(l) **FE**

7.5   Find two matrices A, B, of size 3×3, such that $AB \neq BA$.

7.6   Suppose that you are modeling a nonfatal infectious disease. You assume that the people in the population you are modeling are either susceptible to infection or infected. The following flow diagram shows the rates at which individuals flow from one category to the other.

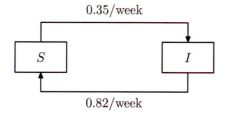

Suppose that at time $t = 0$, 20% of the population is infected. Construct a table to display proportions of susceptible and infected after 1 week, 2 weeks, 3 weeks, 5 weeks, 10 weeks, 20 weeks, and 50 weeks. Use Matlab to assist in your calculations.

7.7   Suppose that you are modeling the ecological succession of a parcel of land that is entirely grassland. After studying the surrounding land, you find that the grassland will eventually become populated with shrubs and then with trees. Because of human intervention, each year some of the land containing trees will be reverted to grassland and landscape populated by shrubs. The following flow diagram represents the dynamics of the ecological succession with classes for grass (G), shrubs (S), and trees (T).

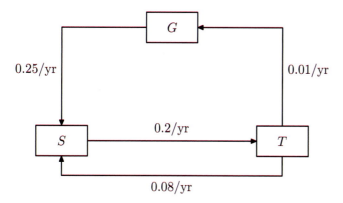

Using Matlab, create a time plot showing how the proportion of the land parcel in each class changes over 25 years. The horizonal axis should show time in years for $t = 0$ to $t = 25$. The vertical axis should show the proportion of land. Three different curves should be plotted, one for each class.

7.8   Make a matrix equation from the following three linear equations.

$$x_1(t + 1) = 0.4x_1(n) + 0.8x_2(t) + 0.1x_3(t)$$

$$x_2(t + 1) = 0.9x_1(t)$$

$$x_3(t + 1) = 0.7x_2(t).$$

7.9   Can this system of equations below be represented by a matrix equation? Why or why not?

$$x_1(t + 1) = 0.4x_1(t)\left(1 - \frac{x_1(t)}{100}\right) - 0.1x_1(t)x_2(t)$$

$$x_2(t + 1) = 0.1x_1(t)x_2(t).$$

7.10   (From [29]) A dietitian prepares a diet specifying the amounts of four basic food groups that a patient should eat: (I) meats and fish, (II) fruits and vegetables, (III) breads and starches, and (IV) milk products. Amounts are given in "exchanges" that represent 1 ounce (meat), 1/2 cup (fruits and vegetables), 1 slice (bread), 8 ounces (milk), or other suitable measurements. The exchanges for breakfast, lunch, and dinner for each group are given in the following table.

Exchanges	Group I	Group II	Group III	Group IV
Breakfast	2	1	2	1
Lunch	3	2	2	1
Dinner	4	3	2	1

The amounts of fat, carbohydrates, and protein (in appropriate units) in each food group are given in the following table.

	Fat	Carbohydrates	Protein
Group I	5	0	7
Group II	0	10	1
Group III	0	15	2
Group IV	10	12	8

There are 8 calories per exchange of fat, 4 calories per exchange of carbohydrates, and 5 calories per exchange of protein.

(a) Create a 3 × 4 matrix representing the exchange information for breakfast, lunch, and dinner for the four food groups. Create a 4 × 3 matrix representing the fat, carbohydrate, and protein content of each of the four food groups. Create a 3 × 1 matrix summarizing the calories per fat/carbohydrate/protein information. Label the matrices **X**, **Y**, and **Z**, respectively.

(b) Find the product **XY**. What do the entries of this matrix represent?

(c) Find the product **YZ**. What do the entries of this matrix represent?

(d) Find the products (**XYZ**) and **X**(**YZ**) and verify that they are equal. What do the entries of this matrix represent?

Long-Term Dynamics or Equilibrium

The previous chapters in this unit focused on how a biological process that can be in one of a few states (such as disease states of susceptible, exposed, infected, and immune) can be modeled to project what fraction of a population changes states over short periods of time. Our objective in this chapter is to extend these ideas to long time periods. In the landscape change case, we may want to know the long-term implications of disturbances, such as fire, storms, and insect outbreaks, on the vegetation across the landscape. We essentially want to take the matrix multiplication approach we developed earlier and apply it over many time periods. An important question in landscape ecology is whether, overall, the fraction of the landscape in each class is always changing or whether the landscape stabilizes. That is, does the landscape eventually have classes or types each of which occupies a certain fraction of the total landscape and then stays near these values?

   You can certainly think of simple cases in which a landscape has been completely urbanized so that there is no longer any forest or in which a region is dammed so that area that was forest is now underwater in a lake. In these cases, it is easy to intuit what happens in the long term, but in many real-world situations, such as recurring fires, it is not easy to tell what happens if fires don't occur too often, particularly from the perspective of the whole landscape. For example, about 10% of the land area of the nonpolar regions of the planet are maintained in savannas (a grassy or herbaceous layer with a thin overstory canopy of trees) because of disturbances, such as fire and foraging by herbivores. So our objective in this chapter is to point out that matrix methods can be used to determine the long-term structure of a landscape. Furthermore, we can determine whether this long-term structure is stable in the face of perturbations, such as a short period of changed environmental conditions that lead to higher fire frequencies and thus a changed landscape or a period of harvesting in which forest is removed. A biological system that stays mostly unchanged over time is said to be at equilibrium, and it is stable if it returns to this equilibrium following a small change in the system. This is closely related to the idea of homeostasis in physiology, which we will discuss later in the text.

8.1 Notion of an Equilibrium

Recall the succession models of Section 6.3. After a long enough time period, if you make some additional calculations, you will see that the fraction of the landscape in each class (submerged, saturated, and dry) will approach some constant. In the simplest model (Example 6.3), all the land will move to the "climax" state, in this case, the dry state. Thus, the vector that describes the landscape would eventually become very close to

$$\begin{bmatrix} 0 \\ 0 \\ 1 \end{bmatrix}.$$

However, when we add more complexity to the system (as in Example 6.6), each class of the system will eventually approach some fixed fraction. What we want to do now is define a mathematical method to find these "final" states: the jargon for this is that we are finding an "eigenvector" for the system. In the models presented in this chapter, the *eigenvector* is a vector whose elements tell us what fraction of the system each class will have after a long time. This is also called the *long-term equilibrium state* of the system.

8.2 Eigenvectors

Suppose that we are modeling a nonfatal disease for which either an individual is susceptible to acquiring the disease (S), is infected with the disease (I), or has recovered from the disease and is susceptible again (we built such a model in Example 6.6 for a common cold spreading among a dormitory population). Suppose that in this model, each day 10% of the susceptibles become infected and 20% of the infected recover and become susceptible again. Then, the transfer matrix $\mathbf{T}$ that models the daily change in this population is

$$\mathbf{T} = \begin{bmatrix} 0.9 & 0.2 \\ 0.1 & 0.8 \end{bmatrix}.$$

(Review Example 6.6 for how such a matrix is constructed.) Suppose that at time $t = 0$, we have 297 susceptible individuals and three infected individuals, and we assume that no new individuals enter the population and no one in the population dies or leaves. Thus, the population size remains at a constant size of 300 individuals. Then, the vector

$$\mathbf{x}(0) = \begin{bmatrix} 297 \\ 3 \end{bmatrix}$$

describes the structure of the population at time $t = 0$.

What do we expect to happen over time? How many individuals do we expect to be infected after 50 days? 100 days? 365 days?

Recall that to get the population structure after 1 day, $t = 1$, we would multiply

$$\mathbf{x}(1) = \mathbf{T}\mathbf{x}(0).$$

If we then wanted to know the population structure after 2 days, $t = 2$, we would multiply again:

$$\mathbf{x}(2) = \mathbf{T}\mathbf{x}(1) = \mathbf{T}\left(\mathbf{T}\mathbf{x}(0)\right) = \mathbf{T}^2\mathbf{x}(0).$$

In general, we have

$$\mathbf{x}(1) = \mathbf{T}\mathbf{x}(0) = \begin{bmatrix} 267.9 \\ 32.1 \end{bmatrix}$$

$$\mathbf{x}(2) = \mathbf{T}\mathbf{x}(1) = \mathbf{T}^2\mathbf{x}(0) = \begin{bmatrix} 247.5 \\ 52.5 \end{bmatrix}$$

$$\mathbf{x}(3) = \mathbf{T}\mathbf{x}(2) = \mathbf{T}^3\mathbf{x}(0) = \begin{bmatrix} 233.3 \\ 66.7 \end{bmatrix}$$

$$\vdots$$

$$\mathbf{x}(t) = \mathbf{T}\mathbf{x}(t-1) = \mathbf{T}^t\mathbf{x}(0),$$

Thus,

$$\mathbf{x}(50) = \begin{bmatrix} 0.9 & 0.2 \\ 0.1 & 0.8 \end{bmatrix}^{50} \begin{bmatrix} 297 \\ 3 \end{bmatrix} = \begin{bmatrix} 200 \\ 100 \end{bmatrix},$$

and

$$\mathbf{x}(51) = \begin{bmatrix} 0.9 & 0.2 \\ 0.1 & 0.8 \end{bmatrix}^{51} \begin{bmatrix} 297 \\ 3 \end{bmatrix} = \begin{bmatrix} 200 \\ 100 \end{bmatrix}.$$

It appears that after 50 days, the proportion of the population in each class (or compartment) is no longer changing. We will refer to this characteristic as the *equilibrium*, that is, the situation in which the proportion or number in each class remains constant after some period of time. The vector representing the proportion or number in each class at equilibrium is known as an *eigenvector*. In the disease model we just considered,

$$\begin{bmatrix} 200 \\ 100 \end{bmatrix}$$

is an eigenvector. Now, just because the population structure is at equilibrium does not mean that individuals are no longer moving between the susceptible and infected classes. There is still movement between the classes; however, the movement is such that the proportion in each class remains constant.

We would like to develop a mathematical method for determining the eigenvector without having to take large powers of matrices. Notice that at equilibrium, $\mathbf{x}(t) = \mathbf{x}(t-1)$. Thus,

$$\mathbf{x}(t) = \mathbf{T}\mathbf{x}(t-1)$$
$$= \mathbf{T}\mathbf{x}(t)$$
$$\begin{bmatrix} x_1 \\ x_2 \end{bmatrix} = \begin{bmatrix} 0.9 & 0.2 \\ 0.1 & 0.8 \end{bmatrix} \begin{bmatrix} x_1 \\ x_2 \end{bmatrix}. \tag{8.1}$$

Recall from Chapter 6 that a matrix multiplication equation such as this can be written as the system of equations

$$x_1 = 0.9x_1 + 0.2x_2. \tag{8.2}$$

$$x_2 = 0.1x_1 + 0.8x_2. \tag{8.3}$$

Notice that we can rewrite Equation (8.2) as

$$0 = -0.1x_1 + 0.2x_2 \tag{8.4}$$

and that we can rewrite Equation (8.3) as

$$0 = 0.1x_1 - 0.2x_2. \tag{8.5}$$

This shows that the Equations (8.2) and (8.3) are equivalent (since Equations (8.4) and (8.5) are multiples of each other). Since the two equations are equivalent, we essentially have one equation and two unknowns; therefore, we will not be able to uniquely determine both values (only one in terms of the other). Thus, we will use one of the equations to solve for x_1 in terms of x_2 and then choose a value for x_2.

If we solve Equation (8.2) for x_1 in terms of x_2, we get

$$0.1x_1 = 0.2x_2$$
$$x_1 = 2x_2. \tag{8.6}$$

Let us choose $x_2 = 1$; then $x_1 = 2x_2 = 2$. So, an eigenvector is

$$\begin{bmatrix} 2 \\ 1 \end{bmatrix}.$$

At this point, you should be thinking, "Doesn't this mean an eigenvector is dependent on what value I choose for x_2?" Indeed, it does. If we had instead chosen $x_2 = 50$, then $x_1 = 2x_2 = 100$, and the eigenvector would be

$$\begin{bmatrix} 100 \\ 50 \end{bmatrix}.$$

An eigenvector for this system is any point (x_1, x_2) on the line $0 = -0.1x_1 + 0.2x_2$. Since the eigenvector is not unique, we will express the eigenvector in a normalized form (a form that is unique). To **normalize** any vector, sum the values of each element of the vector and then divide each element of the vector by that sum. So, if normalizing

$$\begin{bmatrix} 2 \\ 1 \end{bmatrix},$$

the sum of the elements is 3, and the normalized eigenvector is

$$\begin{bmatrix} 2/3 \\ 1/3 \end{bmatrix}.$$

If normalizing

$$\begin{bmatrix} 100 \\ 50 \end{bmatrix},$$

the sum of the elements is 150, and the normalized eigenvector is

$$\begin{bmatrix} 100/150 \\ 50/150 \end{bmatrix} = \begin{bmatrix} 2/3 \\ 1/3 \end{bmatrix}.$$

Notice that no matter what value we choose for x_2, the normalized eigenvector will always be the same.

Now, recall that our population of susceptible and infected individuals was originally comprised of 300 individuals. So at equilibrium, how many individuals are susceptible, and how many are infected? To answer this question, we simply multiply the normalized eigenvector by the total population size, 300:

$$300 \begin{bmatrix} 2/3 \\ 1/3 \end{bmatrix} = \begin{bmatrix} 200 \\ 100 \end{bmatrix}.$$

This is what we expected on the basis of our earlier investigations.

Now, let us apply this method to find the normalized eigenvalue to the ecological succession model that we constructed and investigated in Examples 6.5 and 7.5.

Example 8.1 (Ecological Succession Equilibrium)

Recall that in Examples 6.5 and 7.5, we developed a model for wetland ecological succession consisting of three class: submerged wetlands, saturated but nonsubmerged land, and dry land. The model was defined by the matrix equation

$$\mathbf{v}(t+1) = \begin{bmatrix} 0.94 & 0.02 & 0.01 \\ 0.05 & 0.86 & 0.06 \\ 0.01 & 0.12 & 0.93 \end{bmatrix} \mathbf{v}(t).$$

Find the eigenvector that describes the composition of the wetlands at equilibrium.

Solution: We need to find u, s, and d at equilibrium when

$$\begin{bmatrix} u \\ s \\ d \end{bmatrix} = \begin{bmatrix} 0.94 & 0.02 & 0.01 \\ 0.05 & 0.86 & 0.06 \\ 0.01 & 0.12 & 0.93 \end{bmatrix} \begin{bmatrix} u \\ s \\ d \end{bmatrix}, \tag{8.7}$$

where u is the submerged class, s is the saturated by nonsubmerged class, and d is the dry class. We can represent Equation (8.7) as a system of equations:

$$u = \frac{47}{50}u + \frac{1}{50}s + \frac{1}{100}d. \tag{8.8}$$

$$s = \frac{1}{20}u + \frac{43}{50}s + \frac{3}{50}d. \tag{8.9}$$

$$d = \frac{1}{100}u + \frac{3}{25}s + \frac{93}{100}d. \tag{8.10}$$

Here, each of the coefficients has been rewritten as a fraction (to simplify the algebra). Notice that the sum of Equations (8.8) and (8.9) gives an equation that is the same as multiplying Equation (8.10) by -1. Since the third equation can be written as a sum of the first two, we gain no additional information from the third equation. This means that, we essentially have two equations and three variables. Thus, we must write two of the variables in terms of the third variable and then pick a value for the third variable. Here, we will write u and s in terms of d.

(Continued)

First, we solve Equation (8.8) for u in terms of s and d:

$$\frac{3}{50}u = \frac{1}{50}s + \frac{1}{100}d$$

$$u = \frac{1}{3}s + \frac{1}{6}d. \tag{8.11}$$

Next, we substitute Equation (8.11) into Equation (8.9):

$$s = \frac{1}{20}\left(\frac{1}{3}s + \frac{1}{6}d\right) + \frac{43}{50}s + \frac{3}{50}d$$

$$= \frac{1}{60}s + \frac{1}{120}d + \frac{43}{50}s + \frac{3}{50}d$$

$$= \frac{263}{300}s + \frac{41}{600}d$$

$$\frac{37}{300}s = \frac{41}{600}d$$

$$s = \frac{41}{74}d. \tag{8.12}$$

Now we can substitute Equation (8.12) into Equation (8.11) to get an equation for u in terms of only d:

$$u = \frac{1}{3}\left(\frac{41}{74}d\right) + \frac{1}{6}d$$

$$= \frac{41}{222}d + \frac{1}{6}d$$

$$= \frac{13}{37}d. \tag{8.13}$$

Let us choose $d = 74$; then

$$u = \frac{13}{37} \cdot 74 = 26, \quad \text{and} \quad s = \frac{41}{74} \cdot 74 = 41.$$

Thus, we have the eigenvector

$$\begin{bmatrix} 26 \\ 41 \\ 74 \end{bmatrix}$$

that, when normalized, is

$$\begin{bmatrix} 26/141 \\ 41/141 \\ 74/141 \end{bmatrix} \approx \begin{bmatrix} 0.184 \\ 0.291 \\ 0.525 \end{bmatrix}.$$

8.3 Stability

In this section, we ask what would happen if we started with a different initial condition, a different $\mathbf{v}(0)$.

Consider the model that we developed for the spread of a nonfatal disease with two classes S and I described by Equation (8.1) at equilibrium. We saw that if there was an initial population of 300 individuals with only 3 that were infected, at equilibrium there were 200 susceptible individuals and 100 infected individuals. That is, an eigenvector for the system is

$$\begin{bmatrix} 200 \\ 100 \end{bmatrix}.$$

What would happen if we started with a different ratio of susceptible to infected individuals? Would we still reach the same equilibrium?

Consider starting with

$$\mathbf{x}(0) = \begin{bmatrix} 10 \\ 290 \end{bmatrix},$$

that is, starting with most of the population infected. Then

$$\mathbf{x}(1) = \begin{bmatrix} 0.9 & 0.2 \\ 0.1 & 0.8 \end{bmatrix} \begin{bmatrix} 10 \\ 290 \end{bmatrix} = \begin{bmatrix} 67.0 \\ 233.0 \end{bmatrix}$$

$$\mathbf{x}(2) = \begin{bmatrix} 0.9 & 0.2 \\ 0.1 & 0.8 \end{bmatrix}^2 \begin{bmatrix} 10 \\ 290 \end{bmatrix} = \begin{bmatrix} 106.9 \\ 193.1 \end{bmatrix}$$

$$\mathbf{x}(3) = \begin{bmatrix} 0.9 & 0.2 \\ 0.1 & 0.8 \end{bmatrix}^3 \begin{bmatrix} 10 \\ 290 \end{bmatrix} = \begin{bmatrix} 134.8 \\ 165.2 \end{bmatrix}$$

$$\vdots$$

$$\mathbf{x}(10) = \begin{bmatrix} 0.9 & 0.2 \\ 0.1 & 0.8 \end{bmatrix}^{10} \begin{bmatrix} 10 \\ 290 \end{bmatrix} = \begin{bmatrix} 194.6 \\ 105.4 \end{bmatrix}$$

$$\vdots$$

$$\mathbf{x}(20) = \begin{bmatrix} 0.9 & 0.2 \\ 0.1 & 0.8 \end{bmatrix}^{20} \begin{bmatrix} 10 \\ 290 \end{bmatrix} = \begin{bmatrix} 199.8 \\ 100.2 \end{bmatrix}$$

$$\vdots$$

$$\mathbf{x}(70) = \begin{bmatrix} 0.9 & 0.2 \\ 0.1 & 0.8 \end{bmatrix}^{70} \begin{bmatrix} 10 \\ 290 \end{bmatrix} = \begin{bmatrix} 200 \\ 100 \end{bmatrix}.$$

Thus, we can see that, with a very different initial condition, we still approach the same equilibrium.

What if we started with a different population size? What do you expect would happen? For example, say that we started with a population of 30 individuals with only 1 infected at the initial time:

$$\mathbf{x}(1) = \begin{bmatrix} 0.9 & 0.2 \\ 0.1 & 0.8 \end{bmatrix} \begin{bmatrix} 29 \\ 1 \end{bmatrix} = \begin{bmatrix} 26.3 \\ 3.7 \end{bmatrix}$$

$$\mathbf{x}(2) = \begin{bmatrix} 0.9 & 0.2 \\ 0.1 & 0.8 \end{bmatrix}^2 \begin{bmatrix} 29 \\ 1 \end{bmatrix} = \begin{bmatrix} 24.4 \\ 5.6 \end{bmatrix}$$

$$\mathbf{x}(3) = \begin{bmatrix} 0.9 & 0.2 \\ 0.1 & 0.8 \end{bmatrix}^3 \begin{bmatrix} 29 \\ 1 \end{bmatrix} = \begin{bmatrix} 23.1 \\ 6.9 \end{bmatrix}$$

$$\vdots$$

$$\mathbf{x}(10) = \begin{bmatrix} 0.9 & 0.2 \\ 0.1 & 0.8 \end{bmatrix}^{10} \begin{bmatrix} 29 \\ 1 \end{bmatrix} = \begin{bmatrix} 20.3 \\ 9.7 \end{bmatrix}$$

$$\vdots$$

$$\mathbf{x}(20) = \begin{bmatrix} 0.9 & 0.2 \\ 0.1 & 0.8 \end{bmatrix}^{20} \begin{bmatrix} 29 \\ 1 \end{bmatrix} = \begin{bmatrix} 20.0 \\ 10.0 \end{bmatrix}$$

$$\vdots$$

$$\mathbf{x}(70) = \begin{bmatrix} 0.9 & 0.2 \\ 0.1 & 0.8 \end{bmatrix}^{70} \begin{bmatrix} 29 \\ 1 \end{bmatrix} = \begin{bmatrix} 20.0 \\ 10.0 \end{bmatrix}.$$

After enough time has passed, the proportion in each class is the same in both cases: $\frac{2}{3}$ in the susceptible class and $\frac{1}{3}$ in the infected class.

Why does this happen? Recall that when we were solving for the eigenvector, we never utilized the intial condition information. The eigenvector depends only on the model, not on the initial condition. Thus, it does not matter what initial condition we start with: we will always approach the same equilibrium.

If the eigenvector depends only on the model, then we should expect that if we change one of the values in the transfer matrix, it will change the model's normalized eigenvalue.

Example 8.2 (Changing Transfer Rates)

Consider the model that we developed for the spread of a nonfatal disease with two classes S and I described by Equation (8.1) at equilibrium. Suppose now that each day, 30% of the infected individuals recover and become susceptible again (instead of the 20% we had before). Now the model at equilibrium becomes

$$\begin{bmatrix} x_1 \\ x_2 \end{bmatrix} = \begin{bmatrix} 0.9 & 0.3 \\ 0.1 & 0.7 \end{bmatrix} \begin{bmatrix} x_1 \\ x_2 \end{bmatrix}.$$

Find the normalized eigenvalue for this model. How does changing the recovery rate of the infected class change the normalized eigenvalue? Does it make biological sense?

(Continued)

Solution: Our model can be written as the system of equations:

$$x_1 = 0.9x_1 + 0.3x_2. \tag{8.14}$$

$$x_2 = 0.1x_1 + 0.7x_2. \tag{8.15}$$

Notice that we can rewrite Equations (8.14) and (8.15) as

$$0 = -0.1x_1 + 0.3x_2. \tag{8.16}$$

$$0 = 0.1x_1 - 0.3x_2. \tag{8.17}$$

Since Equation (8.16) can be written as -1 times Equation (8.17), the two equations are equivalent. Thus, we will use one of the equations to solve for x_1 in terms of x_2 and then choose the value of x_2. If we solve Equation (8.14) for x_1 in terms of x_2, we get

$$x_1 = 3x_2.$$

Let us choose $x_2 = 1$; then $x_1 = 3$. Thus, an eigenvector of this model is

$$\begin{bmatrix} 3 \\ 1 \end{bmatrix}.$$

The normalized eigenvector for this system is

$$\begin{bmatrix} 3/4 \\ 1/4 \end{bmatrix}.$$

Thus, we see that when we increase the recovery rate of the infected class, the normalized eigenvector has a smaller proportion in the infected class. This does indeed make biological sense and you should guess what happens to the equilibrium structure of the population as the recovery rate continues to increase.

8.4 Matlab Skills

In Matlab, we use the function `eig` to find eigenvalues (which we will discuss in the following chapter) and eigenvectors. Since the output of this function will not be completely understood until the material of the next chapter is covered, we postpone demonstrating how to find eigenvectors in Matlab until the end of the next chapter.

8.5 Exercises

8.1   Could the matrix below represent a transfer matrix? Why or why not?

$$\begin{bmatrix} 0.7 & 0.2 \\ 0.3 & 0.6 \end{bmatrix}$$

8.2   Write a matrix of size 3×3 that is a transfer matrix with two of its entries being 0.

8.3   Suppose that an eigenvector of a transfer matrix is

$$\begin{bmatrix} 20 \\ 10 \end{bmatrix}.$$

Write the normalized eigenvector for this matrix.

8.4   Suppose that an eigenvector of a transfer matrix is

$$\begin{bmatrix} 0.2 \\ 0.3 \end{bmatrix}.$$

Write the normalized eigenvector for this matrix.

8.5   Suppose that the normalized eigenvector of a transfer matrix in a population model with 2 classes is

$$\begin{bmatrix} 0.4 \\ 0.6 \end{bmatrix}.$$

Explain the meaning of that vector in terms of the equilibrium structure of the population after a long time.

For each transfer matrix in Exercises 8.6 and 8.7, find the normalized eigenvector that would describe the equilibrium structure of the system (using that matrix).

8.6 $\begin{bmatrix} 0.7 & 0.9 \\ 0.3 & 0.1 \end{bmatrix}$

8.7 $\begin{bmatrix} 0.7 & 0.4 \\ 0.3 & 0.6 \end{bmatrix}$

8.8   Suppose that you are modeling a nonfatal infectious disease. You assume the people within the population that you are modeling are either susceptible to infection or infected. The following flow diagram shows the rates at which individuals flow from one category to the other.

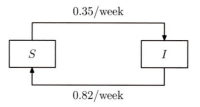

We formed a transfer matrix for this system in Exercise 6.5a.

(a) Find the normalized eigenvector that describes the system's equilibrium structure.
(b) Suppose that the rate of infection increases from 0.35/week to 0.50/week. How does this change the equilibrium structure?
(c) Suppose that the rate of infection decreases from 0.35/week to 0.25/week. How does this change the equilibrium structure?

8.9   (From [16]) Strontium-90 is deposited into pastureland by rainfall. To study how this material is cycled through the ecosystem, we divide the system into four compartments and consider how much Strontium-90 is in each compartment: grasses (G), soil (S), streams (W), and dead organic matter (O). Suppose that the time step is 1 month. A flow diagram displaying the transfer rates of Strontium-90 between compartments is shown below.

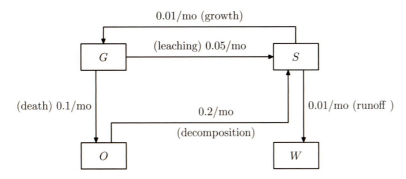

We formed a transfer matrix for this system in Exercise 6.5c.

(a) Find the normalized eigenvector that describes the system's equilibrium structure.
(b) Using Matlab, find approximately how many months pass before the system reaches its equilibrium structure if 100% of the strontium-90 starts in the grass.
(c) Using Matlab, find approximately how many months pass before the system reaches its equilibrium structure if 100% of the strontium-90 starts in the streams.
(d) Using Matlab, find approximately how many months pass before the system reaches its equilibrium structure if 100% of the strontium-90 starts in the dead organic matter.

8.10 (From [16]) Radioisotopes (such as phosphorus-32 and carbon-14) have been used to study the transfer of nutrients in food chains. The flow diagram below shows a compartmental representation of a simple aquatic food chain with phytoplankton (P), zooplankton (Z), and water (W) and the transfer rates of nutrients between these compartments.

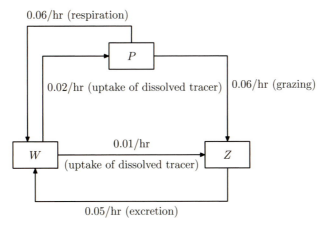

One hundred units (e.g., microcuries) of tracer are dissolved in the water of an aquarium containing a species of phytoplankton and a species of zooplankton.

(a) Construct the transfer matrix that represents the above flow diagram.
(b) Predict the state of the system over the next 6 hours.
(c) Find the normalized eigenvector that describes the system's equilibrium structure.

Leslie Matrix Models and Eigenvalues

The models presented and discussed in Chapters 6, 7, and 8 are limited by the assumption that nothing is created or destroyed. Each area on a landscape may change from one class to another, and each individual in a population changes only from one disease status to another. We did not consider the case of a population in which new individuals are born or individuals die. So in these models, each component (area or individuals) changes only in state. In the landscape succession examples, we never created or destroyed land, but each area on the landscape could change from one state to another. In the disease model, there were no births or deaths, but individuals could move between infected and susceptible states. Much of biology deals with systems in which the entities (cells and individuals) can be born and can die. In this chapter, we discuss models that take into account births and deaths and develop methods to see what happens to the structure of the biological system in this case.

Intuitively, unless we somehow can match up the number of births and deaths regularly, a population will either grow or decline. This should happen no matter how we structure the population, unless there are different birth rates and death rates for the different classes or groups in the population. You may have heard of zero-population growth as arising when births and deaths are exactly matched, but this is difficult to achieve for mammal populations in which there is age structure and the birth rates and death rates vary with age.

The matrix methods we develop in this chapter allow us to project the future changes in a population that is structured, typically by age but more generally by what we call their stage, and to account for the differences in births and mortality for different ages. This is the basis for demographic analysis in humans (and in wildlife populations) that provides an overall growth rate of the population and allows us to project the age structure of the population in a way similar to how we used eigenvalues previously to determine landscape structure. It provides a means to determine whether a wildlife population can be harvested without leading to major declines in the population and guidance as to the magnitude of allowable harvest.

9.1 Leslie Matrix Models

Suppose that we are studying a population of locusts and want to know how their population changes over time. Locust have three stages in their life cycle: (1) egg, (2) nymph or hopper, and (3) adult. Not all locust eggs will survive to become adults. Some will die while they are still eggs. Others will die while they are hoppers. Locusts are able to reproduce (lay eggs) only during the adult stage of their life.

Suppose that in the particular population of locusts we are studying, each year 2% of the eggs survive to become hoppers, 5% of the hoppers survive to become adults, and the adults die soon after they reproduce (i.e., they do not survive to the next year). Additionally, the average female adult locust will produce 1000 eggs before she dies.

Since it is the females that reproduce, we model only the females of the population. Let $x_1(t)$ be the number of female eggs in the locust population at time step t (measured in years), let $x_2(t)$ be the number of female hoppers in the locust population at time step t, and let $x_3(t)$ be the number of female adults in the locust population at time step t. We can write a system of equations to represent how this population changes each year:

$$x_1(t+1) = 1000x_3(t). \tag{9.1}$$
$$x_2(t+1) = 0.02x_1(t). \tag{9.2}$$
$$x_3(t+1) = 0.05x_2(t). \tag{9.3}$$

Equation (9.1) shows that each adult at time step t, that is, $x_3(t)$, produces 1000 eggs, resulting in $1000x_3(t)$ eggs at time step $t+1$. Equation (9.2) shows that 2% of the eggs at time step t, that is, $x_1(t)$, move into the hopper class by time step $t+1$, and Equation (9.3) shows that 5% of the hoppers at time step t, that is, $x_2(t)$, move into the adult class by time step $t+1$.

Can we rewrite our system of Equations (9.1) to (9.3) as a matrix equation? If we rewrite the system of Equations (9.1) to (9.3) as

$$x_1(t+1) = 0x_1(t) + 0x_2(t) + 1000x_3(t),$$
$$x_2(t+1) = 0.02x_1(t) + 0x_2(t) + 0x_3(t), \text{ and}$$
$$x_3(t+1) = 0x_1(t) + 0.05x_2(t) + 0x_3(t),$$

it should become clear that, yes, we can write the system of equations as the matrix equation.

$$\begin{bmatrix} x_1(t+1) \\ x_2(t+1) \\ x_3(t+1) \end{bmatrix} = \begin{bmatrix} 0 & 0 & 1000 \\ 0.02 & 0 & 0 \\ 0 & 0.05 & 0 \end{bmatrix} \begin{bmatrix} x_1(t) \\ x_2(t) \\ x_3(t) \end{bmatrix}$$
$$\mathbf{x}(t+1) = \begin{bmatrix} 0 & 0 & 1000 \\ 0.02 & 0 & 0 \\ 0 & 0.05 & 0 \end{bmatrix} \mathbf{x}(t). \tag{9.4}$$

Now, as in the previous chapters, we are able to find how the population changes over time.

Example 9.1 (Six Years of Locusts)

How does the population of locusts change over the course of 6 years if

(a) there are only 50 adults at the initial time (no eggs and no hoppers)?
(b) there are 50 eggs, 100 hoppers, and 50 adults at the initial time?

__Solution:__ Let

$$\mathbf{A} = \begin{bmatrix} 0 & 0 & 1000 \\ 0.02 & 0 & 0 \\ 0 & 0.05 & 0 \end{bmatrix}.$$

In both cases, we will use the following to find the population structure for each of the 6 years:

$$\mathbf{x}(1) = \mathbf{A}\mathbf{x}(0)$$

$$\mathbf{x}(2) = \mathbf{A}^2\mathbf{x}(0)$$

$$\mathbf{x}(3) = \mathbf{A}^3\mathbf{x}(0)$$

$$\mathbf{x}(4) = \mathbf{A}^4\mathbf{x}(0)$$

$$\mathbf{x}(5) = \mathbf{A}^5\mathbf{x}(0)$$

$$\mathbf{x}(6) = \mathbf{A}^6\mathbf{x}(0).$$

(a) If we start with

$$\mathbf{x}(0) = \begin{bmatrix} 0 \\ 0 \\ 50 \end{bmatrix},$$

then

t	Eggs	Hoppers	Adults
0	0	0	50
1	50,000	0	0
2	0	1000	0
3	0	0	50
4	50,000	0	0
5	0	1000	0
6	0	0	50

(b) If we start with

$$\mathbf{x}(0) = \begin{bmatrix} 50 \\ 100 \\ 50 \end{bmatrix},$$

(*Continued*)

then

t	Eggs	Hoppers	Adults
0	50	100	50
1	50,000	1	5
2	5000	1000	0.05
3	50	100	50
4	50,000	1	5
5	5000	1000	0.05
6	50	100	50

We see that in both cases, the population cycles, but the initial condition determines the cycle.

This type of matrix population model that takes into the account births and the survival rates of each class to the next stage is known as a ***Leslie matrix model***. In general, they have the form $x(t + 1) = Ax(t)$, where

$$
A = \begin{bmatrix}
F_1 & F_2 & F_3 & \cdots & F_n \\
S_1 & 0 & 0 & \cdots & 0 \\
0 & S_2 & 0 & \cdots & 0 \\
\vdots & \vdots & \vdots & \ddots & \vdots \\
0 & 0 & 0 & S_{n-1} & S_n
\end{bmatrix},
$$

where F_i is the average number of female offspring born to an female individual per time step in class i and S_i is the fraction of individuals in class i that survive to class $i + 1$ per time step, for $i < n$. For $i = n$, S_n is the proportion of individuals in class n that survive a subsequent time period (with the same proportion surviving for each future time period). The parameter F_i is known as the ***fecundity rate***, and the parameter S_i is known as the ***survival rate***. Note that the first row gives the equation corresponding to the first stage class x_1 (like eggs or newborns). In Leslie matrix models, we assume that we census the population after reproduction and that only the females are counted. After spending time in the last stage class, individuals die if $S_n = 0$, but if $S_n > 0$ then some fraction of individuals in the last class remain in that class for the next time step. The survival and fecundity rates are assumed to be constant over time. Those rates are independent of the total population. Leslie matrix models do not account for the effects of density-dependent parameters such as overcrowding, environmental disturbances such as natural disasters and weather, seasonal changes, or the harvesting of individuals from the population.

Before we continue on to look at the long-term dynamics of a population modeled by a Leslie matrix, let us consider one more example.

Example 9.2 (American Bison [16])

The Leslie matrix for the American bison is given by

$$A = \begin{bmatrix} 0 & 0 & 0.42 \\ 0.60 & 0 & 0 \\ 0 & 0.75 & 0.95 \end{bmatrix}.$$

The population is divided into calves, yearlings, and adults (age 2 or more). Thus, 95% of the females that reach the age of 2 years survive an additional year and reproduce with the same regularity. If we start a herd with 100 adult females (and an appropriate number of adult males), what will the herd population structure look like for the subsequent 5 years?

Solution: If we start a herd with 100 adult females, then next year's herd structure is given by

$$\begin{bmatrix} 0 & 0 & 0.42 \\ 0.60 & 0 & 0 \\ 0 & 0.75 & 0.95 \end{bmatrix} \begin{bmatrix} 0 \\ 0 \\ 100 \end{bmatrix} = \begin{bmatrix} 42 \\ 0 \\ 95 \end{bmatrix}.$$

Multiplying the prior year's stage structure vector by **A** will give population predictions for subsequent years, which we show in the table below.

Year	Calves	Yearlings	Adults	Total
0	0	0	100	100
1	42	0	95	137
2	40	25	90	155
3	38	24	104	166
4	44	23	117	184
5	49	26	128	203

Notice that the total female population size is growing each year.

9.2 Long-Term Growth Rate (Eigenvalues)

Given a Leslie matrix, we would like to find a stable stage distribution. By this is meant a distribution so that the population remains at that distribution once there. In the case of the locust population, this would mean that the fractions of the population that were eggs, hoppers, and adults would stay the same through time. We saw in Example 9.2 that a Leslie matrix could model a growing population; thus, we do not necessarily want to find vectors $\mathbf{x}(t)$ such that $\mathbf{x}(t + 1) = \mathbf{x}(t)$ as we did in Chapter 8. What we want to do is find a distribution among the stage classes so that the proportion in each class does not change from one time period to the next, though the overall "number" in each class could change. If the overall population increased by a factor of λ, then each class would have increased by a factor of λ if the population were at a stable stage structure. The population will have this property if it satisfies the equation

$$\mathbf{x}(t + 1) = \lambda \mathbf{x}(t), \tag{9.5}$$

where the value λ is known as an ***eigenvalue***.

In Equation 9.5, it is important that this equation holds no matter what time t is used if the population distribution $\mathbf{x}(t)$ is at a stable stage distribution. If $\lambda > 1$, then the population is growing over time when it is at a stable stage distribution. If $\lambda < 1$, then the population is declining over time. If $\lambda = 1$, then the population is remaining constant over time (this is what occurred in the transfer matrices examples). Often, the λ is referred to as the ***long-term growth rate***. This is also called the Malthusian growth rate, after Thomas Malthus, whose treatise, originally published in 1798, *An Essay on the Principle of Population*, pointed out clearly the capacity for populations to increase exponentially.

When population growth is discussed, just as when we talk about the growth of money in a bank account, it is common to state that a population is growing at a rate of 5%, for example. This means that each time step, the population is 5% larger than it was previously, which would correspond to $\lambda = 1 + 0.05 = 1.05$. This is identical to the case of having a bank account with a 5% annual interest rate compounded annually: you would have \$105 at the end of 1 year if you had \$100 in the account at the start of the year. As another example, if a population were declining by 8% per time step, the growth rate per time step would be $\lambda = 1 - 0.08 = 0.92$.

In general, the change in the growth (or decay) rate per time step is

$$\lambda - 1.$$

If $0 < \lambda < 1$, then the population is declining, and the change in the rate per time step is

$$\lambda - 1 < 0.$$

If $\lambda = 0.8$, then
$$\lambda - 1 = 0.8 - 1 = -0.2,$$

and we say that the population is declining by 20% per time step.

If $\lambda > 1$, then the population is increasing, and the change in the growth rate per time step is

$$\lambda - 1.$$

If $\lambda = 1.3$, then
$$\lambda - 1 = 1.3 - 1 = 0.3,$$

and we say that the population is growing by 30% per time step.

All of this discussion of the long-term growth rate applies only when the population is at a stable stage distribution. It may take a number of years or time steps for a population to get close to the stable stage distribution; during this period, the population will typically not be changing by a factor of λ. How long it takes a population to reach a stable stage distribution and its long-term growth rate is one of many questions that mathematical models such as the Leslie matrix model address in the field of demography, which is the study of how a population stage distribution changes (and, in the case of many mammals, such as humans, the important stage distribution is based on age).

If we combine Equation (9.5) with $\mathbf{x}(t + 1) = \mathbf{A}\mathbf{x}(t)$, where $\mathbf{A}$ is our Leslie matrix, then we get

$$\mathbf{A}\hat{\mathbf{x}} = \lambda\hat{\mathbf{x}}, \tag{9.6}$$

where $\hat{\mathbf{x}}$ is the population vector after some period of time. If we can solve Equation (9.6) for λ, then we can determine how our population is growing or declining. This is rather simple for a

2×2 Leslie matrix. We seek to solve Equation (9.6) for λ and the corresponding nonzero vector

$$\hat{\mathbf{x}} = \begin{bmatrix} x_1 \\ x_2 \end{bmatrix}$$

with either or both of x_1 and x_2 not being zero.

Suppose that our Leslie matrix is

$$\mathbf{A} = \begin{bmatrix} a & b \\ c & d \end{bmatrix}.$$

Equation (9.6) can be rewritten as $(\mathbf{A} - \lambda \mathbf{I})\,\hat{\mathbf{x}} = 0$, or

$$\begin{bmatrix} a - \lambda & b \\ c & d - \lambda \end{bmatrix} \begin{bmatrix} x_1 \\ x_2 \end{bmatrix} = \begin{bmatrix} 0 \\ 0 \end{bmatrix}.$$

We can rewrite this matrix equation as a system of equations:

$$(a - \lambda)x_1 + bx_2 = 0. \tag{9.7}$$

$$cx_1 + (d - \lambda)x_2 = 0. \tag{9.8}$$

Multiply Equation (9.7) by c and Equation (9.8) by $a - \lambda$ to get

$$c(a - \lambda)x_1 + cbx_2 = 0. \tag{9.9}$$

$$c(a - \lambda)x_1 + (a - \lambda)(d - \lambda)x_2 = 0. \tag{9.10}$$

Subtract Equation (9.9) from Equation (9.10) to obtain

$$\big[(a - \lambda)(d - \lambda) - bc\big]x_2 = 0,$$

which implies that

$$x_2 = 0 \quad \text{or} \quad (a - \lambda)(d - \lambda) - bc = 0. \tag{9.11}$$

If $x_2 = 0$, then we know that $x_1 \neq 0$, and Equations (9.7) and (9.8) give

$$(a - \lambda)x_1 = cx_1 = 0,$$

and thus

$$\lambda = a \quad \text{and} \quad c = 0.$$

If $x_2 \neq 0$, then by Equation (9.11),

$$(a - \lambda)(d - \lambda) - bc = 0$$

$$\lambda^2 - (a + d)\lambda + (ad - bc) = 0. \tag{9.12}$$

Equation (9.12) is known as the ***characteristic equation*** of the 2×2 matrix $\mathbf{A}$. Solving (9.12) via the quadratic formula yields two solutions:

$$\lambda_1 = \frac{(a + d) + \sqrt{(a + d)^2 - 4(ad - bc)}}{2}, \quad \text{and}$$

$$\lambda_2 = \frac{(a + d) - \sqrt{(a + d)^2 - 4(ad - bc)}}{2}.$$

For any matrix, the eigenvalue with the largest absolute value is known as the ***dominant eigenvalue***. For our matrix models, the dominant eigenvalues corresponds to the long-term growth rates of the systems we are modeling. A Leslie matrix has a unique positive eigenvalue, and its corresponding eigenvector has all positive entries. For a Leslie matrix, this positive eigenvalue is the dominant eigenvalue [12].

Example 9.3 (Finding an Eigenvalue)

Consider the 2×2 Leslie matrix

$$\mathbf{A} = \begin{bmatrix} 1 & 4 \\ 0.5 & 0 \end{bmatrix}.$$

Find the long-term growth rate of the population that this Leslie matrix is modeling.

Solution: To find the long-term growth rate of this population, we find the dominant eigenvalue. First, from the characteristic equation of $\mathbf{A}$, we have

$$\lambda^2 - (1 + 0)\lambda + (1 \cdot 0 - 0.5 \cdot 4) = 0$$

$$\lambda^2 - \lambda - 2 = 0$$

$$(\lambda - 2)(\lambda + 1) = 0$$

$$\lambda = 2, \quad \lambda = -1.$$

The eigenvalue with the largest absolute value is $\lambda = 2$. The long-term growth rate is 2. This means for large times, our population is doubling every time step, that is eventually, $\mathbf{x}(t + 1) = 2\mathbf{x}(t)$. If at time $t = 100$, our population has reached its long-term growth with a population size of 1000 individuals, then at time $t = 101$, we will have a total of 2000 individuals. The change in the growth rate per time step of our population (after a long time) is $\lambda - 1 = 2 - 1 = 1$. This population is growing per time step by 100%.

It turns out there is another method for solving $(\mathbf{P} - \lambda \mathbf{I})\hat{\mathbf{x}} = 0$, where $\mathbf{P}$ is a general 2×2 matrix, but we need a new mathematical concept called the ***determinant***. The determinant of the 2×2 matrix

$$\mathbf{P} = \begin{bmatrix} a & b \\ c & d \end{bmatrix}$$

is

$$\det(\mathbf{P}) = \begin{vmatrix} a & b \\ c & d \end{vmatrix} = ad - cb.$$

Thus, we could also derive the characteristic equation of the 2×2 matrix $\mathbf{P}$ by solving $(\mathbf{P} - \lambda \mathbf{I}) = 0$ for λ:

$$\det(\mathbf{P} - \lambda \mathbf{I}) = 0$$

$$\begin{vmatrix} a - \lambda & b \\ c & d - \lambda \end{vmatrix} = 0$$

$$(a - \lambda)(d - \lambda) - bc = 0$$

$$ad - a\lambda - d\lambda + \lambda^2 - bc = 0$$

$$\lambda^2 - (a + d)\lambda + (ad - bc) = 0.$$

The straight vertical lines around the matrix elements above are the notation for the determinant of the matrix. The reason we introduce this second method is that it is easily expandable to larger matrices. For square matrices larger than 2×2, we use a method known as "expansion by minors." For this method, we pick a row along which we will expand. Here, we will use the example of expanding along the first row of a 3×3 matrix

$$\mathbf{A} = \begin{bmatrix} a & b & c \\ d & e & f \\ g & h & i \end{bmatrix}.$$

First, we take the first entry in the first row and multiply it by the determinant of the matrix that remains, called a minor, when we cross out the first row and the first column. Next, multiply -1 by the second entry in the first row by the determinant of the matrix that remains when we cross out the first row and the second column. Now, multiply the third entry in the first row by the determinant of the matrix that remains when we cross out the first row and third column. Finally, to get the determinant of the matrix $\mathbf{A}$, we sum each of the products together. Thus,

$$\det(\mathbf{A}) = a \begin{vmatrix} e & f \\ h & i \end{vmatrix} - b \begin{vmatrix} d & f \\ g & i \end{vmatrix} + c \begin{vmatrix} d & e \\ g & h \end{vmatrix}.$$

Notice that the definition of the determinant of a 3×3 matrix depends on the determinants of three 2×2 matrices. Likewise, the definition of the determinant of a 4×4 matrix is similarly defined and depends on the determinants of four 3×3 matrices.

Example 9.4 (Eigenvalue of a Transfer Matrix)

Verify that the dominant eigvenvalue of the transfer matrix $\mathbf{A}$ is 1, where

$$\mathbf{A} = \begin{bmatrix} 0.95 & 0 & 0 \\ 0.05 & 0.88 & 0 \\ 0 & 0.12 & 1 \end{bmatrix}.$$

Solution: To find the eigenvalues, we need to solve $\det(\mathbf{A} - \lambda \mathbf{I}) = 0$ for λ. Since $\mathbf{A}$ is a 3×3 matrix, we will use the expansion of minors technique:

$$0 = \det(\mathbf{A} - \lambda \mathbf{I})$$

$$0 = \begin{vmatrix} 0.95 - \lambda & 0 & 0 \\ 0.05 & 0.88 - \lambda & 0 \\ 0 & 0.12 & 1 - \lambda \end{vmatrix}$$

$$0 = (0.95 - \lambda) \begin{vmatrix} 0.88 - \lambda & 0 \\ 0.12 & 1 - \lambda \end{vmatrix} - 0 \begin{vmatrix} 0.05 & 0 \\ 0 & 1 - \lambda \end{vmatrix} + 0 \begin{vmatrix} 0.05 & 0.88 - \lambda \\ 0 & 0.12 \end{vmatrix}$$

$$0 = (0.95 - \lambda) [(0.88 - \lambda)(1 - \lambda) - 0] - 0 + 0$$

$$0 = (0.95 - \lambda)(0.88 - \lambda)(1 - \lambda)$$

$$\lambda = \{0.95, 0.88, 1\}.$$

The eigenvalue of the largest absolute value is $\lambda = 1$. Thus, the dominant eigenvalue of the transfer matrix $\mathbf{A}$ is 1.

Example 9.5 (Eigenvalues of a Leslie Matrix)

Find the long-term growth rate of the Leslie matrix

$$A = \begin{bmatrix} 0 & 4 & 3 \\ 0.5 & 0 & 0 \\ 0 & 0.25 & 0 \end{bmatrix}.$$

Solution: To find the eigenvalues, we need to solve $\det(\mathbf{A} - \lambda \mathbf{I}) = 0$ for λ. Since $\mathbf{A}$ is a 3×3 matrix, we will use the expansion of minors technique:

$$0 = \det(\mathbf{A} - \lambda \mathbf{I})$$

$$0 = \begin{vmatrix} -\lambda & 4 & 3 \\ 0.5 & -\lambda & 0 \\ 0 & 0.25 & -\lambda \end{vmatrix}$$

$$0 = -\lambda \begin{vmatrix} -\lambda & 0 \\ 0.25 & -\lambda \end{vmatrix} - 4 \begin{vmatrix} 0.5 & 0 \\ 0 & -\lambda \end{vmatrix} + 3 \begin{vmatrix} 0.5 & -\lambda \\ 0 & 0.25 \end{vmatrix}$$

$$0 = -\lambda(\lambda^2 - 0) - 4(-0.5\lambda - 0) + 3\left((0.5)(0.25) - 0\right)$$

$$0 = -\lambda^3 + 2\lambda + \frac{3}{8}$$

$$0 = \lambda^3 - 2\lambda - \frac{3}{8}$$

$$0 = \left(\lambda - \frac{3}{2}\right)\left(\lambda^2 + \frac{3}{2}\lambda + \frac{1}{4}\right)$$

$$0 = \left(\lambda - \frac{3}{2}\right)\left(\lambda - \frac{3 - \sqrt{5}}{4}\right)\left(\lambda - \frac{3 + \sqrt{5}}{4}\right)$$

$$\lambda = \left\{\frac{3}{2}, \frac{-3 + \sqrt{5}}{4}, \frac{-3 - \sqrt{5}}{4}\right\}.$$

The eigenvalue with the largest absolute value is $\lambda = \frac{3}{2}$. The long-term growth rate is $\frac{3}{2}$. Thus, for large times, $\mathbf{x}(t + 1) = \frac{3}{2}\mathbf{x}(t)$. The change in the growth rate per time step of the population (after a long time) is $\lambda - 1 = \frac{3}{2} - 1 = \frac{1}{2}$; that is, the population is growing by 50% per time step.

From Example 9.5, we can see that solving the cubic equations resulting from the expansion of minors technique for the determinant of a 3×3 matrix is a bit more involved than solving the quadratic equation resulting from the determinant of a 2×2 matrix. In general, we will use Matlab to find eigenvalues of square matrices larger than 2×2 (see the "Matlab Skills" section at the end of this chapter for details).

In Examples (9.3), (9.4), and (9.5), all the roots of the polynomials derived from $\det(\mathbf{A} - \lambda\mathbf{I}) = 0$ were real numbers. However, it is possible to have complex roots to these polynomials. When there are complex roots, the resulting population structure will exhibit oscillations. Consider Example 9.1, where we saw a locust population oscillating every 3 years.

Example 9.6 (Eigenvalues of an Oscillating Population)

Recall the Leslie matrix of a locust population presented in Example 9.1:

$$A = \begin{bmatrix} 0 & 0 & 1000 \\ 0.02 & 0 & 0 \\ 0 & 0.05 & 0 \end{bmatrix}.$$

Find the eigenvalues for this Leslie matrix.

Solution: To find the eigenvalues, we need to solve the equation $\det(\mathbf{A} - \lambda\mathbf{I}) = 0$ for λ. Since $\mathbf{A}$ is a 3×3 matrix, we will use the expansion of minors technique:

$$0 = \det(\mathbf{A} - \lambda\mathbf{I})$$

$$0 = \begin{vmatrix} -\lambda & 0 & 1000 \\ 0.02 & -\lambda & 0 \\ 0 & 0.05 & -\lambda \end{vmatrix}$$

$$0 = -\lambda \begin{vmatrix} -\lambda & 0 \\ 0.05 & -\lambda \end{vmatrix} - 0 \begin{vmatrix} 0.02 & 0 \\ 0 & -\lambda \end{vmatrix} + 1000 \begin{vmatrix} 0.02 & -\lambda \\ 0 & 0.05 \end{vmatrix}$$

$$0 = -\lambda(\lambda^2 - 0) - 0 + 1000(0.001 - 0)$$

$$0 = -\lambda^3 + 1$$

$$0 = \lambda^3 - 1$$

$$0 = (\lambda - 1)(\lambda^2 + \lambda + 1)$$

$$\lambda = \left\{ 1, \frac{-1 + i\sqrt{3}}{2}, \frac{-1 - i\sqrt{3}}{2} \right\}.$$

Which eigenvalue is dominant? Let $\lambda_1 = 1$, $\lambda_2 = \frac{-1+i\sqrt{3}}{2} = -\frac{1}{2} + i\frac{\sqrt{3}}{2}$, and $\lambda_3 = \frac{-1-i\sqrt{3}}{2} = -\frac{1}{2} - i\frac{\sqrt{3}}{2}$. Then

$$|\lambda_1| = 1$$

$$|\lambda_2| = \sqrt{\left(-\frac{1}{2}\right)^2 + \left(\frac{\sqrt{3}}{2}\right)^2} = \sqrt{\frac{1}{4} + \frac{3}{4}} = 1$$

$$|\lambda_3| = \sqrt{\left(-\frac{1}{2}\right)^2 + \left(\frac{\sqrt{3}}{2}\right)^2} = \sqrt{\frac{1}{4} + \frac{3}{4}} = 1$$

(Continued)

In this case, where all the eigenvalues have the same absolute value, the dominant eigenvalue is the positive real eigenvalue, $\lambda = 1$. However, because the complex eigenvalues have the same absolute value as the dominant eigenvalue, the population structure never attains an equilibrium proportion structure. Instead, the population will oscillate, as we saw in Example 9.1.

9.3 Long-Term Population Structure (Corresponding Eigenvectors)

Now that we have determined how to calculate the long-term growth rate of a population modeled by a Leslie matrix, we would like to determine the long-term population structure, that is, the proportion of the population that is in each class. We do this by finding the eigenvector that corresponds with the dominant eigenvalue of the matrix.

Again, we return to Equation (9.6):

$$\mathbf{A}\hat{\mathbf{x}} = \lambda\hat{\mathbf{x}}.$$

Once we know the value of the dominant eigenvalue, we plug that in for λ and solve the system of equations corresponding to the equation $\mathbf{A}\hat{\mathbf{x}} = \lambda\hat{\mathbf{x}}$. Remember that for our eigenvector to represent the proportions of the population, we normalize it so that the sum of its components is 1.

Example 9.7 (Finding an Eigenvector)

Recall the Leslie matrix from Example 9.3:

$$\mathbf{A} = \begin{bmatrix} 1 & 4 \\ 0.5 & 0 \end{bmatrix}.$$

Find the long-term population structure for this population represented by the Leslie matrix **A**.

Solution: In Example 9.3, we found that the dominant eigenvalue of **A** is 2. To find the long-term population structure, we must find the eigenvector associated with the dominant eigenvalue. We do this by solving

$$\mathbf{A}\hat{\mathbf{x}} = 2\hat{\mathbf{x}}.$$

The corresponding system of equations is

$$1x_1 + 4x_2 = 2x_1. \tag{9.13}$$
$$0.5x_1 + 0x_2 = 2x_2. \tag{9.14}$$

(Continued)

We can solve either Equation (9.13) or Equation (9.14) for x_1 in terms of x_2 to get the relationship

$$x_1 = 4x_2.$$

Let us choose $x_2 = 1$; then $x_1 = 4$. Thus, an eigenvector is

$$\begin{bmatrix} 4 \\ 1 \end{bmatrix}.$$

We normalize this eigenvector to get the long-term population structure:

$$\begin{bmatrix} 4/5 \\ 1/5 \end{bmatrix}.$$

Thus, for large times, we can expect to find 80% of the population in the first class and 20% of the population in the second class.

Recall from Chapter 8 how we found the normalized eigenvector of the transfer matrix

$$\mathbf{T} = \begin{bmatrix} 0.9 & 0.2 \\ 0.1 & 0.8 \end{bmatrix},$$

which described the spread of a nonfatal disease. We solved the system of equations that came from $\hat{\mathbf{x}} = \mathbf{T}\hat{\mathbf{x}}$. If we calculate the eigenvalues of $\mathbf{T}$,

$$0 = \lambda^2 - (0.9 + 0.8)\lambda + (0.9 \cdot 0.8 - 0.2 \cdot 0.1)$$
$$0 = \lambda^2 - 1.7\lambda + 0.7$$
$$0 = (\lambda - 1)(\lambda - 0.7)$$
$$\lambda = \{1, 0.7\},$$

we find that the dominant eigenvalue for $\mathbf{T}$ is $\lambda = 1$. In fact, for all transfer matrices (matrices in which the values in any column sum to 1), the dominant eigenvalue is 1. Thus, when we were calculating the normalized eigenvector, we were actually using the equation $\lambda\hat{\mathbf{x}} = \mathbf{T}\hat{\mathbf{x}}$, where λ is the dominant eigenvalue.

Note that we are only introducing the ideas of eigenvalues and eigenvectors, which can be explored with careful mathematical analysis for more general matrices. Matrix models giving oscillations in the populations have been mentioned only briefly. The properties of matrices with negative eigenvalues and corresponding eigenvectors have not been explored here. The eigenvectors for the dominant eigenvalues for the population models in this unit have positive components. Note that an eigenvector corresponding to the eigenvalue $\lambda = -1$ in Example 9.3 is

$$\begin{bmatrix} -0.5 \\ 1 \end{bmatrix}.$$

9.4 Matlab Skills

In Matlab, the `eig` function calculates both the eigenvalues and their associated eigenvectors. If we have defined a matrix A, then the command

```
eig(A)
```

will produce a column vector of eigenvalues. If A is a 3×3 matrix, then the column vector will be 3×1. To calculate the eigenvectors as well, we need to tell Matlab that we need the second output from the `eig` function. The command

```
[v,lambda] = eig(A)
```

will output two matrices: the second, named `lambda`, contains the eigenvalues along the diagonal and the first, named `v`, contains the eigenvectors that correspond to each of the eigenvalues. The first column of `v` contains the eigenvector that corresponds with the eigenvalue shown in `lambda(1,1)`. The second column of `v` contains the eigenvector that corresponds with the eigenvalue shown in `lambda(2,2)` and so on with the other columns of `v`.

Thus, if we wanted to find the eigenvalues and corresponding eigenvectors of

$$\mathbf{A} = \begin{bmatrix} 1 & 4 \\ 0.5 & 0 \end{bmatrix},$$

we could use the following commands in the command window:

```
──────────── Command Window ────────────
>> A = [1 4; 0.5 0]
A =
     1.0000     4.0000
     0.5000          0

>> eig(A)
ans =
     2
    -1

>> [v,lambda] = eig(A)
v =
     0.9701    -0.8944
     0.2425     0.4472

lambda =
     2      0
     0     -1
```

We see that the eigenvalues are $\lambda = \{2, -1\}$, matching what we found in Example 9.3. The dominant eigenvalue is $\lambda = 2$, which is shown in `lambda(1,1)`. Note that Matlab always displays the dominant eigenvalue in the first row, first column. We look to the first column of `v` to get the eigenvector corresponding to the dominant eigenvalue. Matlab tells us that this is

$$\begin{bmatrix} 0.9701 \\ 0.2425 \end{bmatrix}.$$

You might notice that this eigenvector does not match the one we found in Example 9.7. However, recall that eigenvectors are not unique; rather, only the normalized eigenvector is unique. Let us normalize the eigenvector corresponding to the dominant eigenvalue. To do this in Matlab, we divide the eigenvector by the sum of the entries in the eigenvector, which we can obtain by using the function sum:

```
─────────────────────── Command Window ───────────────────────
>> v(:,1)/sum(v(:,1))
ans =
    0.8000
    0.2000
```

Now, this normalized eigenvector matches the normalized eigenvector for the dominant eigenvalue we found in Example 9.7.

Now that we know how to find eigenvalues and normalized eigenvectors, we would like to see how eigenvectors change when we modify values in the transfer or Leslie matrices. For this, we will make use of loops.

Suppose that we know that the Leslie matrix

$$\mathbf{A} = \begin{bmatrix} 0 & 2 & 3 \\ 0.5 & 0 & 0 \\ 0 & 0.25 & 0 \end{bmatrix}$$

models a population that contains three age classes: hatchlings, juveniles, and adults. Suppose that we wanted to explore how changing the juvenile fecundity changes the long-term population structure (i.e., the eigenvector associated with the dominant eigenvalue). We can construct a *for* loop, to loop through the juvenile fecundity values we want to evaluate and then display the results in a table or graph. Let us look at the following set of juvenile fecundity values:

$$a_{1,2} = \{0.1, 0.2, 0.5, 1.5, 2, 2.5, 3, 4\}.$$

Below is an m-file that finds the dominant eigenvalue and corresponding normalized eigenvector for **A** given each juvenile fecundity value. The m-file constructs a table of results and plots the long-term population structure against the juvenile fecundity:

```
─────────────────────── LeslieFecundity.m ───────────────────────
1    % Filename: LeslieFecundity.m
2    % M-file to
3    %        - Find the dominant eigenvalue and corresponding
4    %          eigenvector for an array of juvenile fecundities
5    %        - Print out a table of the results
6    %        - Print out results graphically
7
8    % Print out the header for the table
9    fprintf('Juvenile    Dominant          Proportion Structure    \n')
10   fprintf('Fecundity Eigenvalue Hatchlings  Juveniles  Adults\n')
11   fprintf('------------------------------------------------------\n')
12
13   % Array of juvenile fecudity values
14   f = [0.1 0.2 0.5 1.5 2 2.5 3 4];
15
16   % Start the loop
17   for i = 1:length(f) % loop thru length of f vector
18
```

```
19    % Construct Leslie matrix with appropriate juvenile fecundity
20    A = [   0    f(i) 3 ;
21          0.5     0  0 ;
22            0   0.25  0 ];
23
24    % Get eigenvalues and eigenvectors
25    [v,lambda] = eig(A);
26
27    % Find dominant eigenvalue in lambda
28    % and make note of position in lambda matrix
29    lambdavector = max(lambda); % array of eigenvalues
30    for j = 1:length(lambdavector)
31      if max(max(abs(lambda))) == abs(lambdavector(j))
32        loc = j; % position of dom eig in lambda matrix
33        deig = lambda(j,j); % dominant eigenvalue
34      end
35    end
36
37    % Normalize eigenvector associated with dominant eigenvalue
38    normv(:,i) = v(:,loc)/sum(v(:,loc));
39    % This creates a matrix called normv in which the
40    % i^th column corresponds to the dominant eigenvector
41    % associated with the i^th juvenile fecundity value
42
43    % Print these values
44    fprintf('   %3.1f   ',f(i)) % juvenile fecundity value
45    fprintf('     %5.3f   ',deig) % dominant eigenvalue
46    fprintf('   %5.3f   ',normv(1,i)) % hatchling proprotion at eq
47    fprintf('      %5.3f   ',normv(2,i)) % juvenile proprotion at eq
48    fprintf('    %5.3f   ',normv(3,i)) % adult proprotion at eq
49    fprintf('\n') % new line
50  end
51
52  % Generate graph
53  h = normv(1,:);  % vector of hatchling proportions at eq
54  j = normv(2,:);  % vector of juvenile proportions at eq
55  a = normv(3,:);  % vector of adult proprotions at eq
56  plot(f,h,'r.-',f,j,'g.-',f,a,'b.-')
57  legend('hatchlings','juveniles','adults')
58  xlabel('Juvenile Fecundity')
59  ylabel('Equilibrium Structure')
```

When this m-file is run in the command window, the output looks like the following:

```
——————————————— Command Window ———————————————
>> LeslieFecundity
   Juvenile    Dominant       Equilibrium Structure
   Fecundity  Eigenvalue  Hatchlings  Juveniles  Adults
   -----------------------------------------------------
      0.1        0.744       0.527       0.354     0.119
      0.2        0.767       0.536       0.350     0.114
      0.5        0.836       0.563       0.337     0.101
      1.5        1.052       0.630       0.299     0.071
      2.0        1.151       0.654       0.284     0.062
      2.5        1.245       0.675       0.271     0.054
      3.0        1.335       0.692       0.259     0.049
      4.0        1.500       0.720       0.240     0.040
```

The graphical output of this m-file is shown in Figure 9.1.

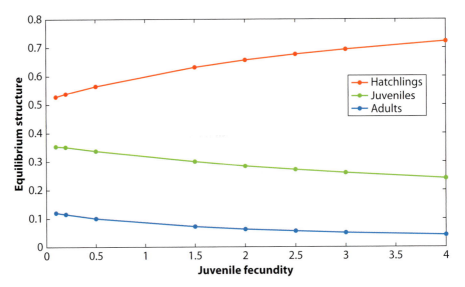

Figure 9.1 Plot generated from the m-file LeslieFecundity.m.

9.5 Exercises

For the Leslie matrices in Exercises 9.1 to 9.6,

(a) find the dominant eigenvalue,
(b) find the normalized eigenvector corresponding to the dominant eigenvalue, and
(c) interpret the values found in (a) and (b) biologically.

9.1 $\begin{bmatrix} 0.7 & 0.9 \\ 0.6 & 0 \end{bmatrix}$

9.2 $\begin{bmatrix} 0.5 & 0.9 \\ 0.4 & 0 \end{bmatrix}$

9.3 $\begin{bmatrix} 2 & 4 \\ 0.1 & 0.2 \end{bmatrix}$

9.4 $\begin{bmatrix} 0.2 & 0.7 \\ 0.7 & 0.2 \end{bmatrix}$

9.5 $\begin{bmatrix} 0.7 & 0.2 \\ 0.2 & 0.4 \end{bmatrix}$

9.6 $\begin{bmatrix} 0.1 & 1.2 \\ 0.4 & 0.3 \end{bmatrix}$

9.7 The Leslie matrix for an insect population is

$$\mathbf{A} = \begin{bmatrix} 0 & 20 & 0 \\ 0.1 & 0 & 0 \\ 0 & 0.2 & 0 \end{bmatrix}.$$

The age classes are 0 to 1 weeks, 2 to 3 weeks, and 4 to 5 weeks, and only insects 2 to 3 weeks old can reproduce. Notice that the time step here is 2 weeks.

(a) If

$$\mathbf{x}(0) = \begin{bmatrix} 0 \\ 100 \\ 0 \end{bmatrix}$$

specifies the initial age structure, project the population forward 2 months (or 4 time periods). Complete these calculations by hand.

(b) Show that $\mathbf{A}^3 = 2\mathbf{A}$ and that $\mathbf{A}^4 = 2\mathbf{A}^2$. Conclude that the population in each age class doubles every month.

(c) Find the long-term growth rate of the population. Is the population growing or declining?

9.8 Cattle on a large ranch are divided into calves, yearlings, and adults. Data indicate that 70% of the calves survive the first year to become yearlings, while 80% of the yearlings mature into adults. In addition, 90% of adults survive a given year, and an adult female produces a single calf each year. Construct the Leslie matrix, assuming that we census the population after reproduction and count only females.

9.9 A population of insects is divided into two stage categories. The females in the first category or stage produce on average 2 female offspring and have a survival rate of 25% to stage 2. The females in stage 2 produce on average 5 female offspring and die after one time period in stage 2.

(a) If the population were to exist for a long time with the above-mentioned conditions staying the same, what would the long-term growth rate of the population be per time period? If at time 150 the population of insects had reached the long-term growth rate with 1 million females present, approximately how many females would be present at time 151?

(b) the population were to exist for a long time with the above-mentioned conditions staying the same, approximately what fraction of the population would be in each of the two stages?

9.10 A population fits the Leslie matrix model with three life stages. The dominant eigenvalue is computed to be 0.3, and a corresponding eigenvector is

$$\begin{bmatrix} 3 \\ 7 \\ 2 \end{bmatrix}.$$

If at time step 400 the population has reached the long-term growth rate with 1200 individuals, how many individuals will be in each class at time step 401?

9.11 A population fits the Leslie matrix model with four life stages. The dominant eigenvalue is computed to be 1.2, and a corresponding eigenvector is

$$\begin{bmatrix} 3 \\ 7 \\ 10 \\ 2 \end{bmatrix}.$$

If at time step 500 the population has reached the long-term growth rate with 6400 individuals, how many individuals will be in each class at time step 501?

9.12 (From [29]) Suppose that the female animals in a population live 3 years. The first year, they are immature and do not reproduce. The second year, they are adolescent and reproduce at a rate of 0.8 female offspring per female individual. The last year, they are adults and produce 3.5 female offspring per female individual. Further suppose that 80% of the first-year females survive to become second-year females and that 90% of second-year females survive to become third-year females. All third-year females die. We are interested in modeling only the female portion of this population.

(a) Construct the Leslie matrix.
(b) Find the long-term growth rate for the population. Is the population growing or declining?
(c) Suppose that a population of 100 first-year females are released into a study area along with a sufficient number of males for reproductive needs. Track the female population over 10 years. Construct a table to display the population structure over the 10 years.
(d) Find the long-term structure of the population.

9.13 (From [29]) In an attempt to save the endangered northern spotted owl, the U.S. Fish and Wildlife Service imposed strict guidelines for the use of 12 million acres of Pacific Northwest forest. This decision led to a national debate between the logging industry and environmentalists. Mathematical ecologists have created a mathematical model to analyze population dynamics of the northern spotted owl by dividing the female owl population into three categories: juvenile (up to 1 year old), subadult (1 to 2 years old), and adult (over 2 years old). The female owl population can be modeled by the Leslie matrix model:

$$\begin{bmatrix} j(t+1) \\ s(t+1) \\ a(t+1) \end{bmatrix} = \begin{bmatrix} 0 & 0 & 0.33 \\ 0.18 & 0 & 0 \\ 0 & 0.71 & 0.94 \end{bmatrix} \begin{bmatrix} j(t) \\ s(t) \\ a(t) \end{bmatrix}.$$

(a) If there are currently 4000 female northern spotted owls made up of 900 juveniles, 500 subadults, and 2600 adults, use Matlab to determine the total number of female owls for each of the next 5 years. Round each answer to the nearest whole number.
(b) Using Matlab, determine the long-term growth rate of the population. What can we conclude about the long-term survival of the northern spotted owl?
(c) Notice that only 18% of the juveniles become subadults. Assuming that, through better habitat management, this number could be increased to 40%, rework part (b). What can we conclude about the long-term survival of the northern spotted owl if the habitat is improved enough to increase juvenial survival?

9.14 (From [47]) Showing that a model is correct can be difficult. Often it is much easier to show that a model is incorrect. Consider the following model that has been proposed [8] to describe the growth of redwoods (*Sequoia sempervirens*). Redwoods frequently live 1000 years or more. The model classifies redwoods into three stages: 0 to 200 years (young), 200 to 800 years (mature), and older than 800 years (old). A current census in a particular stand finds that currently there are 1696 young redwoods, 485 mature redwoods, and 82 old redwoods. The transition between stages over a 50-year period is given by the matrix

$$\begin{bmatrix} 12 & 26 & 6 \\ 0.30 & 0.92 & 0 \\ 0 & 0.18 & 0.67 \end{bmatrix}.$$

Trace this stand of trees through five time steps (five 50-year periods for a total of 250 years) and explain how we know that this model cannot possibly be correct.

Unit 2 Student Projects

Land Classification

Throughout this unit, we will be breaking down systems into classes. In this project, each student will classify the types of land found in his or her hometown. This project provides an inquiry-based introduction to the material in Chapter 6.

Instructor Tasks

1. Give the students a brief demonstration on how to use Google Earth.

Student Tasks

1. Find your hometown on Google Earth and set the altitude to 5000 ft. See if you can spot the building in which you grew up.
2. Print out the Google Earth image.
3. Classify the land into three to six different categories. In a rural area, you might use forest, farmland, housing, water, and so on. In an urban area, you might use large buildings, small buildings, greenspaces, water, and so on. Get creative!
4. Develop a scheme for deciding what proportion of your printed image (map) belongs to each category.
5. How do you think these proportions might have changed over time?

Predator-Prey Models

Recall the logistic difference equation presented in the "Matlab Skills" section in Chapter 5:

$$x_{t+1} = x_t + rx_n(1 - x_n),$$

where x_t is the density of the population at time step t and r is the intrinsic growth rate of the population. Suppose that the population modeled by this logistic difference equation is a prey species for another species and that we want to model the dynamics of both species through time. Let y_t represent the density of the predator population at time step t; then we can model the predator-prey dynamics using the discrete difference equations

$$x_{t+1} = \underbrace{x_t}_{\text{Density at time step } t} + \underbrace{rx_t(1-x_t)}_{\text{Growth term}} - \underbrace{\beta x_t y_t}_{\text{Loss due to predation}} \tag{9.6}$$

$$y_{t+1} = \underbrace{y_t - s y_t}_{\text{Exponential decay term}} + \underbrace{\alpha x_t y_t}_{\text{Growth from prey consumption}}, \tag{9.7}$$

where $0 < s < 1$ is the intrinsic decay rate, $\beta > 0$ measures the impact of predation on the prey population and has units of $\frac{1}{\text{predator density}}$, and $\alpha > 0$ measures the impact of prey consumption on the predator population and has units of $\frac{1}{\text{prey density}}$. Note that the density of the prey population in time step $t+1$ depends on both the density of the prey population at time step t and the density of the predator population at time step t. Likewise, the density of the predator population at time step $t+1$ depends on both the density of the predator population at time step t and the density of the prey population at time step t. This dependence of the variable x on the variable y and vice versa means that these two equations are "coupled" together, and we refer to them as a *system of discrete difference equations*.

1. Suppose that the predator population is not present, that is, $y_t = 0$ for all t. Show that Equation (9.16) reduces to the logistic difference equation. Use the Matlab file `LogisticDifferenceEqn.m` from Section 5.7 to simulate the prey population (in the absence of the predator) when $x_0 = 0.2$ and (a) $r = 0.2$, (b) $r = 0.8$, and (c) $r = 1.2$. What is the density of the prey population after 50 time steps in each case?

2. Suppose that the prey population is not present, that is, $x_t = 0$ for all t. Show that Equation (9.17) reduces to a geometric sequence. What is the fate of the predator population for any value $0 < s < 1$?

3. Modify the Matlab file `LogisticDifferenceEqn.m` from Section 5.7 to simulate both the prey and the predator populations. Use the new file to simulate the dynamics of the system for 100 time steps under the following conditions:
   (a) $r = 0.6$, $s = 0.2$, $\alpha = 2.5$, $\beta = 0.5$, $x_0 = 0.3$, $y_0 = 0.3$
   (b) $r = 0.6$, $s = 0.8$, $\alpha = 2.5$, $\beta = 0.5$, $x_0 = 0.3$, $y_0 = 0.3$
   (c) $r = 0.6$, $s = 0.8$, $\alpha = 2.0$, $\beta = 0.5$, $x_0 = 0.3$, $y_0 = 0.3$
   (d) $r = 0.6$, $s = 0.8$, $\alpha = 1.5$, $\beta = 0.5$, $x_0 = 0.3$, $y_0 = 0.3$
   (e) $r = 0.6$, $s = 0.8$, $\alpha = 1.0$, $\beta = 0.5$, $x_0 = 0.3$, $y_0 = 0.3$
   In each case, describe the dynamics of the predator and prey populations over time. How are cases (a) to (c) qualitatively different from (d) and (e)?
   Write a one-page report, describing your procedures and results. ■

American Bison Matrix Model

The Leslie matrix for the American bison is given by

$$\mathbf{A} = \begin{bmatrix} 0 & 0 & 0.42 \\ 0.60 & 0 & 0 \\ 0 & 0.75 & 0.95 \end{bmatrix}.$$

The population is divided into calves, yearlings, and adults (age 2 years or more). We would like to see how the whole population and the individual classes (calves, yearlings, and adults) change over time and how those results vary when we change the various parameters in the Leslie matrix for the bison population. Thus, if the vector

$$\mathbf{x}(0) = \begin{bmatrix} x_1 \\ x_2 \\ x_3 \end{bmatrix}$$

represents the American bison population at time $t = 0$, where x_1 is the number of calves, x_2 is the number of yearlings, and x_3 is the number of adult bison, then

$$\mathbf{x}(t) = \mathbf{A}^t \mathbf{x}(0)$$

gives the bison population at time t. Suppose that, at $t = 0$, there are 42 calves, 0 yearlings, and 95 adults.

1. Using the Matlab skills you learned in this unit, write an m-file (see Appendix A for description of writing and running an m-file) that allows you to answer each of the below.
   (a) Find the number of calves, yearlings, and adults at times

$$t = \{1, 2, 3, 4, 5, 10, 20, 25, 50, 75\}.$$

   (b) Make a graph that has times $t = 0$ to 100 on the horizontal axis and population size along the vertical axis. Plot the calf, yearling, and adult population sizes for every time step between 0 and 100.
   (c) Change the adult bison fecundity to 0.2, 0.42, 1.0, and 1.4. How do these changes affect the bison population equilibrium structure? Create one graph that has adult fecundity along the horizontal axis and equilibrium population structure along the vertical axis with a data set plotted for each population class (calves, yearlings, and adults).
   (d) Change the calf survival rate to 0.3, 0.5, 0.6, 0.7, and 0.85. How do these changes affect the bison population equilibrium structure? Create one graph that has calf survival rate along the horizontal axis and equilibrium population structure along the vertical axis with a data set plotted for each population class (calves, yearlings, and adults).
   (e) Change the adult survival rate to 0, 0.3, 0.5, 0.75, 0.95, and 0.99. How do these changes affect the bison population equilibrium structure? Create one graph that has adult survival rate along the horizontal axis and equilibrium population structure along the vertical axis with a data set plotted for each population class (calves, yearlings, and adults).
2. Using Matlab, try to discover any parameter changes that cause the population to die out by $t = 200$. You do not need to write an m-file for this exploration. What conclusions can you make about the possibility of the extinction of the American bison based on this exploration?
3. Write a three-page report (not including figures) that describes the analysis you performed and summarizes the findings of your analysis. Your report should have four sections:
   (a) Introduction: Describe the problem put forth and any hypotheses you had about the results of the experiment.
   (b) Methods: Describe the mathematics you employed to solve the described problem.

(c) Results: Describe in words and with graphs the results of your analysis. In this section, include any tables or plots you produced along with descriptions of what is presented in those tables and plots. Also, describe any parameter changes to the Leslie matrix that led to the extinction of the population.

(d) Conclusion: What conclusions can you draw from your results? Comment on any hypotheses you made in light of your findings. Discuss the possible future of the American bison population based on your results. Would you make any policy recommendations based on your results?

4. Write a paragraph that describes how comfortable you feel writing code in Matlab after this project and how much you feel this project has contributed to your understanding of matrix modeling and the equilibrium structure material covered in this unit.

5. Turn in a copy of your report along with a copy of your m-file and your comments on writing Matlab code (from item 4). ■

3
UNIT

Probability

This unit deals with unpredictability, a factor that occurs in every area of the life sciences and, of course, in every area of our lives (including weather, traffic conditions, catching the flu, the sex and genetics of our offspring, and when we die). It might be useful in some circumstances to ignore the aspects of biology that we cannot predict, meaning that we take a purely "deterministic" view. In the deterministic view, we assume that there is no unpredictability in our measurements or in our capacity for determining the future course of events. We saw one example of this in the Leslie matrix models for the age structure of a population. Given the initial age structure of the population as a vector x_0 and a Leslie matrix P, then, at the next time period, the age structure will be Px_0. In the Leslie matrix models, given the present population structure, we can precisely predict the population structure one time period later and, in fact, at any future time n, as we saw by calculating $P^n x_0$. In using this deterministic model, we are deciding to ignore uncertainty about many aspects of the population's future. For example, we assume that the survivorships and fecundities in the matrix P are fixed and do not vary through time.

The Leslie matrix model also assumes that all individuals in the population reproduce and survive exactly consistently according to the elements of the Leslie matrix and that this does not vary in any way. In one sense, this implies that every individual in the population produces exactly the same number of offspring as every other individual of the same age. Clearly, this assumption would not hold in reality, since not only do individuals differ in the capacity to survive and reproduce (the individual differences, which, if heritable, are the basis for the process of natural selection), but these may vary through time and space as environmental conditions change.

The area of mathematics that provides us with methods to account for unpredictability is called "probability." This is closely related to the area of statistics, which applies probability to questions such as how likely it is that the outcomes of two experiments will differ or be the same (e.g., whether the outcomes of the experiments differ "significantly") and how we can best design experiments to evaluate some hypothesis.

This unit will only cover basic probability and models. See [20] for a wonderful, readable book that covers some material from this part of the course in more detail than we do here. In addition to the field of statistics, probability can be applied to many subdisciplines of biology for which the concepts of probability presented here are essential. We will illustrate the concepts in this unit with examples from gambling, genetics, and drug testing. Some areas of biology in which probabalistic methods have been very helpful include the following:

Genomics

This is the application of probability to analyze genetic sequences, determine differences between these sequences, and compare sequences between different individuals/species. The techniques to do this utilize computational methods that are part of the general area of "bioinformatics." The methods also are applied in the use of genetic information in "phylogenetics," the study of the history of life and the relationships between taxa over time and space.

Population Genetics

This uses probability to analyze the genetic structure of a population and how it changes over time, just as we have analyzed the changes in age or stage structure using the Leslie model. Probability ideas are used because of the inherent uncertainty in the process by which mating and assortment of genetic material occurs in many populations. It is not possible to determine exactly which egg and sperm cells will combine, and therefore what the next generation will look like is unpredictable. Think of the simple situation of two birds landing on an island, mating, and founding a population on the island. If by chance a genetically determined characteristic of the parents is not passed on to the offspring, then that characteristic will be completely lost from future generations unless a mutation returns it or an immigrant arrives with that characteristic and interbreeds. The jargon for this is the "founder effect." The limited genetic material contained in the small number of founders of a new population acts to constrain the future genetic composition of the population.

Disease Spread

Epidemiology deals with how diseases spread within and between populations. The initial phase of this is unpredictable. For example, a case of hepatitis A in Knoxville, Tennessee, in 2003 spread unpredictably among some individuals who ate at a restaurant and not to others who also ate there. Similarly, harmful *E. coli* infections might affect some but not all individuals who eat improperly handled meat. Determining why some individuals are affected by exposure to a disease and others are not is a primary question in epidemiology. An associated issue is the application of methods to control the spread of a disease, including vaccination, quarantine, and culling for livestock diseases. Of interest is determining the circumstances under which it is likely that a disease will spread in the absence of a vaccination for that disease. Public health agencies need to consider the benefits of instituting a vaccination program against the costs of delivering the vaccine and its possible side effects.

Vision

Some organisms, including many nocturnal ones, such as owls, have the capacity to perceive images and movement extremely well under very low light conditions. In this situation, very few photons reach the cells in the retina, which responds and passes neural signals on to the brain. Photons hit the retina in an unpredictable manner. If a cell of the retina does not produce a signal, it may be because the field of view is indeed dark, or it may be that by chance a photon did not arrive at the retinal cell from that field of view. The difficulty of perception of an image under low light is enhanced due to the low number of of photons hitting the retina in these conditions. This is part of the problem of signal detection: determining what a signal is in a system with "noise": a central problem in sensory perception.

Probability of Events

To begin our exploration of the role of uncertainty in biological application, we'll define the mathematical formulation associated with different outcomes of an "experiment." You can think of this as a laboratory experiment, such as changing the temperature of a lizard habitat and how that changes the sex of lizard offspring or how food availability affects the number of birds in a clutch. In these cases and many others, there are discrete possible outcomes from an experiment (e.g., offspring are male or female, or the number of eggs in a female's nest is zero, one, two, etc.) that are variable and may not be able to be predicted with certainty. An alternative is that some characteristics are uncertain and can take on a continuous range of values, such as the weight or height of a newborn. We will consider methods to mathematically analyze uncertainty in observations of continuous variables, such as height and weight, in Chapter 25.

Perhaps the simplest case of an observation with discrete outcomes is typified by a coin toss. One side or the other of the coin is facing up after the coin lands. Many situations in biology can be viewed as arising from this situation: the gender of offspring, being infected with the flu or not, or whether a species is present or absent on an island. We will explore here the methods to describe what happens in this situation. Examples include determining the probability that a group of individuals has a certain number of males and females, the number of different species present in a location, or the number of birds that successfully fledge in a clutch. We will illustrate the concepts using classic data from Mendel's experimental genetics on peas.

Many experiments or observations do not have only two possible outcomes. We will use tossing dice to illustrate this, using a six-sided die as one case in which we keep track of which side of the die faces up. This need not be limited to six outcomes (you can think of tossing a die with as many sides as you would like). The number of offspring over the life span, which species are present on an island, and an individual's blood type are all examples of situations in which the outcome has more than two possibilities. Again, we will consider ways to analyze the probability that one of these particular outcomes happens when an experiment or observation is repeated a number of times. So we will be able to compute the probability that a certain number of seeds grow into plants with each of several different colors of flower, as in Mendel's experiments.

10.1 Sample Spaces and Events

In this section, we will confine our attention to experiments for which we can list all possible outcomes. The complete list of possible outcomes for an experiment $\mathcal{E}$ is called the *sample space* for $\mathcal{E}$. Suppose that experiment $\mathcal{E}$ has outcomes $e_1, e_2, \ldots, e_n$; then the set $S = \{e_1, e_2, \ldots, e_n\}$ is called the *sample space* for $\mathcal{E}$, and the elements of the set S are called *elementary events*.

Example 10.1 (Coin Flipping)

Suppose that a coin is flipped once. What is the sample space? What is the sample space if the coin is flipped twice?

Solution: When the coin is flipped, there are two possible elementary events: H (heads) and T (tails). This assumes that the only aspect of the toss we are interested in is which side of the coin is facing up and not in other aspects, such as in which direction the "face" side of the coin is oriented. Thus, in the experiment where the coin is flipped once, the sample space is

$$S_1 = \{H, T\}.$$

Flipping a coin twice is an experiment consisting of two actions. In the first action, the first coin is flipped with sample space S_1. In the second action, the coin is flipped again, and again it has sample space S_1. To consider all possible outcomes when flipping the coin twice, we construct a tree diagram:

$$H \qquad T$$
$$H \quad T \quad H \quad T$$

Starting from the top of the tree, we see that there are four distinct paths to the bottom of the tree. These four paths describe the four possible outcomes of flipping the coin twice, so the sample space is

$$S_2 = \{HH, HT, TH, TT\}.$$

Notice that HT and TH are treated as separate cases because we are keeping track of the order of the tosses. If we were flipping two coins simultaneously and could not differentiate between the two coins (e.g., they are not of different colors or dates of manufacture), then the sample space would be

$$S_3 = \{HH, HT, TT\}$$

since we cannot differentiate between the outcomes HT and TH.

Given a sample space S, a subset E of the sample space (denoted $E \subset S$) is called an ***event***. An event can contain none, some, or all of the elementary events e_i of S.

Example 10.2 (Coin Flipping Events [16])

Suppose that a coin is flipped twice. What elementary events are in the event $E =$ "first and second flips match"?

Solution: From Example 10.1, we know that the sample space is

$$S = \{HH, HT, TH, TT\}.$$

The elemenary events that correspond to "first and second flips match" are HH and TT. Thus,

$$E = \{HH, TT\}.$$

Example 10.3 (Rolling Dice)

Suppose that a single die is rolled. What is the sample space? Suppose that two dice are rolled. What is the sample space? What elementary events are in the event "dice sum to 7"?

Solution: In the experiment where a single die is rolled, the sample space is

$$S_1 = \{1, 2, 3, 4, 5, 6\}.$$

In the experiment where two dice are rolled, the sample space is

$$
S_2 = \left\{
\begin{array}{llllll}
(1,1), & (1,2), & (1,3), & (1,4), & (1,5), & (1,6), \\
(2,1), & (2,2), & (2,3), & (2,4), & (2,5), & (2,6), \\
(3,1), & (3,2), & (3,3), & (3,4), & (3,5), & (3,6), \\
(4,1), & (4,2), & (4,3), & (4,4), & (4,5), & (4,6), \\
(5,1), & (5,2), & (5,3), & (5,4), & (5,5), & (5,6), \\
(6,1), & (6,2), & (6,3), & (6,4), & (6,5), & (6,6)
\end{array}
\right\}.
$$

The event "dice sum to 7" would be

$$E = \{(1,6), (2,5), (3,4), (4,3), (5,2), (6,1)\}.$$

In Example 10.3, there were two actions. Action 1 was rolling the first die, and action 2 was rolling the second die. For action 1, there were six possible outcomes: $\{1, 2, 3, 4, 5, 6\}$. Likewise, for action 2, there were the same six possible outcomes. We saw in Example 10.3 that this resulted in $6 \times 6 = 36$ total elementary events for the experiment of rolling two dice. This is what is known as the *multiplication principle* in probability.

> ### Multiplication Principle
>
> If action 1 in an experiment $\mathcal{E}1$ can be performed m ways and action 2 in experiment $\mathcal{E}2$ can be performed n ways, then a new experiment $\hat{\mathcal{E}}$ consisting of action 1 followed by action 2 can be performed mn ways. Thus, the sample space of $\hat{\mathcal{E}}$ contains mn elementary events.
>   This can be extended to as many successive actions as needed.

We can illustrate the multiplication principle with a tree diagram as well. Suppose that action 1 has four possible outcomes $\{A, B, C, D\}$ and that action 2 has three possible outcomes $\{1, 2, 3\}$.

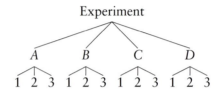

We see that at the bottom of the tree, we have $4 \times 3 = 12$ total elementary events in this experiment, giving the sample space of

$$S = \{A1, A2, A3, B1, B2, B3, C1, C2, C3, D1, D2, D3\}.$$

Notice that to form any particular elementary event in S, we follow a particular path down the tree. For example, to get elementary event $B3$, we follow the branch down to B and then the branch from B down to 3.

Example 10.4 (Mendel's Pea Plants)

In the mid-1800s, Gregor Mendel, now considered the "father of modern genetics," conducted experiments with pea plants to determine the basic principles underlying the passing on of traits from one generation to the next, that is the heredity of traits. One trait that Mendel examined was flower color; he studied pea plants that flowered in either purple or white. Suppose that Mendel had three pea plant seeds each of which might grow into a plant with either purple or white flowers. If all three seeds were planted, what would the sample space of all possible outcomes be? List the elements in the event that there are "exactly two seeds that produce plants with purple flowers."

(Continued)

Solution: For each plant, there are two possibilities: P (purple) and W (white). We can form a tree diagram to enumerate all the possibilities:

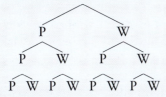

Thus, there are $2 \cdot 2 \cdot 2 = 8$ possible outcomes. Furthermore, the event "exactly two seeds that produce plants with purple flowers" consists of $\{PPW, PWP, WPP\}$.

10.2 Probability of an Event

Now that we have defined events and sample spaces, we need a way to define the probability of an event given that we know the probabilities of the elementary events in that event.

Let experiment $\mathcal{E}$ have sample space $S = \{e_1, \ldots, e_n\}$. The probability of each elementary event e_i in S is $p_i = \omega(e_i)$, where ω is a function satisfying

1. $0 \le p_i \le 1$.
2. $\sum_{i=1}^{n} p_i = p_1 + p_2 + \cdots + p_n = 1$.

Since ω is a function that assigns a probability to each elementary event in S, ω is called a *probability function*. We use the symbol ω because we can think of the probability function as a weighting function; that is, each elementary event is weighted, and the sum of all the weights is 1. Note that the weight for each elementary event may be different.

Let $E = \{s_1, s_2, \ldots, s_k\}$ be an event where $0 \le k \le n$ and each s_i is an elementary event in S (note that the event E can be made up of none, some, or all of the elementary events in S). The *probability of E* is

$$P(E) = \omega(s_1) + \omega(s_2) + \cdots + \omega(s_k),$$

the sum of the probabilities of the elementary events of E. Note that we use an uppercase P to denote the probability function of an event.

A sample space is called *uniform* or *equiprobable* if equal weights are assigned to each elementary event $e_1, \ldots, e_n$, that is, $\omega(e_i) = \frac{1}{n}$ and $P(S) = 1$. Furthermore in this equiprobable case, if event $E = \{s_1, s_2, \ldots, s_k\}$, where $0 \le k \le n$ and each s_i is an elementary event in S, then

$$P(E) = \omega(s_1) + \omega(s_2) + \cdots + \omega(s_k) = \frac{1}{n} + \frac{1}{n} + \cdots + \frac{1}{n} = \frac{\text{no. of elements in } E}{n}.$$

Example 10.5 (More on Mendel's Pea Plants)

Recall Example 10.4. Now suppose that Mendel has those same three seeds and wants to know the probability that *exactly* two out of the three plants will produce purple flowers and the probability that *at least two* of the three plants will produce purple flowers.

Solution: In Example 10.4, we saw that if all three seeds were planted and each seed could produce either a purple flowering plant or a white flowering plant, there were a total of eight possible outcomes (thus, the sample space contains eight elementary events), and that the set $\{PPW, PWP, WPP\}$ corresponded to the event "exactly two seeds that produce plants with purple flowers." If the sample space is uniform, then the probability of each elementary event is $\frac{1}{8}$, and thus the probability of the event $E_1 =$ "exactly two seeds that produce plants with purple flowers" is

$$P(E_1) = P(PPW) + P(PWP) + P(WPP) = \frac{1}{8} + \frac{1}{8} + \frac{1}{8} = \frac{3}{8}.$$

The event $E_2 =$ "at least two seeds that produce plants with purple flowers" contains all the elementary events of E_1 but also contains the elementary event PPP because if all three seeds turn out to be plants with purple flowers, then at least two of these are plants with purple flowers. Thus, $E_2 = \{PPP, PPW, PWP, WPP\}$. Furthermore, if the sample space is uniform, then

$$P(E_2) = P(PPP) + P(PPW) + P(PWP) + P(WPP) = \frac{4}{8} = \frac{1}{2}.$$

We will consider the case where the sample space is nonuniform later in this chapter in Example 10.11.

Example 10.6 (Rolling Three Dice)

If three dice are thrown, find the probability that the sum of the three dice is six.

Solution: Each of the three dice has six possible outcomes: $\{1, 2, 3, 4, 5, 6\}$. By the multiplication principle, there are a total of $6 \times 6 \times 6 = 216$ elementary events. Since we are rolling dice, we will assume that each elementary event is equally likely. Now, let E be the event that the three dice sum to six. We can use a tree diagram to determine how many elementary events are in E. Each path from the top to a bottom node represents one way in which the three dice can sum to six:

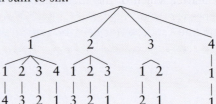

(*Continued*)

We see that there are 10 paths from the top to a bottom node (we determine this by counting the number of bottom nodes). Thus,

$$P(E) = \frac{10}{216} \approx 0.046.$$

Probability Examples from Genetics

Some basic terminology in genetics is given below, but for more detail about any of these terms, consult a text on genetics.

Gene

Genes are genetic material on a chromosome that code for a trait. In many cases, a trait is not determined by a single gene or even by material on a single chromosome. For example, human eye color is determined by genes on two different chromosomes. A gene may be determined by the genetic information at a single location, or *locus*, or it may be determined by genetic material at several locations, or *loci*.

Allele

An allele is a form of a gene at a single locus. For example, in humans, there is an eye color gene on chromosome 15 with two possible alleles: B (brown) and b (blue). The gene for common blood types in humans has three alleles: A, B, and O. Some alleles are dominant over others and are known as *dominant alleles*. Alleles that are not dominant are known as *recessive alleles*. For genes with only two alleles, we will typically use uppercase letters to denote dominant alleles and lowercase letters to denote recessive alleles. For example, the allele for brown eyes, B, is dominant, while the allele for blue eyes, b, is recessive. A gene typically contain pairs of alleles where one allele is inherited from the father and the other from the mother.

Genotype

A genotype is the set of alleles that an organism carries. For diploid organisms such as humans, with pairs of chromosomes, each gene contains one or more pairs of alleles and may have alleles at different loci. The genotype is denoted as a pair (or pairs) of letters that represent the pair (or pairs) of alleles for that particular gene. For example, since the eye color gene at one locus has two alleles, B (brown) and b (blue), there are four possible genotypes: BB, Bb, bB, and bb. In the case of the common blood type gene, since there are three alleles, A, B, and O, there are nine possible genotypes: AA, BB, OO, AB, BA, AO, OA, BO, and OB. In the case in which it doesn't matter which of the alleles came from which parent, there would be only six different genotypes for blood type: AA, BB, OO, AB, AO, and BO. Genes that have two dominant alleles are known as *homozygous dominant*, while genes with two recessive alleles are called *homozygous recessive*. Genes that have different alleles at a locus are known as *heterozygous*.

Phenotype

A phenotype is the physical expression of a trait as determined by the interaction between genetics and developmental or environmental influences. For traits determined by a gene at a single locus, if a gene is homozygous dominant or heterozygous, then the dominant allele will

Table 10.1. Blood type genotypes and their corresponding phenotypes

Phenotype	Genotypes
A	AA
	AO, OA
B	BB
	BO, OB
AB	AB, BA
O	OO

be expressed. In the case of a homozygous recessive gene, the recessive allele is expressed. In the simplest case of eye color, if you have the genotype *BB*, you will have brown eyes since both alleles are for brown eyes. If you have the genotype *bb*, you will have blue eyes since both alleles are for blue eyes. However, if you have the genotype *Bb*, with one allele for brown and one allele for blue, then the dominant allele (*B*) will mask the recessive allele (*b*), and you will have the phenotype for brown eyes. This is a simplified view of eye color, which is controlled by multiple genes, and we will discuss more realistic genetic determinants of eye color in Chapters 11 and 14. In the case of the common blood type gene, the allele *O* is recessive, while the other two are dominant. This arises because alleles *A* and *B* each result in the production of their own antigens, while allele *O* is inactive. Thus, the phenotypes are determined by the antigens produced (see Table 10.1). In the case where the genotype is *AB* or *BA*, neither allele dominates the other, and instead an additional blood type (type *AB*) is formed.

Punnett Square

A Punnett square (named after Reginald C. Punnett) is a diagram that shows the potential genotypes resulting from a mating for which the genotype of each of the parents is known. Figure 10.1 shows the possible genotypes of offspring produced by a woman with brown eyes (genotype *Bb*) and a man with blue eyes (genotype *Bb*).

Carriers of Recessive Alleles

There are many genetic diseases, such as sickle-cell anemia and albinism, that are expressed in the phenotype only if an individual has two recessive alleles. In these cases, an individual is referred to as a "carrier" of the genetic disease if his or her genotype for the disease contains one dominant and one recessive allele. For example, the hemoglobin gene has dominant allele *S* and recessive allele *s*. A person is a carrier for sickle-cell anemia if he or she has genotype *Ss*.

Now we will consider some examples from genetics involving probabilities of events.

		♀	
		B	*b*
♂	*b*	*bB*	*bb*
	b	*bB*	*bb*

Figure 10.1 An example of a Punnett square in which a brown-eyed female with genotype *Bb* and a blue-eyed male of genotype *bb* mate.

Example 10.7 (Albinism)

The gene determining albinism can have dominant allele A or recessive allele a. A set of parents both have genotype Aa, so each is a carrier of the defective allele a. Find the probability that their child will be (a) an albino or (b) a carrier.

Solution: We could draw a tree diagram to solve this problem, but it is more typical to use a Punnett square:

$$
\begin{array}{c|c|c}
 & A & a \\
\hline
A & AA & Aa \\
\hline
a & aA & aa \\
\end{array}
$$

(with ♀ above the columns and ♂ A, a to the left of the rows)

We assume that each of the four outcomes shown in the Punnett square is equally likely. Thus, $P(AA) = \frac{1}{4}$, $P(Aa) = \frac{1}{4}$, $P(aA) = \frac{1}{4}$, and $P(aa) = \frac{1}{4}$.

  (a) The event that an offspring is albino is $E_1 = \{aa\}$. Thus, $P(E_1) = \frac{1}{4}$.
  (b) The event that an offspring is a carrier is $E_2 = \{aA, Aa\}$. Thus,

$$
P(E_2) = P(aA) + P(Aa) = \frac{1}{4} + \frac{1}{4} = \frac{1}{2}.
$$

In the example above, we write Aa and aA as separate genotypes. However, it is typical that the order of the alleles is ignored and that both are referred to as Aa. From this point on in the text, when we talk about the heterozygous genotypes, we will write the dominant allele first and the recessive allele second.

Example 10.8 (Blood Type)

A father and a mother have blood types AO and AB, respectively. What is the probability that their child has blood (a) type A, (b) type B, (c) type AB, and (d) type O?

Solution: First, let us make a Punnett square of the possible offspring genotypes produced by this husband and wife:

$$
\begin{array}{c|c|c}
 & A & B \\
\hline
A & AA & AB \\
\hline
O & OA & OB \\
\end{array}
$$

(with ♀ above the columns and ♂ A, O to the left of the rows)

(Continued)

(a) The phenotype of type A blood corresponds to genotypes AA and AO. We can write this as if the event E_1 is "type A"; then $E_1 = \{AA, AO\}$ and

$$P(E_1) = P(AA) + P(AO) = \frac{1}{4} + \frac{1}{4} = \frac{1}{2}.$$

(b) If E_2 is "type B," then $E_2 = \{BB, BO\}$ and

$$P(E_2) = P(BB) + P(BO) = 0 + \frac{1}{4} = \frac{1}{4}.$$

(c) If E_3 is "type AB," then $E_3 = \{AB\}$ and

$$P(E_3) = P(AB) = \frac{1}{4}.$$

(d) If E_4 is "type O," then $E_4 = \{OO\}$ and

$$P(E_4) = P(OO) = 0.$$

10.3 Combinations and Permutations

To calculate the total numbers of events in our sample space, we use the concepts of combinations and permutations. In the sample spaces considered so far, the order of actions constituting an elementary event matters. That is, HT is not the same as TH in the coin flipping example. Furthermore, the outcome of the first action does not change the possible outcome of subsequent actions. If the first coin flipped is H, there are still the same two options for the second coin: H or T. It is not always the case that the outcome of the first action does not affect the outcome of subsequent actions.

Suppose that five cards are dealt from a standard deck of 52 cards and that the order in which they were dealt is noted. The first card dealt could be any one of the 52 cards in the deck. For the second card, there is already one card missing from the deck (the first card dealt), so there are only 51 possible cards that could be dealt second. For the third card, there are now 2 cards missing from the deck, so there are only 50 possible cards that could be dealt third. Likewise, there are 49 possible cards that could be dealt fourth and 48 possible cards that could be dealt fifth. By the multiplication principle, this deal can be made $52 \times 51 \times 50 \times 49 \times 48$ ways. We can express this multiplication in terms of factorials:

$$52 \times 51 \times 50 \times 49 \times 48 = \frac{52!}{(52 - 5)!}.$$

In general, if k objects are selected from n objects and the order of selection is noted, then this action can be accomplished in

$$P[n, k] = \frac{n!}{(n - k)!}$$

ways (note that 0! is defined to be 1). This expression gives the number of *permutations* of k objects selected from n objects.

Example 10.9 (Genome [39])

Gene order refers to the permutation of genome arrangements. Vesicular stomatitis virus (VSV) is a prototype RNA virus that encodes five genes (N-P-M-G-L). These five genes can occur in any gene order. How many different gene orders (i.e., permutations) are there?

Solution: There are a total of $n = 5$ genes to choose from. We must place these genes in $k = 5$ slots where the order matters. Thus, there are

$$P[5, 5] = \frac{5!}{(5 - 5)!} = 5! = 120$$

different gene orders.

In many card games, the order in which the cards appear in the hand dealt is not relevant. When k objects are selected from n objects without regard to the order of their selection, the number of ways this can be accomplished is

$$C[n, k] = \frac{P[n, k]}{k!} = \frac{n!}{k!(n - k)!}.$$

This expression gives the number of *combinations* of k objects selected from n objects. This expression can be used to count the number of events when the order of terms in the elementary events does not matter.

Example 10.10 (Mendel's Pea Plants Revisited)

In Example 10.4, we calculated the number of elementary events in the event "exactly two seeds that produce plants with purple flowers." We can now use the combinations function to perform that calculation.

Solution: There are three seeds; thus, $n = 3$. We want to know how many possible outcomes have exactly two of the plants producing purple flowers; thus, $k = 2$:

$$C[3, 2] = \frac{3!}{2!(3 - 2)!} = \frac{3 \cdot 2 \cdot 1}{(2 \cdot 1)(1)} = 3.$$

This is the same answer that we obtained in Example 10.4.

The example above may make it seem like it is more work to use the combinations function than it is to enumerate all the possibilities and just select the ones that match the criteria. However, suppose that Mendel now has 300 seeds and wants to know how many possible outcomes have exactly 200 of the plants producing purple flowers. It would take quite some time to enumerate all 2^{300} possibilities, and it would then be ever so tedious to go through that massive list and select all of the outcomes with exactly 200 purple flowering plants. In this case, it would be much faster to use the combinations formula, especially since most scientific calculators will quickly compute the value for you:

$$C[300, 200] = \frac{300!}{200!(300 - 200)!} \approx 4.16 \times 10^{81}.$$

10.4 Binomial Experiments

The simplest experiment that we might describe is one in which something happens or does not happen. You either get the flu this year or do not, a seed germinates this season or does not, or you either get a placebo when participating in a trial of a new drug or you get the drug being evaluated. These are all examples of a *Bernoulli experiment*, named after one of the founders of probability, and are like a toss of a single coin on a table in which it lands either heads up or heads down. Many observations in biology fit this type of experiment. You can also think of the outcome of an experiment as "success" or "failure." If we assume that success occurs a fraction p of the time, this means that when we repeat the experiment many times, approximately a proportion p of the outcomes will be a success. Of course, the probability of a failure is $1 - p$ since success and failure are the only two possibilities.

When performing an experiment several times, such as tossing a coin 10 times, we often want to know what the probability is that we get a certain number of outcomes of one of the two possibilities (e.g., the probability that in 10 coin tosses we get exactly nine heads and one tail). The assumption here is that each outcome of the experiment is independent; that is, it does not matter what the outcome was in earlier experiments. In this case, there is a simple way to calculate the probability of a certain number of successes, called the *binomial probability*. Suppose that the probability of a success is p and that we perform the experiment n times. Then the probability of having exactly k successes and $n - k$ failures is

$$B[n; k] = \binom{n}{k} p^k (1 - p)^{n-k} \quad \text{for } k = 0, 1, \ldots, n.$$

This is called the *binomial probability distribution*. The formula arises since there are exactly $\binom{n}{k}$, called "n choose k," ways to assign the k successes among n experiments, and for each of these, the probability of exactly k successes and $n - k$ failures is $p^k (1 - p)^{n-k}$. In this,

$$\binom{n}{k} = \frac{n!}{k!(n - k)!},$$

which counts the number of ways we can distribute exactly k objects in n spaces. Notice that

$$\binom{n}{k} = C[n, k].$$

Example 10.11 (A Binomial Experiment with Mendel's Pea Plants)

Recall in Example 10.5 that we calculated the probability that at least two of the three seeds would produce purple flowering pea plants. In the example, we assumed that each of the eight outcomes was equally likely. However, it may be the case that each outcome is not equally likely. Suppose that Mendel knows that for each of his three seeds, there is a 75% chance that the seed will produce a plant with purple flowers and a 25% chance that the seed will produce a plant with white flowers. Calculate the probability that exactly two of the seeds produce plants with purple flowers and the probability that at least two of the seeds produce plants with purple flowers.

Solution: Since there are only two outcomes for each seed and we know the probability of success (purple flowers, $p = 0.75$), we can use the binomial probability function to calculate the probability that exactly two of the seeds will produce plants with purple flowers. Since there are three seeds, $n = 3$, and since we want exactly two of those seeds to produce plants with purple flowers, $k = 2$,

$$B[3; 2] = C[3, 2] \left(\frac{3}{4}\right)^2 \left(1 - \frac{3}{4}\right)^{3-2} = \frac{3!}{2!(3-2)!} \left(\frac{9}{16}\right) \left(\frac{1}{4}\right) = \frac{27}{64}.$$

Now, to calculate the probability that at least two of the seeds produce plants with purple flowers, we can take the sum of the probability that exactly two of the seeds produce plants with purple flowers and the probability that exactly three of the seeds produce plants with purple flowers. We have already calculated the first probability. For the second probability, $p = 0.75$, $n = 3$, and $k = 3$. Thus,

$$B[3; 3] = C[3, 3] \left(\frac{3}{4}\right)^3 \left(1 - \frac{3}{4}\right)^{3-3} = \frac{3!}{3!(3-3)!} \left(\frac{27}{64}\right) (1) = \frac{27}{64},$$

and therefore

$$B[3; 2] + B[3; 3] = \frac{27}{64} + \frac{27}{64} = \frac{54}{64} = \frac{27}{32}.$$

10.5 Matlab Skills

Given a certain experiment $\mathcal{E}$, we would like to utilize Matlab to find the probability of an event $E \subset S$, where S is the sample space for the experiment $\mathcal{E}$. To do this, we will write code in Matlab that simulates the experiment. Then we will repeat the experiment many times, noting the outcome of the experiment each time. If we run the experiment 1000 times and the outcomes corresponding to event E occur 126 times, then we will say that $P(E) = 126/1000 = 0.126 = 12.6\%$.

Generating Random Numbers

Before we can successfully simulate an experiment, we need a way to introduce randomness or unpredictability into our simulations. Matlab can produce "random" numbers. Imagine that

Table 10.2. The Matlab code in the left-hand column will produce a random number in corresponding interval listed in the right-hand column

Interval	Code
(0,1)	rand(1)
(1,2)	rand(1)+1
(−1,0)	rand(1)-1
(0,10)	10*rand(1)
(0,0.5)	(0.5)*rand(1)
(−1,1)	2*rand(1)-1
(−10,10)	2*10*rand(1)-10

you are tossing a dart at a line segment between 0 and 1 and that the dart has an equal chance of landing at any point. You don't have good aim, so the dart can hit near the end of the line near 0 just as likely as it can hit in the middle near .5 or near the end at 1. Matlab uses a numerical method to do this that is good but not perfect, and so officially these are called "pseudorandom numbers." For our purposes, we'll abbreviate this and call them random numbers. The function rand(x), where x is a positive integer, will return an $x \times x$ matrix of random numbers between 0 and 1, that is, in the open interval $(0, 1)$. These numbers are said to be chosen according to a "uniform random distribution" on the interval $(0, 1)$. Thus, rand(1) will return a single random number between 0 and 1, and rand(4) will return a 4×4 matrix of random numbers between 0 and 1. If you want a nonsquare matrix of random numbers, use rand(x,y), where x and y are positive integers. For example, rand(1,2) will return a 1×2 matrix of random numbers between 0 and 1.

Suppose that we do not want random numbers between 0 and 1. Table 10.2 shows some examples of how to modify the rand function in order to obtain random numbers in various intervals.

Often when conducting an experiment, we want to choose between one of several possible outcomes. For example, if we are flipping a coin, there are two possible outcomes. If we are rolling a standard die, there are six possible outcomes. To pick among the various choices, we will use a combination of the rand function, the ceil and floor functions, and a code structure known as an *if statement*. An *if statement* has the following format:

```
if condition
  % what to do if condition is true
else
  % what to do if condition is false
end
```

Suppose that you wanted to simulate the flipping of a coin. You want to use the function rand to generate two possibilities. First, we generate a random number, x, in the interval $(0, 2)$. Next, take floor(x) to round the value down. If the x is in the interval $(0,1)$, floor will round it down to 0. If the x is in the interval $[1, 2)$, floor will round it down to 1. Next, we can use the *if statement* to say that if floor(x) is equal to 0, then it is a heads. Otherwise, if floor(x) is equal to 1, then it is a tails. Here is what that code would look like:

```
                              ──── coinflip.m ────
1   │  % M-file for flipping a coin
2   │
3   │  % Generate a random number in (0,2)
4   │  x = 2*rand(1);
5   │
6   │  % Determine if heads or tails
7   │  if floor(x) == 0
8   │    fprintf('Your coin landed head side up!\n')
9   │  else
10  │    fprintf('Your coin landed tail side up!\n')
11  │  end
```

Notice that to test the equality of the condition that `floor(x)` is equal to 0, we use a double equals sign, `==`. This different notation is used to differentiate between testing an equality (as we did above in the *if statement*) and setting a variable equal to a value (as we did when we set `x` equal to a random number between 0 and 2). Try copying this m-file and then running it several times to see that sometimes you get heads and other times tails.

Suppose that you wanted to simulate rolling a die. Here we will not need to use the *if statement* since all we need is the generated number:

```
                              ──── dieroll.m ────
1   │  % M-file for rolling a die
2   │
3   │  % Generate a random number in (0,6)
4   │  x = 6*rand(1);
5   │
6   │  % Print out result
7   │  fprintf('You rolled a %d.\n',ceil(x))
```

Again, try copying this m-file and then running it several times to see that you roll different values. Notice that in the m-file for rolling the die, we used the function `ceil` instead of `floor`. This is because we wanted to round up to the values 1, 2, 3, 4, 5, and 6. If we had used `floor`, the random number would have been rounded down to one of the values 0, 1, 2, 3, 4, or 5.

Writing an M-file as a Function

As we will see shortly, it is often valuable to write our own functions in Matlab. These functions will work the same way as the built-in functions in Matlab, such as `polyfit`, `floor`, `mean`, and so on. However, when we write our own functions, we can have as many inputs and outputs as we want, and we can define the function to perform any number of operations desired. Functions in Matlab are m-files where the first line of executable code starts with the word `function`.

Suppose that you wanted to flip a coin *n* times, where *n* is some integer value. We could write this as a function where the input is the number of times you wish to flip the coin and the output is a 1×2 vector that contains the number of times heads was flipped and the number of times tails was flipped. The m-file for this function is shown below and is called coinflips.m:

```
                              ──── coinflips.m ────
1   │  % function outcome = coinflips(N)
2   │  %
3   │  % Input:
```

```
4     %   N = number of times to flip the coin
5     %
6     % Output:
7     %   outcome = array with structure
8     %       [# heads flipped, # tails flipped]
9     %
10    function outcome = coinflips(N)
11
12    % Get N random numbers in (0,2)
13    x = 2*rand(1,N);
14
15    % Initialize heads and tails counters to zero
16    Hcount = 0;
17    Tcount = 0;
18
19    % For each random number, decide if it is a head or tail
20    for i = 1:N
21        if floor(x(1,i))==0
22            Hcount = Hcount + 1;
23        else
24            Tcount = Tcount + 1;
25        end
26    end
27
28    % Record outcome
29    outcome = [Hcount Tcount];
```

In coinflips.m, first notice the block of commenting before the line

```
function outcome = coinflips(N)
```

This commenting is what will appear if you type the following into the command window:

```
─────────────────── Command Window ───────────────────
>> help coinflips
```

Next, notice that we create a vector of N random numbers (N being the input of the function), one for each coin flip. Since we want to keep track of how many times a head is flipped and how many times a tail is flipped, we create counters called Hcount and Tcount and set them initially to 0. Next, we use a loop to go through each of the coin flips. For each coin flip, we use an *if statement* to decide whether a head or a tail was flipped. After we have considered each coin flip (i.e., after we are done with the loop), we construct our output vector, outcome.

Suppose that we wanted to simulate 1000 coin flips. Now that we have our coinflips(N) function, we can simply type the following into the command window:

```
─────────────────── Command Window ───────────────────
>> coinflips(1000)
```

If we run this same command multiple times, the output of the function will vary each time. Below is the output when one of the authors ran this command 10 times:

```
                              ── Command Window ──
>> coinflips(1000)
ans =
    479    521

>> coinflips(1000)
ans =
    492    508

>> coinflips(1000)
ans =
    480    520

>> coinflips(1000)
ans =
    496    504

>> coinflips(1000)
ans =
    514    486

>> coinflips(1000)
ans =
    492    508

>> coinflips(1000)
ans =
    494    506

>> coinflips(1000)
ans =
    528    472

>> coinflips(1000)
ans =
    485    515

>> coinflips(1000)
ans =
    493    507
```

Estimating the Probability of an Event

Once we have a way to simulate a certain experiment multiple times, we would like to estimate the probability of a certain event occurring.

Returning to the coin flipping example, suppose that we wanted to know the probability of flipping a head. We could run our coinflips function with $N = 1000$, see how many times heads occurred, and then divide by 1000. However, if we did this, we would get a slightly different probability each time. For example, the 10 outputs shown above would correspond to the probabilities 0.479, 0.492, 0.480, 0.496, 0.514, 0.492, 0.494, 0.528, 0.485, and 0.493 (for flipping a heads).

Another approach to estimating the probability of flipping a head would be to take the average of the 10 different outputs and divide that number by 1000. If we did this, we would get a probability of 0.4943. We could extend this and take the average of 500 different outputs and divide that number by 1000. However, we would not want to enter `coinflips(1000)` into the command window 500 times. Thus, let us write an m-file that does this for us:

```
                          ──────── ProbHeadFlip.m ────────
1    % Filename: ProbHeadFlip.m
2    % M-file to compute probability of flipping a heads
3
4    % Number of times to collect output
5    n = 500;
6
7    % Initialize heads count
8    Hcount = 0;
9
10   % Run coinflips n times
11   for i = 1:n
12       % Coinflips will run 1000 experiments each time
13       output = coinflips(1000);
14
15       % sum up number of heads outputs
16       Hcount = Hcount + output(1,1);
17   end
18
19   % Compute average number of heads
20   Haverage = Hcount/1000;
21
22   % Estimate probability of flipping heads
23   ProbH = Haverage/n;
24
25   % Print out answer
26   fprintf('P(Heads) = %6.4f\n',ProbH);
```

When one of the authors ran this file five times in the command window, the outputs were as follows:

```
                          ──────── Command Window ────────
>> ProbHeadFlip
P(Heads) = 0.4993

>> ProbHeadFlip
P(Heads) = 0.5006

>> ProbHeadFlip
P(Heads) = 0.5000

>> ProbHeadFlip
P(Heads) = 0.4996

>> ProbHeadFlip
P(Heads) = 0.5006
```

We can see that these values are a better estimate of the true probability of flipping heads, $P(H) = 0.5$, than if we used only one output from the `coinflips` function with $N = 1000$.

If we change the value of n in the `ProbHeadFlip.m` file to $n = 10,000$, the answers will be even closer to the true probability. However, the larger we make n, the longer the file will take to run, and at some point we must decide that our estimate is good enough.

Biological Example of Probability Estimation: Albinism

Here we will consider an example similar to Example 10.7.

Suppose we know that both John and Jane are carriers for albinism. What is the probability that their child will be a carrier for albinism? What is the probability that a child of theirs will have albinism?

SIMULATE GENERATION OF AN OFFSPRING GENOTYPE

We can use Matlab to simulate the "experiment" of John and Jane having a child. First, we construct a function in Matlab for which the input is the genotype of each parent (with respect to albinism) and the output is the genotype of the offspring. For the genotype with respect to albinism, there are three possibilities: (1) heterozygous (Aa or aA), (2) homozygous dominant (AA), or (3) homozygous recessive (aa). In our Matlab function, we will use numbers to represent each of these cases:

```
———————————— OneChild.m ————————————
% function child = OneChild(mom,dad)
%
% Inputs:
%    mom = genotype of mother**
%    dad = genotype of father**
% ** For genotypes use
%        1 for heterozygous, Aa or aA
%        2 for homozygous dominant, AA
%        3 for homozygous recessive, aa
%
% Output:
%    child = genotype of offspring
function child = OneChild(mom,dad)


% Within this function we will use
%    0 to represent a recessive allele
%    1 to represent a dominant allele

% Determine allele inherited from mother
if mom == 1
    %if mom heterozygous, randomly choose between two alleles
    childallele1 = floor(2*rand(1));
elseif mom == 2
    %if mom homozygous dominant, then child inherits A
    childallele1 = 1;
else
    %if mom homozygous recessive, then child inherits a
    childallele1 = 0;
end

% Determine allele inherited from father
if dad == 1
```

```
34        %if dad heterozygous, randomly choose between two alleles
35        childallele2 = floor(2*rand(1));
36    elseif dad == 2
37        %if dad homozygous dominant, then child inherits A
38        childallele2 = 1;
39    else
40        %if dad homozygous recessive, then child inherits a
41        childallele2 = 0;
42    end
43
44  . % Determine the genotype of the child
45    if (childallele1==1 && childallele2==1)
46        child = 2;
47    elseif (childallele1==0 && childallele2==0)
48        child = 3;
49    else
50        child = 1;
51    end
```

Notice in the OneChild function the use of elseif in the *if statements* to test a second condition. In this case, when determining the allele inherited from the mother, first the code checks if the mother is heterozygous. If she is, then we use the rand function to determine the allele that is inherited from the mother. If she is not heterozygous, then the code goes to the next condition and checks if the mother is homozygous dominant. If she is, then the child will inherit a dominant allele. Finally, if the mother is not heterozygous and not homozygous dominant, then the code goes to the else condition. This assumes that if the mother is not heterozygous and not homozygous dominant, then the only option left is that she is homozygous recessive, in which case the child inherits a recessive allele. This process is repeated for choosing an allele from the father.

Next, notice that when the genotype of the child is determined (in the third and last *if statement* structure), we use the && notation to indicate that the condition in the *if statement* that must be satisfied is actually two conditions both of which must be satisfied. We use the || notation to indicate that the condition in the *if statement* is true if either of the conditions is satisfied. So, if our code reads something like

```
if (condition A && condition B)
   % something here
else
   % something here
end
```

then the *if statement* is true if condition A AND condition B hold. Likewise, if our code reads something like

```
if (condition A || condition B)
   % something here
else
   % something here
end
```

then the *if statement* is true if condition A OR condition B holds.

ESTIMATING PROBABILITIES

Now that we have a function for the "experiment" of generating the genotype of an offspring, we would like to calculate the probability of different events. We will find these probabilities by running the experiment 1000 times, tabulating the results, and then taking the average of the 1000 experimental runs. This is similar to how we estimated the probability of flipping a head in the coin flipping experiment. Below is the code to complete these estimations, named `Albinism.m`:

──────────────── Albinism.m ────────────────

```
1    % Filename: Albinism.m
2    % M-file for determining the probability of child with two albinism
3    %    carrying parents
4    % The possible outcomes for the child are
5    %    (a) an albinism carrier (Aa,aA)
6    %    (b) have no albinism alleles (AA)
7    %    (c) albino (aa)
8    % This file finds the probability of each outcome.
9
10   % Set how many times to run the experiment
11   N = 1000;
12
13   % Set variables to sum up probabilities
14   A = 0;      % sum of heterozygous probabilities
15   B = 0;      % sum of homozygous dominant probabilities
16   C = 0;      % sum of homozygous recessive probabilities
17
18   for i = 1:N
19       % Set counters
20       a = 0;   % heterozygous counter
21       b = 0;   % homozygous dominant counter
22       c = 0;   % homozygous recessive counter
23
24       for j = 1:N
25           % Get child's genotype if both parents are carriers
26           child = OneChild(1,1);
27
28           % Increase appropriate counter
29           if child == 1
30               a = a + 1;
31           elseif child == 2
32               b = b + 1;
33           else
34               c = c + 1;
35           end
36       end
37
38       % Add the probabilities for this set of N experiments
39       %   to the running sum
40       A = A + a/N;
41       B = B + b/N;
42       C = C + c/N;
43   end
44
45   % Print out results
46   fprintf('P(carrier of albinism) = %6.4f\n',A/N);
47   fprintf('P(has albinism) = %6.4f\n',C/N);
48   fprintf('P(no recessive alleles) = %6.4f\n',B/N);
```

When one of the authors ran this file five times in the command window, the outputs were as follows:

```
──────────────────────────────── Command Window ────────────────────────────────
>> Albinism
P(carrier of albinism) = 0.5001
P(has albinism) = 0.2501
P(no recessive alleles) = 0.2498

>> Albinism
P(carrier of albinism) = 0.4996
P(has albinism) = 0.2501
P(no recessive alleles) = 0.2503

>> Albinism
P(carrier of albinism) = 0.5003
P(has albinism) = 0.2495
P(no recessive alleles) = 0.2502

>> Albinism
P(carrier of albinism) = 0.4995
P(has albinism) = 0.2500
P(no recessive alleles) = 0.2505

>> Albinism
P(carrier of albinism) = 0.5004
P(has albinism) = 0.2497
P(no recessive alleles) = 0.2500
```

10.6 Exercises

10.1   Give the appropriate sample spaces for each of the following experiments.

(a) Outcomes when three coins are flipped.
(b) Outcomes when two coins are flipped and a die is rolled.
(c) The sex of children in a family with five children.
(d) The genotype of an offspring when each parent is heterozygous.
(e) The genotype of an offspring when one parent is heterozygous and the other is homozygous recessive.
(f) The genotype of an offspring when one parent is homozygous dominant and the other is homozygous recessive.
(g) An island contains three species of spiders (species *A*, *B*, and *C*) with many individuals present of each species. Two spiders are collected from the island, chosen one at a time.
(h) An individual's blood type is determined (*A*, *B*, *AB*, or *O*), as is their Rh-factor (+ or −).

10.2   Consider the experiment where two dice are rolled. Find the probability of the following events.

(a) "sum is odd"
(b) "doubles"
(c) "sum of 5"
(d) "sum of 8"

10.3   A mother and father each have blood type *AB*. Find the probability that their child has the following blood type.

   (a) *AB*
   (b) *A*
   (c) *B*

10.4   Suppose that a family has five children. Find the probability that there are four girls and one boy.

10.5   Suppose that a family has five children. Find the probability that there are exactly three girls and two boys.

10.6   Using only four genes denoted by (N-P-M-G), how many different gene orders are there?

10.7   Suppose that you want to plant four trees in a plot and can choose from 10 different species. How many ways can be chosen to plant the trees in the plot?

10.8   Suppose that there are five types of toppings to put on a certain pizza. If you want to have only two toppings, how many choices do you have?

10.9   In plants and animals that sexually reproduce, each parent gives half of its genes to the offspring. Somatic cells contain two unique copies of each chromosome, and the two copies of each chromosome form a chromosome pair, or *homologous chromosomes*. During the process of *meiosis*, the somatic cell will divide into *gametes*, or sex cells, each of which contains only one copy of each chromosome.

   (a) The common fruit fly has four different chromosomes; thus, in each somatic cell, there are eight chromosomes arranged in four pairs. How many unique gametes can one fruit fly form?
   (b) Humans have 23 different chromosomes; thus, in each somatic cell, there are 46 chromosomes arranged in 23 chromosome pairs. How many unique gametes can one human form?

10.10 *Achondroplasia* is a type of dwarfism caused by the presence of the dominant allele *D* on the gene. However, genotype *DD* is not viable and usually causes death before birth. Thus, a dwarf will have genotype *Dd*, while a nondwarf will have genotype *dd*. If a dwarf and nondwarf mate, what is the probability that their first child will be a nondwarf?

10.11 In Example 10.4, we saw that Mendel's pea plants had two colors of flowers: purple and white. On the gene for flower color, the purple allele, *P*, is dominant, while the white allele, *p*, is recessive.

   (a) If a heterozygous pea plant with purple flowers is crossed with a pea plant with white flowers, what proportion of pea plants in the next generation will have white flowers?

   There are some flowering plants for which the combination of two alleles can express three different phenotypes. For example, for some plants, flower color is determined by two alleles, *R* and *r*, where *RR* flowers are red, *Rr* flowers are pink, and *rr* flowers are white.

   (b) If a plant with red flowers is crossed with a plant with white flowers, what proportion of plants in the next generation will have pink flowers?
   (c) If a plant with pink flowers is crossed with a plant with pink flowers, what proportion of plants in the next generation will have pink flowers?

10.12 A child with blood type O has a mother with blood type A and a father with blood type B.

(a) What are the genotypes of the parents?
(b) List all possible genotypes of children from these two parents.
(c) What is the probability that a child from these two parents has blood type O?

10.13 One of the genes for eye color in humans has two alleles, B and b. The brown allele, B, is dominant over the blue allele, b, and so genotypes BB and Bb have brown eyes, while genotype bb has blue eyes. Suppose that a child has brown eyes and that the child's father has blue eyes.

(a) What is the genotype of the child?
(b) What are the possible genotypes of the mother?
(c) Suppose that the child grows up and mates with a blue-eyed individual. What is the probability that their offspring will have brown eyes?

10.14 The *National Vital Statistics Report for 2006* [32] reported that there were a total of 2,426,264 deaths in the United States in 2006. The following table shows the 15 most common causes of death and their numbers.

Cause	Number of Deaths
Diseases of heart	631,636
Malignant neoplasms	559,888
Cerebrovascular diseases	137,119
Chronic lower respiratory diseases	124,583
Accidents (unintentional injuries)	121,599
Diabetes mellitus	72,449
Alzheimer's disease	72,432
Influenza and pneumonia	56,326
Nephritis, nephrotic syndrome, and nephrosis	45,344
Septicemia	34,234
Intentional self-harm (suicide)	33,300
Chronic liver disease and cirrhosis	27,555
Essential hypertension and hypertensive renal disease	23,855
Parkinson's disease	19,566
Assault (homicide)	18,573
All other causes (residual)	447,805

Convert the table to read in probabilities, that is, the probability that a randomly selected death was of a certain cause. Verify that the probabilities sum to 1.

10.15 Using the code for OneChild.m as an example, write a function to simulate the generation of the blood type (A, B, AB, or O) of an offspring given the blood genotypes of the offspring's parents.

Probability of Compound Events

In Chapter 10, we concentrated on determining the probability of a single event. For example, we found the probability that an offspring has a certain genotype if the genotypes of the parents are known. Many experiments or observations involve cases in which there are several different characteristics under consideration and we wish to determine the probability that a certain combination of events occurs. These are called compound events. An example we consider first is the case of human blood types that include not just the classification as A, B, AB, and O and but the Rh classification as well. For example, we'll consider how to determine the probability that someone is of blood type A and is Rh-positive. For the case of human characteristics, we'll see how this can be expanded to consider the probability that individuals in your city have a certain mix of eye color, hair color, and blood type; whether they are diabetic; and so on. All of these are discrete characteristics, and we will consider how to calculate the probability that a randomly chosen individual has a certain set of these characteristics. You can think of this as finding the frequency in the population (e.g., of the people in your city) having a particular mixture of traits, such as the frequency of blue-eyed, brown-haired individuals. We'll see that there is a nice geometric interpretation for finding the probability that two events both happen or that either of them happens. The concepts developed here can be extended to consider continuous traits, such as body weight, height, and cholesterol level, which we will discuss in Chapter 25.

11.1 Compound Events

One example of compound events is blood typing with the Rh classification of blood. As we saw in Chapter 10, there are four different blood types based on the alleles that determine antigen production: A, B, AB, and O. However, blood types can also be positive or negative (this is how we get blood types like A positive or B negative). It is the Rh classification of blood that determines whether it Rh is positive or negative. Rh classification of blood types is determined by two alleles, R and r, where R is the dominant allele. The genotype rr produces a negative Rh classification, while the genotypes RR and Rr produce positive Rh classifications. Given the blood types and Rh classifications of an offspring's parents, we could ask about the probability of a child is Rh-positive and has type B blood. We could consider other possible combinations of these two characterizations of blood type.

Let $\mathcal{E}$ be some experiment with sample space $S = \{e_1, e_2, \ldots, e_n\}$. We will represent events as subsets of S. Let E and F be two possible events in S and e be a particular outcome of experiment $\mathcal{E}$. A *compound event* is an event in S that involves some combination of E, F, not E, and not F. Here are the various possibilities:

E or F

The event "E or F" will occur if the outcome e is either in event E or in event F or in both E and F. We denote this as $E \cup F$, where the symbol $\cup$ denotes the *union* of E and F. This can be seen pictorially in Figure 11.1(a).

Example 11.1 (Blood Typing: $E \cup F$)

Suppose that our sample space consists of blood types with Rh factor:

$$S = \begin{cases} e_1 = \text{"type A and Rh-positive"} \\ e_2 = \text{"type A and Rh-negative"} \\ e_3 = \text{"type B and Rh-positive"} \\ e_4 = \text{"type B and Rh-negative"} \\ e_5 = \text{"type AB and Rh-positive"} \\ e_6 = \text{"type AB and Rh-negative"} \\ e_7 = \text{"type O and Rh-positive"} \\ e_8 = \text{"type O and Rh-negative"} \end{cases}.$$

Additionally, suppose that $E = $ "type B blood" and $F = $ "Rh-positive." The elementary events $e_3 = $ "type B and Rh-positive," $e_4 = $ "type B and Rh-negative," $e_1 = $ "type A and Rh-positive," $e_5 = $ "type AB and Rh-positive," and $e_7 = $ "type O and Rh-positive" are all in $E \cup F$. That is, $E \cup F = \{e_1, e_3, e_4, e_5, e_7\}$.

E and F

The event "E and F" will occur if the outcome e is in both event E and in event F. We denote this as $E \cap F$, where the symbol $\cap$ denotes the *intersection* of E and F. This can be seen pictorially in Figure 11.1(b). Notice that *and* is more restrictive than *or*.

Example 11.2 (Blood Typing: $E \cap F$)

Suppose that we have the same sample space S as in Example 11.1 and that events $E = $ "type B blood" and $F = $ "Rh-positive." Then $E \cap F = \{e_3\}$.

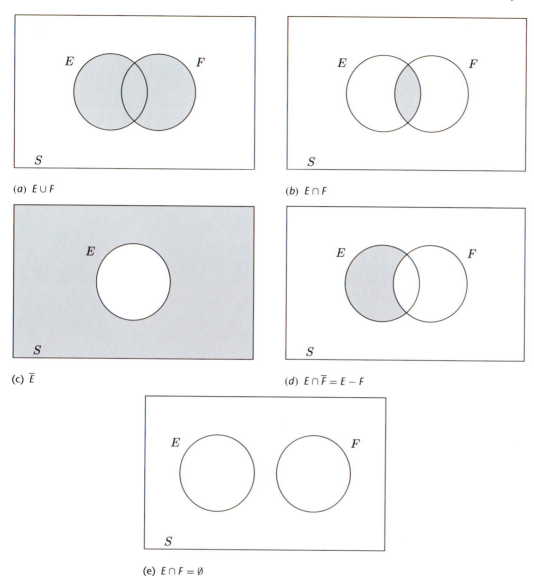

Figure 11.1 Venn diagrams of the various types of compound events: (a) E or F, (b) E and F, (c) not E, (d) E but not F, and (e) events E and F are mutually exclusive; that is, they do not intersect.

Not E

The event "not E" will occur if the outcome e is not in event E. We denote this as $\overline{E}$. The set $\overline{E}$ is often referred to as the ***complement*** of the set E. A visual representation of $\overline{E}$ can be seen in Figure 11.1(c).

Example 11.3 (Blood Typing: $\overline{E}$)

Suppose that we have the same sample space S as in Example 11.1 and that event $E =$ "type B blood." Then $\overline{E} = \{e_1, e_2, e_5, e_6, e_7, e_8\}$.

E but not F

The event "E but not F" will occur if the outcome e is in event E but not in event F. We denote this as $E \cap \overline{F}$ or $E - F$. This can be seen pictorially in Figure 11.1(d).

Example 11.4 (Blood Typing: $E \cap \overline{F}$)

Suppose that we have the same sample space S as in Example 11.1 and that events $E =$ "type B blood" and $F =$ "Rh-positive." Then $E \cap \overline{F} = \{e_4\}$.

Mutually Exclusive

Let $\emptyset$ represent the **null or empty set**, that is, an impossible event. We call events E and F **mutually exclusive** if $E \cap F = \emptyset$. A visual representation of mutual exclusivity is shown in Figure 11.1(e).

Example 11.5 (Blood Typing: $E \cap F = \emptyset$)

Suppose that we have the same sample space S as in Example 11.1 and that events $E =$ "type B blood" and $F =$ "type O blood." A person cannot have both type B blood and type O blood; thus, these events are mutually exclusive. Thus, $E \cap F = \emptyset$.

Note how intersection and union work with "not":

$$\overline{A \cap B} = \overline{A} \cup \overline{B} \text{ and } \overline{A \cup B} = \overline{A} \cap \overline{B}.$$

11.2 Finding the Probability of a Compound Event

In Chapter 10, we computed the probability of an event E by adding the probability of each elementary event in E. In this section, we discover how to compute the probability of compounds events via a set of probability axioms and properties.

Probability Axioms

Let S be a sample space. Then

1. $P(S) = 1$.
2. If E is an event in S, then $0 \leq P(E) \leq 1$.
3. If A and B are mutually exclusive events in S (i.e., $A \cap B = \emptyset$), then

$$P(A \cup B) = P(A) + P(B) \text{ and } P(A \cap B) = 0.$$

Example 11.6 (Eye Color)

Suppose that we know the following about the probability of an individual's eye color:

- $P(\text{brown eyes}) = \frac{1}{2}$
- $P(\text{blue eyes}) = \frac{1}{8}$
- $P(\text{green eyes}) = \frac{3}{8}$

What is the probability that the individual has blue or green eyes?

Solution: Since the individual can have only one of these three eye colors, each of the above events (brown eyes, blue eyes, and green eyes) is mutually exclusive. Thus,

$$P(\text{blue} \cup \text{green}) = P(\text{blue}) + P(\text{green}) = \frac{1}{8} + \frac{3}{8} = \frac{4}{8} = \frac{1}{2}.$$

Properties with Complements

Let E and F be events in a sample space.

1. $P(\overline{E}) = 1 - P(E)$.
2. $P(E - F) = P(E \cap \overline{F}) = P(E) - P(E \cap F)$.

The first property comes directly from our axioms. We know that

$$P(E \cup \overline{E}) = P(S) = 1$$

and that

$$E \cap \overline{E} = \emptyset.$$

Combining these two facts, we get

$$P(E \cup \overline{E}) = 1$$
$$P(E) + P(\overline{E}) = 1$$
$$P(\overline{E}) = 1 - P(E).$$

Notice that this property also implies that $P(A) = 1 - P(\overline{A})$.
Notice that

$$E = (E \cap F) \cup (E \cap \overline{F})$$

and that $(E \cap F)$ and $(E \cap \overline{F})$ are mutually exclusive. Thus, from the third axiom, we get

$$P(E) = P(E \cap F) + P(E \cap \overline{F})$$
$$P(E) - P(E \cap F) = P(E \cap \overline{F}).$$

Rearranging this equation gives the second property above involving $E \cap \overline{F}$.

Example 11.7 (At Least One Girl)

A family has three children. What is the probability that there is at least one girl among the three children?

Solution: If the family has three children and each child can be a boy or a girl, there are a total of $2 \times 2 \times 2 = 8$ possible ways this family could be formed. Let event $A =$ "at least one girl." We could list all the ways there could be at least one girl among three children, or we could look at the complement of A. That is, $\overline{A} =$ "all three children are boys." Now, there is only one way to have all three boys; thus, $P(\overline{A}) = \frac{1}{8}$. Thus,

$$P(A) = 1 - P(\overline{A}) = 1 - \frac{1}{8} = \frac{7}{8}.$$

From Example 11.7, we see that it is sometimes easier to answer a probability question by thinking about the complement of the event rather than thinking about the event itself.

Example 11.8 (A Little Wager)

Suppose that a friend asked you to make a little wager. She hands you two dice and says, "If you can roll double sixes at least once in 21 rolls, I'll buy you a pizza. However, if you don't roll the double sixes, then you have to buy me a pizza." Do you take the wager?

Solution: In each roll, there are $6 \times 6 = 36$ combinations. So, in 21 rolls, there are 36^{21} combinations (a very large number!). Let event $A =$ "do not get $(6, 6)$ in 21 rolls." Thus, $\overline{A} =$ "get at least one $(6, 6)$ in 21 rolls."

In one roll, there is only one way for $(6, 6)$ to occur, so there are 35 ways for $(6, 6)$ not to occur. Do this 21 times. There are 35^{21} ways for a $(6, 6)$ not to occur in 21 rolls. Thus,

$$P(\overline{A}) = 1 - P(A) = 1 - \frac{35^{21}}{36^{21}} = 0.4466.$$

Chances are less than one-half that you roll a double six at least once in 21 rolls!

Next, notice that $E \cup F$ is the union of the mutually exclusive events F and $E \cap \overline{F}$. Thus, using the third axiom and the previous probability law, we can write

$$P(E \cup F) = P(F) + P(E \cap \hat{F})$$
$$= P(F) + P(E) - P(E \cap F),$$

which is the first property with union and intersection.

Properties with Union and Intersection

Let E and F be events in a sample space.

1. $P(E \cup F) = P(E) + P(F) - P(E \cap F)$.
2. $P(\overline{E} \cap \overline{F}) = 1 - P(E \cup F)$.

Notice that if $E \cap F = \emptyset$, then the first property above reduces to the third probability axiom.

We can rearrange the first property above to get

$$P(E \cap F) = P(E) + P(F) - P(E \cup F). \qquad (11.1)$$

Now, by applying properties of complements twice and substituting in Equation (11.1), we can write

$$
\begin{aligned}
P(\overline{E} \cap \overline{F}) &= P(\overline{E}) - P(\overline{E} \cap F) \\
&= [1 - P(E)] - [P(F) - P(E \cap F)] \\
&= 1 - P(E) - P(F) + P(E \cap F) \\
&= 1 - P(E) - P(F) + [P(E) + P(F) - P(E \cup F)] \\
&= 1 - P(E \cup F),
\end{aligned}
$$

which is the second property with union and intersection.

Example 11.9 (Dice Rolling [16])

In rolling two dice, what is the probability of rolling a sum of 6 or doubles?

Solution: Let events $A =$ "sum of 6" and $B =$ "doubles." If we want to find $P(A \cup B)$, then we need to find $P(A)$, $P(B)$, and $P(A \cap B)$. In rolling two dice, there are $6 \times 6 = 36$ possible outcomes. The elementary events in each event are

- $A = \{(1,5), (2,4), (3,3), (4,2), (5,1)\}$,
- $B = \{(1,1), (2,2), (3,3), (4,4), (5,5), (6,6)\}$, *and*
- $A \cap B = \{(3,3)\}$.

Thus,

$$P(A \cup B) = P(A) + P(B) - P(A \cap B) = \frac{5}{36} + \frac{6}{36} - \frac{1}{36} = \frac{10}{36} = \frac{5}{18}.$$

Example 11.10 (Blood Typing in South America)

A study of a tribe in South America revealed that 75% were of blood type A and the rest had blood type O, 60% were Rh-negative, and 30% were both Rh-positive and blood type A. Find the probabilities of a person being

(a) type A or Rh-positive,
(b) type A and Rh-negative,
(c) Rh-positive but not type A, and
(d) type O and Rh-negative.

<u>Solution:</u> We can use a Venn diagram to help us figure out these probabilities.

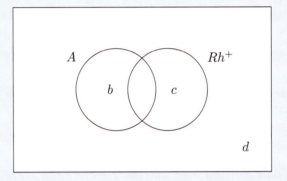

Let events A = "type A" and Rh^+ = "Rh-positive." We are given that $P(A) = 0.75$, $P(A \cap Rh^+) = 0.30$, and $P(\overline{Rh^+}) = 0.60$. From this, we can immediately devise that $P(\overline{A}) = 0.25$ and $P(Rh^+) = 0.40$ (by properties with complements). The desired events are shown in the Venn diagram. Combining properties with complements, unions, and intersections, we obtain the following:

(a) $P(A \cup Rh^+) = P(A) + P(Rh^+) - P(A \cap Rh^+) = 0.75 + 0.4 - 0.3 = 0.85$.
(b) $P(A \cap \overline{Rh^+}) = P(A) - P(A \cap Rh^+) = 0.75 - 0.3 = 0.45$. (See region b in the Venn diagram).
(c) $P(\overline{A} \cap Rh^+) = P(Rh^+) - P(A \cap Rh^+) = 0.4 - 0.3 = 0.1$. (See region c in the Venn diagram).
(d) $P(\overline{A} \cap \overline{Rh^+}) = 1 - P(A \cup Rh^+) = 1 - 0.85 = 0.15$. (See region d in the Venn diagram).

Properties with unions of two sets can be generalized. For example, let A, B, and C be events in sample space S then

$$P(A \cup B \cup C) = P(A) + P(B) + P(C) - [P(A \cap B) + P(A \cap C) + P(B \cap C)] + P(A \cap B \cap C).$$

Note that if A, B, and C are mutually exclusive so that they have no outcomes in common between any two or more of them, then

$$P(A \cup B \cup C) = P(A) + P(B) + P(C).$$

Example 11.11 (Beetles)

Suppose that we have a population of beetles in which 30% have wings and the rest are wingless. You select a beetle at random and record whether it has wings and then put it back. You do this three times. What is the probability that on at least one sample, you obtained a winged beetle?

Solution: You might think that the chance on each selection is 0.3; thus, over three tries, the chance of getting at least one beetle is $0.3 + 0.3 + 0.3 = 0.9$. However this certainly cannot be correct which you can see by considering what would happen if you selected four beetles for which a similar reasoning would give 1.2 for the probability. This is not good because by our second probability axiom, any event must have a probability that is between 0 and 1.

What is wrong with this thinking?

"At least one winged beetle" means selecting a winged beetle on one of the three tries, two of the three tries, or all three tries.

Let A = winged beetle on try 1, B = winged beetle on try 2, and C = winged beetle on try 3.

$$P(A \cup B \cup C) = P(A) + P(B) + P(C) - [P(A \cap B) + P(A \cap C) + P(B \cap C)]$$
$$+ P(A \cap B \cap C)$$
$$= 0.3 + 0.3 + 0.3 - [0.3 \times 0.3 + 0.3 \times 0.3 + 0.3 \times 0.3] + 0.3 \times 0.3 \times 0.3$$
$$= 0.657.$$

Great! But this involved several calculations.

Let us use properties of complements to reduce the number of calculations we must make. P(no winged beetle on one try)=0.7, and over three tries this gives the probability of no winged beetle to be $0.7^3 = 0.343$. Then P(at least one winged beetle) $= 1 - 0.343 = 0.657$. The latter method requires fewer calculations.

11.3 Probability Viewed as Darts Tossed at a Dart Board

One way to visualize an experiment is to think of the sample space as a dart board, in which you toss darts at the board and assume that any dart you toss has an equal chance of landing anywhere on the board. We also assume that any dart actually hits the board somewhere so that the board represents the entire set of possible outcomes of the experiment. The outcome of the experiment is the location that the dart hits on the board. Then, if we view the dartboard as having regions on it (such as a standard round board with circles of various radii starting from the small center circle), each region corresponds to an event in the sample space of the experiment of tossing a dart. The probability of an event is simply the probability that the dart lands inside the region that corresponds to this event. We can estimate this probability by tossing a large number of darts, N, and counting the number that land inside the region, M. Then the probability of the event is M/N. Another way to estimate the probability that a dart lands inside a particular region (because we assume that a dart is equally likely to land anywhere on the dart board) is simply by taking the area of the region, R, and dividing it by the total area of the board, A. If we toss a very large number of darts, we expect M/N to be close to R/A. In fact, one way to test the assumption that a dart is equally likely to hit any point on the board is to see if M/N gets closer and closer to R/A as we toss more and more darts; that is, we let N get bigger and bigger.

Standard dart boards have points assigned to darts landing in different regions. This means that you assign a number to the outcome of an experiment (the dart toss). This is an example of a *random variable*: it assigns a number to each possible outcome of an experiment. Note that, in this case, there is only one number assigned to where a dart lands (remember that this is what a function does), but that many different dart tosses can have the same number assigned if all of them land in the same region. This means that the function (random variable) is not invertible. That is, if someone tells you the point he received from a toss of the dart, you cannot tell where his dart landed exactly; all you can tell is the region in which it landed.

Many of the basic rules of probability can be thought of in terms of dart tosses. The probability that a dart lands in one of two regions, C or D, is simply the sum of the areas of these two regions if they don't overlap (i.e., they are disjoint sets). If C and D overlap, though, to find the probability that a dart lands in one or the other of these two regions if we add the areas of C and D, we will have counted *twice* the area that is in the intersection of C and D. This is why we have $P(C \cup D) = P(C \text{ or } D) = P(C) + P(D) - P(C \cap D)$. Also, the probability that event C doesn't occur is the area of the dartboard outside of C divided by the total area of the dartboard, or $1 - \text{Area}(C)/A$.

11.4 Matlab Skills

Estimating the Probability of Compound Events

Estimating the probability of compound events is similar to estimating the probability of a single event, only now we must keep track of multiple events.

Consider Example 11.9, in which two dice were rolled and we want to know the probability of rolling doubles or a sum of 6. We can easily simulate the rolling of two dice; however, now we must keep track of every time we roll doubles (i.e., the two numbers rolled are equal) and every time we roll a sum of 6. Otherwise, the probability of the event "doubles or a sum of 6" is computed in the same fashion as in Section 10.5. Below is an example of an m-file that computes the probability of rolling doubles or a sum of 6:

DiceRolling.m

```
1   % Filename: DiceRolling
2   % M-file to compute probability of rolling doubles or a sum of 6.
3
4   % Number of times to run experiment
5   N = 1000;
6
```

```
7     % Initialize sum of probabilities
8     C = 0;
9
10    for i = 1:N
11          % Set event counter to zero
12          c = 0;
13
14          for j = 1:N
15                % Roll two dice x and y
16                x = ceil(6*rand(1));
17                y = ceil(6*rand(1));
18
19                % If roll doubles or sum of 6, increase counter.
20                % Otherwise do nothing.
21                if (x==y || (x+y)==6)
22                    c = c + 1;
23                end
24          end
25
26          % Add on prob of event after rolling N times
27          C = C + c/N;
28    end
29
30    % Print out result
31    fprintf('P(doubles OR sum of 6) = %6.4f\n',C/N);
```

Generating Genotypes with Multiple Allele Pairs

In this chapter, we looked at genetic traits that are determined by more than one gene, such as blood type with Rh factor (one gene for blood type, another gene for Rh factor) and eye color (a brown/blue gene and a blue/green gene; for more on eye color genes, see Student Projects for this Unit for the Eye Color Project).

Another example of this comes from the classical pea plant experiments by Gregor Mendel. Mendel did many of his experiments on the pea plant *Pisum sativum*. In these experiments, he traced many traits through several generations of pea plants. Two of those traits were whether the seeds produced were spherical or wrinkled and whether the seeds produced were green or yellow. As it turns out, these two traits for *Pisum sativum* seeds are determined by two separate genes. The gene for spherical/wrinkled seeds can have the *S* allele (dominant) for spherical seeds and the *s* allele (recessive) for wrinkled seeds. The gene for yellow/green seeds can have the *Y* allele (dominant) for yellow seeds and *y* allele (recessive) for green seeds.

If we want to determine the probability of an event such as $E =$ "offspring has yellow wrinkled seeds," we need to be able to simulate the experiment of two parent plants (each with two seed genes) producing an offspring with two seed genes, assuming that the two genes are independent. Below is the Matlab function OneChild2Genes that completes this task. This function is similar to the OneChild function in Section 10.5. The inputs are the genotypes of each parent (for both genes), and the output is the genotype of the offspring:

―――――――――――――――――――― OneChild2Genes.m ――――――――――――――――――――

```
1     % function child = OneChild2Genes(mom,dad)
2     %
3     % Inputs:
4     %    mom = genotype of mother**
```

```
5    %     array containing genoypte of first & second genes
6    %  dad = genotype of father**
7    %     array containing genoypte of first & second genes
8    % ** For genotypes use
9    %     1 for heterozygous, Aa or aA
10   %     2 for homozygous dominant, AA
11   %     3 for homozygous recessive, aa
12   %
13   % Output:
14   %  child = genotype of offspring array containing
15   %     genotype for first gene and second genen
16   function child = OneChild2Genes(mom,dad)
17
18   % Within this function we will use
19   %    0 to represent a recessive allele
20   %    1 to represent a dominant allele
21
22   % Loop through both genes
23   for i = [1 2]
24       % determine allele for gene i from mother
25       if mom(i) == 1
26           %if mom heterozygous, randomly choose between two alleles
27           childallele1 = floor(2*rand(1));
28       elseif mom(i) == 2
29           %if mom homozygous dominant, then child inherits A
30           childallele1 = 1;
31       else
32           %if mom homozygous dominant, then child inherits a
33           childallele1 = 0;
34       end
35
36       % determine allele for gene i from father
37       if dad(i) == 1
38           %if dad heterozygous, randomly choose between two alleles
39           childallele2 = floor(2*rand(1));
40       elseif dad(i) == 2
41           %if dad homozygous dominant, then child inherits A
42           childallele2 = 1;
43       else
44           %if dad homozygous dominant, then child inherits a
45           childallele2 = 0;
46       end
47
48       % Determine the genotype of the child for gene i
49       if (childallele1 == 1 && childallele2==1)
50           child(i) = 2;
51       elseif (childallele1==0 && childallele2==0)
52           child(i) = 3;
53       else
54           child(i) = 1;
55       end
56   end
```

Notice the differences in this function from the OneChild function. The function OneChild2Genes now takes in 1×2 vectors of genotype information for each parent and then loops through each gene. The code contained inside the loop is the same as the code contained in OneChild except in OneChild2Genes, the ith gene is dealt with through each pass of the loop.

11.5 Exercises

11.1   Consider the following using the sample space $\{1, 2, 3, 4, 5, 6\}$ and events $A = \{1, 3, 5\}$ and $B = \{1, 2, 3, 4\}$.

(a) Find $A \cup B$ and $A \cap B$.
(b) Find $\overline{A}$.
(c) Find $\overline{A \cup B}$.
(d) Are A and B mutually exclusive?

11.2   In the Venn diagram shown below, A is the event "female" and B is the event "left-handed." Explain the meaning of the areas labeled a, b, and c in the diagram.

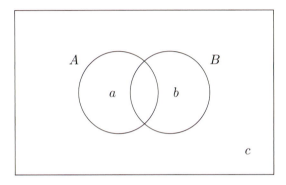

11.3   (From [16]) In the Venn diagram shown below, M is the event "male," H is the event "a heavy smoker," and E is the event "developed emphysema."

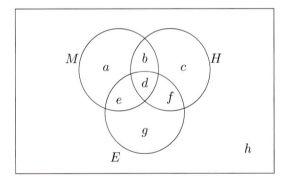

(a) Express in words the event labeled e.
(b) Locate the event "a female with emphysema who is not a heavy smoker."
(c) Locate the event $\overline{M} \cap \overline{H} \cap \overline{E}$ in the Venn diagram.

11.4   For a roll of two dice, let A be the event "a sum of 10," B the event "at least one 6," and C the event "doubles." Explain the meaning of the following events in words and list the corresponding elementary events.

(a) $A \cup C$
(b) $A \cap B$
(c) $C \cap \overline{B}$

11.5 A coin is tossed three times. Let B be the event "at least one tail" and C the event "a head on the third toss." List all elementary events in each of the following events.

(a) $\overline{B}$

(b) $B \cap C$

(c) $C - B$

11.6 Assume that $P(A \cap B) = 0.2$, $P(A) = 0.4$, and $P(\overline{A} \cap \overline{B}) = 0.3$. Find $P(B)$.

11.7 Assume that $P(A) = 0.4$, $P(B) = 0.4$, and $P(A \cup B) = 0.7$. Find $P(\overline{A} \cap \overline{B})$.

11.8 Which of the following pairs of events are mutually exclusive?

(a) For a roll of two dice, "a sum of 9" and "at least one 2"

(b) For a person to be blood type AB and $Rh-$negative

(c) For a single card selected from a standard deck to be "red" and "a spade"

11.9 (From [16]) In the population of Dry Gulch, 75% of the population are cowboys and 90% beer drinkers. Only 5% stay away from beer and horses.

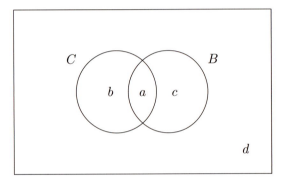

(a) Suppose that C is the event "cowboy" and that B is the event "beer drinker." Express, in words, the events a, b, c, and d.

(b) Find the proportion of the population that are beer-drinking cowboys.

11.10 For infections of humans by *Mycobacterium tuberculosis*, which causes the disease tuberculosis, 35% of infections are resistant to the drug rifampin, 50% are resistant to the drug ethambutol, and 15% are resistant to both drugs. In the Venn diagram below, the event R reperesents those with tuberculosis who are resistant to rifampin, and the event E represents those with tuberculosis who are resistant to ethambutol.

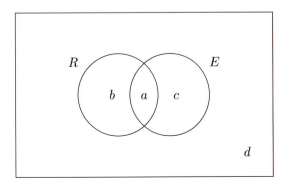

(a) Express in words event *a*. What proportion of humans infected with
    Mycobacterium tuberculosis are in event *a*?
(b) What fraction of tuberculosis infections are resistant to only rifampin or ethambutol
    but not both? Express this event in terms of *a*, *b*, *c*, and *d*.

11.11 A meerkat study classified individuals as helping to feed pups (*F*) or not and helping to
guard the colony (*G*) or not. The study also distinguished whether male (*M*) or female
($\overline{M}$) meerkats did each of these jobs.

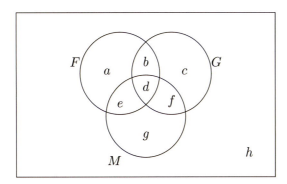

It was found that 12% of all meerkats were males that helped feed pups and guard,
25% were females that helped feed pups and guard, 6% were males that fed pups but
did not guard, and 8% were females that guarded but did not feed pups. Overall, 58%
of the meerkats helped feed pups, 55% helped guard, and 50% were males.

(a) What percentage of meerkats studied were males that did not feed pups or guard?
    Indicate what event this is on the Venn diagram.
(b) What percentage of meerkats studied were females that did not feed pups or guard?
    Indicate what event this is on the Venn diagram.

Conditional Probability

Often in biology, situations arise in which information can be gained by making observations or producing an experimental result that might change the conclusion or outcome of an observation. An important practical example occurs in testing for a disease in which some test (possibly on blood or saliva) is used to indicate whether an individual has the disease or is a carrier. Often, such testing is not for the disease directly but rather for some trait that is indicative of the disease; for example, a high cholesterol level might be an indicator of potential heart disease. However, these tests are not perfect. For example, a urine-based pregnancy test may indicate that a woman is not pregnant, but she may be pregnant, particularly if the test is performed soon after implantation of an egg. Performing a test provides some additional information, but no test is completely accurate. In this chapter, we will show how to account for this, including how to update the probability that someone has a disease if there is additional information from a test that is performed. Similarly, if there is a genetic component to a disease and we have information about an individual's siblings, we can obtain additional information that modifies whether an individual is at risk of having the disease. To answer these questions, we will first consider the notion of conditional probability, that is, finding the probability that an event occurs given that you know that some other event occurs. An example of this is finding the probability that someone indeed does have a disease when a test is performed that indicates that they do not have the disease.

The ideas in this chapter underlie an entire approach to data analysis in science called Bayesian methods. The notion is that science proceeds by updating our understanding of the world through experiments that revise our prior understanding. So, performing a large number of blood tests for a particular disease updates our information about the prevalence of that disease and thus enhances our ability to accurately determine the fraction of the population tested who have the disease. Also in this chapter, we will discuss the notion of independence, in which knowledge about the outcome of an experiment provides no additional information about a different observation.

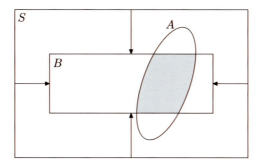

Figure 12.1 The shaded area is the event $A \cap B$.

12.1 Conditional Probability

For an event B with $P(B) > 0$, we define the ***conditional probability*** of event A given B, meaning the probability of event A occurring given that event B has occurred, by

$$P(A|B) = \frac{P(A \cap B)}{P(B)}. \tag{12.1}$$

When we know that a certain event B has occurred, the sample space reduces to B, so we look for where event A overlaps event B. See Figure 12.1.

   Notice that this ensures that $P(B|B) = 1$. The conditional probability of the complement of an event is given by

$$P(\overline{A}|B) = 1 - P(A|B).$$

However, in general, $P(A|\overline{B}) \neq 1 - P(A|B)$.

   Let us look at two examples of conditional probability. The first example investigates the probability of being a carrier for Tay-Sachs disease.

Example 12.1 (Tay-Sachs Disease [16])

Tay-Sachs disease is a serious disorder of the nervous system that usually results in death by age 2 or 3. Affected individuals have genotype tt, while normal (nonaffected) individuals have genotype Tt or TT.

   Judy has a little brother with Tay-Sachs disease and is worried that she may carry the recessive allele. What is the probability of this?

Solution: First, notice that both of Judy's parents must be Tt if her little brother has the disease. We assume here that neither parent can be tt since each lived to adulthood.

		♀	
		T	t
♂	T	TT	Tt
	t	tT	tt

(Continued)

Since Judy does not have the disease, we eliminate tt as a possibility for her genotype. Looking at the Punnett square, we intuitively guess that the probability should be $\frac{2}{3}$. Let us show that this is true by using the definition of conditional probability.

Let $A =$ "Judy is Tt" and $B =$ "Judy is not tt." Then

$$P(A|B) = \frac{P(A \cap B)}{P(B)}.$$

Looking at the Punnett square, we see that three out of the four possible genotypes for Judy are not tt. Thus, $P(B) = \frac{3}{4}$. The event $A \cap B$ corresponds to Judy being genotype Tt. Since two of the four possible genotypes are Tt, $P(A \cap B) = \frac{2}{4} = \frac{1}{2}$. Therefore,

$$P(A|B) = \frac{1/2}{3/4} = \frac{1}{2} \times \frac{4}{3} = \frac{2}{3}.$$

Thus, the probability that Judy is a carrier for Tay-Sachs disease given that she is normal (i.e., not tt) is $\frac{2}{3}$.

This next example considers drug testing in which a control, in this case a placebo pill, is used. Given the results of the testing, by using conditional probability, we can ask, "What is the probability the drug will be effective given that you actually take the drug and not the placebo?"

Example 12.2 (Drug Testing [16])

A test for a new sleeping pill involved 200 individuals in which 100 of the individuals were given the sleeping pill and the other 100 received a sugar pill. The results of the test are shown in the following table.

	Improved sleep	Did not sleep better
Sleeping pill	71	29
Sugar pill	58	42

What is the probability that if you take the sleeping pill you will sleep better?

Solution: Let event $A =$ "took sleeping pill" and event $B =$ "slept better." We want to find the probability that you will sleep better *given* that you took the sleeping pill, that is, $P(B|A)$. To use the definition of conditional probability, we need to find $P(A \cap B)$ and $P(A)$. The event $A \cap B$ corresponds to individuals who took the sleeping pill and slept better; 71 of the 200 tested individuals fall into this category. Thus, $P(A \cap B) = 71/200 = 0.355$. Since 100 of the tested individuals were given the sleeping pill, $P(A) = 100/200 = 0.5$. Thus,

$$P(B|A) = \frac{P(A \cap B)}{P(A)} = 0.355/0.5 = 0.71.$$

Thus, the chance that you will sleep better if you take the sleeping pill is 71%.

Notice that by multiplying both sides of Equation (12.1) by $P(B)$, we can get an expression for $P(A \cap B)$:

$$P(B)P(A|B) = \frac{P(A \cap B)}{P(B)}P(B)$$

$$P(B)P(A|B) = P(A \cap B).$$

As above, we can obtain two expressions for $P(E \cap F)$ using the definition of conditional probability. Let E and F be events in sample space S. If $P(F) \neq 0$ and $P(E) \neq 0$, then

$$P(E \cap F) = P(E|F)P(F)$$

and

$$P(E \cap F) = P(F|E)P(E).$$

Example 12.3 (Beetle Sampling without Replacement)

We have a population of 150 beetles. Thirty percent have wings, and the rest are wingless. You select a beetle at random and record whether it has wings and do not put it back. Then you select another beetle. What is the probability of the following?

(a) The first and second beetle both have wings.
(b) The second beetle has wings (regardless of whether the first beetle had wings).

Solution: Let event A = "first beetle is winged" and event B = "second beetle is winged."

(a) We want to compute $P(A \cap B)$. We should decide to use $P(A \cap B) = P(A|B)P(B)$ or $P(A \cap B) = P(B|A)P(A)$. To determine this, we need to see what information we are given. We are told that of the 150 beetles, 30% have wings. That is, 45 of the 150 have wings. Thus, for the first beetle picked, there is a 30% chance that it will have wings. Therefore, $P(A) = 0.3$. Now, we can see that we will want to use $P(A \cap B) = P(B|A)P(A)$, so we need to find $P(B|A)$. The event $B|A$ is the event that the second beetle has wings, given that the first beetle has wings. If there were 45 winged beetles to start with and the first beetle picked had wings, then of the remaining 149 beetles left to choose from, 44 have wings. Thus, $P(B|A) = \frac{44}{149} = 0.295$, and

$$P(A \cap B) = P(B|A)P(A) = 0.295 \times 0.3 = 0.0885.$$

Thus, the chance that both the first and the second beetles you pick have wings is 8.85%.

(*Continued*)

(b) Here, we want to compute $P(B)$. Recall that we can think of event B as

$$B = (A \cap B) \cup (\overline{A} \cap B)$$

and, by the third probability axiom,

$$P(B) = P(A \cap B) + P(\overline{A} \cap B).$$

In part (a), we computed $P(A \cap B) = 0.0885$. Thus, we need only compute $P(\overline{A} \cap B)$. Since $P(A) = 0.3$, by the property of complements, $P(\overline{A}) = 1 - 0.3 = 0.7$, where event $\overline{A} =$ "first beetle is *not* winged." If there are $150 - 45 = 105$ wingless beetles to start with and the first beetle picked does not have wings, then of the remaining 149 beetles to choose from, there are still 45 that do have wings. Thus, $P(B|\overline{A}) = \frac{45}{149} = 0.302$, and

$$P(\overline{A} \cap B) = P(B|\overline{A})P(\overline{A}) = 0.302 \times 0.7 = 0.211.$$

Therefore, we have

$$P(B) = P(A \cap B) + P(\overline{A} \cap B) = 0.0885 + 0.211 = 0.2995.$$

Thus, the chance that the second beetle has wings (regardless of whether the first beetle has wings) is 29.95%.

12.2 Independence

Assuming that both $P(A)$ and $P(B)$ are positive, when the occurrence of event B does not change the probability that event A will occur and vice versa so that the occurrence of event A does not change the probability that event B occurs, we call events A and B **independent**. Mathematically, we say that events A and B are independent if $P(A|B) = P(A)$ and $P(B|A) = P(B)$.

If A and B are independent, the probability of the intersection takes a special form.

Property of Independent Events

If $P(A) > 0$, $P(B) > 0$, and A and B are independent events, then

$$P(A \cap B) = P(A)P(B).$$

Notice that when two events are mutually exclusive, the probability of either one or the other event or both occurring is the sum of the probabilities of each event. When two events are independent, the probability of both events occurring is the product of the probabilities of each event. Two events being mutually exclusive is very different from those two events being independent since the probability of both events happening, for mutually exclusive events, is zero.

If A and B are independent events, then we can compute the conditional probability:

$$P(A|B) = \frac{P(A \cap B)}{P(B)} = \frac{P(A)P(B)}{P(B)} = P(A).$$

So, we see that when events A and B are independent, knowing the event B has occurred does not change the probability of event A occurring. Notice that if $P(A|B) = P(A)$, then

$$P(B|A) = \frac{P(A \cap B)}{P(A)} = \frac{P(A|B)P(B)}{P(A)} = \frac{P(A)P(B)}{P(A)} = P(B).$$

In calculating probabilities, if we can show that either $P(A|B) = P(A)$ or $P(B|A) = P(B)$ then we can assume that events A and B are independent.

Recall that if events A and B are mutually exclusive, then $P(A \cap B) = 0$. Thus, if $P(B) \neq 0$, by the definition of conditional probability,

$$P(A|B) = \frac{P(A \cap B)}{P(B)} = 0.$$

Can events A and B be both independent *and* mutually exclusive? Suppose that they are; then $P(A|B) = P(A)$ and $P(A|B) = 0$, and thus $P(A) = 0$. Furthermore, $P(B|A) = P(B)$ and $P(B|A) = 0$, and thus $P(B) = 0$. Therefore, if events A and B are both independent and mutually exclusive, then $P(A) = 0$ and $P(B) = 0$, which is a contradiction since independent events are assumed to have positive probability. We will not be considering events that are both independent and mutually exclusive.

Example 12.4 (Independence in Dice Rolling)

Two dice are tossed one at a time. Let

A = "6 on first die,"
B = "sum of 7," and
C = "sum of 8."

Which events are independent? Which events are mutually exclusive?

Solution: Clearly, B and C are mutually exclusive with nonzero probabilities and therefore are not independent.

There are six possible outcomes on a die and only one way to roll a 6. Thus, $P(A) = \frac{1}{6}$.

In rolling two dice, there are $6 \times 6 = 36$ possible outcomes. There are six ways to get a "sum of 7": $\{(1, 6), (2, 5), (3, 4), (4, 3), (5, 2), (6, 1)\}$. Thus, $P(B) = \frac{6}{36} = \frac{1}{6}$. There are five ways to get a "sum of 8": $\{(2, 6), (3, 5), (4, 4), (5, 3), (6, 2)\}$. Thus, $P(C) = \frac{5}{36}$. Event $A \cap B$ can occur only with a 6 on the first die and a 1 on the second die. Thus, $P(A \cap B) = \frac{1}{36}$. Event $A \cap C$ can occur only with a 6 on the first die and a 2 on the second die. Thus, $P(A \cap C) = \frac{1}{36}$.

$$P(A|B) = \frac{P(A \cap B)}{P(B)} = \frac{1/36}{6/36} = \frac{1}{6}$$

(Continued)

and

$$P(B|A) = \frac{P(A \cap B)}{P(A)} = \frac{1/36}{1/6} = \frac{1}{6}.$$

Thus, A and B are independent (and not mutually exclusive).

$$P(A|C) = \frac{P(A \cap C)}{P(C)} = \frac{1/36}{5/36} = \frac{1}{5}.$$

Thus, A and C are not independent and not mutually exclusive.

When is it reasonable to assume two events are independent? Some examples are:

1. When the result of the second sample is independent of the first which arises if the first sample is placed back into the sampling pool
2. When the sex of the second child is independent of that of the first child.
3. When there is random selection of alleles on genes located on separate chromosomes (Mendel's second law of independent assortment)
4. When the genotypes of nonrelated individuals are independent due to the individuals being chosen from completely different groups

First, we look at a modified version of the beetle sampling example for which the first beetle chosen from the population is placed back into the population before sampling the second beetle.

Example 12.5 (Beetle Sampling with Replacement)

Suppose that we have a population of 150 beetles. Forty-five of the beetles have wings, and the rest are wingless. You select a beetle at random and record whether it has wings and then put it back amongst the other beetles. Then you select another beetle. What is the probability of the following?

(a) The first and second beetle both have wings.
(b) The first beetle has wings, and the second beetle does not have wings.

Solution : Let events $A =$ "first beetle has wings," $B =$ "second beetle has wings," and $C =$ "second beetle does not have wings." The probability that the first beetle has wings is $P(A) = \frac{45}{150}$. Since we place the first beetle back into the population, the conditions return to what they were before selecting the first beetle: 45 winged beetles and 105 wingless beetles. Then,

(a) $P(B) = \frac{45}{150}$ and $P(A \cap B) = P(A)P(B) = \frac{45}{150} \frac{45}{150} = \frac{3}{10} \frac{3}{10} = \frac{9}{100}$ or 9%.

(b) $P(C) = \frac{105}{150}$ and $P(A \cap C) = P(A)P(C) = \frac{45}{150} \frac{105}{150} = \frac{3}{10} \frac{7}{10} = \frac{21}{100}$ or 21%.

Thus, there is a 9% chance that both the first and the second beetles will have wings and a 21% chance that the first beetle will have wings while the second beetle is wingless.

Next, we will return to the example of blood typing and Rh classification. Since the genes that determine blood type and Rh classification are separate, the events of having particular genotypes with respect to each gene are independent.

Example 12.6 (Blood Typing)

Suppose that Jacob has O^+ blood with genotype OO/Rr and that Anna has B^- blood with genotype BO/rr. What is the probability that a child of Jacob and Anna will have (a) B^+ blood and (b) O^- blood?

Solution: Let us start by making a Punnett square for each gene.

	♀	
	B	O
♂ O	OB	OO
O	OB	OO

The probability of Jacob and Anna's child being type B is $P(B) = \frac{2}{4} = \frac{1}{2}$ and being type O is $P(O) = \frac{1}{2}$.

	♀	
	r	r
♂ R	Rr	Rr
r	rr	rr

The probability of their child being Rh-positive is $P(Rh^+) = \frac{2}{4} = \frac{1}{2}$ and being Rh-negative is $P(Rh^-) = \frac{1}{2}$.

Thus,

(a) the probability that a child of Jacob and Anna's is B^+ is
$P(B \cap Rh^+) = P(B)P(Rh^+) = \frac{1}{2} \cdot \frac{1}{2} = \frac{1}{4}$, and

(b) the probability that a child of Jacob and Anna's is O^- is
$P(O \cap Rh^-) = P(O)P(Rh^-) = \frac{1}{2} \cdot \frac{1}{2} = \frac{1}{4}$.

Next, we look at an example of the probability of two possible carriers for Tay-Sachs disease having a child with Tay-Sachs disease.

Example 12.7 (Tay Sachs Disease [16])

Jack and Judy each had brothers afflicted with Tay-Sachs disease. From Example 12.1, the probability of being a carrier if your sibling has Tay-Sachs disease is $\frac{2}{3}$. What is the probability that Jack and Judy have a child with Tay-Sachs disease?

Solution: The only way the child can be tt is if both Jack and Judy are Tt. Thus, let events A = "Judy is Tt," B = "Jack is Tt," and C = "child is tt." Since A and B are reasonably assumed to be independent, $P(A \cap B) = P(A)P(B) = \frac{4}{9}$, $P(C|A \cap B) = \frac{1}{4}$, and $P(C) = P(C \cap (A \cap B)) = P(C|A \cap B)P(A \cap B) = \frac{1}{4} \times \frac{4}{9} = \frac{1}{9}$.

It should be noted that the independence property can be extended to multiple independent events. For example, if A, B, and C are independent events, then

$$P(A \cap B \cap C) = P(A)P(B)P(C).$$

Example 12.8 (Family Planning)

Assume that the births of boys and girls are equiprobable and independent.

(a) What is the probability that in a family with two children, both the children girls?
(b) What is the probability that in a family with three children, all the children are girls?
(c) What is the probability that in a family with six children, all the children are the same sex?
(d) What is the probability that in a family with six children, there is at least one girl?

Solution:

(a) Let events A = "first child is a girl" and B = "second child is a girl." These are independent events; thus,

$$P(A \cap B) = \frac{1}{2} \times \frac{1}{2} = \frac{1}{4} = 0.25.$$

Thus, the probability that in a family with two children both will be girls is 0.25 or 25%.

(b) Let events A = "first child is a girl," B = "second child is a girl," and C = "third child is a girl." These are independent events; thus,

$$P(A \cap B \cap C) = \frac{1}{2} \times \frac{1}{2} \times \frac{1}{2} = \frac{1}{8} = 0.125.$$

Thus, the probability that in a family with three children all three will be girls is 0.125 or 12.5%.

(Continued)

(c) To compute the probability that all the children are of the same sex we need to compute the probability that all the children are girls and the probability that all the children are boys. That is,

$$P(\text{same sex}) = P(\text{all girls}) + P(\text{all boys}).$$

Let events $A_1, A_2, \ldots, A_6$ be the events that the first, second, $\ldots$, sixth (respectively) child is a girl. Then

$$P(\text{all girls}) = P(A_1 \cap A_2 \cap \cdots \cap A_6) = \left(\frac{1}{2}\right)^6.$$

Similarly,

$$P(\text{all boys}) = P\left(\overline{A_1} \cap \overline{A_2} \cap \cdots \cap \overline{A_6}\right) = \left(\frac{1}{2}\right)^6.$$

Thus,

$$P(\text{same sex}) = \left(\frac{1}{2}\right)^6 + \left(\frac{1}{2}\right)^6 = \frac{1}{32} = 0.03125.$$

Therefore, the probability that in a family of six children all the children are of the same sex is 0.03125 or 3.125%.

(d) Let $A =$ "at least one girl." Then $\overline{A} =$ "no girls," that is, all boys. Then

$$P(\overline{A}) = \left(\frac{1}{2}\right)^6,$$

and thus

$$P(A) = 1 - \left(\frac{1}{2}\right)^6 = 0.9844.$$

Therefore, in a family with six children, there is 98.44% chance that at least one of the children will be a girl.

12.3 Matlab Skills

To estimate the conditional probability $P(A|B)$ of an experiment $\mathcal{E}$, we simulate the experiments (as in Chapters 10 and 11), but now we must keep track of two events, $A \cap B$ and B, so that we can compute

$$P(A|B) = \frac{P(A \cap B)}{P(B)}.$$

Biological Example of Conditional Probability: Tay-Sachs Disease

Consider Example 12.1, in which Judy has a little brother who has Tay-Sachs disease and is worried that she might be a carrier of the disease. From the example, we know that each of Judy's parents is a carrier for Tay-Sachs disease.

GENERATING THE OFFSPRING'S GENOTYPE

We can use the OneChild function developed in Section 10.5 to simulate the "experiment" of the generation of Judy's genotype. Recall that the function OneChild takes in two integer values from the set $\{1, 2, 3\}$ representing the genotype of the mother and the father, where

- 1 = heterozygous genotype, Aa or aA;
- 2 = homozygous dominant, AA; and
- 3 = homozygous recessive, aa.

The output of the file is an integer value from the set $\{1, 2, 3\}$, which represents the genotype of the offspring.

COMPUTING THE CONDITIONAL PROBABILITY

Using the OneChild function, we can repeat the experiment many times and compute the probability. However, now we must keep track of how many times two different events occur. Let A = "Judy is heterozygous" and $\overline{B}$ = "Judy is not homozygous recessive." Notice $A \cap \overline{B} = A$. Thus,

$$P(A|\overline{B}) = \frac{P(A \cap \overline{B})}{P(\overline{B})} = \frac{P(A)}{P(\overline{B})}.$$

Therefore, in order to compute the probability that Judy is a carrier given that she does not have the disease, we must keep track of the number of experiments that produce a heterozygous genotype and the number of experiments that produce a homozygous recessive genotype. The code to do this is shown below and is called TaySachs.m:

──────── TaySachs.m ────────

```
1   % Filename: TaySachs.m
2   % M-file for determing P(A|not B) where
3   %     A = offspring is a carrier
4   %     B = offspring does have TaySachs
5   % given that both parents are carriers
6
7   % Set how many times to run the experiment
8   N = 1000;
9
10  % Set variables to sum up the probabilities of events
11  A = 0;        % probability of Aa or aA
12  B = 0;        % probability of aa
13
14  % Run set of N experiments N times
15  for i = 1:N
16      % Set counters
17      a = 0;    % heterozygous counter
18      b = 0;    % homozygous recessive counter
19
20      % Run experiment N times
21      for j = 1:N
22          % Get child's genotype if both parents are carriers
23          child = OneChild(1,1);
24
25          % Increase appropriate counter
26          if child == 1
27              a = a + 1;
```

```
28        elseif child == 3
29              b = b + 1;
30        else
31              % do nothing
32        end
33    end
34
35    % Add the probabilities for this set of N experiments
36    % to the running sum
37    A = A + a/N;
38    B = B + b/N;
39    end
40
41    % Estimated Probability of Event A
42    PA = A/N;
43
44    % Estimated Probability of Event NOT B
45    PB = 1 - B/N;
46
47    % Print out results
48    fprintf('P(carrier|no disease)=%6.4f\n',PA/PB)
```

Notice that since the code inside of the loops kept track of when a homozygous recessive genotype was generated and we wanted the probability of when this did not happen, at the end we calculated

$$P(\overline{B}) = 1 - P(B),$$

where $B =$ "Judy is homozygous recessive."

When one of the authors ran TaySachs.m five times in the command window, the outputs were as follows:

```
─────────────── Command Window ───────────────
>> TaySachs
P(carrier|no disease)=0.6670

>> TaySachs
P(carrier|no disease)=0.6671

>> TaySachs
P(carrier|no disease)=0.6664

>> TaySachs
P(carrier|no disease)=0.6661

>> TaySachs
P(carrier|no disease)=0.6677
```

Biological Example of Conditional Probability: Drug Testing

Recall Example 12.2, in which a new sleeping pill is being tested. Of the 200 individuals participating in the study, 100 are given that new sleeping pill, and the remaining 100 are given a sugar pill (a placebo). The results of the study are given in the following table.

	Improved sleep	Did not sleep better
Sleeping pill	71	29
Sugar pill	58	42

In Example 12.2, we asked: What is the probability that if you take the sleeping pill, you will sleep better? Let A = "took sleeping pill" and B = "slept better." We want to estimate

$$P(B|A) = \frac{P(A \cap B)}{P(A)}.$$

To do this, first we need to simulate an experiment in which we randomly choose one of the 200 study participants and then check whether they took the sleeping pill or sugar pill and whether they had improved sleep. Once we have a Matlab function to simulate this experiment, we can repeat this experiment several times, keeping track of the events $A \cap B$, "took sleeping pill AND had improved sleep," and A, "took sleeping pill." By keeping track of these events, we can estimate the probability of sleeping better given that you took the sleeping pill.

SIMULATING THE EXPERIMENT

To simulate the experiment, we will construct a 200×2 matrix where each row represents an individual who participated in the study, the first column represents whether the individual took a sugar pill or sleeping pill, and the second column represents whether the individual had improved sleep. In the first column, we will use 0 to represent "took sugar pill" and 1 to represent "took sleeping pill." In the second column, we will use 0 to represent "did not have improved sleep" and 1 to represent "did have improved sleep." Note that this Matlab function will have no input. However, the output will be a 1×2 matrix, specifically the row of the 200×2 matrix representing the individual who was selected:

```
                        DrugTesting.m
1    % function out = DrugTesting
2    %   Inputs: none
3    %
4    %   Output:
5    %      out = 1x2 matrix representing results for 1 individual
6    %         column 1: 0 = sugar pill, 1 = sleeping pill
7    %         column 2: 0 = no improved sleep, 1 = improved sleep
8    function out = DrugTesting
9
10   % Create matrix of drug testing results
11   Results = zeros(200,2);
12
13   % Set first 1/2 of column 1 to "took sleeping pill"
14   Results(1:100,1) = ones(100,1);
15   % Set first 71 of column 2 to "improved sleep"
16   Results(1:71,2) = ones(71,1);
17   % Now rows 1-71 = "took sleeping pill" & "had improved sleep"
18   % and rows 72-100 = "took sleeping pill" & "no improved sleep"
19
20   % Set rows 101-158 of column to to "improved sleep"
21   Results(101:158,2) = ones(58,1);
22   % Now rows 101-158 = "took sugar pill" & "had improved sleep"
23   % and rows 159-200 = "took sugar pill" & "no improved sleep"
24
25   % Randomly pick one of the 200 individuals
```

```
26    x = ceil(200*rand(1));   % generates integers from 1 to 200
27
28    % Create function output
29    out=Results(x,:);
```

Notice that the `DrugTesting` function is equally likely to choose any of the 200 participants.

COMPUTING THE CONDITIONAL PROBABILITIES

Now that we have a function to conduct the experiment of choosing of the drug study participants and to display their results, we can replicate this experiment many times and keep track of the results to estimate various probabilities. We will write an m-file to estimate the probability that a participants had improved sleep given that they took the sleeping pill:

─────── SleepingPillEfficacy.m ───────

```
1    % Filename: SleepingPillEfficacy.m
2    % M-file to estimate probability of sleeping better given that you take
3    %   a sleeping pill.
4
5    % Set how many times to run the experiment
6    N = 1000;
7
8    % Set variables to sum up the probabilities of events
9    A = 0;         % probability took sleeping pill
10   B = 0;         % probability took sleeping pill + had improved sleep
11
12   % Run set of N experiments N times
13   for i = 1:N
14       % Set counters
15       a = 0;    % took sleeping pill counter
16       b = 0;    % took sleeping pill + had improved sleep counter
17
18       % Run experiment N times
19       for j = 1:N
20           % Select one participant
21           x = DrugTesting;
22
23           % If participant took sleeping pill
24           if x(1,1)==1
25               a = a + 1;
26
27               % If participant also had improved sleep
28               if x(1,2)==1
29                   b = b + 1;
30               end
31           end
32       end
33
34       % Add the probabilities for this set of N experiments
35       % to the running sum
36       A = A + a/N;
37       B = B + b/N;
38   end
39
40   % Print out result
41   fprintf('P(improved sleep|took sleeping pill)=%6.4f\n',(B/N)/(A/N));
```

Notice the nested if statements. First, we check to see if the particular participant selected took the sleeping pill. If they did, then we increase the "took sleeping pill" counter and check to see if they also had improved sleep. If they also had improved sleep, then we additionally increase the "took sleeping pill + had improved sleep" counter.

When one of the authors ran `SleepingPillEfficacy.m` five times in the command window, the outputs were as follows:

```
──────────────────── Command Window ────────────────────
>> SleepingPillEfficacy
P(improved sleep|took sleeping pill)=0.7112

>> SleepingPillEfficacy
P(improved sleep|took sleeping pill)=0.7100

>> SleepingPillEfficacy
P(improved sleep|took sleeping pill)=0.7104

>> SleepingPillEfficacy
P(improved sleep|took sleeping pill)=0.7098

>> SleepingPillEfficacy
P(improved sleep|took sleeping pill)=0.7098
```

12.4 Exercises

12.1   Decide if events A and B are independent using conditional probability.

  (a) Two dice are tossed. Let $A =$ "sum of 8" and $B =$ "both numbers are even."
  (b) Select a single card from a standard deck. Let $A =$ "a heart" and $B =$ "an ace."
  (c) A couple have known blood genotypes AB and BO. Let $A =$ "their child has genotype BO" and $B =$ "their child has blood type B."

12.2   (From [16]) One thousand entering college freshmen were polled to study the relationship between hair and eye color. Results are shown in the table below. If event $A =$ "light hair" and event $B =$ "light eyes," estimate $P(A|B)$, $P(B|A)$, $P(\overline{A}|B)$, and $P(\overline{B}|\overline{A})$.

	Light hair	Dark hair
Light eyes	350	120
Dark eyes	230	300

12.3   (From [16]) In a study of survival rates of elk on a game preserve, a large group of young elk were followed over a 10-year period. Eighty percent were alive after 5 years, while 50% were still living after 10 years. Estimate the probability that a 5-year-old elk will live an additional 5 years.

12.4   The cure rate for a certain disease is 80%. Two people have recently become infected with this disease. Assume that each person who becomes cured is independent of the other person who becomes cured. Find the probability that

  (a) both will be cured,
  (b) neither will be cured, and
  (c) only one of the two will be cured.

12.5   Let A and B be two independent events with $P(A) = P(B) = 0.5$. Find $P(A \cap B)$.

12.6   (From [16]) An examination of Dick's family history has determined that the probability that he carries the recessive allele c responsible for cystic fibrosis is $\frac{1}{6}$. Similarly, Jane's chance of being Cc is $\frac{1}{3}$.

(a) Find the probability that both Dick and Jane are carriers of the c allele.
(b) Find the probability of their producing a child with cystic fibrosis.

12.7   (From [16]) Mary is a 25-year-old woman with two children. She has just learned that her father has Huntington's chorea, a serious degenerative disease of the nervous system that often does not manifest itself until late in life. This affliction is caused by the presence of a single dominant allele, D. Thus, Mary's father is now known to have genotype Dd. Assuming that Mary's mother and husband are both genotype dd (which is justified by the rareness of the disease), find the probability that

(a) Mary has Huntington's chorea and
(b) neither of Mary's children has Huntington's chorea, given that Mary has the disease.

12.8   (From [16]) In cattle, coat color is determined by two alleles, R and r, and the possible coat colors are red (RR), white (rr), and roan (Rr). In addition, horned cattle have genotype hh, while hornless cattle are either HH or Hh. A hornless roan cow is mated to a horned roan bull, and a horned white calf results. Assuming that the genes are unlinked (this, they are independent), find the probability that the next calf from this mating will be a horned roan.

12.9   (From [16]) In parakeets, the color of the feathers is determined by the combined effect of two unlinked (independent) genes. Let A and a be the alleles on one gene and B and b be the alleles on the second gene. Shown in the table below are the colors resulting from various genotypes. Suppose that a yellow female and a blue male have produced a white offspring.

(a) What are the genotypes of the parents?
(b) If this pair were to have many offspring, what proportions would you expect to be green, yellow, blue, and white?
(c) Answer question (b), assuming that both parents are Aa/Bb.

Phenotype	Green	Yellow	Blue	White
Genotypes	AA/BB	aa/BB	AA/bb	Aa/bb
	Aa/BB	aa/Bb		aa/bb
	AA/Bb			
	Aa/Bb			

12.10   (From [49]) Assume that a couple is equally like to have a boy or a girl. A family has three children. Find the probability of the following events.

(a) "all children are girls"
(b) "at least one boy"
(c) "at least two girls"
(d) "at most two boys"

12.11   (From [49]) Assume that 20% of a very common insect species in your study area is parasitized. Assume that insects are parasitized independently of each other. If you collect 10 specimens of this species, what is the probability that at least one specimen in your sample is parasitized?

12.12   (From [49]) Assume that the probability that an insect of a certain species lives more than 5 days is 0.1. Find the probability that in a sample of 10 insects of this species, at least one insect will still be alive after 5 days.

Sequential Events

In the previous chapter, we saw how to account for additional information about the outcome of an experiment using conditional probability. This was framed in the simple situation of either-or types of outcomes. For example, a test for a disease is positive or negative, or an individual has a disease or does not have it. We will see several more examples of drug testing in this chapter, including how to account for the errors associated with these tests. We will also explore situations in which there can be a sequence of observations or experiments for which what occurs first can affect the outcome of what happens second, which affects what happens third, and so on. So these sequences are not independent events. The example that we will emphasize is that of inherited human genes. A human pedigree shows several generations of a genetically related family tree. Human pedigrees can be considered as sequential experiments in which genetic information is transferred from generation to generation. Your genes are determined by your parents' genes, which are determined by their parents' genes. The standard notation for human pedigree charts is shown in Table 13.1. Here, we will develop the extension of the conditional probabilities in the previous chapter to account for the multiple outcomes that arise in sequential situations, such as pedigrees, to look at the probabilities of passing on genetic traits through multiple generations.

Suppose that we know that both members of a couple are carriers for a certain disease. What is the probability that their granddaughter has the disease or is a carrier for the disease? Before we can answer these types of questions, we need to introduce some more probability theory.

13.1 Partition Theorem

Suppose that we have a sample space S that is divided into two regions B_1 and B_2 such that these two regions are mutually exclusive and $B_1 \cup B_2 = S$ (see Figure 13.1(a)). This is known as a *partition* of the sample space. To find the probability of some event A in this sample space S, we use the partition property.

Table 13.1. Human pedigree notation.

Symbol	Meaning
□	Male homozygous dominant
○	Female homozygous dominant
■	Male homozygous recessive
●	Female homozygous recessive
◧	Male heterozygous
◐	Female heterozygous

Partition Property

If B_1 and B_2 form a partition of the sample space S, then

$$P(A) = P(A|B_1)P(B_1) + P(A|B_2)P(B_2).$$

We can think of the property of partitions as "the probability of event A is the probability of event A given event B_1 weighted by the probability of event B_1 plus the probability of event A given event B_2 weighted by the probability of event B_2."

To generalize the partition property, the sample space S can be partitioned into n regions, $B_1, B_2, \ldots, B_n$ such that each B_i is mutually exclusive from the others and $B_1 \cup B_2 \cup \cdots \cup B_n = S$ (see Figure 13.1(b)).

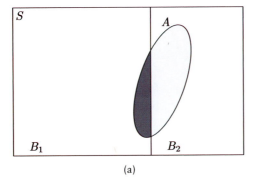
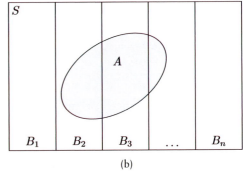

(a)   (b)

Figure 13.1 (a) The sample space S is partitioned into events B_1 and B_2. The darker-shaded region corresponds to $A \cap B_1$. The lighter-shaded region corresponds to $A \cap B_2$. The probability of event A is $P(A) = P(A|B_1)P(B_1) + P(A|B_2)P(B_2)$. **(b)** The sample space S is partitioned into events B_1, B_2, $\ldots$, B_n. The probability of event A is $P(A) = P(A|B_1)P(B_1) + \cdots + P(A|B_n)P(B_n)$. Notice that for some k, we can have $P(A|B_k) = 0$. For example, above, $P(A|B_1) = 0$ and $P(A|B_n) = 0$.

Example 13.1 (Albinism [16])

Consider the genetic pedigree shown in Figure 13.2. Both of Jay's parents are carriers for albinism. Jay's sister has the disorder, but Jay does not (though it is unknown if he is a carrier). Jay's wife, Mary, has no history of albinism in her family, so we can assume that she is homozygous dominant with respect to the albinism gene. What is the probability that Jay and Mary's child will be a carrier of the disorder?

Solution: Let event A = "child is a carrier." To find $P(A)$, we can partition the sample space into B_1 = "Jay is heterozygous" and B_2 = "Jay is homozygous dominant." Given that Jay's mother and father are both heterozygous, the Punnett square showing the possible genotypes for Jay is

		$\male$	
		A	a
$\female$	A	AA	Aa
	a	aA	aa

Since Jay is not an albino, $P(B_1) = \frac{2}{3}$ and $P(B_2) = \frac{1}{3}$.

Given that Jay is heterozygous, the Punnett square displaying the possible genotypes for a child of Jay and Mary's is

		$\female$	
		A	A
$\male$	A	AA	AA
	a	aA	aA

Thus, $P(A|B_1) = \frac{1}{2}$.

Given that Jay is homozygous dominant, the Punnett square displaying the possible genotypes for a child of Jay and Mary's is

		$\female$	
		A	A
$\male$	A	AA	AA
	A	AA	AA

Thus, $P(A|B_2) = 0$.

(Continued)

By the partition property, we have

$$P(A) = \left(\frac{1}{2}\right)\left(\frac{2}{3}\right) + (0)\left(\frac{1}{3}\right) = \frac{1}{3}.$$

Thus, the probability that a child of Jay and Mary's could be a carrier of albinism is roughly 0.33 or 33%.

Generalized Partition Property

If $B_1, B_2, \ldots, B_n$ forms a partition of the sample space S, then the probability of event A occurring is

$$P(A) = \sum_{i=1}^{n} P(A|B_i)P(B_i).$$

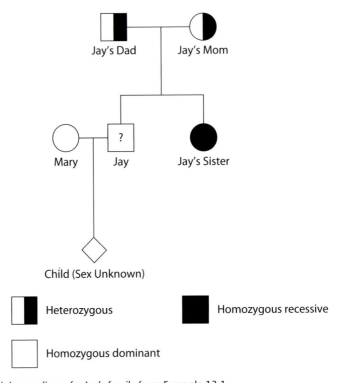

Figure 13.2 The albinism pedigree for Jay's family from Example 13.1.

Example 13.2 (Pedigree)

Suppose that we know that the genotypes of Billy's maternal grandparents for a certain genetic disease are Aa and Aa and that the genotype of Billy's father is AA (no disease). What is the probability that Billy is a carrier for the disease?

Solution: Using a Punnett square, we can show that there are three possibilities for the genotype of Billy's mother.

- $B_1 =$ "mom is Aa"
- $B_2 =$ "mom is aa"
- $B_3 =$ "mom is AA"

These are three mutually exclusive events that partition the sample space. Furthermore,

- $P(B_1) = \frac{1}{2}$
- $P(B_2) = \frac{1}{4}$
- $P(B_3) = \frac{1}{4}$

Let event $A =$ "Billy is a carrier (Aa)." We can use the generalized partition property to find $P(A)$:

$$P(A) = P(A|B_1)P(B_1) + P(A|B_2)P(B_2) + P(A|B_3)P(B_3)$$
$$= \left(\frac{1}{2}\right)\left(\frac{1}{2}\right) + (1)\left(\frac{1}{4}\right) + (0)\left(\frac{1}{4}\right)$$
$$= \frac{1}{4} + \frac{1}{4} + 0$$
$$= \frac{1}{2}.$$

Thus, the probability that Billy is a carrier of the disease is 0.5 or 50%.

How would Example 13.2 have been different if we had considered a disease like Tay-Sachs disease? Recall that individuals having Tay-Sachs disease usually do not live beyond the age of 4. Thus, if both of Billy's maternal grandparents were carriers of Tay-Sachs disease, that is, if both were Tt, then the Punnett square showing the possible genotypes for Billy's mother would be as follows:

		♀	
		T	t
♂	T	TT	Tt
	t	tT	tt

However, since Billy's mother survived to reproduce, we can assume that she does not have Tay-Sachs disease; that is, her genotype is not tt. Thus, if

- $B_1 =$ "mom is Tt,"
- $B_2 =$ "mom is tt," and
- $B_3 =$ "mom is TT,"

then

- $P(B_1) = \frac{2}{3}$,
- $P(B_2) = 0$, and
- $P(B_3) = \frac{1}{3}$.

Consequently, the probability that Billy is a carrier for Tay-Sachs disease is $\frac{1}{3}$.

13.2 Bayes' Theorem

We would like a way to define $P(B|A)$ in terms of the reverse conditional probability $P(A|B)$. The mathematical tool we have for this is known as Bayes' theorem.

The definition of conditional probability tells us that

$$P(B|A) = \frac{P(B \cap A)}{P(A)},$$

and rearranging by algebra gives

$$P(B \cap A) = P(A|B)P(B).$$

Combining these together, we get the formula in Bayes' theorem.

Bayes' Theorem

If events A and B are in sample space S such that $P(A) \neq 0$, then

$$P(B|A) = \frac{P(A|B)P(B)}{P(A)}.$$

If the sample space has a partition, then this result can be generalized.

Bayes' Theorem with Partition

If S is partitioned by mutually exclusive events $B_1, B_2, \ldots, B_n$, then

$$P(B_k|A) = \frac{P(A|B_k)P(B_k)}{P(A)},$$

---

where

$$P(A) = \sum_{i=1}^{n} P(A|B_i)P(B_i).$$

Let us look at a few examples in which we can use Bayes' theorem to determine the conditional probability of an event.

Example 13.3 (Head Injuries [16])

Among individuals who have sustained head injuries, X-rays reveal that only about 6% have skull fractures. Nausea is a standard symptom of a skull fracture and occurs in 98% of all skull fracture cases. With other types of head injuries, nausea is present in about 70% of all cases. Suppose that an individual who has just suffered a head injury is not nauseous. Find the probability that he has a skull fracture.

<u>**Solution:**</u> Let events F = "skull fracture" and N = "nausea." Since only 6% of head injuries result in skull fractures,

$$P(F) = 0.06 \text{ and } P(\overline{F}) = 0.94.$$

Given individuals who have skull fractures, 98% have nausea. Given those who do not have skull fractures, 70% have nausea. Thus,

$$P(N|F) = 0.98 \text{ and } P\left(N|\overline{F}\right) = 0.70.$$

Applying Bayes' theorem with a partition of, F and $\overline{F}$ gives

$$P\left(F|\overline{N}\right) = \frac{P\left(\overline{N}|F\right)P(F)}{P\left(\overline{N}|F\right)P(F) + P\left(\overline{N}|\overline{F}\right)P\left(\overline{F}\right)}$$

$$= \frac{(1-0.98)(0.06)}{(1-0.98)(0.06) + (1-0.7)(0.94)}$$

$$= 0.004.$$

Thus, it is very unlikely that a person having just suffered a head injury and having no nausea has a skull fracture (the probability is less than .01).

The *sensitivity* of a medical test is defined as the probability that a test will be positive given that a person has the disease for which he or she is are being tested. The *sensitivity* is usually represented by

$$P(+|D)$$

with + for a positive test result and D for the disease being present.

The *specificity* of a medical test is the probability that a test will be negative given that the person does not have the disease. The *specificity* is usually represented by

$$P(-|\overline{D})$$

with $-$ for a negative test result and $\overline{D}$ for the disease not being present.

From the *specificity* and the property of complements (used on $+,-$), the probability of a "false positive" is

$$P(+|\overline{D}) = 1 - P(-|\overline{D}).$$

Similarly, from the *sensitivity*, the probability of a "false negative" is

$$P(-|D) = 1 - P(+|D).$$

But note that one cannot obtain $P(+|\overline{D})$ directly from the sensitivity since $P(+|\overline{D}) \neq 1 - P(+|D)$.

Consider the following drug testing example, in which we consider the validity of a drug test under different conditions by calculating the probabilities of various outcomes.

Example 13.4 (Drug Testing)

Suppose that a drug test for an illegal drug is such that it is 98% accurate in the case of a user of that drug (i.e., it produces a positive result with probability 0.98 in the case that the tested individual uses the drug) and 90% accurate in the case of a nonuser of the drug (i.e., it is negative with probability 0.90 in the case that the person does not use the drug). Suppose that 10% of the entire population uses this drug.

(a) What is the probability of a false positive with this test (i.e., the probability of obtaining a positive drug test given that the person tested is a nonuser)?
(b) What is the probability of obtaining a false negative for this test (i.e., the probability that the test is negative but that the individual tested is a user)?
(c) You test someone, and the test is positive. What is the probability that the tested individual uses this illegal drug?

Solution: Let events $+$ be "drug test is positive for an individual," $-$ be "drug test is negative for an individual," and A be "the person tested has the drug being tested for present in his or her system." Since 10% of the population uses the drug in question,

$$P(A) = 0.10 \text{ and } P\left(\overline{A}\right) = 0.90.$$

From the accuracy of the drug test, the sensitivity and the specificity are

$$P(+|A) = 0.98 \text{ and } P\left(-|\overline{A}\right) = 0.90.$$

(Continued)

(a) First, we want to find the probability of obtaining a positive drug test given that the person tested is a nonuser, that is, $P\left(+|\overline{A}\right)$. Using the property of complements, we have

$$P\left(+|\overline{A}\right) = 1 - P\left(-|\overline{A}\right) = 1 - 0.90 = 0.10.$$

Thus, the chance of a false positive is 10%.

(b) Next, we want to find the probability of obtaining a negative drug test given that the person tested is a user, that is, $P(-|A)$. Again,

$$P(-|A) = 1 - P(+|A) = 1 - 0.98 = 0.02.$$

Thus, the chance of a false negative is 2%.

(c) Finally, we want to find the probability that a tested individual has the drug in his or her system given that he or she tested positive; that is, we want to find $P(A|+)$. According to Bayes' theorem,

$$P(A|+) = \frac{P(+|A)P(A)}{P(+)}.$$

Using the property of partitions, we know that

$$P(+) = P(+|A)P(A) + P\left(+|\overline{A}\right)P\left(\overline{A}\right)$$
$$= (0.98)(0.10) + (0.10)(0.90)$$
$$= 0.188.$$

Thus,

$$P(A|+) = \frac{P(+|A)P(A)}{P(+)} = \frac{(0.98)(0.10)}{0.188} = 0.521.$$

Thus, the probability that the person who tested positive actually has the drug in his or her system is 0.521 or 52.1%.

Let us look at another medical testing scenario where there is the possibility of false positives and false negatives.

Example 13.5 (Genetic Testings [16])

Suppose that there is a disease caused by the presence of two recessive alleles, aa, and that a genetic test has been developed to determine whether a normal individual is a carrier, Aa, or a noncarrier, AA. Suppose that among known carriers, the test correctly identifies the presence of the a allele 80% of the time. Among known AA individuals, the test indicates

(*Continued*)

the presence of the a allele 30% of the time. What is the probability that a person is a carrier if both of his or her parents are carriers given that the test given to this person indicates a positive result (that is, the allele a is present)?

Solution: Since the test was developed for genotypes AA and Aa, we can assume that if a person is taking this test, then he or she does not have the disease; that is, the person does not have genotype aa.

Let event C = "person is a carrier (Aa)," $\overline{C}$ = "person is not a carrier (AA)," + be "test result is positive," and − be "test result is negative."

Since the tested individual's parents are both carriers, the Punnett square showing the possible genotypes for the individual are

		♀	
		A	a
♂	A	AA	Aa
	a	aA	aa

Since the individual does not have the disease, the probabilities of being a carrier and a noncarrier are, respectively,

$$P(C) = \frac{2}{3} \text{ and } P(\overline{C}) = \frac{1}{3}.$$

Since the test sensitivity is 80% for carriers and the false positive chance is 30% for noncarriers,

$$P(+|C) = 0.80 \text{ and } P(+|\overline{C}) = 0.30.$$

Notice that we can partition the sample space into the mutually exclusive events C and $\overline{C}$. Applying Bayes' theorem, we find that

$$P(C|+) = \frac{P(+|C)P(C)}{P(+|C)P(C) + P(+|\overline{C})P(\overline{C})}$$

$$= \frac{(0.80)\left(\frac{2}{3}\right)}{(0.80)\left(\frac{2}{3}\right) + (0.30)\left(\frac{1}{3}\right)}$$

$$= 0.842.$$

Thus, the probability that the individual who tested positive is a carrier is 0.842 or 84.2%.

13.3 Exercises

13.1 For three events A, B, and C, let B and C form a partition of the sample space. If $P(A|B) = 0.7$, $P(A|C) = 0.3$, $P(B) = 0.8$, and $P(C) = 0.2$, find $P(A)$.

13.2 For two events A and C given that $P(A|C) = 0.7$, $P(A) = 0.3$, and $P(C) = 0.1$, find $P(C|A)$.

13.3 For two events A and B, with $P(A) = 0.8$, $P(B|A) = 0.3$, and $P(B|\overline{A}) = 0.4$, find the following.

(a) $P(A|B)$
(b) $P(\overline{A}|B)$

13.4 (From [29]) For mutually exclusive events R_1, R_2, and R_3, we have $P(R_1) = 0.05$, $P(R_2) = 0.6$, and $P(R_3) = 0.35$. Also, $P(Q|R_1) = 0.40$, $P(Q|R_2) = 0.30$, and $P(Q|R_3) = 0.60$. Find the following.

(a) $P(R_1|Q)$
(b) $P(R_2|Q)$
(c) $P(R_3|Q)$
(d) $P(\overline{R_1}|Q)$

13.5 If two fair dice are rolled, find the probabilities of the following results.

(a) A sum of 10, given that the sum is greater than 5.
(b) A "double" (two identical numbers), given that the sum is 12.
(c) A double, given that the sum is 11.

13.6 For a test for a certain disease, given that the sensitivity of that test is 0.8 and the specificity of that test is 0.3, find

(a) the probability that the test is positive, given that the disease is not present, and
(b) the probability that the test is negative, given that the disease is present.

13.7 For a test for a certain disease, given that the sensitivity of that test is 0.9 and the specificity of that test is 0.5, find

(a) the probability of a false positive and
(b) the probability of a false negative.

13.8 A screening test for a disease has a sensitivity of 0.80. The test gives a positive result in 10% of all cases when the disease is not present. Assume that the prevalence of the disease is 1 in 1000. If the test is administered to a randomly chosen individual, what is the probability that the result is positive?

13.9 (From [49]) Suppose that you have a batch of red-flowering pea plants of which 40% are of genotype CC and 60% of genotype Cc with the C allele for red-flowers being dominant. You pick one plant at random and cross it with a white-flowering pea plant of genotype cc. Find the probability that the offspring of this cross will have white flowers.

13.10 (From [49]) Suppose that you have a batch of red- and white-flowering pea plants where all three genotypes, CC, Cc, and cc, are equally represented. The allele C for red flowers is dominant. You pick one plant at random and cross it with a white-flowering pea plant. What is the probability that the offspring will have red flowers?

13.11 (From [49]) If a person is treated with a placebo, there is a 10% chance the person will feel instant relief. In a clinical trial, half the subjects are treated with a new drug; the other half receive the placebo. If an individual from this trial is chosen at random, what is the probability that the person will have experienced instant relief from the placdbo?

13.12 In a certain area, ten percent of the rabbits in a population are slow and their chances of being captured by a fox are 0.6. Among faster rabbits, there is a 20% chance of being caught. Find the probability that a rabbit will be caught.

13.13 (From [16]) Consider the following pedigree.

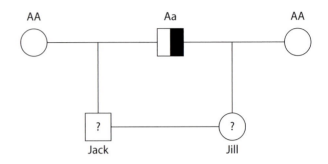

Suppose that Jack and Jill have a child.

(a) Find the probability that their child will be *Aa*.
(b) Find the probability that their child will be *aa*.

13.14 (From [16]) Shown below is a pedigree for a family with Huntington's chorea. This genetic disease is caused by the presence of a single dominant allele, *D*, and has the unusual property that it does not manifest itself until age 35 or later.

(a) Find the probability that Mary has Huntington's chorea.
(b) Find the probability that neither of Mary's daughters has Huntington's chorea.

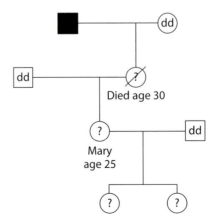

13.15 (From [29]) The sensitivity and specificity for breast cancer screening during a clinical breast cancer examination by a trained expert are approximately 0.54 and 0.94, respectively.

(a) If 2% of women in the United States have breast cancer, find the probability that a woman who tests positive during a clinical breast examination actually has breast cancer.

(b) Suppose that a positive clinical examination has occurred and that the woman is directed to receive a mammogram. Suppose that for this woman, the sensitivity and specificity for the mammogram are 0.89 and 0.93, respectively. Given that both the clinical exam and the mammogram result in a positive test, what is the probability that the woman has breast cancer? Assume that the mammogram and clinical examination are independent events.

13.16 (From [16]) Voice analyzers are used to detect tiny modulations in a person's voice when that person is lying. To test the effectiveness of this device, an individual was instructed to lie on half of the responses. When a lie was told, the voice analyzer indicated "lie" in 90% of all cases. When the truth was told, the device indicated "lie" in only 5% of the cases. Based on these results, find the probability that the person is lying given that the device indicates "lie."

13.17 (From [30]) The probability that a person with certain symptoms has hepatitis is 0.8. The blood test used to confirm this diagnosis gives positive results for 90% of people with the disease and 5% of those without the disease. What is the probability that an individual who has the symptoms and who reacts positive to the test actually has hepatitis?

13.18 (From [30]) A 1995 study by the Harvard School of Public Health reported that 86% of male students who live in a fraternity house are binge drinkers. The study also reported that 71% of fraternity members who do not reside in a fraternity house are binge drinkers and that 45% of male students who do not belong to a fraternity are binge drinkers. Suppose that 10% of U.S. male students live in a fraternity house, 15% belong to a fraternity but do not live in a fraternity house, and 75% do not belong to a fraternity.

(a) What is the probability that a randomly selected male student is a binge drinker?

(b) If a randomly selected male student is a binge drinker, what is the probability that he lives in a fraternity house?

Population Genetics Models

The examples we have considered on human pedigrees have allowed us to estimate the probabilities that certain genetic traits are passed down through a family. Another scale at which we can consider genetic traits being passed on from one generation to the next is by looking not at families but rather across an entire population of interbreeding individuals. In this case, we consider the changes in the structure of the population through time by considering what fraction of the population in each generation has a certain genotype. This approach serves as the basis for the study of the impacts of evolutionary forces, including natural selection and mutation, on the genetics of populations. The approach was formulated mathematically through what is called the "modern synthesis," which formally connected the fields of Mendelian genetics (we saw many simple examples of this in considering Mendel's pea experiments), with the ideas of Charles Darwin on evolution by natural selection.

Population genetics is the study of the changes through time in the genetics of a population and considers the mechanisms responsible for these changes. The key variables are the frequencies at which different alleles occur in the population, and the key assumption we will make here is that there is random mating, meaning that there is no tendency for individuals of one genotype to preferentially mate with those of another (or the same) genotype. From this, we will see that, similar to the long-term equilibrium stable age structure that arises in the Leslie population model, there is a stable equilibrium genetic structure of the population that arises. We will consider here only a simple case of natural selection in which one genotype has a higher probability of surviving than another, and we will see how to formulate this in terms of diseases that have detrimental effects on the survival of offspring.

In this chapter, our goal is to investigate questions such as the following:

- Does a given allele maintain itself in the population over time?
- What are the impacts of different mechanisms that can lead to evolution, meaning changes in genotype frequency in the population?

14.1 Hardy-Weinberg Equilibrium

Suppose that we want to follow a genetic trait determined by alleles A and a from one generation to the next. For now, we assume these are not sex linked. Let

- $P_0 = P(AA)$ be the current frequency of AA in the population,
- $Q_0 = P(Aa)$ be the current frequency of Aa in the population, and
- $R_0 = P(aa)$ be the current frequency of aa in the population.

Thus,

$$P_0 + Q_0 + R_0 = 1.$$

In genetics, it is common to ask what the frequency of the A or a allele is. Let

- $p_0 = P(A)$ be the current frequency of the A allele, and
- $q_0 = P(a)$ be the current frequency of the a allele.

Thus,

$$p_0 + q_0 = 1.$$

We would like a way to relate the probabilities P_0, Q_0, and R_0 to p_0 and q_0. In the current generation, each heterozygous individual has one A allele and one a allele. We can think of this as half of the frequency Q_0 contributes is to p_0 and the other half contributes to q_0. Since each homozygous dominant individual has only A alleles, the entire frequency of P_0 contributes to p_0. Likewise, since each homozygous recessive individual has only a alleles, the entire frequency of R_0 contributes to q_0. Thus, we have the following two equations:

$$p_0 = P_0 + \frac{1}{2}Q_0.$$

$$q_0 = \frac{1}{2}Q_0 + R_0.$$

Notice that we still have

$$p_0 + q_0 = P_0 + \frac{1}{2}Q_0 + \frac{1}{2}Q_0 + R_0 = P_0 + Q_0 + R_0 = 1.$$

As long as there is a way to distinguish between at least two of the genotypes in a population, we can determine the allele frequency within that population.

Example 14.1 (Sickle Cell Anemia Allele Frequency [16])

Hemoglobin is an iron-rich protein that gives blood its red coloring. Hemoglobin is found in every red blood cell and is the compound that allows the cell to carry oxygen to different parts of the body. The gene that produces hemoglobin will produce normal healthy hemoglobin, known as hemoglobin-A, when it contains two A alleles. The hemoglobin gene

(Continued)

has a recessive allele, S, that causes the production of hemoglobin-S. A red blood cell containing hemoglobin-S will have a crescent or sickle shape (instead of the regular disc shape) and is not able to deliver as much oxygen throughout the body.

School-aged children in Ghana were screened for sickle cell anemia, and the following results were found.

Genotype	AA	AS	SS
Frequency	0.834	0.161	0.005

What are the allele frequencies?

Solution: We have that $P_0 = 0.834$, $Q_0 = 0.161$, and $R_0 = 0.005$. Thus,

$$p_0 = P_0 + \frac{1}{2}Q_0 = 0.834 + (0.5)(0.161) = 0.9145, \text{ and}$$

$$q_0 = 1 - p_0 = 1 - 0.9145 = 0.0855.$$

Thus, 91.45% of the alleles in the population are A, and the remaining 8.55% are S.

Given that we know the genotype frequencies in the current generation, how can we find the genotype frequencies in the subsequent generation? We can form the frequencies of the next generation using a Punnett square. We assume that in a population, the females with a p frequency of A alleles and a q frequency of a alleles randomly mate with the males with a p frequency of A alleles and a q frequency of a alleles (see Figure 14.1).

We can see from Figure 14.1 that the subsequent generation has frequencies of p^2 homozygous dominant individuals, $2pq$ heterozygous individuals, and q^2 homozygous recessive individuals. Thus,

$$P_1 = p^2,$$
$$Q_1 = 2pq, \text{ and}$$
$$R_1 = q^2.$$

Notice that

$$P_1 + Q_1 + R_1 = p^2 + 2pq + q^2 = (p + q)^2 = 1.$$

Assuming that in the first generation we had started with allele frequencies of p_0 and q_0, what would the allele frequencies in the subsequent generation be? We would have genotype

		♀	
		A (p)	a (q)
♂	A (p)	AA (p^2)	Aa (pq)
	a (q)	Aa (pq)	aa (q^2)

Figure 14.1 A population with frequency p of A alleles and frequency q of a alleles where random mate selection occurs. The Punnett square shows that the next generation has frequencies p^2 homozygous dominant individuals, $2pq$ heterozygous individuals, and q^2 homozygous recessive individuals.

frequencies $P_1 = p_0^2$, $Q_1 = 2p_0q_0$, and $R_1 = q_0^2$. Thus, the allele frequencies would be

$$p_1 = P_1 + \frac{1}{2}Q_1 = p_0^2 + \frac{1}{2}2p_0q_0 = p_0^2 + p_0q_0 = p_0(p_0 + q_0) = p_0 \times 1 = p_0, \text{ and}$$

$$q_1 = \frac{1}{2}Q_1 + R_1 = \frac{1}{2}2p_0q_0 + q^2 = p_0q_0 + q_0^2 q_0(p_0 + q_0) = q_0 \times 1 = q_0.$$

Therefore, the allele frequencies in the subsequent generation are the same as in the first generation. Since the allele frequencies are the same as in the previous generation we have

$$P_1 = p_0^2, \quad Q_1 = 2p_0q_0, \quad \text{and} \quad R_1 = q_0^2.$$

Since these allele frequencies and genotype frequencies remain the same over the generations, we drop the subscript and have

$$p = P + \frac{1}{2}Q, \quad q = \frac{1}{2}Q + R, \quad p + q = 1,$$

$$P = p^2, \quad Q = 2pq, \quad R = q^2, \text{ and } P + Q + R = 1.$$

The phenomenon of allele and genotype frequencies remaining the same over generations is known as the **_Hardy-Weinberg equilibrium_**. We should note here that this equilibrium of allele frequencies and, consequently, genotype frequencies depends on the assumption of random mating. Thus, if individuals select their mates on the basis of physically visible traits such as eye color, complexion, or height, then the Hardy-Weinberg equilibrium would not hold for the alleles of that gene. However, it is reasonable to assume that individuals do not select mates on the basis of traits such as blood type or whether they have sickle cell anemia. For the majority of gene loci in humans, the random mating hypothesis is reasonable.

Example 14.2 (Navajo Indians in New Mexico [16])

Of the Navajo Indians in New Mexico, 77.7% have blood type O, and 22.3% have blood type A. What is the frequency of the genotype AO?

Solution: Let p represent the frequency of allele A and q represent the frequency of allele O. We want to find $P(AO) = 2pq$. Since 77.7% have blood type O, which is given by genotype OO, $q^2 = 0.777$, and $q = 0.881$. Then $p = 1 - q = 0.119$. Thus, the frequency of allele A is 11.9%. Now,

$$Q = P(AO) = 2pq = 2(0.119)(0.881) = 0.210.$$

Thus, 21.0% of the Navajo Indian population in New Mexico have the blood genotype AO.

Example 14.3 (Sickle Cell Anemia [82])

The incidence rate of sickle cell anemia in Africa is 4%, whereas the incidence rate of sickle cell anemia among African Americans is 0.25%. Find the proportion of each of these populations that are carriers of the sickle cell gene.

Solution: Let us first consider the population in Africa. Let q be the frequency of the recessive allele, S. Then $q^2 = 0.04$, and thus $q = 0.2$. Then

$$p = 1 - q = 1 - 0.2 = 0.8.$$

Thus,

$$Q = P(AS) = 2pq = 2(0.8)(0.2) = 0.32.$$

Therefore, 32% of the population in Africa carries the allele of sickle cell anemia.
   Next, let us consider the population of African Americans. Here, $q^2 = 0.0025$, and thus $q = 0.05$. Then

$$p = 1 - q = 1 - 0.05 = 0.95.$$

Thus,

$$Q = P(AS) = 2pq = 2(0.95)(0.05) = 0.095.$$

Therefore, only 9.5% of the African American population are carriers of the sickle cell anemia allele.

14.2 Hardy-Weinberg Selection Model

Suppose that in a population, the probability that an individual survives to reproduce depends on his or her genotype for a given trait. How might this affect the Hardy-Weinberg equilibrium?
   Suppose that we are considering a gene with alleles A and a, where A is the dominant allele. Let event S = "survive to reproductive age." Next, let

$$s_1 = P(S|AA),$$
$$s_2 = P(S|Aa), \text{ and}$$
$$s_3 = P(S|aa).$$

Notice that if we are considering a genetic disease that is fatal prior to reaching reproductive age, such as Tay-Sachs disease, then $s_3 = 0$.
   Let p_0 and q_0 be the allele frequencies of A and a, respectively, among the current generation. Then

$$P(AA) = p_0^2, \quad P(Aa) = 2p_0q_0, \quad \text{and} \quad P(aa) = q_0^2$$

for the current generation. First, we compute the probability that an individual is of a given genotype given that he or she reaches reproductive age. Thus, we will know the frequencies of genotypes of the current generation that actually reproduce. By the partition theorem,

$$P(S) = P(S|AA)P(AA) + P(S|Aa)P(Aa) + P(S|aa)P(aa)$$
$$= s_1 p_0^2 + 2s_2 p_0 q_0 + s_3 q_0^2$$

By Bayes' theorem,

$$P(AA|S) = \frac{P(S|AA)P(AA)}{P(S)} = \frac{s_1 p_0^2}{s_1 p_0^2 + 2s_2 p_0 q_0 + s_3 q_0^2},$$

$$P(Aa|S) = \frac{P(S|Aa)P(Aa)}{P(S)} = \frac{2s_2 p_0 q_0}{s_1 p_0^2 + 2s_2 p_0 q_0 + s_3 q_0^2}, \quad \text{and}$$

$$P(aa|S) = \frac{P(S|aa)P(aa)}{P(S)} = \frac{s_3 q_0^2}{s_1 p_0^2 + 2s_2 p_0 q_0 + s_3 q_0^2}.$$

What are the allele frequencies p_1 and q_1 for the next generation? The new allele frequencies among the children of the survivors of the current generation are

$$p_1 = P(AA|S) + \frac{1}{2}P(Aa|S)$$

$$= \frac{s_1 p_0^2 + s_2 p_0 q_0}{s_1 p_0^2 + 2s_2 p_0 q_0 + s_3 q_0^2}, \quad \text{and}$$

$$q_1 = \frac{1}{2}P(Aa|S) + P(aa|S)$$

$$= \frac{s_2 p_0 q_0 + s_3 q_0^2}{s_1 p_0^2 + 2s_2 p_0 q_0 + s_3 q_0^2}.$$

This can be extended to many generations into the future using difference equation ideas we developed in Chapter 5

$$q_{n+1} = q_n \frac{s_2 p_n + s_3 q_n}{s_1 p_n^2 + 2s_2 p_n q_n + s_3 q_n^2}. \tag{14.1}$$

Note that for a fatal disease such as Tay-Sachs disease, for which $s_3 = 0$, this equation reduces to

$$q_{n+1} = q_n \frac{s_2}{s_1(1 - q_n) + 2s_2 q_n}. \tag{14.2}$$

Example 14.4 (Future Allele Frequencies for Tay-Sachs Disease)

An individual with Tay-Sachs disease rarely lives past 4 years. Thus, for Tay-Sachs disease, $s_3 = 0$. In 1964, the Tay-Sachs incidence rate among the Jewish population of New York City was 0.00023 [48]. Assume that individuals who do not have Tay-Sachs disease have

(Continued)

a 90% chance of surviving and reproducing each generation. For the Jewish population in New York City, answer the following questions:

(a) What were the allele frequencies in 1964?
(b) What would you predict the allele frequencies to be in 2030 given that there are about 22 years between generations?
(c) What would you predict as the frequency of carriers in 2030?

Solution: Let p represent the frequency of the dominant allele, T, and q represent the frequency of the recessive allele, t. Since individuals born with Tay-Sachs disease do not live to reproduce, we will use the equation

$$q_{n+1} = q_n \frac{s_2}{s_1(1 - q_n) + 2s_2 q_n}$$

to predict the allele frequencies of future generations.

(a) Let q_0 be the allele frequency of t in 1964. Then $P(tt) = q_0^2 = 0.00023$; thus, $q_0 = 0.0152$. Then $p_0 = 0.9848$.
(b) Given that there is about 22 years between generations and generation 0 is in 1964, generation $n = 1$ occurs at about year 1986, generation $n = 2$ occurs at about year 2008, and generation $n = 3$ occurs at about year 2030. Now, we can estimate q_n for generations $n = 1, 2, 3$. Since $s_1 = s_2 = 0.9$, we can simplify the equation for q_{n+1} to

$$q_{n+1} = q_n \frac{s}{s(1 - q_n) + 2s q_n} = \frac{q_n}{(1 - q_n) + 2q_n} = \frac{q_n}{1 + q_n}.$$

$$q_1 = \frac{0.0152}{1 + 0.0152} = 0.0150$$

$$q_2 = \frac{0.0150}{1 + 0.0150} = 0.0148$$

$$q_3 = \frac{0.0148}{1 + 0.0148} = 0.0146$$

Thus, by 2030, we would expect to have allele frequencies $q_3 = 0.0146$ and $p_3 = 1 - q_3 = 0.9854$.

(c) By 2030, we would expect to have the frequency of carriers to be

$$Q_3 = P(Aa) = 2p_3 q_3 = 0.0144.$$

Thus, we would expect 1.44% of the Jewish population of New York City to be carriers of Tay-Sachs disease. The calculated values of genotype and allele frequencies at each generation are given in the below table.

(*Continued*)

Generation, n	0	1	2	3
Year	1964	1986	2008	2030
q	0.0152	0.0150	0.0148	0.0146
p	0.9848	0.9850	0.0952	0.0954
P	0.9698	0.9702	0.9706	0.9710
Q	0.0299	0.0296	0.0292	0.0288
R	0.0003	0.0002	0.0002	0.0002

14.3 Exercises

14.1 The allele b occurs with a frequency of 0.8 in a population of clams. Give the frequency of genotypes BB, Bb, and bb using the Hardy-Weinberg model.

14.2 In a Hardy-Weinberg situation, you have sampled a population in which the percentage of the homozygous recessive genotype (aa) is 36%. What is the frequency of the A allele? What is the frequency of the genotype Aa?

14.3 In a Hardy-Weinberg situation, suppose that 98 out of 300 individuals in a population express the dominant phenotype. What percentage of the population would you predict to be heterozygotes?

14.4 In a Hardy-Weinberg situation, suppose that 98 out of 300 individuals in a population express the recessive phenotype. What percentage of the population would you predict to be heterozygotes?

14.5 Within a population of butterflies, the color brown (B) is dominant over the white (b). Twenty percent of all the butterflies are white.

(a) Find the percentage of butterflies in this population that are heterozygous.
(b) Find the frequency of the homozygous dominant individuals.

14.6 (From [16]) Allele A is dominant over allele a, and so genotypes AA and Aa are identical in appearance. Suppose that there are four times as many $A\_$ individuals (meaning with at least one A allele) as aa genotypes, that is, $P + Q = 4R$. Use the Hardy-Weinberg model to estimate the allele frequency of allele a and predict the proportions of various genotypes in the population.

14.7 (From [16]) Use the Hardy-Weinberg model to estimate the proportion of carriers in a given population. Each genetic disease is due to the presence of two recessive alleles.

(a) Cystic fibrosis occurs in approximately 1 out of every 1600 births in the United States.
(b) In the Jewish population of northern Europe, Tay-Sachs disease has an incidence rate of 1 in 6000.
(c) Albinism among the Indians on the San Blas Islands off Panama occurs with probability 1/130.

14.8 (From [16]) Let p, q, and r denote the frequencies of alleles A, B, and O, respectively, in a given population for the standard blood groups.

(a) Assuming the random mating hypothesis, show the following.

    i. $P(\text{"blood type } O\text{"}) = r^2$
    ii. $P(\text{"blood type } A\text{"}) = p^2 + 2pr$
    iiii. $P(\text{"blood type } B\text{"}) = q^2 + 2qr$
    iv. $P(\text{"blood type } AB\text{"}) = 2pq$

(b) A study of blood types in France found the frequencies shown in the table below. Using the results from (a), find the frequencies of alleles A, B, and O.

Blood type	O	A	B	AB
Frequency	0.441	0.435	0.090	0.034

(c) In France's population, what fraction of those with blood type A carry the O allele?

14.9   (From [16]) For Huntington's chorea, it has been estimated that $s_2 \approx 0.75 s_1$ and $s_3 = 0$, where s_1, s_2, and s_3 are the survival rates for dd, Dd, and DD genotypes, respectively. Recall, that Huntington's chorea is caused by the presence of a single D allele.

(a) If q_n is the frequency of the rare D allele, let $A = d$ and $a = D$ in the selection model. Show that

$$q_{n+1} = \frac{3q_n}{4 + 2q_n}.$$

(b) if $q_0 = 0.006$, find $q_1, q_2, \ldots, q_5$.

14.10   In a given population, only the A and B alleles are present in the blood; there are no individuals with type O blood or with O alleles in this particular population. If 100 people have type A blood, 75 have type AB blood, and 25 have type B blood, what are the allele frequencies of this population (i.e., what are p and q)?

Unit 3 Student Projects

Human Eye Color

We have seen in examples and exercises throughout this unit that one hypothesis for how human eye color is determined assumes that eye color is controlled by one gene with two possible alleles, *B* (brown) and *b* (blue), where brown is dominant. However, we know that there is a greater variety of eye color than just brown and blue.

Currently, at minimum, three genes are involved in determining eye color, and it is likely that more are involved. However, of those three genes, two are much better understood than the third. The first gene is called "*bey2*" and is located on chromosome 15. This gene can contain alleles *B* for brown and *b* for blue, of which *B* is dominant. The second gene is called "*gey*" and is located on chromosome 19. This gene can contain alleles *G* for green and *b* for blue, of which *G* is dominant. Because these two genes are fairly well understood, one common model for eye color distinguishes between three color groups, brown, blue, and green, as determined by the alleles present on the *bey2* and *gey* genes. The way that eye color is determined is given in Table 14.1.

Student Tasks

In this project, you will use Matlab to simulate eye color phenotypes of babies that Cordelia and her husband, Jay, could produce. Cordelia has brown eyes, and her father has blue eyes.

Prep Work

Answer the following questions before continuing with the project.

1. What are the possible allele combinations that, Cordelia's mother can have on the *bey2* gene?
2. What is the only possible combination of alleles that, Cordelia can have on the *bey2* gene?
3. Suppose that Cordelia has a *Gb* on the *gey* gene and that Cordelia's husband, Jay, has green eyes with a *Gb* on the *gey* gene. Construct two Punnett squares

Table 14.1. Eye color genotypes determined by the *bey2* and *gey* genes shown with their corresponding phenotypes.

Eye Color	Genotypes		Descriptions
	Bey2	*Gey*	
Brown	Bb	bb	At least one B on *bey2*
	BB	bb	
	Bb	Gb	
	BB	Gb	
	Bb	GG	
	BB	GG	
Green	bb	Gb	bb on *bey2* and
	bb	GG	At least one B on *gey*
Blue	bb	bb	bb on both *bey2* and *gey*

(one for each gene), showing the possible genotypes for each gene, of potential children of Cordelia and Jay.

4. What are the possible phenotypes of potential children of Jay and Cordelia?
5. What is the probability that Jay and Cordelia's first child will have brown eyes? Green eyes? Blue eyes?

Utilizing Matlab to Run Probability Experiments

In the prep work, we calculated by hand the probabilities of Cordelia and Jay having a blue-eyed child, a green-eyed child, and a brown-eyed child. Suppose that the probability of having a blue-eyed child was $x/100$ or $x\%$. Then, theoretically, if Cordelia and her husband have 100 children, approximately x of them should have blue eyes. We will use Matlab to conduct an experiment like this to test if this is correct.

1. Using the Matlab skills you have learned in this unit, write an m-file (see Appendix A for a description of writing and running an m-file) that does the following.
   (a) Write a function in an m-file (call it `eyecolor.m`) that will randomly select an allele from Cordelia and an allele from Jay for both the *bey2* and the *gey* genes (you will need to make use of the `rand` function, `if` statements, and `for` loops for this). Use the resulting combinations of alleles on each gene to determine the eye color of Cordelia and Jay's child. The input of your function should be the genotypes of Cordelia and Jay for each gene. The output of your function should be the resulting eye color.
   (b) In another m-file (call it `hundredbabies.m`), write code that will call the `eyecolor` function 100 times. Keep track of each result by using three counters (one for each eye color); for example, each time the resulting eye color is blue, you will increase the blue eye color counter by 1. Use Matlab to determine how many times the eye color was blue, green, and brown. You should run this file several times and see if the results vary.
   (c) Currently, your `hundredbabies.m` file should run the experiment of creating 100 of Cordelia and Jay's children and to determine how many have brown, green, and blue eyes. Now modify your `hundredbabies.m` file so that the experiment is run 600 times (you will need another `for` loop to accomplish this). Compute the average proportion of 100 children, over the 600 experiments, that have each eye color.

2. Using the m-files you have written and the genotypes for Cordelia and Jay determined in the Prep Work section, determine how many of Cordelia and Jay's children would have brown, green, and blue eyes if they had 100 children.

3. Repeat the previous step supposing that Jay's genotype on the *gey* gene was *GG* instead of *Gb*, as is stated in the Prep Work section.

4. Repeat the previous step supposing that Cordelia's genotype on the *gey* gene was *GG* and that Jay's genotype on the *gey* gene was *Gb*.

5. Repeat the previous step supposing that Cordelia and Jay have genotypes different from the one you have tested in the previous three steps.

6. Compose a three-page report that describes the analysis you performed and that summarizes the findings of your analysis. Your report should have four sections:

   (a) Introduction: Describe the problem put forth and address why one might want to use a computer simulation to conduct the experiment of Cordelia and Jay having 100 children.

   (b) Methods: Describe how your m-files are structured and written and explain how these files are used to calculate probabilities.

   (c) Results: Describe in words the results of your analysis. In this section, include a table of the results you obtained using different genotypes for Cornelia and Jay. Describe how changing the genotypes of Cordelia and Jay changed the proportion of their 100 children that had brown, green, and blue eyes.

   (d) Conclusion: What conclusions can your draw from your results? Describe how you might use the files you have written to determine the possible eye colors of your own potential children.

7. Turn in a copy of your report along with your m-files.   ∎

American Bison Population Extinction

Recall the American bison Leslie matrix model from Unit 2 Student Projects. Let the Leslie matrix for the American bison population be given by

$$\mathbf{A} = \begin{bmatrix} 0 & 0 & F \\ 0.60 & 0 & 0 \\ 0 & 0.75 & 0.60 \end{bmatrix}.$$

The population is divided into calves, yearlings, and adults (age 2 years or more). Suppose that the population starts at time 0 with 42 calves, 0 yearlings, and 95 adults. In the Unit 2 Student Project, we changed various parameters in the Leslie matrix to discover what parameter changes lead to the extinction of the population by $t = 200$. In this project, we will suppose that the adult fecundity F varies from year to year according to a uniform distribution over the range $(0, 1.7)$. To simulate this, we will use Matlab to randomly generate the value of F at each time step (each year) using the rand function. That is,

```
F = 1.7*rand(1);
```

Given that F is randomly chosen at each time step, each simulation of the population over the next 200 years will be different. In the project, your task is to determine what proportion of those simulations lead to the extinction of the American bison population by $t = 200$.

Student Tasks

1. Using the Matlab skills you have learned in this unit and the previous unit, write an m-file to determine the proportion of simulations in which the American bison population goes extinct. This is known as the extinction probability. Build this model using the following steps.

   (a) First, assume that $F = 1.1$. Now write an m-file that utilizes a *for-loop* to determine the number of calves, yearlings, and adults each year (assume that the time step is 1 year). After the *for-loop*, use an *if statement* to determine if the population is extinct.

   (b) Next, within the *for-loop* you have created, add the line `F=1.7*rand(1)` to randomly choose the value for F each year. Run this a few times and check if each simulation is producing different results.

   (c) Now, you currently have a loop that runs one 200-year simulation. Encase that loop inside another *for-loop* to run the simulation N times. Make sure that you keep track of how many of the simulations lead to extinction of the bison population. You will need a counter for this. (See some of the m-files in Chapter 10 for examples of multiple loops and the use of counters.)

2. Use the code that you have written to run your file for $N = 10, 20, 50, 100, 200, 500, 1000$.

3. Suppose that F varies from year to year according to a uniform distribution, with range $(0, 3)$. Change F in your m-file accordingly and run your file for $N = 10, 20, 50, 100, 200, 500, 1000$.

4. Suppose that F varies from year to year according to a uniform distribution, with range $(0, 1)$. Change F in your m-file accordingly and run your file for $N = 10, 20, 50, 100, 200, 500, 1000$.

5. Compose a three-page report (not including figures or tables) that describes the analysis you performed and that summarizes the findings of your analysis. Your report should have four sections:

   (a) Introduction: Describe the problem put forth and any hypothesis you had about the results of the experiment.

   (b) Methods: Describe the mathematics you employed to solve the described problem. Specifically, describe how you used Matlab to incorporate probability into the Leslie matrix model for the American bison.

   (c) Results: Describe in words and with tables the results of your analysis. In this section, include any tables of results that you produced along with descriptions of what is presented in those tables. Also, describe how changing the number of simulations run affected the proportion of simulations in which the American bison went extinct. Likewise, describe how changing the distribution from which the adult fecundity value F was chosen affected the proportion of simulations in which the American bison went extinct.

   (d) Conclusions: What conclusions can you draw from your results? Comment on any hypotheses you made in light of your findings. Discuss the possible future of the American bison population based on your results. Would you make any policy recommendations based on your results?

6. Write a paragraph that describes how comfortable you feel writing code in Matlab after this project and how much you feel this project has contributed to your understanding of matrix modeling and probability. ∎

4
UNIT

Limits and Continuity

In the first three units of this book, we have investigated a variety of mathematical approaches to biological questions. A common theme throughout has been an emphasis on problems that could be expressed in a "discrete-time" manner. Thus, we analyzed the growth rate of cell populations in which we assumed that the population at the next time step depended only on the population at the current time step (first-order difference equations), or we looked at the age structure of a population from one time step to another (Leslie matrix population model) or analyzed the drug concentration in the bloodstream just after giving a dose and how this concentration changed from one dose to the next.

However, in many biological situations, it is more appropriate to consider the variables of interest continuously rather than discretely. For example, the concentration of a drug in the bloodstream may vary through time, and we may want to know how this concentration changes not only after giving a dose but also in the time period between doses (we made an assumption about this previously when we assumed that it declined exponentially). Many physiological processes do not move in "jumps" but rather change continuously through time (e.g., body temperature, body weight, brain activity, enzyme concentrations, etc.).

The area of mathematics that deals with variables that change continuously rather than in jumps is called calculus (which is short for differential and integral calculus). Many would consider the development of calculus to be one of the greatest conceptual achievements of the human mind. The ideas of calculus permeate all of modern science and have proven to be useful in applications to areas as diverse as neurobiology, economics, forensic science, ecology, and epidemiology. Our goal in this portion of the text is to help you learn about the conceptual foundations of calculus, provide you with some of the standard "tools" used to apply calculus in science, and do this in a biological context so that you will see how biological questions may be addressed using this wonderful conceptual gift.

The first concept we will discuss is one we have already seen: the idea of a limit. We developed this when thinking about populations varying through discrete time (generation by generation) in which we said that a population's size had a long-term limit if its size moved closer and closer to a particular value after a large number of generations. We called this the steady-state population size or the population's long-term equilibrium population size. Of course, this did not apply only to populations, as we saw the same idea in the example of a drug's concentration in the body (measured just after a periodic dose is given) and saw that this concentration approached a limit after a large number of regular, periodic doses. In the Leslie matrix examples for age-structured population growth, we saw that the population growth rate from one generation (or time period) to the next changed but eventually became closer and closer to the largest eigenvalue of the population projection matrix.

We will now consider the same idea of a limit, but rather than letting a quantity (generations or number of doses) get very large, we will let a variable get close to some fixed number and see the impact of that variable on a function. For example, suppose that blood flow rate through the heart (measured in cubic centimeters per second) is a function of a drug's concentration in the blood. We might want to know what the limit of the blood flow rate is when the concentration of the drug approaches some value.

All of this relates to the issue of what we will call "continuity": whether a function (such as blood flow rate) changes smoothly as we vary the drug concentration or whether there is a sudden "jump" or shift in flow rate at some concentration. A simple example of this is a drug concentration that becomes lethal at some level (e.g., the person's heart stops, and so the blood flow rate drops suddenly to zero). This would not be a continuous change in blood flow and should be an outcome to be avoided! The next few chapters formalize the ideas of limit and continuity.

Limits of Functions

Biology deals with many processes that change in time or space. These include the number of fish in a stream, the rate at which biochemical reactions occur, the metabolic rate of an individual, the feeding rate of individuals, the death rates of populations, the elimination rate of a drug, and the rate of change in lineages. As this list indicates, rates occur throughout biology, expressing the way that some measurable phenomenon changes with time (or space, as you would observe by following the changes in number of fish as you moved downstream).

You are already very familiar with average rates since this is used, for example, when your pulse rate is taken. You measure how many beats occur in a minute, and this gives you a rate per minute (typically in the range of 60 to 80 if you are at rest). If you have never measured this yourself, you should do so and observe how the pulse rate changes after both mild and stronger exercise. The measurement you make is an average rate over the time period you measure it. You might try the experiment of measuring your pulse rate not over only 1 minute but also over 2 minutes and then over 30 seconds, 10 seconds, and 5 seconds. Do you get consistent results? For example, if the results are consistent, then the number of pulses you measure over 1 minute should be exactly 12 times the number you obtained in 5 seconds.

Of course, it is difficult to count what a pulse is when the time period is very short, such as a fraction of a second. This is true for heart-beats because there is a natural period (about 1 Hertz, or 1 cycle per second, when at rest) for heart rate. Trying to measure biological processes with natural cycles over shorter time periods than their natural cycle lengths is not feasible. However, there are many biological processes for which it makes sense to measure them over shorter and shorter time periods, at least down to some level.

Many biochemical reactions occur on time scales much faster than seconds, as do the basic life processes of photosynthesis and respiration. For these processes, it would make sense to measure their rates in shorter and shorter time periods, assuming that you had the instruments to do so. For example, the rate of a leaf's photosynthesis is by definition the rate at which the leaf takes in CO_2, though there are some complications here since respiration also occurs in leaves, causing CO_2 to be produced. Plant physiologists, therefore, usually refer to the photosynthesis rate of a leaf as the net photosynthesis rate since it takes into account the uptake rate of CO_2 as well as the rate at which the leaf releases CO_2 through respiration. These rates can change quite rapidly, often within fractions of seconds. If you want to estimate how rapidly photosynthesis responds to environmental changes, such as the light change that occurs when the sun is covered by a cloud (and the light level on a leaf changes quickly), then you might record how photosynthesis changes when you measure the net uptake rate of CO_2 over shorter and shorter time periods.

It is the objective of this section to present the mathematics involved when you measure some biological process, such as the photosynthetic rate of a leaf, over very short time periods. This is similar in some respects to what you saw in previous sections when you looked at how a biological process behaved after a long time period (letting the number of time steps go to infinity), only here we are looking at very short rather than very long time periods. After we develop the mathematics for this, we will see how we can use measurements made over very short time periods to eventually determine how the total process responds over long time periods. In the photosynthesis example, this means that we will be able to determine the total amount of CO_2 taken up by a leaf over a long time period by knowing how the leaf responds over short time periods. Since CO_2 uptake by a leaf is directly related to the energy available for plant growth, when summed over all the leaves on a plant, this affects the crop yields that are available from agricultural systems.

Note that although we are using photosynthesis as the example process here, this same approach applies to processes such as the elimination of drugs from the body. If we can estimate the rate at which a drug is metabolized or excreted over very short time periods, we will see that we can use this information to estimate how much drug is metabolized or excreted over long time periods and thus estimate the amount of drug left in the body at any time. As you might have already guessed, the method we will use to do this is essentially to "sum up" the changes over short time periods to get the total change (in drug in the body) over long time periods.

15.1 Limit of a Function

Recall in Chapter 5 that you were introduced to the concept of a *limit* by considering limits of sequences. Now we will develop limits of functions.

We write

$$\lim_{x \to a} f(x) = L$$

and say that "the *limit* of $f(x)$, as x approaches a, equals L" if we can make the values of $f(x)$ arbitrarily close to L (as close to L as we like) by taking x to be sufficiently close to a but not equal to a. The precise mathematical definition of a limit is given in Appendix B.

For example, if $f(x)$ represents the photosynthetic rate of a leaf at time x, then the limit as x approaches a will give the photosynthetic rate exactly at time a. If $f(x)$ gives the dissolved oxygen content of a stream at location x along the stream, then the limit as x approaches a will give the dissolved oxygen content at location a.

Example 15.1 (Photosynthetic Capacity of Leaves [26, 73])

Recall in Example 5.3 that we estimated the limit of the photosynthetic rates for leaves higher on a soybean plant as compared to leaves lower on the plant. The shape of the curve

(Continued)

of the data for both the upper and the lower leaves looks like a type of *Michaelis-Menten curve*, which has the form

$$P = \frac{abI}{aI + b} - d, \tag{15.1}$$

where a, b, and d are positive constants, I is the light level, and P is the photosynthetic rate. The light level I has units of μmol of photons per m^2 of leaf area per second, or μmol m^{-2} s^{-1}. The photosynthetic rate P has units of μmol of O_2 per m^2 of leaf area per second, where here the measure of photosynthetic rate is the rate at which O_2 is produced by the leaf area rather than the rate at which CO_2 is absorbed.

Technically, a Michaelis-Menten curve has the form $P = \frac{abI}{aI+b}$. Given this form, if a and b are positive constants, then P will always be positive for positive I. However, the data for Example 5.3 show some negative P values for positive I values (this indicates that at low light levels, the soybean plant is actually losing CO_2 and absorbing rather than releasing O_2). Thus, we subtract a constant d from the standard Michaelis-Menten curve, where $-d$ is the *dark respiration rate*, that is, the respiration rate of the soybean plant when there is no light (i.e., $P(0) = -d$).

If we fit Equation (15.1), using a nonlinear least-squares regression (recall in Chapter 3 that we covered linear regressions), to each of the data sets in Table 5.2, we get the following equations:

$$P_L(I) = \frac{1.62I}{0.195I + 8.29} - 3.13 \qquad \text{for the lower leaves.}$$

$$P_U(I) = \frac{1.84I}{0.132I + 14.0} - 3.03 \qquad \text{for the upper leaves.}$$

In full sunlight at sea level, the light level, measured in PPFD (photosynthetic photon flux density), is approximately 2000 μmol m^{-2} s^{-1}. What is the limit of P_L and P_U as the light level I approaches 2000 μmol m^{-2} s^{-1}?

Solution: To determine the limit of each function, let us examine the values of P_L and P_U as the values of I get very close to 2000.

I	P_L	P_U
1000.000	4.82529	9.60036
1500.000	4.93486	10.0167
1750.000	4.96672	10.1407
1900.000	4.98993	10.2003
1990.000	4.98993	10.2320
1999.000	4.99069	10.2350
1999.900	4.99077	10.2353
1999.990	4.99078	10.2353
1999.999	4.99078	10.2353

(Continued)

It appears that the lower leaves approach a photosynthetic oxygen evolution rate of approximately 4.99 μmol m^{-2} s^{-1} and that the upper leaves approach a rate of approximately 10.24 μmol m^{-2} s^{-1}. Thus, as light level increases, the upper leaves are producing oxygen, O_2, at a faster rate than the lower leaves. As we noted earlier, the differences between the photosynthetic rates of the leaves have been found to arise because of differences in the chloroplast density and the thickness of the leaves from the different locations in the plant.

These estimates of the maximum photosynthetic rate for upper- and lower-canopy soybean leaves are derived from the fitted equations for P_L and P_U. The data from which these fitted equations were derived included oxygen evolution rates measured only up to light levels of approximately 1000 μmol m^{-2} s^{-1}. So, without taking measurements of photosynthetic rates at higher light levels, we cannot say for certain that the estimated maximum photosythetic rates at full sunlight we derived above are accurate. They are the best estimates that can be made given the available data, but there are additional constraints on photosynthetic rates at very high light levels that may constrain the maximum rates to be below those that we estimated.

In the previous example, we found the limit of the functions $P_L(I)$ and $P_U(I)$ by examining function values as we increased the value of I closer and closer to 2000. Why not also examine the function's values as we *decrease* the value of I closer and closer to 2000? In the example above, approaching the value 2000 "from the right" (along the number line) does not make sense because the maximum value of I is 2000. However, if we were interested in the finding the limit of $P_L(I)$ and $P_U(I)$ at $I = 800$, then it would make sense to examine function values as we allow values to get closer and closer to 800 μmol m^{-2} s^{-1} from the left and the right (along the number line).

In general, we can determine the limit of a function $f(x)$ at the value $x = a$ by examining function values as we increase x closer and closer to a and decrease x closer and closer to a. Consider the following example, which examines the limit of a quadratic function as x approaches 2.

Example 15.2 (Thinking About Limits)

Consider the function $f(x) = x^2$. We know that $f(2) = (2)^2 = 4$, but is 4 the limit of $f(x)$ as $x \to 2$? Let us look at values of x approaching 2 to find out.

x	$f(x) = x^2$	x	$f(x) = x^2$
1.9	3.61	2.1	4.41
1.99	3.9601	2.01	4.0401
1.999	3.996001	2.001	4.004001
1.9999	3.99960001	2.0001	4.00040001

Notice that as x approaches the value 2, approaching both "from the left" with values that are smaller than 2 and "from the right" with values are are larger than 2, the function values approach 4. Thus, we see that as $x \to 2$, $f(x) \to 4$. Therefore, 4 is the limit of $f(x)$ as $x \to 2$.

Notice that by our definition of a limit, the function value need not exist at the point at which we are trying to find the limit; that is, $f(a)$ need not exist (have a value). Let's look at an example.

Example 15.3 (Finding a Limit)

Find $\lim\limits_{t \to 0} \frac{\sqrt{t^2+9}-3}{t^2}$.

<u>Solution:</u> Notice that the function $g(t) = \frac{\sqrt{t^2+9}-3}{t^2}$ cannot be evaluated at 0 (since the denominator would be 0). Thus, the function $g(t)$ is not defined at $t = 0$; that is, $g(0)$ does not exist. However, this does not prevent $g(t)$ from having a limit as $t \to 0$. When we look at the graph of $g(t)$, we can see that though the function is not defined at $t = 0$, it appears that the function values approach some point as we let $t \to 0$ from the left and from the right.

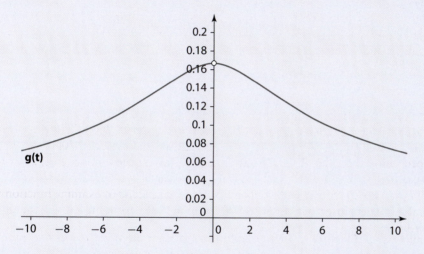

From the graph, however, it is unclear what value the function is approaching as t approaches 0. So let us look at a table of values.

t	$f(t) = \frac{\sqrt{t^2+9}-3}{t^2}$
± 1.0	0.16228
± 0.5	0.16553
± 0.1	0.16662
± 0.05	0.16666
± 0.01	0.16667˙

Recall that $0.16666\cdots = \frac{1}{6}$. Since the function values approach $0.16666\ldots$ as $t \to 0$, the limit is $\frac{1}{6}$.

Notice that if you use a small enough value for t (very close to 0) when computing function values on your calculator, you find that the function evaluates to 0. This occurs because $\sqrt{t^2+9}$ is very close to 3 when t is small. In fact, when t is sufficiently small (like 0.000001), the calculator's value for $\sqrt{t^2+9}$ is $3.0000\ldots$, so when we subtract 3 from this, we get 0 in the numerator.

(Continued)

Notice that we can also show that the limit is $\frac{1}{6}$ by rationalizing the numerator.

$$\frac{\sqrt{t^2+9}-3}{t^2} \times \frac{\sqrt{t^2+9}+3}{\sqrt{t^2+9}+3} = \frac{t^2+9-9}{t^2\left(\sqrt{t^2+9}+3\right)} = \frac{t^2}{t^2\left(\sqrt{t^2+9}+3\right)} = \frac{1}{\sqrt{t^2+9}+3}$$

As $t \to 0$, the numerator remains the constant 1, and the denominator approaches $\sqrt{9}+3 = 6$. Thus,

$$\lim_{t\to0} \frac{\sqrt{t^2+9}-3}{t^2} = \lim_{t\to0} \frac{1}{\sqrt{t^2+9}+3} = \frac{1}{6}.$$

Are there times when the limit does not exist? Yes. Let's look at an example.

Example 15.4 (Limit Does Not Exist)

Consider $\lim_{x\to0} \frac{|x|}{x}$. Suppose that δ is some very small positive number, i.e., close to 0. Then, when $x = -\delta$, we have $\frac{|x|}{x} = \frac{\delta}{-\delta} = -1$. When $x = \delta$, we have $\frac{|x|}{x} = \frac{\delta}{\delta} = 1$. No matter how small δ becomes, the function is -1 on the left of 0 and 1 on the right of 0. Since these two values do not match, the limit does not exist.

15.2 Limit Properties

In some biological situations, we are concerned with mixtures of functions. Estimating the revenue from harvesting a crop requires computing a product of the yield of the crop, $Y(r)$, which depends on a measure of rainfall, r, and the income received per unit yield $I(r)$, which might also depend on rainfall since rainfall could affect the overall supply of the crop. Thus, revenue is the product $Y(r)I(r)$. If $C(r)$ gives the cost of harvesting the yield $Y(r)$, then the ratio $C(r)/Y(r)$ gives the cost per unit yield for growing this crop. In many situations, the functions that are of interest to us are made up of sums, differences, products, and ratios of other functions. We now point out the basic rules for taking limits for these types of mixtures of functions.

Limit Properties

If L, M, a, and c are real numbers and $\lim_{x\to a} f(x) = L$ and $\lim_{x\to a} g(x) = M$, then

1. $\lim_{x\to a} c = c$
2. $\lim_{x\to a} [f(x) + g(x)] = L + M$
3. $\lim_{x\to a} [f(x) - g(x)] = L - M$
4. $\lim_{x\to a} [f(x)g(x)] = LM$
5. $\lim_{x\to a} [f(x)/g(x)] = L/M$, provided that $M \neq 0$

Let us apply some of these principles.

Example 15.5 (Factoring to Simplify)

Find $\lim\limits_{x \to 2} \frac{x^2-4}{x-2}$.

Solution: Notice that if we apply limit property 5, we find that the limit has the indeterminate form $\frac{0}{0}$. However, if we simplify using algebra, then

$$\frac{x^2-4}{x-2} = \frac{(x-2)(x+2)}{x-2} = x+2 \text{ for } x \neq 2.$$

Note that the original function $(x^2 - 4)/(x - 2)$ is equal to the linear function $x + 2$ everywhere except for $x = 2$. This is because the original function is not defined at $x = 2$ (since we cannot divide by 0). Recall that by the definition we have for a limit, we do not need to know the value of the function at the point we are approaching (in this case, $x \to 2$). Instead, we are interested in the value that the function is *approaching* as $x \to 2$. Since the function is equal to the linear function $x + 2$ at all points except $x = 2$ (the point we are approaching), the limit of our original function $(x^2 - 4)/(x - 2)$ as $x \to 2$ is equal to the limit of the linear function $x + 2$ as $x \to 2$, that is,

$$\lim\limits_{x \to 2} \frac{x^2-4}{x-2} = \lim\limits_{x \to 2} x + 2 = 4.$$

Example 15.6 (Getting a Common Denominator to Simplify)

Find $\lim\limits_{x \to 1} \left[\frac{1}{x-1} - \frac{2}{x^2-1} \right]$.

Solution: If we attempt to apply limit property 2, we find

$$\lim\limits_{x \to 1} \left[\frac{1}{x-1} - \frac{2}{x^2-1} \right] = \lim\limits_{x \to 1} \frac{1}{x-1} - \lim\limits_{x \to 1} \frac{2}{x^2-1}.$$

Let us consider each limit in turn. What happens to the fraction $1/(x-1)$ as $x \to 1$? First, notice that the function $1/(x - 1)$ is undefined at $x = 1$.

Next, suppose that x is smaller than 1 but approaching 1 (i.e., approaching from the left). As the value of x gets closer and closer to 1, the value of the denominator gets smaller and smaller but is a negative number. Thus, the value of the fraction $1/(x - 1)$ gets larger and larger but is a negative number. Thus, we would say that as x approaches 1 from the left, the function diverges to negative infinity, denoted $-\infty$.

(Continued)

Now suppose that x is larger than 1 but approaching 1 (i.e., approaching from the right). Now, as the value of x gets closer and closer to 1, the value of the denominator gets smaller and smaller but is a positive number. Thus, the value of the fraction gets larger and larger, and we would say that as x approaches 1 from the right, the function diverges to infinity, denoted ∞.

So, we have two problems here: the function $1/(x-1)$ does not approach a finite value as $x \to 1$, and the function values do not have the same behavior when you approach from the left as when you approach from the right.

Now consider the second function $2/(x^2-1)$. For this function, as $x \to 1$ from either the left or the right, the denominator gets smaller and smaller. Thus, the value of the fraction gets larger and larger, either positive or negative, depending on whether x approaches 1 from the left or from the right.

To apply limit property 2 to find the limit of the difference of two functions, the limits of each of those two functions must be real numbers. This is not the case here. So, instead, we will use algebra to rewrite the difference of fractions as one fraction. Thus, we will find a common denominator to simplify

$$
\frac{1}{x-1} - \frac{2}{x^2-1} = \frac{1(x^2-1) - 2(x-1)}{(x-1)(x^2-1)}
$$

$$
= \frac{x^2 - 1 - 2x + 2}{(x-1)[(x-1)(x+1)]}
$$

$$
= \frac{x^2 - 2x + 1}{(x-1)^2(x+1)}
$$

$$
= \frac{(x-1)(x-1)}{(x-1)^2(x+1)}
$$

$$
= \frac{1}{x+1} \quad \text{for } x \neq 1.
$$

Thus,

$$
\lim_{x \to 1} \left[\frac{1}{x-1} - \frac{2}{x^2-1} \right] = \lim_{x \to 1} \frac{1}{x+1} = \frac{1}{2}.
$$

In each of the examples so far, we let x approach a finite number. As you saw in the landscape-change examples in Unit 2 and the case of a regular sequence of drug dosages in Chapter 5, we may be interested in the long-term outcome of a biological process. Similar to letting n go to ∞ in these discrete model situations, here we can allow x to approach ∞ or $-\infty$. This means that x will take on larger and larger positive or negative values, respectively. We use the notation

$$
\lim_{x \to \infty} f(x) = L
$$

to mean that as x gets larger, $f(x)$ approaches L. Likewise, we use the notation

$$
\lim_{x \to -\infty} f(x) = M
$$

to mean that as x gets smaller with negative values, $f(x)$ approaches M.

All of the limit properties 2 to 5 are also valid if $x \to \infty$ or $x \to -\infty$. Next, we have two more limit properties that involve $x \to \pm\infty$ (i.e., $x \to \infty$ or $x \to -\infty$).

More Limit Properties

For $n > 0$ and $r > 0$,

6. $\displaystyle\lim_{x \to \pm\infty} \frac{1}{x^n} = 0$

7. $\displaystyle\lim_{x \to \infty} e^{-rx} = 0$

If $\displaystyle\lim_{x \to -\infty} f(x) = L$ or $\displaystyle\lim_{x \to \infty} f(x) = L$, then we say that the function $f(x)$ has a *horizontal asymptote*, $y = L$. The equation representing a population with logistic growth has a horizontal asymptote. This equation is derived from a differential equation, which we explore in depth in Chapter 27. The *logistic growth equation* for a population is given by

$$P(t) = \frac{KP_0 e^{rt}}{K + P_0(e^{rt} - 1)} \quad \text{for } t > 0,$$

where

- $P(t) =$ population size at time t,
- $K =$ carrying capacity of the population (the maximum population size the environment can sustain),
- $P_0 =$ the population size at time 0 (this must be given), and
- $r =$ the intrinsic growth rate of the population (how rapidly a population grows over time), $r > 0$.

Notice in Figure 15.1 that as time goes to infinity, the population size gets infinitesimally close to its carrying capacity. Mathematically, it never actually reaches the carrying capacity, but biologically, since we deal only with whole individuals (not fractional individuals), we know that we can actually reach the carrying capacity. Here, the carrying capacity is a horizontal

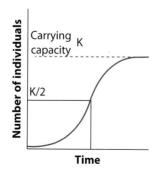

Figure 15.1 Graph showing logistic growth. Notice the dashed line ($P = K$) indicating the carrying capacity K. The population approaches this carrying capacity as time grows large. Thus, $\lim_{t \to \infty} P(t) = K$, where $P(t)$ is the population size at time t.

asymptote described by the line $P = K$. We can verify that $P = K$ is a horizontal asymptote by evaluating the limit

$$\lim_{t \to \infty} P(t).$$

However, notice that

$$\lim_{t \to \infty} P(t) = \lim_{t \to \infty} \frac{KP_0 e^{rt}}{K + P_0(e^{rt} - 1)} = \frac{\infty}{\infty}.$$

The fraction $\frac{\infty}{\infty}$ is an **indeterminate form**, which essentially means "to be determined." Thus, in order to evaluate the limit, we must use some clever algebra to change the form of the function (without actually changing the function) so that we do not obtain an indeterminate form. In this case, we will multiply by 1 in a clever way. Note that multiplying by 1 will not change the value of the function. Notice that

$$\frac{KP_0 e^{rt}}{K + P_0(e^{rt} - 1)} \times \frac{e^{-rt}}{e^{-rt}} = \frac{KP_0}{Ke^{-rt} + P_0(1 - e^{-rt})}.$$

Thus, using limit property 7, we find

$$\lim_{t \to \infty} P(t) = \lim_{t \to \infty} \frac{KP_0 e^{rt}}{K + P_0(e^{rt} - 1)}$$

$$= \lim_{t \to \infty} \frac{KP_0}{Ke^{-rt} + P_0(1 - e^{-rt})}$$

$$= \frac{KP_0}{K(0) + P_0(1 - 0)} = \frac{KP_0}{P_0} = K.$$

Thus, we have shown that the function $P(t)$ has a horizontal asymptote of $P = K$.

Note that $\frac{\infty}{\infty}$ is not the only indeterminate form. The full list of indeterminate forms is given below.

Indeterminate Expressions

The following expressions are indeterminate

$$\frac{\infty}{\infty}, \frac{0}{0}, \infty - \infty, \infty \times 0, 0^0, \infty^0, \text{ and } 1^\infty.$$

Let us consider another biological example of a horizontal asymptote.

Example 15.7 (Maximum Photosynthetic Capacity of Leaves [22, 73])

In Example 15.1, the function

$$P = \frac{abI}{aI + b} - d,$$

where a, b, and d are positive constants, is a model representing the photosynthetic rate P (measured in μmol of O_2 per m^2 of leaf area per second) at light level I (measured in

(Continued)

μmol of photons per m^2 of leaf area per second). Does this function have a horizontal asymptote? If so, what does the horizontal asymptote represent biologically?

Solution: We can determine if this function has a horizontal asymptote by taking the limit of the function $P(I)$ as $I \to \infty$.

We start by writing the limit as the difference of two limits using limit property 3:

$$\lim_{I \to \infty} \frac{abI}{aI + b} - d = \lim_{I \to \infty} \frac{abI}{aI + b} - \lim_{I \to \infty} d.$$

Notice that

$$\lim_{I \to \infty} \frac{abI}{aI + b} = \frac{\infty}{\infty}.$$

In this case, we multiply by 1 in the form of $\frac{1/I}{1/I}$, and thus

$$\lim_{I \to \infty} \frac{abI}{aI + b} - d = \lim_{I \to \infty} \frac{abI}{aI + b} - \lim_{I \to \infty} d \qquad \text{(Limit property 3)}$$

$$= \lim_{I \to \infty} \frac{abI}{aI + b} \times \frac{1/I}{1/I} - \lim_{I \to \infty} d \qquad \text{(Multiply by 1 in a clever way)}$$

$$= \lim_{I \to \infty} \frac{ab}{a + \frac{b}{I}} - \lim_{I \to \infty} d$$

$$= \frac{\lim\limits_{I \to \infty} ab}{\lim\limits_{I \to \infty} a + \lim\limits_{I \to \infty} \frac{b}{I}} - \lim_{I \to \infty} d \qquad \text{(Limit properties 2 and 5)}$$

$$= \frac{ab}{a + 0} - d \qquad \text{(Limit properties 1 and 5)}$$

$$= b - d.$$

Thus, $P(I)$ has a horizontal asymptote at $P = b - d$. Since $P(I)$ is an increasing function, we can think of the horizontal asymptote as an upper bound on the value of the photosynthetic rate P. That is, the photosynthetic rate of a plant will never be larger than $b - d$. In the case of the soybean plant, in Example 15.1 we saw that for the lower leaves of the plant, $b = 8.29$ and $d = 3.13$, while for the upper leaves of the plant, $b = 14.0$ and $d = 3.03$. Thus, for the soybean plant, the photosynthetic rate will never exceed

$$b - d = 8.29 - 3.13 = 5.16 \ \mu\text{mol m}^{-2} \text{ s}^{-1}$$

for the lower leaves and

$$b - d = 14.0 - 3.03 = 11.0 \ \mu\text{mol m}^{-2} \text{ s}^{-1}$$

for the upper leaves. Note that each calculation has been rounded to the appropriate number of significant digits.

(Continued)

Full sunlight at sea level has a light level (measured in PPFD) of approximately 2000 μmol m^{-2} s^{-1}. Thus, it may seem a bit unrealistic to consider the limit of $P(I)$ as $I \to \infty$. Consider the difference between the upper bound $b - d$ and the value of $P(2000)$ in order to determine if the estimate of maximum photosynthetic rate obtained by taking the limit is very different from the estimated value at full sunlight:

$$P_{\max} - P(2000) = (b - d) - \left(\frac{2000ab}{2000a + b} - d \right)$$

$$= b - \frac{2000ab}{2000a + b}$$

$$= \frac{b(2000a + b) - 2000ab}{2000a + b}$$

$$= \frac{b^2}{2000a + b}.$$

Recall from Example 15.1 that for the lower leaves of the soybean plant, $a = 0.195$, while for the upper leaves of the soybean plant, $a = 0.132$. Thus,

$$P_{\max} - P(2000) = \frac{(8.29)^2}{2000(0.195) + 8.29} = 0.173 \, \mu \, \text{mols} \, \text{m}^{-2} \, \text{s}^{-1} \text{ for the lower leaves.}$$

$$P_{\max} - P(2000) = \frac{(14.0)^2}{2000(0.132) + 14.0} = 0.705 \, \mu \, \text{mols} \, \text{m}^{-2} \, \text{s}^{-1} \text{ for the upper leaves.}$$

This illustrates a difficulty in using a single equation for photosynthesis. It is not perfect, but it is still useful in describing the general dependence of photosythesis on light level and the relative differences between the photosynthetic responses of different types of leaves.

Example 15.8 (Approaching Infinity and Negative Infinity)

Find $\lim\limits_{x \to \infty} \dfrac{3x}{\sqrt{x^2 + 8}}$ and $\lim\limits_{x \to -\infty} \dfrac{3x}{\sqrt{x^2 + 8}}$.

Solution: Divide the top and bottom by x and recall that $x = \sqrt{x^2}$ for $x > 0$:

$$\lim_{x \to \infty} \frac{3x}{\sqrt{x^2 + 8}} \times \frac{\frac{1}{x}}{\frac{1}{\sqrt{x^2}}} = \lim_{x \to \infty} \frac{3}{\sqrt{1 + \frac{8}{x^2}}} = \frac{3}{\sqrt{1 + 0}} = 3.$$

(*Continued*)

Thus, $y = 3$ is a horizontal asymptote for the function $f(x) = \frac{3x}{\sqrt{x^2+8}}$ for $x > 0$.

To find $\lim\limits_{x \to -\infty} \frac{3x}{\sqrt{x^2+8}}$, recall that $x = -\sqrt{x^2}$ for $x < 0$. Then, if we divide the top and bottom by x, we get

$$\lim_{x \to -\infty} \frac{3x}{\sqrt{x^2+8}} \times \frac{\frac{1}{x}}{\frac{1}{-\sqrt{x^2}}} = \lim_{x \to \infty} \frac{3}{-\sqrt{1+\frac{8}{x^2}}} = \frac{3}{-\sqrt{1+0}} = -3.$$

Thus, $y = -3$ is also a horizontal asymptote for $x < 0$.

Example 15.9 (Arctic Foxes [29, 54])

The age-weight relationship of female Arctic foxes caught in Svalbard, Norway, can be estimated by the function

$$M(t) = 3102e^{-e^{-0.022(t-56)}},$$

where t is the age of the fox in days and $M(t)$ is the weight of the fox in grams.

(a) Use $M(t)$ to estimate the largest size that a female fox can attain.
(b) Estimate the age, in days, of a female fox when she has reached 80% of her maximum weight.

Solution:

(a) First, notice that

$$\lim_{t \to \infty} e^{-0.022(t-56)} = \lim_{t \to \infty} e^{(0.022)(56)}e^{-0.022t} = e^{(0.022)(56)} \times 0 = 0$$

since $\lim\limits_{t \to \infty} e^{-rt} = 0$, where r is a positive constant. Thus,

$$\lim_{t \to \infty} M(t) = 3102e^0 = (3102)(1) = 3102.$$

Thus, the maximum weight that a female Arctic fox can attain according to this model is 3102 grams.

(Continued)

(b) A female Arctic fox will be at 80% of her maximum weight when she is $(0.8)(3102) = 2481.6$ grams. To find the corresponding age, we solve $M(t) = 2481.6$ for t:

$$2481.6 = M(t)$$
$$(0.8)(3102) = 3102e^{-e^{-0.022(t-56)}}$$
$$0.8 = e^{-e^{-0.022(t-56)}}$$
$$\ln(0.8) = -e^{-0.022(t-56)}$$
$$-\ln(0.8) = e^{-0.022(t-56)}$$
$$\ln(-\ln(0.8)) = -0.022(t-56)$$
$$\frac{-1}{0.022}\ln(-\ln(0.8)) = t - 56$$
$$t = \frac{-1}{0.022}\ln(-\ln(0.8)) + 56 \approx 124 \text{ days.}$$

Thus, when the female Arctic fox is about 124 days old, she will weigh approximately 80% of her maximum weight.

15.3 Matlab Skills

If we have a definition of a function f, Matlab can calculate the value of f at a specific value x. Recall that to define a "function" in Matlab, we must create an m-file that defines that function. For example, if we needed to work with the function

$$f(x) = \frac{\sqrt{x^2 + 9} - 3}{x^2},$$

we could create the following m-file to define that function:

f.m
```
1   % Creates the function f(x)
2   % Input: x (a real number)
3   % Output: y
4
5   function y = f(x)
6   y = (sqrt(x^2+9)-3)/(x^2);
```

Now suppose that we want to find the limit of $f(x)$ as $x \to 2$ or as $x \to 0$. With a program like Matlab, we cannot find the exact limit; however, we can estimate that limit numerically. Notice that for

$$\lim_{x \to 0} \frac{\sqrt{x^2 + 9} - 3}{x^2},$$

though a limit exists (which we showed in Example 15.3), the value $f(0)$ does not exist. Thus, to find the limit, we will look at the values of $f(x)$ for x close to 0 but $x \neq 0$. We can approximate the limit by looking at values $f(x+h)$ and $f(x-h)$, where h becomes increasingly small. Additionally, we will consider values close to $x = 0$ on both sides of $x = 0$ (see the discussion of right and left limits in Chapter 16):

```
                              limit.m
1   % Find the limit of a function f
2   % Make sure the file f.m is contained in the same folder as this file
3
4   % Input: x = value at which you want to find the limit
5   %        delta = how close do you want to get to the value x
6
7   % Output: LL = limit from the left side
8   %         RL = limit from the right side
9
10  function [LL,RL] = limit(h,x)
11
12  % Left limit
13  LL = f(x-h);
14
15  % Right limit
16  RL = f(x+h);
```

Now, if we want to estimate the left and right limits as $x \to 0$, we would use the following commands in the command window:

```
                          Command Window
>> [LL,RL] = limit(1,0)
LL =
    0.1623
RL =
    0.1623

>> [LL,RL] = limit(0.1,0)
LL =
    0.1666
RL =
    0.1666

>> [LL,RL] = limit(0.01,0)
LL =
    0.1667
RL =
    0.1667

>> [LL,RL] = limit(0.001,0)
LL =
    0.1667
RL =
    0.1667
```

Notice that as we let the input for h get smaller, and smaller the left and right limits seem to approach $0.16666\ldots$, or $\frac{1}{6}$. Thus, given the numerical approximations, we estimate the limit as $x \to 0$ to be $\frac{1}{6}$.

Now, if we want to find the limits as $x \to 2$, we would use the following commands in the command window:

```
────────────────── Command Window ──────────────────
>> [LL,RL] = limit(1,2)
LL =
     0.1623
RL =
     0.1381

>> [LL,RL] = limit(0.1,2)
LL =
     0.1526
RL =
     0.1501

>> [LL,RL] = limit(0.01,2)
LL =
     0.1515
RL =
     0.1513

>> [LL,RL] = limit(0.001,2)
LL =
     0.1514
RL =
     0.1514

>> [LL,RL] = limit(0.0001,2)
LL =
     0.1514
RL =
     0.1514
```

Notice that as we let the input for h get smaller and smaller, the left and right limits approach 0.1514. Thus, given these numerical approximations, we estimate the limit as $x \to 2$ to be 0.1514.

If we now want to consider a different function, such as

$$f(x) = \frac{1}{(x-2)^2},$$

all we need to do is change the file f.m appropriately:

```
────────────────────── f.m ──────────────────────
1   % Creates the function f(x)
2   % Input: x (a real number)
3   % Output: y
4
5   function y = f(x)
6   y = 1/(x-2)^2;
```

Now, what do we find as $x \to 2$?

```
──────────────── Command Window ────────────────
>> [LL,RL] = limit(1,2)
LL =
      1
RL =
      1

>> [LL,RL] = limit(0.1,2)
LL =
   100.0000
RL =
   100.0000

>> [LL,RL] = limit(0.01,2)
LL =
    1.0000e+004
RL =
    1.0000e+004

>> [LL,RL] = limit(0.001,2)
LL =
    1.0000e+006
RL =
    1.0000e+006

>> [LL,RL] = limit(0.0001,2)
LL =
    1.0000e+008
RL =
    1.0000e+008

>> [LL,RL] = limit(0.00001,2)
LL =
    1.0000e+010
RL =
    1.0000e+010
```

Here, it appears that the smaller we make the value of h, the larger the value of $f(x)$ becomes. Thus, as $x \to 2$, we see that

$$\frac{1}{(x-2)^2}$$

is growing without bound. For this function, it would be a fair estimation to say that the limit does not exist as $x \to 2$.

15.4 Exercises

15.1   Compute each of the following limits, if it exists.

(a) $\displaystyle\lim_{x \to 1} \frac{x^2 - 1}{x - 1}$

(b) $\displaystyle\lim_{x \to -3} \frac{x^2 - x + 12}{x + 3}$

(c) $\displaystyle\lim_{x\to -3} \frac{x^2 - x - 12}{x + 3}$

(d) $\displaystyle\lim_{x\to 0} \frac{x^2 + x}{x}$

(e) $\displaystyle\lim_{h\to 0} \frac{(3 + h)^2 - 9}{h}$

(f) $\displaystyle\lim_{t\to 0} \frac{\sqrt{2 - t} - \sqrt{2}}{t}$

(g) $\displaystyle\lim_{x\to 1} \left[\frac{2}{1 - x^2} + \frac{1}{x - 1} \right]$

(h) $\displaystyle\lim_{s\to 0} \left[\frac{1}{s\sqrt{s + 1}} - \frac{1}{s} \right]$

(i) $\displaystyle\lim_{x\to 2} \frac{4x + 1}{2x - 1}$

(j) $\displaystyle\lim_{x\to 2} \sqrt{4x + 1}$

(k) $\displaystyle\lim_{y\to 3} (y^3 + 2)(y^2 - 5y)$

(l) $\displaystyle\lim_{u\to -2} \sqrt{u^4 - 3u + 6}$

(m) $\displaystyle\lim_{t\to 0} \frac{\sqrt{3 + t} - \sqrt{3}}{t}$

15.2   Compute $\displaystyle\lim_{x\to\infty} f(x)$ for each of the following functions $f(x)$.

(a) $f(x) = \dfrac{3x - 2}{x + 1}$

(b) $f(x) = \dfrac{x^2 - 1}{x^2 + 1}$

(c) $f(x) = \dfrac{3x}{\sqrt{x^2 + 8}}$

(d) $f(x) = \dfrac{3x}{x^2 + 1}$

15.3   Compute $\displaystyle\lim_{x\to -\infty} g(x)$ for each of the following functions $g(x)$.

(a) $g(x) = \dfrac{2x}{\sqrt{x^2 - 4}}$

(b) $g(x) = \dfrac{3x}{x^2 - 2}$

(d) $g(x) = \dfrac{x^3 + 2}{x^3 - 2}$

(c) $g(x) = \dfrac{3x - 2}{5 - x}$

15.4   (From [29]) Suppose that a small pond normally contains 12 units of dissolved oxygen in a fixed volume of water. Suppose also that at time $t = 0$, a quantity of organic waste is introduced into the pond, reducing the oxygen concentration $f(t)$ in the pond. After t weeks, the oxygen concentration in the pond is given by

$$f(t) = \frac{12t^2 - 15t + 12}{t^2 + 1}.$$

As time goes on, what will be the ultimate concentration of oxygen? Will it return to 12 units?

15.5   (From [29]) The concentration of a drug in a patient's bloodstream h hours after it was injected is given by

$$A(h) = \frac{0.17h}{h^2 + 2}.$$

Find and interpret $\lim\limits_{h \to \infty} A(h)$.

15.6   (From [29, 37]) Researchers have developed a mathematical model that can be used to estimate the number of teeth $N(t)$ at time t (days of incubation in the egg prior to hatching) for *Alligator mississppiensis*, where

$$N(t) = 71.8e^{-8.96e^{(-0.0685t)}}.$$

(a) Find $N(65)$, the number of teeth of an alligator that hatched after 65 days of incubation.

(b) Find $\lim\limits_{t \to \infty} N(t)$ and use this value as an estimate of the upper bound on the number of teeth of a newborn alligator. Does this estimate differ significantly from the estimate of part (a)?

15.7   (From [29, 50]) To develop strategies to manage water quality in polluted lakes, biologists must determine the depth of sediments and the rate of sedimentation. It has been determined that the depth of sediment $D(t)$ (in centimeters) with respect to time (in years before 1990) for Lake Coeur d'Alene, Idaho, can be determined by the equation

$$D(t) = 155 \left(1 - e^{-0.0133t}\right).$$

(a) Find $D(20)$. What does this value represent biologically?

(b) Does $D(t)$ have a horizontal asymptote? If so, give a biological interpretation of the horizontal asymptote.

15.8   In Exercise 5.1, data were presented for light responses for soybean leaves after a 30-day period of scarce rainfall. The data were divided into two sets, one set representing the light response for leaves on the upper portion of the soybean plant and the other set representing light response for leaves on the lower portion of the plant. The following Michaelis-Menten types of curves are used to model the data:

$$P_L(I) = \frac{2.74I}{0.107I + 25.5} - 2.65 \qquad \text{for the lower leaves.}$$

$$P_U(I) = \frac{3.47I}{0.0924I + 37.6} - 3.30 \qquad \text{for the upper leaves.}$$

Determine the equation for the horizontal asymptotes for $P_L(I)$ and $P_U(I)$. What do these horizontal asymptotes represent biologically?

15.9   (From [29, 52]) In an article on the local clustering of cell surface receptors, the research analyzed limits of the function

$$C = F(S) = \frac{1 - S}{S(1 + kS)^{f-1}},$$

where C is the concentration of free ligand in the medium, S is the concentration of free receptors, f is the number of functional groups of cells $(f > 1)$, and k is a positive constant. Find each of the following limits.

(a) $\lim\limits_{S \to 1} F(S)$

(b) $\lim\limits_{S \to 0} F(S)$

15.10   (From [33]) When a stimulus is applied to a person, a small amount of time passes before the person reacts. This is because it takes time for the stimulus to reach the brain, for the excitation of the neurons in the brain to reach a threshold level to signal the release of a response, and for the response to reach the appropriate muscles and organs. The time it takes for the excitation of the neurons in the brain to reach a threshold level to signal the release of a response is known as the excitation factor, $\varepsilon(t)$, which depends on time t (measured in milliseconds). The threshold level is denoted h_E. Thus, when $\varepsilon(t) \geq h_E$, the response is released to travel through the nervous system to the appropriate muscles and organs. It can be shown that the relationship between the excitation factor and time is given by

$$t = \frac{1}{k_2} \ln \left[\frac{k_1 E}{k_1 E - k_2 \varepsilon(t)} \right],$$

where k_1 and k_2 are positive constants and E is the intensity of excitation produced by a stimulus of constant intensity.

(a) Solve the equation above explicitly for $\varepsilon(t)$.
(b) Using your solution from (a), verify that if $t_2 > t_1$, then $\varepsilon(t_2) > \varepsilon(t_1)$. This shows that the function $\varepsilon(t)$ is an increasing function.
(c) Compute $\varepsilon_\infty = \lim\limits_{t \to \infty} \varepsilon(t)$. This value represents an upper bound on the value for $\varepsilon(t)$ since $\varepsilon(t)$ is an increasing function.
(d) The signal to release the response will occur only when $\varepsilon(t) \geq h_E$. In order for this to be feasible, we must have $\varepsilon_\infty > h_E$. Show that if $k_1 E > k_2 h_E$, then $\varepsilon_\infty > h_E$.

15.11   Write your own Matlab function f.m for

$$f(x) = x^2.$$

Then, using limit.m, estimate

(a) $\lim\limits_{x \to -1} x^2$

(b) $\lim\limits_{x \to 1} x^2$

(c) $\lim\limits_{x \to 0} x^2$

15.12 Write your own Matlab function f.m for

$$f(x) = \ln x.$$

Then, using limit.m, estimate

(a) $\lim\limits_{x \to 3} \ln x$

(b) $\lim\limits_{x \to 1} \ln x$

(c) $\lim\limits_{x \to 0} \ln x$

When computing

$$\lim\limits_{x \to 0} \ln x$$

using limit.m, why does the left limit evaluate to an imaginary number?

15.13 Find an example of a function f and a point x for which the Matlab program limit.m will return different values for the left and right limits no matter how small you make delta. Write your own Matlab function f.m for the function you chose and use the program limit.m to estimate the left and right limit values for $f(x)$.

16

Limits of Continuous Functions

Our observations of biological processes such as photosynthesis are inherently constrained by the tools we use to measure them. Sometimes the methods being used are able to detect small changes of these processes over short periods of time. There are devices that can measure the rapid response of leaf photosynthetic rate to changes in light level, and these instruments can respond almost instantly to changes in the flux (rate of movement) of CO_2 into the leaf. It is the net flux of CO_2 into a leaf (or O_2 out of the leaf) that is a measure of photosynthesis at the leaf level. Using these tools allows us to draw a graph through time of the leaf's response to light or temperature changes and provides a direct way to estimate the amount of carbon fixed through photosynthesis as the environment varies. In such a graph, we may see rapid decreases in photosynthesis as light suddenly decreases, but the change will not be instantaneous. There is no immediate change in CO_2 uptake by a soybean leaf from 22 to 8 μmol m^{-2} s^{-1} when light suddenly drops from full sunlight to 20% of full sunlight [22]. Rather, all values of photosynthetic rate between 22 and 8 μmol m^{-2} s^{-1} occur over the approximately 2 minutes it takes for the leaf's photosynthetic rate to adjust to this changed light environment. The mathematical term we use to describe this continual, gradual change is "continuity," and the fact that the photosynthetic rate takes on all values between 22 and 8 μmol CO_2 m^{-2} s^{-1} arises from the "intermediate value theorem" described in Section 16.3.

In contrast to the above situation, sometimes we do not have the tools or the ability to observe how a biological process might change gradually. If we can observe a process only occasionally, we cannot say with certainty how it changes over the time between observations. In this case, it is simplest to assume a linear change over the time interval between observations or assume that it stays at the previous observed level over the time interval and then changes to the new level only at the end of the time interval. In the latter case, the graph of the process would have sudden "jumps," and the mathematical description is that the graph of the process is "discontinuous" at the times that it jumps.

Dealing with biological processes that do not suddenly jump but change in a consistent manner is generally simpler than having to be concerned about sudden "surprises." So it is helpful to be careful about defining what we mean by continuous variation in a process, which requires considering how the measured process is affected on both sides of a factor that affects it. For example, consider a drug that affects heart rate, such as digitalis or beta-blockers. We may be interested in knowing how heart rate is affected by a drug concentration just below level a as the drug concentration increases to a as compared to how heart rate is affected by drug concentration just above a as the drug concentration decreases to a. We may be interested in whether this depends on possible different concentrations a. To define "continuity" of a function at a point a in its domain, we need to define one-sided limits so that the limit of our function can be determined as one approaches a from the left (e.g., with values below a) and the limit of our function as one approaches a from the right (e.g., with values above a).

16.1 Right and Left Limits

We write

$$\lim_{x \to a^-} f(x) = L$$

and say that the ***left-hand limit*** of $f(x)$ as x approaches a (or the limit of $f(x)$ as x approaches a from the left) is equal to L if we can make the values of $f(x)$ arbitrarily close to L by taking x to be sufficiently close to a with $x < a$.

We can write a similar definition for the **right-hand limit** of $f(x)$ as x approaches a, where $x > a$:

$$\lim_{x \to a^+} f(x) = L.$$

> *Limit Properties*
>
> 8. $\lim_{x \to a} f(x) = L$ if and only if $\lim_{x \to a^-} f(x) = L$ and $\lim_{x \to a^+} f(x) = L$.

The above says that the limit of a function exists at a point $x = a$ if we get the same value of the function when we let x get closer and closer to a from values above a that we do when we let x get closer and closer to a from values below a. If this doesn't happen, then there is some kind of jump of the function $f(x)$ at the point $x = a$, and we say that $\lim_{x \to a} f(x)$ is not defined.

From the idea that if $\lim_{x \to a} f(x)$ is not defined we have some kind of jump in function $f(x)$, we are able to define a function as continuous at a point if it doesn't have a jump there. For example, if you are drawing a graph of the function $f(x)$ by hand, then $f(x)$ is continuous at $x = a$ if you don't have to lift your pencil off the paper when you come to the value of the function at $x = a$. If you do have to lift your pencil off the paper, then we say that $f(x)$ is discontinuous at $x = a$. A biological example of a discontinuity is pulse rate at the onset of a sudden heart attack: essentially, pulse rate drops instantly at the time of the attack. Another example is drug concentration in the blood, which essentially changes instantly at the time of the drug's administration, at least in the local blood near the site of the infusion.

16.2 Continuity

For a in the domain of the function f, we say that f is ***continuous*** at a point a if $\lim_{x \to a} f(x) = f(a)$. Notice that this statement implies three things:

1. $\lim_{x \to a^-} f(x)$ exists
2. $\lim_{x \to a^+} f(x)$ exists
3. $\lim_{x \to a^-} f(x) = \lim_{x \to a^+} f(x) = f(a)$

If a function is continuous at a, then the limits from the right and the left have the same value, $f(a)$.

If a function is continuous at every point in its domain, we say that the entire function is ***continuous***. If the function is not continuous at some point, we say that it is ***discontinuous*** at that point. There are several types of discontinuities.

If as $x \to a^+$ the function values $f(x)$ get larger without bound, then we denote this behavior by

$$\lim_{x \to a^+} f(x) = \infty.$$

This limit does not exist as a finite number, but we use this notation to indicate the behavior of the function. We see this behavior in part (a) of Figure 16.1 since the function values "blow up" (i.e., get very large) as $x \to 0^+$. If as $x \to a^+$ the function values $f(x)$ get smaller without bound (with negative values), then we say

$$\lim_{x \to a^+} f(x) = -\infty.$$

Similar notation holds for $x \to a^-$, $x \to \infty$, and $x \to -\infty$. Note that in part (b) of Figure 16.1, both the right-hand limit and the left-hand limit exist at $x = 1$, but they are not equal because of the jump in the graph there.

Most functions representing biological processes are continuous. But if there were points of discontinuity, they would be one of the following two types.

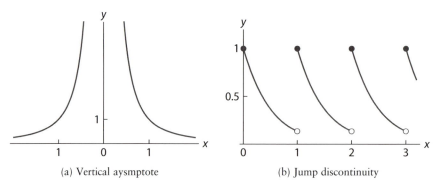

(a) Vertical aysmptote (b) Jump discontinuity

Figure 16.1 Examples of discontinuous functions.

Types of Discontinuities

1. A function has an ***infinite discontinuity*** at the point *a*, if $\lim_{x \to a^-} f(x) = \pm\infty$ and
$\lim_{x \to a^+} f(x) = \pm\infty$ (i.e., the limits do not exist) and either $f(a)$ is not defined or
$f(a) = c$, some finite number. In this case, we say that there is a vertical asymptote
at $x = a$.

2. A function has a ***jump discontinuity*** at the point *a* if $\lim_{x \to a^-} f(x) = L$ and
$\lim_{x \to a^+} f(x) = M$ and $L \neq M$. It may be that either $L = f(a)$ or $M = f(a)$.

In some biological situations, it may appear that a function suddenly jumps and that it is discontinuous, but this may arise from constraints on the measurement process. Figure 16.2 shows the photosynthetic rate *P*, measured as CO_2 uptake per unit leaf area per unit time, for a soybean leaf on the upper part of a plant grown under normal conditions. This graph shows the changes in *P* through time when the light level is suddenly decreased at 10 minutes and then increased back to the original level after 7 minutes at the lower light level. Note that *P* appears to suddenly drop when light is reduced, but this is not really a discontinuity because measurements of *P* are made approximately every minute. So we cannot tell from the measurements whether there is a discontinuity in *P* at the time that light is reduced or whether there is a rapid decrease in *P* that we would be able to see as continuous if we made observations more rapidly than every minute. It is reasonable to suspect that *P* is continuous in time based on the observations made when *I* is increased, for which a gradual increase occurs in *P* back to the photosynthetic rate that was present before the light was reduced, taking approximately 8 minutes to return to this level. However, if you were to make a mathematical model for leaf response to light changes, the data would indicate that a reasonable first model would have a sudden, discontinuous drop in *P* when light is decreased and a gradual, continuous increase in *P* when light is increased. One reason this might matter for soybean growth and crop yield is that there can be sudden light decreases and increases throughout the day as clouds pass by.

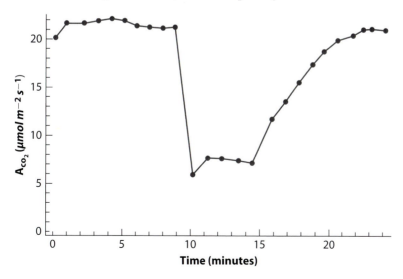

Figure 16.2 Photosynthetic rate of the upper leaves of a soybean plant (measured in μmol of CO_2 per m^2 of leaf area per second) over a 25-minute period. Data from [22].

Some examples of continuous functions are the following:

1. Power functions: $f(x) = x^\kappa$, where κ is any real number.
2. Polynomial functions (by extension of the fact that power functions are continuous).
3. Exponential functions: $f(x) = a^x$.
4. Logarithmic functions: $f(x) = \log_a x$ for $x > 0$, $a > 0$.
5. Trigonometric functions: $f(x) = \sin x$, $\cos x$.

As we have already seen, each of these functions appears in many biological applications. Exponential functions arise in population growth, power functions describe allometric relationships, and trigonometric functions describe many periodic phenomena that arise from seasonal or diurnal factors. Since many models in biology are built up by combining functions of these forms, many functions in biological applications are continuous. Particular cases in which discontinuities may arise, however, are when a function has a denominator that can be 0. This is one reason to be careful about models that include ratios: these can "blow up" in value when the denominator becomes small in value if the numerator does not also become small.

Let us now determine whether some functions are continuous.

Example 16.1 (Continuous Function?)

Where is the function $f(x) = x^3 - 3x + 1$ continuous?

Solution: $f(x)$ is a polynomial function and therefore continuous everywhere, that is, for all real numbers x.

Example 16.2 (Continuous Function?)

Where is the function $f(x) = \frac{1}{x-2}$ continuous?

Solution: Notice that this example is similar to a portion of Example 15.6. What happens to the fraction $1/(x - 2)$ as $x \to 2$? First, notice that the function $1/(x - 2)$ is undefined at $x = 2$.

Next, suppose that x is smaller than 2 but approaching 2 (i.e., approaching from the left). As the value of x gets closer and closer to 2, the value of the denominator gets smaller and smaller but is a negative number. The value of the fraction $1/(x-2)$ gets smaller than any number. When $x = 1.999$, the value of the fraction is -1000. Thus, we would say that as x approaches 2 from the left, the function diverges to negative infinity, and we write this as

$$\lim_{x \to 2^-} \frac{1}{x - 2} = -\infty.$$

(Continued)

Now, suppose that x is larger than 2 but approaching 1 (i.e., approaching from the right). Now, as the value of x gets closer and closer to 2, the value of the denominator gets smaller and smaller but is a positive number. Thus, the value of the fraction gets larger and larger, and we would say that as x approaches 2 from the right, the function diverges toward infinity, and we write this as

$$\lim_{x \to 2^+} \frac{1}{x-2} = \infty.$$

For any other value of x that we choose, say, $x = a$, where $a \neq 2$, we find that the limit evaluates to

$$\lim_{x \to a} \frac{1}{x-2} = \frac{1}{a-2}.$$

Thus, the function $f(x)$ is continuous for all real numbers except 2. We write $f(x)$ is continuous on $(-\infty, 2) \cup (2, \infty)$. We say that "$f$ has an infinite discontinuity at $x = 2$."

Example 16.3 (Continuous Function?)

Where is the function $f(x) = \frac{x^2 - x - 2}{x - 2}$ continuous?

<u>Solution:</u> Notice that

$$\frac{x^2 - x - 2}{x - 2} = \frac{(x - 2)(x + 1)}{x - 2} = x + 1 \;\; \text{for } x \neq 2.$$

So, this function looks like the line $y = x + 1$, except that it is not defined at the point $x = 2$ (See Figure 16.3). We know that a line is a polynomial of degree 1 and that polynomials are continuous everywhere on the real line. So, this function is continuous everywhere but at the point of discontinuity ($x = 2$). We write $f(x)$ is continuous on $(-\infty, 2) \cup (2, \infty)$. We say that f has a removable discontinuity at $x = 2$.

Example 16.4 (Continuous Function?)

Where is the function $f(x) = \begin{cases} 2 & \text{if } x \geq 0 \\ 1 & \text{if } x < 0 \end{cases}$ continuous?

<u>Solution:</u> Notice that

$$\lim_{x \to 0^-} f(x) = 1 \;\; \text{and} \;\; \lim_{x \to 0^+} f(x) = 2.$$

We write $f(x)$ is continuous on $(-\infty, 0) \cup (0, \infty)$. We say that f has a jump discontinuity at $x = 0$.

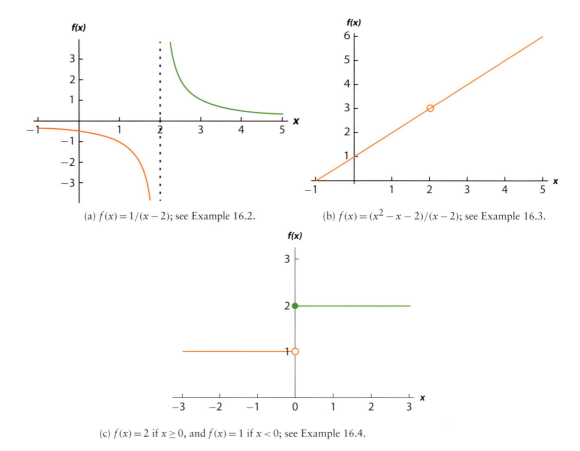

(a) $f(x) = 1/(x-2)$; see Example 16.2.

(b) $f(x) = (x^2 - x - 2)/(x - 2)$; see Example 16.3.

(c) $f(x) = 2$ if $x \geq 0$, and $f(x) = 1$ if $x < 0$; see Example 16.4.

Figure 16.3 Examples of functions with a discontinuity.

Continuity Properties

If $f(x)$ and $g(x)$ are continuous at $x = a$, then so are the functions:

1. $f(x) + g(x)$
2. $f(x) - g(x)$
3. $f(x)g(x)$
4. $f(x)/g(x)$ if $g(a) \neq 0$

To understand property (a), suppose that f and g are continuous at $x = a$. Then, we know that $\lim_{x \to a} f(x) = f(a)$ and $\lim_{x \to a} g(x) = g(a)$. From the limit properties from 15.2, each of the above functions is continuous. For example,

$$\lim_{x \to a} f(x)g(x) = [\lim_{x \to a} f(x)] \cdot [\lim_{x \to a} g(x)] = f(a)g(a).$$

The other properties (b) to (d) can be explained in a similar fashion.

Example 16.5 (Discontinuity)

Is the function $f(x) = \begin{cases} \frac{1}{x}\left(\frac{1}{2+x} - \frac{1}{2}\right) & \text{if } x \neq 0 \\ -\frac{1}{4} & \text{if } x = 0 \end{cases}$ continuous at $x = 0$? Why?

Solution:

1. Notice that $f(0) = -\frac{1}{4}$.

2. Notice that

$$\frac{1}{x}\left(\frac{1}{2+x} - \frac{1}{2}\right) = \frac{1}{x} \times \frac{2 - (2 + x)}{2(2 + x)} = \frac{1}{x} \times \frac{-x}{4 + 2x} = \frac{-1}{4 + 2x}.$$

3. Thus, $\lim_{x \to 0} \frac{1}{x}\left(\frac{1}{2+x} - \frac{1}{2}\right) = \lim_{x \to 0} \frac{-1}{4+2x} = -\frac{1}{4}$.

4. Since the limit exists at $x = 0$ and is equal to $f(0)$, the function $f(x)$, by definition, is continuous.

We currently do not have all the techniques we need to compute *all* limits algebraically. Any function may have an indeterminate form as $x \to a$. The indeterminate forms are

$$\infty - \infty, \quad \frac{\infty}{\infty}, \quad \frac{0}{0}, \quad 1^{\infty}, \quad 0^{0}, \quad \text{and} \quad \infty^{0}.$$

Often, we can manipulate the function algebraically to determine the resulting limit. However, there are cases for which we cannot algebraically manipulate our way out of the indeterminate form. Thus, we cannot calculate the limit algebraically, but this does not stop us from trying to calculate the limit numerically.

Example 16.6 (Estimating the Limit Graphically)

Compute $\lim_{x \to 0} \frac{\tan 4x}{x}$.

Solution: As $x \to 0$, we see that the function gives the indeterminate form $\frac{0}{0}$. Since there is no obvious algebraic manipulation to get this function out of indeterminate form, we can use a graphing calculator or Matlab to get an estimate for the limit. After zooming in a bit, we should see that the limit is 4.

16.3 Intermediate Value Theorem

Many biological processes are functions of time. If these processes do not have sudden jumps, then the functions describing them are continuous. The cases in biology for which sudden jumps occur are rare but do occur for threshold phenomena such as arise when a heart suddenly stops beating (i.e., cardiac arrest) or when someone stops breathing because of a sudden electric shock. In the situation of a continuous biological process that does not jump, the intermediate value theorem guarantees that if we measure the biological process at two times and get two different values, then the process must have taken on any value intermediate between the two values that we measured at some intermediate time. Thus, if the photosynthetic rate of a leaf is 2 μmol of CO_2 per m^2 per second at 10:00 a.m. and 3.5 μmol of CO_2 per m^2 per second at 11:00 a.m., then we can be assured that at some time over that hour, the leaf's photosynthetic rate was 2.8 μmol of CO_2 per m^2 per second. The following theorem about continuity provides us some information about the values of biological processes at intermediate times when we measure the process only at particular times.

> *Intermediate Value Theorem*
>
> If f is continuous on a closed interval $[a, b]$ and N is some value between $f(a)$ and $f(b)$, then there exists a value c in the open interval (a, b) such that $f(c) = N$.

An illustration of the intermediate value theorem is shown in Figure 16.4. Note that in Figure 16.4(a), there exists only one value $c \in (a, b)$ such that $f(c) = N$; however, in Figure 16.4(b), there are three such values. It is important to note that even if more than one such values exists, the intermediate value theorem guarantees the existence of at least *one* value $c \in (a, b)$ such that $f(c) = N$.

Finally, note the importance of continuity in the intermediate value theorem. If the function f is not continuous on $[a, b]$, we could simply use a jump discontinuity to ensure that there is some value N for which there is no $c \in (a, b)$ such that $f(c) = N$.

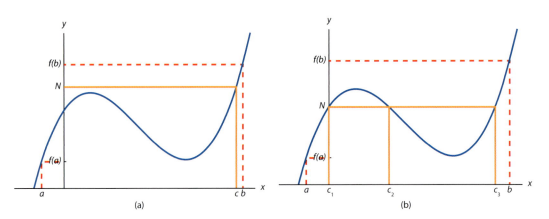

(a)   (b)

Figure 16.4 Intermediate value theorem.

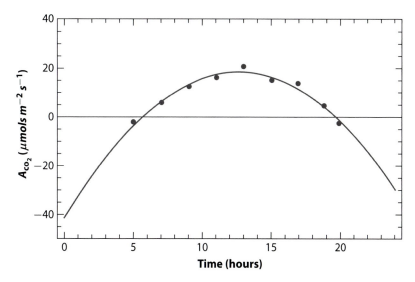

Figure 16.5 Data extracted from figure 2 in [58] showing the photosynthetic rate (measured as the absorption of CO_2 in μmol of CO_2 per m^2 of leaf area per second) over the course of 1 day (24 hours). The points on the graph indicate the actual data; the curve that is shown is a quadratic polynomial least-squares regression of the data and has the formula $P(t) = -0.374t^2 + 9.44t - 41.3$, where t is measured in hours.

Example 16.7 (Photosynthetic Rates over the Course of 1 Day)

Figure 16.5 shows data collected by Rogers et al. [58] of the photosynthetic rate (measured as the absorption of CO_2 in μmol of CO_2 per m^2 of leaf area per second) over the course of 1 day (24 hours) at day 164 in the life cycle of a soybean plant. The data can be approximated by the quadratic function

$$P(t) = -0.374t^2 + 9.44t - 41.3,$$

where P is the photosynthetic rate measured in μmol m^{-2} s^{-1} and t is measured in hours. The formula for the quadratic can be found using a nonlinear least-squares regression with a second-order polynomial.

(a) Is it appropriate to assume that the photosynthetic rate is continuous over the course of 1 day?

(b) Given that we can assume the photosynthetic rate is continuous over the course of 1 day, use the intermediate value theorem to show that at some time during the course of a day, the photosynthetic rate must be 10 μmol m^{-2} s^{-1}.

Solution:

(a) The measurements of photosynthesis were made at only a few times during the day, so we cannot be certain that during the day there were not sudden changes due to

(Continued)

decreases in light level from clouds passing over or other factors. The fitted function for $P(t)$ assumes that we are ignoring any very short-time changes in photosynthesis that occur. This is reasonable if instead of thinking of $P(t)$ as the photosynthetic rate at exactly time t, we think of it as an average photosynthetic rate during a short time period (e.g., 10 minutes) centered at time t. To a first approximation, therefore, if we are not interested in the detailed second-by-second photosynthetic rate, it is reasonable to assume that the daily course of photosynthesis follows the quadratic function given by $P(t)$, which is continuous.

(b) If we assume that the photosynthetic rate is continuous over the course of a day and is approximated by $P(t) = -0.374t^2 + 9.44t - 41.3$, then the estimated photosynthetic rate $P(t)$ is continuous over the closed interval $[0, 24]$. Note that $P(0) = -41.3$ and $P(24) = -30.0$. Thus, we know from the intermediate value theorem that $P(t)$ must have values between -41.3 and -30.0. However, the value in which we are interested, 10, is not contained within the interval $[-30.0, -41.3]$. In order to show that $P(t) = 10$ for some t in the interval $[0, 24]$, we must find some other time t for which $P(t) < 10$ and some other time t for which $P(t) > 10$. We already have two values of t for which $P(t) < 10$, namely, $t = 0$ and $t = 24$. Let us choose $t = 12$, which is appears to be close to where $P(t)$ peaks:

$$P(12) = 18.2 > 10.$$

Now, since $P(t)$ is continuous over $[0, 24]$, it is must be continuous over $[0, 12]$, and thus, by the intermediate value theorem, there is a value of t in the interval $[0, 12]$ such that $P(t) = 10$.

Additionally, since $P(t)$ is continuous over $[0, 24]$, it is must be continuous over $[12, 24]$, and thus, by the intermediate value theorem, there is a value of t in the interval $[12, 24]$ such that $P(t) = 10$.

Thus, there are two times over the course of 1 day when the photosynthetic rate is 10 μmol at least m^{-2} s^{-1}. This can be seen directly from the graph of $P(t)$ by looking at how the horizontal line with photosynthetic rate equal to 10 crosses the graph of $P(t)$ at two points.

16.4 Matlab Skills

Recall that if we want to work with the function $f(x) = x^2$ in Matlab, we can write an m-file, such as f.m below:

———————————— f.m ————————————

```
1   % Creates the function f(x)
2   % Input: x (a real number)
3   % Output: y
4
5   function y = f(x)
6   y = x^2;
```

Now, if we wanted to graph that function, we could write another m-file, such as `graphf.m` below:

```
─────────────── graphf.m ───────────────
1   % Create a graph of f
2   % Make sure the file f.m is contained in the same folder as this file
3   % Inputs: xmin = minimum x value to graph
4   %         xmax = maximum x value to graph
5   % Output: a graph
6
7   function graphf(xmin,xmax)
8
9   % Create 1000 x values to graph
10  n = (xmax-xmin)/1000;
11  x = [xmin:n:xmax];
12
13  % Create 1000 f(x) values to graph
14  for i = 1:length(x)
15      F(i) = f(x(i));
16  end
17
18  % Create plot
19  plot(x,F,'r.')
20  xlabel('x')
21  ylabel('f(x)')
```

The graph in Figure 16.6 is produced when the following command is run in the command window:

```
─────────────── Command Window ───────────────
>> graphf(-5,5)
```

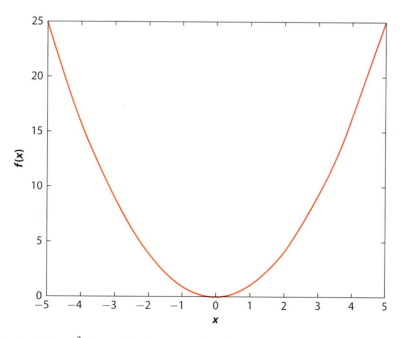

Figure 16.6 Graph of $f(x) = x^2$ drawn using the `graphf` function.

Now, suppose that we want to plot a function that is defined piecewise. What must we change in the f.m and graphf.m files? It turns out that we have to change only the function f.m. Now, how do we define a piecewise function in Matlab?

Suppose that we want to graph the function

$$f(x) = \begin{cases} x^2 & x \leq 2 \\ 3x + 2 & x > 2 \end{cases}. \tag{16.1}$$

Notice that *if $x \leq 2$, then $f(x) = x^2$*, and *if $x > 2$, then $f(x) = 3x + 2$*. We can use *if statements* when defining f.m to correctly describe the piecewise function:

```
─────────────────────────── f.m ───────────────────────────
1   % Creates the function f(x)
2   % Input: x (a real number)
3   % Output: y
4
5   function y = f(x)
6   if x <= 2
7       y = x^2;
8   else
9       y = 3*x+2;
10  end
```

The graph in Figure 16.7 is produced when the following command is run in the command window:

```
──────────────────────── Command Window ────────────────────────
>> graphf(-5,5)
```

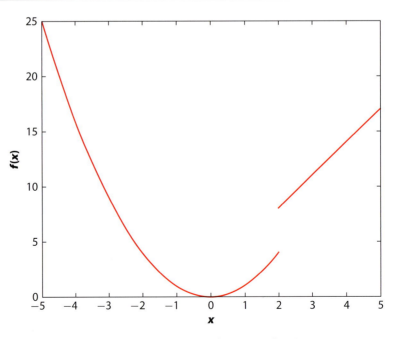

Figure 16.7 Graph of $f(x)$ (from Equation 16.1) drawn using the graphf function.

16.5 Exercises

16.1   Determine where $f(x)$ is continuous.

(a) $f(x) = x^2 - x - 6$

(b) $f(x) = x^4 + 3x^3 - 2x + 1$

(c) $f(x) = \dfrac{x^2 - x - 6}{x + 2}$

(d) $f(x) = \dfrac{x^2 - x - 6}{x - 3}$

(e) $f(x) = x^3 \sin x$

(f) $f(x) = \dfrac{3^x}{x}$

(g) $f(x) = \dfrac{x^2 + 2x + 17}{x^2 - 1}$

16.2   The following functions are not continuous at $x = 1$. Tell which type of discontinuity occurs there.

(a) $f(x) = \dfrac{3}{1 - x}$

(b) $f(x) = \begin{cases} x, & x \le 1 \\ x + 2, & x > 1 \end{cases}$

16.3   Use Matlab to sketch graphs of the following functions. For each function, create the appropriate `f.m` file and then use the `graphf` function to graph the function. Choose appropriate values for `xmin` and `xmax`. If a function is not continuous at a point, state which type of discontinuity it is.

(a) $f(x) = \begin{cases} \dfrac{1}{x - 1}, & x \ne 1 \\ 2, & x = 1 \end{cases}$

(b) $g(x) = \begin{cases} x^3, & x < 0 \\ x^2, & 0 \le x \le 2 \\ 2x, & x > 2 \end{cases}$

(c) $s(t) = \begin{cases} \dfrac{1}{t + \frac{\pi}{2}}, & t < -\frac{\pi}{2} \\ \tan t, & -\frac{\pi}{2} \le t \le \frac{\pi}{2} \\ \dfrac{1}{t - \frac{\pi}{2}}, & t > \frac{\pi}{2} \end{cases}$

16.4   Let

$$f(x) = \begin{cases} x^2 - 2, & x \le 1 \\ 2x - 3, & x > 1 \end{cases}$$

Show that f is continuous on the domain $(-\infty, \infty)$.

16.5   (From [29]) An aggressive tumor is growing according to the function $N_1(t) = 2^t$, where $N_1(t)$ represents the number of cells at time t (in months). The tumor continues to grow for 40 months, at which time the tumor is diagnosed.

(a) How many tumor cells are present when the tumor is diagnosed?

(b) Suppose that the patient receives chemotherapy immediately after the tumor has grown for 40 months and that 99.9% of the cells are instantaneously killed. How many tumor cells are left immediately after the patient receives chemotherapy? Write your answer as 2 to a power; that is, solve $(1 - 0.999)(2^{40}) = 2^x$ for x.

(c) After the chemotherapy, the tumor cells continue to die off according to the function $N_2(t) = 2^{40.03 - .25t}$. Show that $N_2(40)$ is approximately 2^x for the x you found in part (b).

(d) Graph the number of tumor cells before and after chemotherapy. Graph $N_1(t)$ for $0 < t < 40$ and $N_2(t)$ for $40 \le t < \infty$. (Hint: To draw a manageable graph, use $\log_2 N(t)$ for the vertical axis.)

(e) Is the graph in part (d) continuous? If not, then find the value(s) of t where the function is discontinuous.

(f) How many months after the chemotherapy must the patient wait until the tumor contains less than 2^{10} cells?

16.6   (From [29]) During pregnancy, a woman's weight naturally increases. When she delivers, her weight immediately decreases by the approximate weight of the child. Suppose that a 120-pound woman gains 27 pounds during pregnancy, delivers a 7-pound baby, and then, through diet and exercise, loses the remaining weight during the next 20 weeks.

(a) Graph the weight gain and loss during the pregnancy and the 20 weeks following the birth of the baby. Assume that the pregnancy lasted 40 weeks, that delivery occurs immediately after this time interval, and that the weight gain/loss before and after birth is linear (i.e., use equations for lines, $y = mx + b$). Write down the resulting piecewise function.

(b) Is this graph a continuous function? If not, then find the value(s) of t where the function is discontinuous.

16.7   (From [29]) To be continuous or not: that is the question.

(a) When a patient receives an injection, the amount of the injected substance in the body immediately goes up. Comment on whether the function that describes the amount of the drug with respect to time is a continuous function. Justify your reasoning.

(b) When a person takes a pill, the amount of the drug contained in the pill is immediately inside the body. Comment on whether the function that describes the amount of the drug inside the body with respect to time is a continuous function. Justify your reasoning.

16.8   (From [29, 78]) Researchers at Iowa State University and the University of Arkansas have developed a piecewise function that can be used to estimate the body weight (in grams) of a male broiler chicken during the first 56 days of life according to

$$W(t) = \begin{cases} 48 + 3.64t + 0.6363t^2 + 0.00963t^3, & 1 \le t \le 28 \\ -1004 + 65.8t, & 28 < t \le 56, \end{cases}$$

where t is the age of the chicken (in days).

(a) Determine the weight of a male broiler that is 25 days old.

(b) Is $W(t)$ a continuous function?

(c) Use Matlab to graph $W(t)$. Let the horizontal axis range from 1 to 56 and the vertical axis range from 0 to 3000. Comment on the accuracy of the graph.

(d) Comment on why the researchers would use two different functions to estimate the weight of a chicken at various ages. (*Hint*: To get an idea of why, graph $48 + 3.64t + 0.6363t^2 + 0.00963t^3$ for $t \in [1, 56]$ and $-1,004 + 65.8t$ for $t \in [1, 56]$.)

16.9   In Example 16.7, we saw that at day 164 in the life cycle of a soybean plant, the rate of photosynthesis (measured as the absorption of CO_2 in μmol of CO_2 per m^2 of leaf area per second) of the plant over the course of 1 day (24 hours) can be approximated by the quadratic function

$$P(t) = -0.374t^2 + 9.44t - 41.3,$$

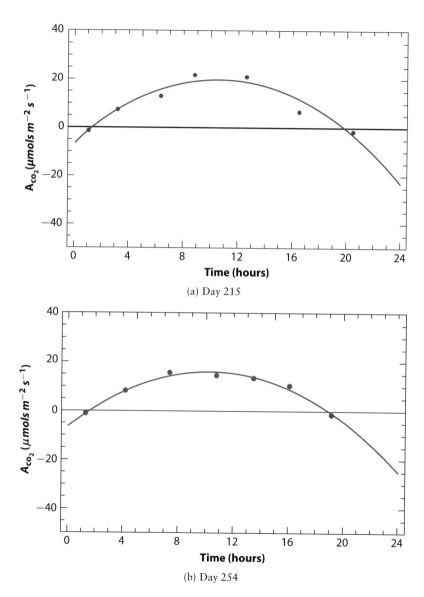

(a) Day 215

(b) Day 254

Figure 16.8 Data extracted from figure 2 in [58] showing the photosynthetic rate (measured as μmol m^{-2} s^{-1} of CO_2) over the course of 1 day (24 hours) for soybean plants at (a) day 215 and (b) day 254 in the soybean life cycle. The points on the graph indicate the actual data; the curve shown is a quadratic polynomial least-squares regression of the data.

where P is the photosynthetic rate measured in μmol m^{-2} s^{-1} and t is measured in hours (see Figure 16.5). Over the life cycle of the soybean plant, the photosynthetic rate over the course of a day changes. The photosynthetic rate of the soybean plant over the course of 1 day can be approximated by

$$P(t) = -0.237t^2 + 5.01t - 6.58 \qquad \text{at 215 days old, and} \qquad (16.2)$$

$$P(t) = -0.214t^2 + 4.38t - 6.23 \qquad \text{at 254 days old.} \qquad (16.3)$$

Use the intermediate value theorem to show that when the soybean plant is 215 days old, there is at least one time during the day when the photosynthesis rate is 10 μmol m^{-2} s^{-1} and that when the soybean plant is 254 days old, there is at least one time during the day when the photosynthetic rate is 5 μmol m^{-2} s^{-1}.

16.10 Fay and Knapp [22] measured the amount of CO_2 absorbed by the upper leaves of soybean plants over a 30-minute period when the plant was grown under dought conditions. During the 30-minute experiment, the soybean plant is initially in full sunlight. After 8 to 10 minutes, the plant is shaded for approximately 8 to 10 minutes, after which it is returned to full sunlight. Below is a graph showing the absorption rate of CO_2 in μmol per m^2 of leaf area per second over the full experiment for the soybean plant under drought conditions. The absorption rate was sampled roughly at 1-minute intervals.

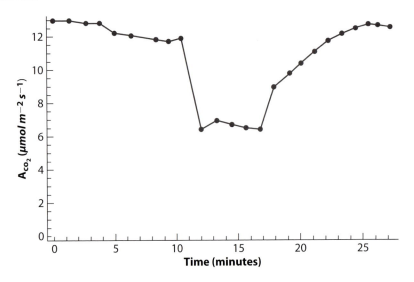

(a) If you were to construct a function to model these data, would that function be continuous? Why or why not?
(b) Consider the time interval from 0 to 15 minutes. Can you guarantee that there exists a time at which the absorption rate was exactly 8 μmol m^{-2} s^{-1}? Keep in mind your answer to (a).

Unit 4 Student Projects

Investigating Biodiversity

This project will use ideas of probability and limits of sequences from the last two units to explore measuring biodiversity. Simpson's index, which was introduced in Chapter 1, can be used to quantify the biodiversity of a group of similar organisms in a certain geographic area. Biodiversity takes into account both species richness (the number of different species in an area) and species evenness (the degree of equitability in the distribution among the species under consideration). Maximum evenness occurs when all the represented species have the same number of individuals. Simpson's Index represents the probability that two individuals selected (one after the other, without replacement) from an area are from different species. When Simpson's Index is close to 1, then that area is considered to be high in biodiversity in the variety of species considered.

By creating several scenarios, we will explore the limits of evenness and richness in this project. What are the restrictions on Simpson's Index when considering evenness alone? Exploring the limits of Simpson's Index in specific cases will illustrate the interplay between evenness and richness.

We will use the data below on species of salamanders in the Great Smoky Mountains National Park. This park is considered to be the "Salamander Capital of the World," with 31 species being found there. These data are only a sample of the data collected by the Citizen Science program at the Great Smoky Mountains Institute at Tremont. The following table provides information on the number of individuals of each species found during sampling periods in each of two locations: Lower Dorsey Stream and Pig Pen Stream.

Salamander Species	Lower Dorsey Stream	Pig Pen Stream
Spotted dusky salamander	7	18
Imitator salamander	6	3
Seal salamander	5	15
Black-bellied salamander	7	11
Desmognathus spp. salamander	4	17
Blue ridged two-line salamander	1	31
Spring salamander	2	1
Northern slimy salamander	0	1
Santeetlah salamander	1	0
Southern red-backed salamander	2	0

1. Write the definition of species evenness and richness. Find (or make up) two simple examples: one with high evenness (but not richness) and one with high richness (but not evenness).
2. For the data in Table 1, suppose that we consider an area having only the imitator and the seal salamanders from the Lower Dorsey Stream. There are 11 total individuals: six imitator salamanders and five seal salamanders. Suppose that you randomly remove an individual from that area and then randomly remove another individual without replacement of the first individual. What is the probability that both of that both of the selected individuals are seal salamanders?
3. What is the probability that both of the individuals selected in part 2 are imitator salamanders?
4. What is the probablity that both of the selected individuals are the same species, that is either imitator salamanders or seal salamanders?
5. What is the probability that the two selected individuals are from two different species?
6. Suppose in a certain area that there are n_1 imitator salamanders and n_2 seal salamanders. In total, there are $n_1 + n_2$ salamanders. Suppose that you again select an individual and then select another individual without replacement. What is the probability that the two selected salamanders are from different species?
7. Suppose that we have a larger data set, with the ith species having n_i individuals. There are S types of species recorded. What would be the probability that two selected individuals chosen without replacement are from different species?

Simpson's Index of diversity is one of several diversity indices. It represents the probability that two individuals randomly selected from a sample without replacement will belong to different species. In a certain area or sample, let

$$D = \sum_{i=1}^{S} \frac{n_i(n_i - 1)}{N(N - 1)},$$

where n_i is the number of individuals in species i, N is the total number of individuals, and S is the number of species. Then Simpson's index of diversity is

$$SID = 1 - D.$$

When SID is close to 1, the sample is considered to be highly diverse.

1. Calculate the Simpson's index SID for the data in the table for the Lower Dorsey Stream. Then calculate the index for the Pig Pen Stream. Which stream area is more biodiverse with respect to salamanders based on these data?
2. Suppose that a plot has 4 species with 100 individuals of each species. Calculate SID for this plot.
3. Suppose that a plot has 4 species and n individuals of each species. Calculate SID for this plot.
4. Using limits of sequences, how large can D become with only 4 species with a large number of individuals in each species? If you let $n \to \infty$, in your answer to the last item, what is the limit of SID, emphasizing only evenness?
5. Suppose that a plot has S species with n individuals in each species. Find SID and simplify as much as possible.

We can view our formula from the last item as giving SID as a function of S and n. We use what we have learned about limits of sequences to investigate how this index changes as n and S become large.

1. How does SID change as n gets large, keeping S fixed? This indicates the influence of evenness.
2. How does SID change as S gets large, keeping n fixed? This indicates the influence of richness.
3. Examining your work so far, which factor, evenness or richness, can bring SID closer to 1?
4. Calculate the limits below and discuss your results.

$$\lim_{S \to \infty} (\lim_{n \to \infty} SID(S, n))$$

$$\lim_{n \to \infty} (\lim_{S \to \infty} SID(S, n))$$

Compose a one-page report, describing your procedures and results.   ■

5
UNIT

Derivatives

A main idea in this unit involves what we can learn from the instantaneous rate of change of a function, which is called the derivative of the function. Instantaneous rates of change of biological processes can provide a great deal of information about the process. If the instantaneous rate of change is positive, then at least at that time we know that the process is increasing in value. Why does knowing a derivative matter in biology aside from telling us whether something we can measure is increasing or decreasing? Using derivatives, we will be able to tell if a population is increasing fast; this would correspond to a large positive rate of change. If a population is decreasing, then the rate of change is negative. If the rate of change is zero, then the population is not changing, so it is at a constant level.

Much of biology deals with life systems that are in some sense stable: you have likely heard of homeostasis, an example being the fairly constant body temperature in "homeotherms," or organisms that maintain fairly constant temperatures, as compared to ectotherms, in which the organism's temperature is greatly affected by external temperature conditions and varies with this. Stability here refers to the idea that certain processes not only stay near the same value (98.6 degrees Fahrenheit for human body temperature) but also return to this value if perturbed (e.g., because of a fever, assuming the individual does not die). Much of the science of disease deals with cases in which the body's physiological capacity to maintain a stable condition is lost. Heart arhythmias, fevers arising from infection, and many disorders of the nervous system are examples of this, and there is a field of dynamical disease that studies how this arises (a great exposition is given in [27]).

So we would like to have a way to tell when a physiological process is homeostatic, and one objective of this text is to provide you with the mathematical machinery to understand this. One part of this is understanding that when a derivative is equal to 0 (we will see that the notation is $f'(x) = 0$), then the quantity being measured is stationary (not changing). Consider the example of $f'(x)$ giving the cell division rate in healthy tissue as a function of some enzyme level.

Then, $f'(x) = 0$ implies that the division rate is constant. Now suppose that there is a genetic mutation that modifies the expression of the protein pathway that produces the enzyme, thus affecting the form of the function $f(x)$ and its derivative. If we had a way to quantify how this occurred, we could potentially investigate the changes in the cell division rate and whether an instability arises that could lead to tumor growth. Since most biological processes involve rates of change, the ideas in this unit on derivatives are key to describing biological processes and analyzing how they might be affected by the factors that modify them.

Rates of Change

It is likely impossible for there to be a living world with no change. There would be no births or deaths of individuals of any species, no breathing or circulatory systems with any rhythms, no movement, no mating, and no aging: it would be a rather dull place. In many ways, a central theme of all of biology is change: sometimes change brought about by normal phenomena, such as births and deaths, and sometimes change brought about by abnormal phenomena, such as the failure of DNA repair mechanisms, leading to cancers. Since change is so central to biology, it is helpful to carefully define it and to develop the tools needed to analyze it. This is the objective of the following chapters, and we start in this chapter by describing how one goes from describing an average rate of change over some time period to an instantaneous rate of change that is the essence of differential calculus. We have already described many biological processes that change: photosynthetic rate with light and temperature, metabolic rate with level of activity, and population size with availability of food resources or level of disease. Some of these rates change with time, and some change with other factors, such as temperature, that may themselves depend on time. Differential calculus provides the tools for developing a science of change. So we will see here how to measure change and how to formally define it, and in later chapters we will go on to analyze the properties of changing biological phenomena based on the factors that cause the changes.

As an example, remember that the rate at which crickets "chirp" depends on the insect's body temperature, which itself depends on the temperature of the environment in which the cricket is placed. So, if $C(T)$ gives the rate at which a cricket chirps when its body temperature is T, then we may be interested in finding out how rapidly the chirping rate (measured as number of chirps per minute) is changing as its body temperature drops near dusk. One way to do this would be to listen for the chirping rate at 6:00 p.m. (say when it is 24 chirps per minute) and again at 6:30 p.m. (when it is 9 chirps per minute), so the rate of decrease of the chirping rate over this half hour is $15/30 = 0.5$ chirps per minute. This means that for each minute between 6:00 p.m. and 6:30 p.m., the chirping rate decreases by 0.5 chirp per minute. It may be difficult to think of "half a chirp," but this means that there is a reduction, on average, over this 30-minute time period of 1 chirp every 2 minutes. The above uses data on the actual chirping rate in time, not body temperature, which is a bit more difficult to measure. What we get from the observations is an estimate of how rapidly the chirping rate decreases over this half hour. In this section, we will discuss how this relates to instantaneous changes in the chirping rate at times between 6:00 p.m. and 6:30 p.m.

17.1 Average Rate of Change

To denote how a quantity y changes from time $t = t_1$ to $t = t_2$ (where $t_2 > t_1$), that is, the *average rate of change* over t_1 to t_2, we use the notation

$$\frac{\Delta y}{\Delta t} = \frac{y(t_2) - y(t_1)}{t_2 - t_1}.$$

Notice that this is the definition of the slope of the line that connects the points $(t_1, y(t_1))$ and $(t_2, y(t_2))$. Additionally, notice that if $\frac{\Delta y}{\Delta t} > 0$, then the quantity y increased from t_1 to t_2 (since $t_2 > t_1$). However, if $\frac{\Delta y}{\Delta t} < 0$, then the quantity decreased from t_1 to t_2.

Let us consider an example.

Example 17.1 (Alaskan Moose [29, 62])

Researchers from the Moose Research Center in Soldotna, Alaska, have developed a mathematical relationship between the age of a captive female moose and its weight. The function is

$$M(t) = 369(0.93)^t \left(t^{0.36}\right),$$

where M is the weight of the moose (in kg) and t is the age (in years) of the moose. Find the average rate of change in the weight of a female moose between the ages of 2 and 3 years.

Solution: On the interval from $t = 2$ to $t = 3$, the average rate of change is

$$\frac{M(3) - M(2)}{3 - 2} = \frac{369(0.93)^3(3)^{0.36} - 369(0.93)^2(2)^{0.36}}{3 - 2} \approx 31.194.$$

Therefore, the average amount of weight gained by a captive female moose between the ages of 2 and 3 is 31.194 kg.

Now let us consider an example for which we have data but no function fitted to those data. We will see how to use the formula for the average rate of change to determine how the data are changing over time.

Example 17.2 (Federal Spending on Stem Cell Research)

The U.S. government provides funding for scientific research through many different agencies. The National Institutes of Health (NIH) provide money for health research through research grants. The table below shows the amount of money the NIH awarded for stem cell research as reported by the NIH [83, 84].

Year	Funding Allocated (in millions US$)
2005	$609
2006	$643
2007	$968
2008	$938
2009	$1044
2010	$1099
2011	$1179
2012	$1374

Find the average rate of change in the amount of funding for stem cell research from (a) 2005 to 2007, (b) 2007 to 2008, and (c) 2007 to 2009.

Solution: Let $A(t)$ represent the amount of money the NIH allocated to stem cell research in year t.

(a) $\dfrac{A(2007) - A(2005)}{2007 - 2005} = \dfrac{968 - 609}{2007 - 2005} = \dfrac{359}{2} = 179.5$

Thus, the amount of funding for stem cell research increased by an average of $179.5 million per year from 2005 to 2007.

(b) $\dfrac{A(2008) - A(2007)}{2008 - 2007} = \dfrac{938 - 968}{2008 - 2007} = \dfrac{-30}{1} = -30.0$

Thus, the amount of funding for stem cell research decreased by an average of $30.0 million per year from 2007 to 2008. Since this is only a 1-year period, we can say that the amount of funding from 2007 to 2008 decreased by $30.0 dollars.

(c) $\dfrac{A(2012) - A(2007)}{2012 - 2007} = \dfrac{1374 - 968}{2012 - 2007} = \dfrac{406}{5} = 81.2$

Thus, the amount of funding for stem cell research increased by an average of $81.2 million per year from 2007 to 2012.

Although the average rate of change has been defined above as a quantity changing with respect to time, this is not the only way to define the rate of change. In fact, for any function $y(x)$, where x does not necessarily represent time, we can define the average rate of change as

$$\frac{\Delta y}{\Delta x} = \frac{y(x_2) - y(x_1)}{x_2 - x_1}.$$

In Section 17.7, we will use Matlab to consider how the number of times a cricket chirps in 1 second changes with respect to temperature. Thus, if N is the number of cricket chirps per second and T is the temperature in degrees Fahrenheit, then

$$\frac{\Delta N}{\Delta T} = \frac{N(T_2) - N(T_1)}{T_2 - T_1}$$

is the average rate of change in chirps per second with respect to temperature, where T_1 and T_2 are two distinct temperatures.

17.2 Estimating Rates of Change for Data

So far, we have considered the average rate of change of a quantity over some time interval. However, it is often useful to know how a quantity is changing at a particular time. To denote how a quantity y is changing at a particular time t, that is, the *instantaneous rate of change* at time t, we use the notation $y'(t)$. Again, though the instantaneous rate of change has been defined here in terms of how a quantity y changes with respect to time, we can define the instantaneous rate of change for a quantity y with respect to another quantity x, or $y'(x)$.

Suppose that we are given a set of data such as in the table below and want to estimate the instantaneous rate of change of the data at t_3.

Time t	t_1	t_2	t_3	t_4	t_5
Data x	x_1	x_2	x_3	x_4	x_5

One way to estimate the rate of change of the data at time t_3 is to average the rates of change from t_2 to t_3 and from t_3 to t_4:

$$x'(t_3) \approx \frac{1}{2} \left(\begin{array}{l} \text{average rate of change} \\ \text{from } t_2 \text{ to } t_3 \end{array} + \begin{array}{l} \text{average rate of change} \\ \text{from } t_3 \text{ to } t_4 \end{array} \right)$$

$$= \frac{1}{2} \left(\frac{x(t_3) - x(t_2)}{t_3 - t_2} + \frac{x(t_4) - x(t_3)}{t_4 - t_3} \right).$$

Example 17.3 (Drug Concentration [34])

The table below gives the concentration of a drug (in $\mu g/cm^3$), $c(t)$, in the blood stream at time t (in minutes). Estimate the rate of change of the concentration of drug at $t = 0.2$, $t = 0.4$, and $t = 0.8$.

t	0	0.1	0.2	0.3	0.4	0.5	0.6	0.7	0.8	0.9	1.0
$c(t)$	0.84	0.89	0.94	0.98	1.00	1.00	0.97	0.90	0.79	0.63	0.41

(*Continued*)

Solution: Let $c'(t)$ represent the instantaneous rate of change of the drug concentration at time t.

$$c'(0.2) \approx \frac{1}{2}\left(\frac{c(0.2) - c(0.1)}{0.2 - 0.1} + \frac{c(0.3) - c(0.2)}{0.3 - 0.2}\right)$$

$$= \frac{1}{2}\left(\frac{0.94 - 0.89}{0.1} + \frac{0.98 - 0.94}{0.1}\right) = \frac{0.5 + 0.4}{2} = 0.45.$$

Thus, we estimate the rate of change at $t = 0.2$ to be 0.45 μg/cm^3/min. Since this is positive, the drug concentration is increasing.

$$c'(0.4) \approx \frac{1}{2}\left(\frac{c(0.4) - c(0.3)}{0.4 - 0.3} + \frac{c(0.5) - c(0.4)}{0.5 - 0.4}\right)$$

$$= \frac{1}{2}\left(\frac{1.00 - 0.98}{0.1} + \frac{1.00 - 1.00}{0.1}\right) = \frac{0.2 + 0}{2} = 0.1.$$

Thus, we estimate the rate of change at $t = 0.4$ to be 0.1 μg/cm^3/min. So the drug concentration is still increasing but at a slower rate than at $t = 0.2$.

$$c'(0.8) \approx \frac{1}{2}\left(\frac{c(0.8) - c(0.7)}{0.8 - 0.7} + \frac{c(0.9) - c(0.8)}{0.9 - 0.8}\right)$$

$$= \frac{1}{2}\left(\frac{0.79 - 0.90}{0.1} + \frac{0.63 - 0.79}{0.1}\right) = \frac{-1.1 + (-1.6)}{2} = -1.35.$$

Thus, we estimate the rate of change at $t = 0.8$ to be -1.35 μg/cm^3/min. So by this time, the drug concentration is decreasing.

17.3 Velocity

One classic example of a rate of change frequently used in the natural sciences is velocity. The *average velocity* of an object, denoted $\bar{v}$, is the ratio of the displacement of the object (measured in length) and the time interval over which that displacement took place, that is,

$$\bar{v} = \frac{\text{change in position}}{\text{change in time}}.$$

If $s(t)$ is the position of an object at time t, then

$$\bar{v} = \frac{\Delta s}{\Delta t} = \frac{s(t_2) - s(t_1)}{t_2 - t_1}.$$

Velocity is the instantaneous rate of change of the position of the object with respect to time and is denoted as $v(t)$ or $s'(t)$. Note that speed and velocity are not exactly the same. Velocity has direction (up/down or left/right or forward/backward, i.e., $\pm$), whereas speed does not. Speed is defined to be the absolute value of velocity.

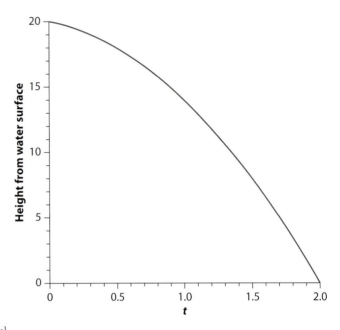

[Photo © Andy Rouse/www.andyrouse.co.uk]

Suppose that you are observing a kingfisher bird while it is hunting. The kingfisher will dive into the water from some height to catch its prey. After several observations of the bird diving from a branch, you determine an equation for its height from the water as a function of time. That function is

$$h(t) = -4t^2 - 2t + 20,$$

where $h(t)$ is the height (in feet) above the water's surface at time t. Notice from the graph that it takes 2.0 seconds for the kingfisher to reach the water's surface.

You want to know the velocity of the bird when it hits the surface. Since the kingfisher reaches the surface of the water at $t = 2$, we could estimate the velocity of the bird by taking the average of the average velocities from $t = 1$ to $t = 2$ and from $t = 2$ to $t = 3$, as in Section 17.2:

$$\begin{aligned}
v(2) &= \frac{1}{2}\left(\frac{h(2) - h(1)}{2 - 1} + \frac{h(3) - h(2)}{3 - 2}\right) \\
&= \frac{1}{2}\left(\frac{0 - 14}{2 - 1} + \frac{-22 - 0}{3 - 2}\right) \\
&= \frac{1}{2}(-14 - 22) = -\frac{36}{2} = -18 \text{ ft/s.}
\end{aligned}$$

Why are the units expressed as feet/second? When "time" is the independent variable, recall that a rate of change describes how something is changing with respect to time. In this case of velocity, we are considering how many feet the kingfisher has moved within the span of 1 second.

The fact that velocity is negative at this time means that the bird is moving toward the ground (i.e., in a negative direction) at a speed of 18 feet/second.

One could argue that the function $h(t)$ describes only the kingfisher bird's height from $t = 0$ to $t = 2$, that is, until it hits the surface of the water, and thus it is unreasonable to use values like $h(3)$. Instead of estimating the velocity of the kingfisher at $t = 2$ by using an average of average velocities, let us take the average velocity of the kingfisher bird over increasingly smaller time intervals ending at $t = 2$.

Time Interval	Average Velocity
$[0, 2]$	$\overline{v} = \frac{0-20}{2-0} = -10$ ft/s
$[1, 2]$	$\overline{v} = \frac{0-14}{2-1} = -14$ ft/s
$[1.5, 2]$	$\overline{v} = \frac{0-8}{2-1.5} = -16$ ft/s
$[1.8, 2]$	$\overline{v} = \frac{0-3.44}{2-1.8} = -17.2$ ft/s
$[1.9, 2]$	$\overline{v} = \frac{0-1.76}{2-1.9} = -17.6$ ft/s
$[1.99, 2]$	$\overline{v} = \frac{0-0.1796}{2-1.99} = -17.96$ ft/s
$[1.999, 2]$	$\overline{v} = \frac{0-0.017996}{2-1.999} = -17.996$ ft/s

It appears that as the time interval decreases, the average velocity is approaching the value -18 ft/s. In fact, if we take the limit of the average velocities as the time interval shrinks to 0, we can determine the exact velocity at $t = 2$:

$$
\begin{aligned}
v(2) &= \lim_{t \to 2} \frac{h(t) - h(2)}{t - 2} \\
&= \lim_{t \to 2} \frac{-4t^2 - 2t + 20 - 0}{t - 2} \\
&= \lim_{t \to 2} \frac{-2(2t^2 + t - 10)}{t - 2} \\
&= \lim_{t \to 2} \frac{-2(2t + 5)(t - 2)}{t - 2} \\
&= \lim_{t \to 2} -2(2t + 5) \\
&= -2\,[2(2) + 5] = -18 \text{ ft/s.}
\end{aligned}
$$

17.4 Photosynthesis

In Unit 4, we encountered examples that examine how the rate of photosynthesis changes with time of day and light level. Recall that the photosynthetic rate is the rate at which a unit area of leaf assimilates carbon dioxide (CO_2) or respires oxygen (O_2) per second. Although the photosynthetic rate is a rate of change (the rate of change of CO_2 being absorbed into the leaf per second), scientists have observed that photosynthetic rate changes with time of day and with light level. Thus, the rate of change of the photosynthetic rate becomes a quantity of interest.

Let us first consider the rate of change of the photosynthetic rate with respect to time of day. Rogers et al. [58] collected data on the photosynthetic rate measured as the absorption of CO_2 in μmol CO_2 m^{-2} s^{-1} (note that this means μmol/m^2/s) over the course of 1 day (24 hours) for different days in the life cycle of a soybean plant. In Example 16.7, we considered the data collected on day 164 in the life cycle and saw that the data could be approximated by the quadratic function

$$P(t) = -0.374t^2 + 9.44t - 41.3,$$

where P is the photosynthetic rate and t is time measured in hours. Figure 16.4 shows the data for times from 5:00 a.m. to 8:00 p.m. and the curve $P(t)$ for $t \in [0, 24]$. Notice that there are times of day for which the photosynthetic rate is negative. This occurs when there is no sunlight and the soybean leaves release CO_2 rather than absorb (or assimilate) CO_2. However, the curve $P(t)$ was obtained on the basis of photosynthetic rates measured during daylight hours. We should not expect that the value of $P(t)$ for evening hours is indicative of actual nighttime CO_2 loss since additional evening measurements would be needed. We will return in Chapter 19 to analyze photosynthetic rate using a different data set that includes some early morning and late evening hours and that fits a different, more realistic curve to the data.

To determine the exact rate of change of the photosynthetic rate at $t = 5$ (5:00 a.m.), we take the limit of the average rate of change

$$\frac{P(t) - P(5)}{t - 5}$$

as the interval between t and 5 shrinks to 0 (i.e., as $t \to 5$). Here we will use the notation $P'(t)$ to denote the instantaneous rate of change of the photosynthetic rate at hour t:

$$P'(5) = \lim_{t \to 5} \frac{P(t) - P(5)}{t - 5}$$

$$= \lim_{t \to 5} \frac{(-0.374t^2 + 9.44t - 41.3) - (-0.364(5)^2 + 9.44(5) - 41.3)}{t - 5}$$

$$= \lim_{t \to 5} \frac{-0.374t^2 + 9.44t + 0.364(5)^2 - 9.44(5)}{t - 5} \qquad \text{(Simplify)}$$

$$= \lim_{t \to 5} \frac{-0.374\left(t^2 - 5^2\right) + 9.44(t - 5)}{t - 5} \qquad \text{(Factor)}$$

$$= \lim_{t \to 5} \frac{(t - 5)\left(-0.374(t + 5) + 9.44\right)}{t - 5} \qquad \text{(Factor using } (a^2 - b^2) = (a - b)(a + b))$$

$$= \lim_{t \to 5} \left(-0.374(t + 5) + 9.44\right) \qquad \text{(Simplify)}$$

$$= -0.374(5 + 5) - 9.44 = 5.70 \ \mu\text{mols } CO_2 \ \text{m}^{-2} \ \text{s}^{-1} \ \text{h}^{-1}.$$

The positive sign of the rate of change at $t = 5$ means that at 5:00 a.m., the photosynthesis rate is increasing. The amount by which it is increasing is 5.70 μmol of CO_2 per m^2 per second per hour. The units for this rate of change are a bit cumbersome when first encountered. It may be helpful

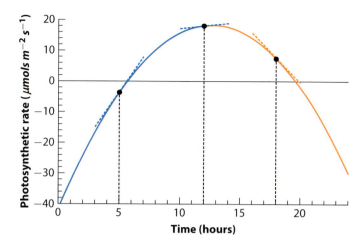

Figure 17.1 Graph of $P(t) = -0.374t^2 + 9.44t - 41.3$. The portions of the graph that are increasing and decreasing are shown as the solid blue and solid orange curves, respectively. Additionally, lines tangent to the curve $P(t)$ at $t = 5$, 12, and 18 are shown with their colors indicating positive slope (blue) or negative slope (orange).

to think of the rate-of-change units as

$$\frac{\text{photosynthetic rate units}}{\text{hour}},$$

where the photosynthetic rate units are μmol CO_2 m^{-2} s^{-1}.

We can make this same calculation for $t = 12$ (12:00 p.m.), and find that $P'(12) = 0.46$ (photosynthetic rate units)/hour. At 12:00 p.m., the photosynthetic rate is still increasing (since $P'(12) > 0$), but the value of the rate of change is much smaller. This means that the photosynthetic rate is increasing less quickly than it was a 5:00 a.m.

The graph of $P(t)$ in Figure 17.1 highlights the region over which $P(t)$ is increasing (solid green curve) and over which region $P(t)$ is decreasing (solid red curve). If the function is increasing at a point, we expect the rate of change to be positive, as we saw when we calculated the instantaneous rate of change at $t = 5$ and $t = 12$. However, if the function is decreasing at a point, we expect the rate of change to be negative. If we calculate the instantaneous rate of change at $t = 18$ (6:00 p.m.) when the function $P(t)$ is decreasing, indeed we find that $P'(18) = -4.02$ (photosynthetic rate units)/hour.

Figure 17.1 also shows dashed lines tangent to the curve $P(t)$ at $t = 5$, 12, and 18. The slopes of these tangent lines are equal to instantaneous rates of change at $t = 5$, 12, and 18, respectively. When the photosynthetic rate, $P(t)$, is increasing or decreasing more quickly, the slope of the tangent line is steeper than when it is increasing or decreasing less quickly.

The photosynthetic rate of plants changes not only with time of day but also with light level. In Example 15.1, we saw that the photosynthetic rate for leaves on soybean plants increases according to a Michaelis-Menten type of curve with respect to light level. In general, if P represents the photosynthetic rate with units μmol CO_2 m^{-2} s^{-1} and I represents light level measured in terms of photosynthetic photon flux density [PPFD] with units μmol photons m^{-2} s^{-1}, then

$$P(I) = \frac{abI}{aI + b} - d,$$

where a, b, and d are positive constants. Using data for soybean leaves on the lower portion of the plant collected by Fay and Knapp [22], we found in Example 15.1 that the photosynthetic rate as a function of light level can be modeled by

$$P(I) = \frac{1.62I}{0.195I + 8.29} - 3.13.$$

What is the rate of change of the photosynthetic rate at the low light level of 200 μmol photons m^{-2} s^{-1}? To determine the exact rate of change of the photosynthetic rate at $I = 200$, we take the limit of the average rate of change

$$\frac{P(I) - P(200)}{I - 200}$$

as the interval between I and 200 shrinks to 0 (i.e., as $I \to 200$). Here we will use the notation $P'(I)$ to denote the instantaneous rate of change of the photosynthetic rate at light level I:

$$P'(200) = \lim_{I \to 200} \frac{P(I) - P(200)}{I - 200}$$

$$= \lim_{I \to 200} \frac{\left(\frac{1.62I}{0.195I + 8.29} - 3.13\right) - \left(\frac{1.62(800)}{0.195(200) + 8.29} - 3.13\right)}{I - 200}$$

$$= \lim_{I \to 200} \frac{\left(\frac{1.62I(0.195(200) + 8.29)}{0.195I + 8.29}\right) - \left(\frac{1.62(200)(0.195I + 8.29)}{0.195(200) + 8.29}\right)}{I - 200} \qquad \text{(Simplify)}$$

$$= \lim_{I \to 200} \frac{\frac{1.62I(0.195(200) + 8.29) - 1.62(200)(0.195I + 8.29)}{(0.195I + 8.29)(0.195(200) + 8.29)}}{(I - 200)} \qquad \text{(Write numerator as one fraction)}$$

$$= \lim_{I \to 200} \frac{1.62I(8.29) - 1.62(200)(8.29)}{(I - 200)(0.195I + 8.29)(0.195(200) + 8.29)} \qquad \text{(Simplify)}$$

$$= \lim_{I \to 200} \frac{1.62(8.29)(I - 200)}{(I - 200)(0.195I + 8.29)(0.195(200) + 8.29)} \qquad \text{(Factor)}$$

$$= \lim_{I \to 200} \frac{1.62(8.29)}{(0.195I + 8.29)(0.195(200) + 8.29)} \qquad \text{(Simplify)}$$

$$= \frac{1.62(8.29)}{(0.195(200) + 8.29)^2} \approx .006 \;\text{(photosynthetic rate units)/(light level units),}$$

where the photosynthetic rate units are μmol CO$_2$ m^{-2} s^{-1} and the light level units are μmol photons m^{-2} s^{-1}.

Notice that the rate of change of the photosynthetic rate with respect to light level is increasing when $I = 200$ since $P'(200) > 0$.

17.5 Other Examples of Rates of Change

Population Growth

Suppose that we describe the number of individuals in a population at time t using the function $N(t)$. The average rate of change of the population over the time interval $[t_1, t_2]$ would be

$$\frac{\Delta N}{\Delta t} = \frac{N(t_2) - N(t_1)}{t_2 - t_1}.$$

The average rate of change of a population describes how much the population increases or decrease per unit time over a given interval of time, $[t_1, t_2]$. We denote the instantaneous rate of change of the population as $N'(t)$. The instantaneous rate of change of the population is often referred to as the *instantaneous growth rate*. However, the population is growing only if $N'(t) > 0$. If $N'(t) < 0$, then the population is declining, and if $N'(t) = 0$, then the population remains constant. The instantaneous growth rate of a population describes how much a population is increasing or decreasing per unit time at a given instant of time.

If we divide the instantaneous growth rate at time t by the number of individuals in the population at time t, that is,

$$\frac{1}{N(t)} N'(t),$$

then we obtain the *instantaneous per capita growth rate*.

Biomass

For some species, it is easier to measure the total mass of the population than it is to measure the number of individuals. One example of this is the floating freshwater aquatic plant *Lemna minor*, a species of duckweed. This plant is one of the smallest known flowering plants. When scientists want to measure the "size" of the population at various times to estimate how quickly a population of *Lemna minor* is growing, they measure the mass of the population by measuring either the fresh-weight mass (the mass when the plants still contain moisture) or the dry-weight mass (the mass after the plants have been allowed to dry). If $W(t)$ denotes the biomass of a population or organism, then $W'(t)$ represents the instantaneous growth rate of the population or organism. Again, the population is growing only if $W'(t) > 0$, and the instantaneous growth rate describes how quickly the population is increasing or decreasing per unit time at a given instant of time.

Drug Absorption into the Body

As we saw in Section 5.6, for many drugs it is assumed that the amount of drug remaining in the body t hours after a dose is administered is $x(t) = be^{-kt}$, where $b > 0$ is the size of the dose given and $k > 0$ is a drug-specific constant. The instantaneous rate of change $x'(t)$ represents the rate of change of the amount of drug in the body at a given time t.

Allometric Properties

In Section 4.3, we discussed the concept of allometry. Recall that two variables x and y are said to be allometrically related if $y = ax^b$, where a and b are real constants. Many aspects of a single organism can be described by an allometric relationship, an example being the relationship between an individual's body weight and brain weight. The number of species on an island is allometrically related to the size of the island by $S = \alpha A^\beta$, where S is the number of species on the island, A is the area of the island, and α and β are constants related to a geographic region. The instantaneous rate of change $S'(A)$ describes how quickly the number of species on an island changes per unit area increase in island size. This allows one to consider the impacts on biodiversity caused by changes in island area arising from climate change.

17.6 Definition of a Derivative at a Point

In Section 17.3, we saw that the instantaneous rate of change of the position with respect to time of the kingfisher bird at $t = 2$ could be expressed as the limit

$$\lim_{t \to 2} \frac{h(t) - h(2)}{t - 2}.$$

In Section 17.4, we saw that the instantaneous rate of change of the photosynthetic rate with respect to time of day at $t = 5$ could be expressed as the limit

$$\lim_{t \to 5} \frac{P(t) - P(5)}{t - 5}.$$

Additionally, we saw that the instantaneous rate of change of the photosynthetic rate with respect to light level at $I = 200$ could be expressed as the limit

$$\lim_{I \to 200} \frac{P(I) - P(200)}{I - 200}.$$

In each case, we took the limit of the average rate of change as the difference between the independent variable of our function and the point of interest decreased to 0. Since we use this same limit again and again, it has a special name: the *derivative*.

In mathematical terms, for a function $f(x)$, the ***derivative of f at the point*** $x = a$ is defined as

$$f'(a) = \lim_{x \to a} \frac{f(x) - f(a)}{x - a}.$$

Thus, the derivative at a point is just the instantaneous rate of change at that point. Note that the fraction

$$\frac{f(x) - f(a)}{x - a}$$

is the slope of the line between the points on the function $(x, f(x))$ and $(a, f(a))$. This fraction is often referred to as a *difference quotient*.

Although it is useful to know the derivative (or instantaneous rate of change) of some quantity at a point $x = a$, it is often more useful to determine a function that describes the rate of change at any value x. We will explore this idea in the next chapter.

17.7 Matlab Skills

Biologists have observed a linear relationship between the temperature and the frequency with which a cricket chirps. The following data were measured for the striped ground cricket [53] and has been placed in ascending temperature order.

Temperature °F	Chirps/sec	Temperature °F	Chirps/sec
69.4	15.4	82.0	17.1
69.7	14.7	82.6	17.2
71.6	16.0	83.3	16.2
75.2	15.5	83.5	17.0
76.3	14.4	84.3	18.4
79.6	15.0	88.6	20.0
80.6	16.0	93.3	19.8
80.6	17.1		

Let $C(T)$ be the number of chirps per second for temperature T. How do we estimate rates of change for these data? We illustrate using two methods.

Method 1: Using Least-Squares Regression

The first method is to fit a function to the data and then use that function to determine average rates of change and estimate instantaneous rates of change. If we plot these data, we can see that they seem to have a linear relationship (see Figure 17.2), and thus we can use the methods developed in Chapter 3 to determine the least-squares regression line for the data. The equation for the regression line is

$$C(T) = 0.211925T + -0.309144.$$

The m-file used to determine this equation is `CricketChirps.m`.

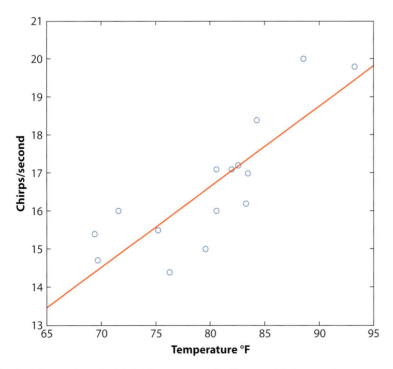

Figure 17.2 Graph of the number of cricket chirps per second with respect to temperature.

Now, if we wanted to find the average rate of change in chirps per second as the temperature rose from 70°F to 80°F, we could use

$$\frac{C(80) - C(70)}{80 - 70}.$$

This calculation is computed in `CricketChirps.m`. This calculation indicates that as the temperature rises from 70°F to 80°F, there is an average increase of 0.211925 chirps per second per °F. If we want to estimate the instantaneous rate of change at any point, we can take the derivative of the least-squares regression line $C'(T) = 0.211925$, which implies that the rate of change in chirps per second with respect to temperature is 0.211925 chirps per second per °F:

```
————————————————— CricketChirps.m —————————————————
1   % Cricket Chirping Data
2   T = [69.4 69.7 71.6 75.2 76.3 79.6 80.6 80.6 82.0 82.6 83.3 83.5 ...
3   84.3 88.6 93.3];
4   C = [15.4 14.7 16.0 15.5 14.4 15.0 16.0 17.1 17.1 17.2 16.2 17.0 ...
5   18.4 20.0 19.8];
6
7   % Plot out data
8   plot(T,C,'bo')
9   xlabel('Temperature \circF')
10  ylabel('Chirps/second')
11  axis([65 95 13 21])
12
13  hold on; % Allows next plot command to print on same graph
14
15  % Plot LSR
16  M = polyfit(T,C,1);
17  X = 65:95;
18  plot(X,polyval(M,X),'r-')
19
20  hold off; % Next plot command will print on a new graph
21
22  % Display the equation
23  fprintf('Eqn for LSR: C(T) = %f T + %f\n',M(1),M(2))
24
25  % Average rate of change in chirps/sec as temp goes from 70 to 80
26  avg = (polyval(M,80)-polyval(M,70))/(80-70);
27  fprintf('[C(80)-C(70)]/[80-70] = %f chirps/sec/degree F\n',avg)
```

When the `CricketChirps` m-file is run in the command window, the following output is obtained:

```
————————————————— Command Window —————————————————
>> CricketChirps
Eqn for LSR: C(T) = 0.211925 T + -0.309144
[C(80)-C(70)]/[80-70] = 0.211925 chirps/sec/degree F
```

Method 2: Estimating Directly from the Data

The second method does not require fitting a function to the data. We will use the data directly to estimate the rate of change at each point (except the end points). Thus, if we consider each temperature value T_i and each chirps per second value C_i for $i = 1, 2, \ldots, 15$, we will compute

$$\text{estimated rate of change at } T_i = \frac{1}{2}\left(\frac{C_i - C_{i-1}}{T_i - T_{i-1}} + \frac{C_{i+1} - C_i}{T_{i+1} - T_i}\right)$$

for $i = 2, 3, \ldots, 14$. If we examine our data set, we will see that there are two data points that have the same temperature: $(T_7, C_7) = (80.6, 16.0)$ and $(T_8, C_8) = (80.6, 17.1)$. If we use the formula above with this current data set, we will encounter an error when computing

$$\frac{C_8 - C_7}{T_8 - T_7}$$

since $T_8 - T_7 = 0$. How should we fix this? One solution is to take the average value of chirps per second at $T = 80.6$. Thus, we replace the two data points with the one data point $(80.6, 16.6)$ (notice this change in the m-file `CricketChirpsD.m`). The m-file `CricketChirpsD.m`, with its updated data set, estimates the rate of change at each data point and then plots these rates of change as well as the original data:

```
                          ─── CricketChirpsD.m ───
1   % Cricket Chirping Data
2   T = [69.4 69.7 71.6 75.2 76.3 79.6 80.6 82.0 82.6 83.3 83.5 84.3 ...
3   88.6 93.3];
4   C = [15.4 14.7 16.0 15.5 14.4 15.0 16.6 17.1 17.2 16.2 17.0 18.4 ...
5   20.0 19.8];
6
7   % Estimate rates of change at each point (not including end points
8   for i = 2:length(T)-1
9      % average rate of change from i-1 to i
10      L = (C(i)-C(i-1))/(T(i)-T(i-1));
11
12      % average rate of change from i to i+1
13      R = (C(i+1)-C(i))/(T(i+1)-T(i));
14
15      % estimate rate of change at i
16      RoC(i) = (L+R)/2;
17   end
18
19   % Plot out data
20   subplot(2,1,1); plot(T,C,'bo')
21   xlabel('Temperature \circF')
22   ylabel('Chirps/second')
23   axis([65 95 13 21])
24
25   % Plot out rate of change information
26   subplot(2,1,2); plot(T(2:length(T)-1),RoC(2:length(T)-1),'go-')
27   xlabel('Temperatee \circF')
28   ylabel('Chirps/second/\circF')
```

When run in the command window, `CricketChirpsD` produces the graph shown in Figure 17.3.

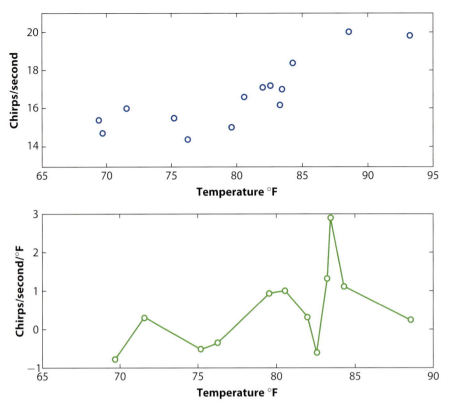

Figure 17.3 The top graph shows the number of cricket chirps per second with respect to temperature. The bottom graph shows the estimated rate of change of cricket chirps per second per °F.

17.8 Exercises

17.1   A ball is thrown upward. The data below give its height y above the ground in meters at time t in seconds. Estimate values for the average velocity on the intervals $[0, 1], [2, 3],$ and $[4, 5]$.

t	0	1	2	3	4	5
y	2	30	47	54	50	35

17.2   A table of data is given below. Estimate values for the average rate of change of y at each of the x-values given (except $x = 0.0$ and $x = 1.0$). Use the average of two difference quotients for each estimate. Where is the rate of change negative? Where is the rate of change the greatest?

x	0.0	0.1	0.2	0.3	0.4	0.5	0.6	0.7	0.8	0.9	1.0
y	17	13	11	9	6	11	15	27	30	22	19

17.3   In [56], Reis et al. used slow-motion videos to collect time-series data on the vertical position of a cat's tongue as it drank water in an effort to better understand the kinematics (i.e., motion) of a cat's tongue while it laps up water. Reis et al. collected time-series data of the vertical position of the tongue through 11 laps of the tongue. The graph below shows the average vertical position of the cat's tongue over one lap of the tongue (i.e., the average position of the tongue at each time over the 11 laps of the tongue).

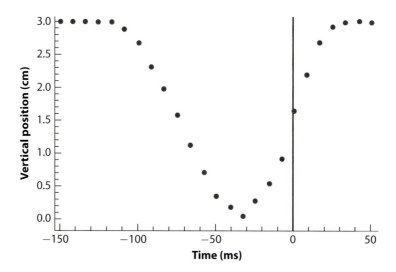

(a) Estimate the average velocity of the cat's tongue as it descends downward.
(b) Estimate the average velocity of the cat's tongue as it retracts into the mouth.

17.4 At a time t seconds after a ball is thrown up in to the air, it is at a height of

$$f(t) = -4.9t^2 + 16t + 1 \text{ meters.}$$

(a) What is the average velocity of the ball during the first second? Give units.
(b) Find the instantaneous velocity of the ball at $t = 1$. Give units.

17.5 In [71], Steury and Murray model the cyclic trends of a hare population using

$$H(t) = \bar{h} + \frac{m}{2} \cos\left(\frac{2\pi t}{10}\right),$$

where $H(t)$ is the density of the hare population measured in hares per hectare at year t, $\bar{h}$ is the average hare density, and m is the difference between high and low hare densities. A particular population of hares has an average density of 1.06 hares/hectare, with a minimum density of 0.31 hares/hectare and a maximum density of 1.81 hares/hectare.

(a) How does this population vary with time? Using Matlab, graph $H(t)$ for 5 years.
(b) Use the graph to determine at what times the population reaches a maximum value. A minimum value. Does the population density maxima and minima occur at the values you expect? Does the model make sense?
(c) Use the graph to decide when the population is growing fastest and when it is decreasing fastest.
(d) How fast is the population changing at $t = 1.5$ (a year and a half after the initial observation)? Is it increasing or decreasing?

17.6   Suppose that the population of a town can be approximated by

$$P(t) = 16,250(0.87)^t,$$

where P is the population of the town t years after 1985 (starting on January 1, 1985).

(a) Find the rate of change of the population of the town on January 1, 2000. Is the population increasing or decreasing?
(b) Find the rate of change of the population of the town on January 1, 2010. Is the rate of change faster or slower than it was in 2000?
(c) Use Matlab to graph $P(t)$. Explain how the rate of change of the population changes over time.
(d) What do you expect to happen to the population in the long term? What mathematical tools can you use to verify your hypothesis?

17.7   (From [34]) A laboratory study investigating the relationship between diet and weight in adult humans found that the weight of a subject, W (in pounds), was a function, $W = f(c)$, of the average number of calories per day, c, consumed by the subject.

(a) Interpret the statements $f(1800) = 155$, $f'(2000) = 0$ in terms of diet and weight.
(b) What are of the units of $f'(c)$?

17.8   (From [34, 77]) When you take a breath, a muscle called the diaphragm reduces the pressure around your lungs, which expand to fill with air. The table below shows the volume of a lung V as a function of the reduction in pressure from the diaphragm P. Pulmonologists (lung doctors) define the compliance of the lung as the derivative of this function, that is, compliance $= V'(P)$.

Pressure Reduction (cm of water)	Volume (liters)
0	0.20
5	0.29
10	0.49
15	0.70
20	0.86
25	0.95
30	1.00

(a) What are the units of compliance?
(b) Write an m-file similar to `CricketChirpsD.m` to estimate the compliance for pressure reductions of 5, 10, 15, 20, and 25 cm of water. Interpret the resulting graph.
(c) When is compliance largest?
(d) Explain why the compliance gets small when the lung is nearly full (around 1 liter).

17.9   The atmospheric carbon dioxide levels in Barrow, Alaska, can be modeled using the function defined by

$$C(t) = 0.04t^2 + 0.6t + 330 + 7.5\sin(2\pi t),$$

where $C(t)$ is the concentration of carbon dioxide in the atmosphere measured in parts per million and t is measured in years since 1960 [81].

(a) Using Matlab, graph C for t from 0 to 40. Note that $t = 40$ corresponds to the year 2000.

(b) Estimate $C'(60.2)$ using points close to $t = 60.2$. What month and year does $t = 60.2$ represent? Interpret the value $C'(60.2)$ in terms of rates of change. Looking at the graph from (a), can the rate of change at this one point in time give a valid indication of the overall trend in atmospheric carbon dioxide levels? Explain.

(c) Write a program similar to `CricketChirpsD.m` where the array of time data is given by

```
t = 0:40;
```

and the array of atmospheric carbon dioxide levels is given by

```
C = 0.04*t.^2 + 0.6*t + 330 + 7.5*sin(2*pi*t);
```

Note that you will need to change `T` to `t` in the appropriate places within the code since Matlab is case sensitive.

(d) When is the rate of change of carbon dioxide the greatest? The smallest?

(e) The function C is the sum of a quadratic function and a sine function. What is the physical significance/interpretation of each function?

17.10 In [55], Reed and Hill measured how the metabolic rate of a person changes after he or she eats a meal. After a person ingests food, the metabolic rate increases for a period of time and then returns to a resting metabolic rate. This phenomenon is known as the thermic effect of food. Reed and Hill show that the thermic effect of food for one person is *approximated* by the equation

$$M(t) = -10.28 + 175.9te^{-t/1.3},$$

where $M(t)$ is the difference between the measured metabolic rate at time t and the resting metabolic rate (measured in kilojoules per hour) and t is the number of hours since the person ate.

(a) If $t = 0$ corresponds to the time the person finished his or her meal, what is the interpretation of $M(0)$?

(b) Using Matlab, graph the function with a horizontal axis of $[0, 6]$ and a vertical axis of $[-20, 100]$. Note that this can be done in Matlab by using the command `axis ([0 6 -20 100])` after using the plot command.

(c) Find the average rate of change of the thermic effect of food during the first 30 minutes after eating.

(d) Find the average rate of change of the thermic effect of food during the first hour after eating. Is the rate of change larger or smaller than in (c)?

(e) Find the average rate of change of the thermic effect of food during the first 6 hours after eating. Why is this average rate of change so much smaller than the others?

(f) Estimate the (instantaneous) rate of change of a person's metabolic rate thirty minutes after eating. Estimate this rate of change using the average rate of change from $t = 0$ to $t = 0.5$, and the average rate of change from $t = 0.5$ to $t = 1$.

Derivatives of Functions

The previous chapter illustrated how we can calculate average rates of change in biological quantities, such as the weight of an individual and the concentration of a drug, over some time period. Using these average rates, we saw how one might define the instantaneous rate of change of a quantity by considering average rates of change over shorter and shorter time periods. We called this the derivative of the function describing the quantity and noted that in some cases we can actually obtain a formula for the derivative by taking a limit of a quotient. In this chapter, we develop this idea further and show how a derivative is related to the slope of a function, how to calculate derivatives, and how derivatives are related to the continuity of a function. Because so many biological processes change through time, we will be able to use derivatives to determine how rapidly a biological process changes and how this rate of change itself varies through time. For example, we will have a way to determine how rapidly a leaf's photosynthetic rate changes when the temperature or light changes and how a cell population's growth rate changes as it depletes the agar growth substance in a petri dish. We will start by considering how a leaf's photosynthetic rate changes over time when the leaf is at a constant light level.

18.1 Concept of a Derivative

The diurnal response of an organism's behavior and physiology to the 24-hour day/night cycle of our planet is a major feature of living systems. Circadian rhythms are the daily repeating patterns of many organisms, including behaviors such as the closing of flowers at night or the opening of stomata in leaves of C3 plants during the day. An interesting property of circadian rhythms is that many are endogenous, arising from an internal biological clock. For example, in 1729, the French scientist Jean-Jacques d'Ortous de Mairan discovered that circadian movements were not dependent on the daily cycling of light and dark even though they were synchronized with it. Photosynthesis is an example of a process with a circadian rhythm. Photosynthesis is the process by which plants convert carbon dioxide and water to sugar and oxygen. Tiny openings in the leaves, called stomata, open and close to regulate the amount of CO_2 taken in and the

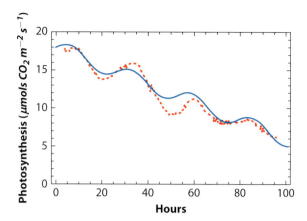

Figure 18.1 Data from [31] on the rate at which carbon dioxide is assimilated in kidney bean leaflets when exposed to constant high light over 100 hours (in red). The blue curve shows a function approximating the data, $A(t) = 18 - 0.2t + e^{-0.0001t} \sin\left(\frac{t}{4.2}\right)$.

amount of water and O_2 exiting. The circadian rhythms of photosynthesis can be investigated by manipulating environmental conditions and measuring the corresponding changes in rhythm.

One means of measuring the rate of photosynthesis in a plant is to measure the amount of carbon dioxide assimilated by the leaf (measured in μmol of CO_2) per square meter of leaf area per second. Hennessey et al. [31] measured the rate at which carbon dioxide is assimilated in kidney bean leaflets when exposed to constant high light over 100 hours (a little over 4 days). The red points in Figure 18.1 show the data collected by Hennessey et al. How might we describe the "shape" of the data? In Unit 1, we used functions to approximate data without regular rhythmic patterns. You will recall that trigonometric functions have regular periodic behavior, and Figure 18.1 shows a blue curve that approximates the shape of the data using a trigonometric function. The equation for the curve is

$$A(t) = 18 - 0.2t + e^{-0.0001t} \sin\left(\frac{t}{4.2}\right), \tag{18.1}$$

where t is in hours and A is in μmol CO_2 m^{-2} s^{-1}. The equation obviously doesn't match the data exactly, but it does mimic several features of the data: a regular periodic change, a decrease in the amplitude of the variation, and a decreasing trend in the photosynthetic rate. We might expect these features to be a general characteristic of CO_2 assimilation for leaves held under constant light and thus be useful in analyzing the processes that cause these changes.

Now that we have a curve approximating the data, how might we describe this curve? We might consider where the graph crosses the x and y axes (if it does at all). We might try and describe how how it goes up and down. We might describe how some parts are steeper than other parts, relating to the rates of change in the assimilation of carbon dioxide. These rates of change could be approximated by the slope at each point of the graph by looking at the slope of the tangent line at each point (a description of the "steepness" at each point, with "+ slope" indicating an increase and "$-$ slope" indicating a decrease).

The **derivative** of a function $f(x)$ is a function representing the **instantaneous rate of change** of the function $f(x)$, that is, how one variable f is changing with respect to another variable x. The derivative function also gives the slope of the tangent line at every point of $f(x)$. In the context of the curve describing the photosynthesis curve, the derivative would describe how the rate of carbon dioxide assimilation is changing over time.

The function used to approximate the photosynthesis data in Figure 18.1 is

$$A(t) = 18 - 0.2t + e^{-0.0001t} \sin\left(\frac{t}{4.2}\right),$$

where t is in hours and A is in μmol CO_2 m^{-2} s^{-1}. Thus, the derivative of $A(t)$ will be a function that represents the rate of change in the rate of carbon dioxide assimilation with respect to time. In the next section, we develop the mathematical definition of a derivative so that we can determine the function representing the derivative of $A(t)$.

Historical Note: Calculus was developed independently by Sir Isaac Newton and Gottfried Wilhelm Leibniz. Newton was studying physical movement and developed the derivative as a function to represent *velocity* (distance per unit time) given a function that describes the position of an object at each time point. A speedometer in a vehicle displays the distance per unit time of that vehicle at any given time point. However, one should note that velocity can be both positive and negative, indicating the direction in which an object is moving (forward or backward). A speedometer displays only the numerical value of the velocity, not the sign (whether the vehicle is moving forward or backward).

18.2 Limit Definition of a Derivative of a Function

In Section 17.6, we defined the derivative of a function at a point. Recall that for a function $f(x)$, the derivative of f at the point $x = a$ is defined as

$$f'(a) = \lim_{x \to a} \frac{f(x) - f(a)}{x - a}.$$

In this section, we define the derivative of a function. The derivative of the function $f(x)$, denoted $f'(x)$, is a function that represents the value of the derivative of f at any point x in the domain of f.

Note that the difference quotient

$$\frac{f(x) - f(a)}{x - a}$$

represents the rate of change of the function between x and a. This difference quotient is also the slope of the line between $(x, f(x))$ and $(a, f(a))$.

To generalize the derivative of f at a point to the derivative of f (in general), we let $h \equiv a - x$ and rewrite the difference quotient as

$$\frac{f(x) - f(a)}{x - a} = \frac{f(x) - f(x + h)}{-h}$$

$$= \frac{f(x + h) - f(x)}{h}.$$

Note that this difference quotient does not depend on a specific x-value. In the definition of the derivative of f at the point $x = a$, we take the limit of the difference quotient as $x \to a$. Note that as $x \to a$, then $h \to 0$, and thus we define the *derivative* of the function f as

$$f'(x) = \lim_{h \to 0} \frac{f(x + h) - f(x)}{h}.$$

Throughout this text, several alternative notations for $f'(x)$ will be used. If $y = f(x)$, then each of the following also represent the function $f'(x)$:

$$\frac{d}{dx}f(x) \quad \frac{d}{dx}y \quad \frac{dy}{dx} \quad y'$$

Let's try obtaining the derivative functions in a few simple cases.

Example 18.1 (Using the Definition of the Derivative)

Find the derivative function of $f(x) = x^2$ and determine the slope of the tangent line at $x = 3$.

Solution: Using the limit definition of a derivative, we have

$$
\begin{aligned}
f'(x) &= \lim_{h \to 0} \frac{(x+h)^2 - x^2}{h} \\
&= \lim_{h \to 0} \frac{x^2 + 2xh + h^2 - x^2}{h} \\
&= \lim_{h \to 0} 2x + h \\
&= 2x.
\end{aligned}
$$

Since the derivative function represents the slope of the tangent line at x, we find that

$$f'(3) = 2(3) = 6.$$

So the slope of the tangent line at $x = 3$ is 6.

Example 18.2 (Equation of the Tangent Line)

Find the equation of the tangent line to the graph of the function $f(x) = x^2$ at $x = 3$.

Solution: Using our work from the above example, $f'(3) = 6$ is the slope of our tangent line. Now we calculate $f(3)$ giving the function value (y-value) of the function at $x = 3$,

$$f(3) = 3^2 = 9.$$

The point $(3, 9)$ is on our tangent line. The equation of the tangent line, using the point-slope form, is

$$y - 9 = 6(x - 3)$$

or

$$y = 6x - 9.$$

Both forms of the tangent line equation are correct.

Example 18.3 (Using the Definition of the Derivative)

Find the derivative function of $f(x) = x^2 + 3$.

Solution: By definition,

$$f'(x) = \lim_{h \to 0} \frac{(x+h)^2 + 3 - x^2 - 3}{h}$$

$$= \lim_{h \to 0} \frac{x^2 + 2xh + h^2 - x^2}{h}$$

$$= \lim_{h \to 0} 2x + h$$

$$= 2x.$$

This makes sense. Notice that the function in this example is simply 3 added to the function in the previous example. Adding 3 to the function only moves the graph up vertically; it does not change the shape of the graph. Thus, adding 3 to the function has no effect on the derivative, which measures the steepness of the graph at each point x.

There is another way to write the definition of a derivative function:

$$f'(a) = \lim_{x \to a} \frac{f(x) - f(a)}{x - a},$$

which gives the derviative of $f(x)$ at $x = a$. You may use whichever version you feel more comfortable with.

Now, let us return to the circadian rhythm function for photosynthesis in kidney bean leaflets described by Equation 18.1. Notice that this function is the sum of a linear function and a function that is the product of an exponential function and sine function. Let us use the limit definition of a derivative to determine how to find the derivative of a function that is the sum of two functions.

Sum of Functions

Let $S(t) = f(t) + g(t)$. Then

$$S'(t) = \lim_{h \to 0} \frac{S(t+h) - S(t)}{h} \qquad \text{(Limit definition of derivative of } S(t))$$

$$= \lim_{h \to 0} \frac{f(t+h) + g(t+h) - f(t) - g(t)}{h} \qquad \text{(Replace } S(t) \text{ with } f(t) + g(t))$$

$$= \lim_{h \to 0} \left(\frac{f(t+h) - f(t)}{h} + \frac{g(t+h) - g(t)}{h} \right) \qquad \text{(Rearrange terms)}$$

$$= \lim_{h \to 0} \frac{f(t+h) - f(t)}{h} + \lim_{h \to 0} \frac{g(t+h) - g(t)}{h} \qquad \text{(Limit property 2 (Section 15.2))}$$

$$= f'(t) + g'(t) \qquad \text{(Limit definition of derivative of } f(t) \text{ and } g(t)).$$

Thus, the derivative of the sum of two functions is the sum of the derivatives of each function; we call this the **derivative sum rule**. Therefore, we can first find the derivative of the linear function and then find the derivative of the other function. So, we now need to know *how* to determine the derivative function of a linear function and the derivative of the product of an exponential and sine function. We start with the linear function.

Some linear functions are constant functions (horizontal lines). We start with this simpler case and then move to the general function for a line $f(x) = mx + b$.

Constant Functions

If $f(x) = c$, then

$$f'(x) = \lim_{h \to 0} \frac{c - c}{h} = 0.$$

Linear Functions

If $f(x) = mx + b$, then

$$f'(x) = \lim_{h \to 0} \frac{m(x + h) + b - mx - b}{h} = \lim_{h \to 0} \frac{mh}{h} = \lim_{h \to 0} m = m.$$

Notice that the constant b does not appear in the derivative of the linear function. If we think of $mx + b$ as the sum of two functions, the linear function $y = mx$ and the constant function $y = b$, we have already seen that the derivative of the constant function will be 0. Thus, the derivative of $y = mx$ is m, the slope of the line. This makes sense because the slope of a line to any point on the line $y = mx$ will be the slope of the line $y = mx + b$, which is m. Also, as we saw in Example 18.3, when a constant is added to a function, it has the effect of shifting the function up or down the vertical axis but does not change the slope of the function at any point.

Now, to take the derivative of

$$A(t) = 18 - 0.2t + e^{-0.0001t} \sin\left(\frac{t}{4.2}\right),$$

we could rewrite the function as $A(t) = f(t) + g(t)$, where $f(t) = 18 - 0.2t$ and $g(t) = e^{-0.0001t} \sin(t/4.2)$. At this point, we can determine that $f'(t) = -0.2$.

Let us now turn our attention to the function $g(t)$. Notice that $g(t)$ is the product of two functions. Let us use the limit definition of a derivative to determine how to find the derivative of the product of two functions.

Product of Functions

Let $P(t) = f(t)g(t)$. Then

$$P'(t) = \lim_{h \to 0} \frac{P(t + h) - P(t)}{h} \qquad \text{(Limit definition of derivative of } P(t)\text{)}$$

$$= \lim_{h \to 0} \frac{f(t + h)g(t + h) - f(t)g(t)}{h} \qquad \text{(Replace } P(t) \text{ with } f(t)g(t)\text{)}$$

Now, we will use a trick to add 0 to the numerator in a clever and helpful way. We will add and subtract the term $f(t)g(t + h)$ in the numerator of the quotient:

$$P'(t) = \lim_{h \to 0} \frac{f(t + h)g(t + h) - f(t)g(t + h) + f(t)g(t + h) - f(t)g(t)}{h}$$

$$= \lim_{h \to 0} \frac{\big[f(t + h) - f(t)\big]g(t + h)}{h} + \frac{f(t)\big[g(t + h) - g(t)\big]}{h} \qquad \text{(Rearrange terms)}$$

$$= \left(\lim_{h \to 0} \frac{f(t + h) - f(t)}{h}\right)\left(\lim_{h \to 0} g(t + h)\right) + f(t)\left(\lim_{h \to 0} \frac{g(t + h) - g(t)}{h}\right)$$

(Limit properties 1, 2, and 4 (Section 15.2))

$$= f'(t)g(t) + f(t)g'(t) \qquad \text{(Limit definition of derivative of } f(t) \text{ and } g(t))$$

We will refer to this derivative property as the ***derivative product rule***. Note the derivative of a product of functions is *not* the product of the derivatives.
Recall that we rewrote

$$A(t) = 18 - 0.2t + e^{-0.0001t} \sin\left(\frac{t}{4.2}\right)$$

as $A(t) = f(t) + g(t)$, where $f(t) = 18 - 0.2t$ and $g(t) = e^{-0.0001t} \sin(t/4.2)$. Now, let us write $g(t)$ as $g_1(t)g_2(t)$, where $g_1(t) = e^{-0.0001t}$ and $g_2(t) = \sin(t/4.2)$. Thus,

$$A'(t) = f'(t) + g_1'(t)g_2(t) + g_1(t)g_2'(t).$$

We next turn our attention to developing rules for the derivatives of exponential and sine functions.
But first, to recap, so far we have developed the following derivative rules:

	Function	Derivative
Sum rule	$S(x) = f(x) + g(x)$	$S'(x) = f'(x) + g'(x)$
Constant function	$f(x) = c$	$f'(x) = 0$
Linear function	$f(x) = mx + b$	$f'(x) = m$
Product rule	$P(x) = f(x)g(x)$	$P'(x) = f'(x)g(x) + f(x)g'(x)$

18.3 Derivatives of Exponential Functions

Exponential functions can be used to describe a variety of biological and physical phenomena, including population growth with unlimited resources, radioactive decay, and Newton's Law of Cooling. To motivate finding the derivative function of an exponential function, let us consider an example of exponential population growth.

Lemna minor is a species of duckweed, a floating freshwater aquatic plant, that can be found in freshwater ponds and slow-moving streams in Africa, Asia, Europe, and North America. The plant is quite small with leaves measuring between 1 and 8 mm long. Because of the plant's small size and high growth rate, researchers studying wastewater management have investigated the use of *Lemna minor* (and other varieties of duckweed) in treating (i.e., purifying) wastewater. In one laboratory study of *Lemna minor*, Vermaat and Hanif [74] found that in domestic wastewater the fresh-weight biomass of the *Lemna minor* (measured in grams per growing container) increased exponentially with time. The data, averaged at each time over several

Time (days)	Plant Fresh Weight (g)
0	0.0416
4	0.0592
8	0.1006
12	0.1717

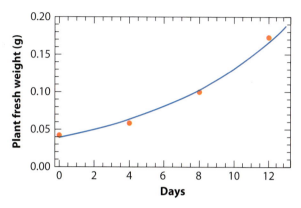

Figure 18.2 Data from [74] of the fresh-weight biomass of *Lemna minor* measured in grams per growing container over time shown in the table of values and as orange points on the graph. Each data point represents the average value over several experiments. The blue curve shows a function approximating the data, $W(t) = 0.0394e^{0.120t}$.

experiments, are shown (in orange) in Figure 18.2. Using regression techniques from Chapter 4, we can determine that the function

$$W(t) = 0.0394e^{0.120t}$$

approximates the data (shown as the blue curve in Figure 18.2), where t is measured in days and W is the fresh-weight biomass of *Lemna minor* per container measured in grams.

What is the rate of change in the biomass of *Lemna minor* in these experiments? Using the approximating function $W(t)$, let us calculate the average rate of change in population size at $t = 2, 4, 6, 8, 10$, and 12. For example, we can approximate the rate of change at $t = 2$ by using the slope of the secant line through $(0, W(0))$ and $(4, W(4))$:

$$W'(2) \approx \frac{W(4) - W(0)}{4 - 0} = 0.0060.$$

The average rate of change at each time t is shown in Table 18.2.

A plot of the average rate of change at each time is shown in Figure 18.3. What type of function might describe the shape of the data in Figure 18.3? It looks as if we could draw an exponential curve through the points in Figure 18.3. Again, using regression techniques from Chapter 4, we can determine that the function

$$W'(t) = 0.00476e^{0.120t}$$

fits the data in Figure 18.3 perfectly (i.e., the curve passes through every data point).

When we compare $W(t)$ and $W'(t)$, what do we notice?

$$W(t) = 0.0394e^{0.120t} \qquad W'(t) = 0.00476e^{0.120t}$$

Table 18.1. Average rates of change of *Lemna minor* fresh-weight biomass at selected times. Note that t has units of days and that $(W(t+2) - W(t-2))/4$ has units of grams per day.

t	$\dfrac{W(t+2) - W(t-2)}{4}$
2	0.0060
4	0.0077
6	0.0097
8	0.0124
10	0.0157
12	0.0200

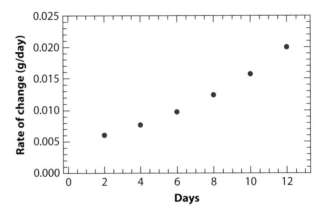

Figure 18.3 Value of the average rate of change of $W(t)$ at $t = 2, 4, 6, 8, 10,$ and 12 days.

The only difference between $W(t)$ and $W'(t)$ is the coefficient in front of the exponential term. Thus, the derivative of $W(t)$ is directly proportional to the original function $W(t)$; that is, there exists a value k such that

$$W'(t) = kW(t). \tag{18.2}$$

To find the value of k, we substitute the equations for $W(t)$ and $W'(t)$ into Equation (18.2):

$$0.00476e^{0.120t} = k\left(0.0394e^{0.120t}\right)$$

$$0.00476 = k(0.0394)$$

$$k = 0.121.$$

Notice that the value of k is approximately the value of the coefficient in the exponent of the exponential term in $W(t)$ and $W'(t)$. In fact, if there were no truncation errors due to using decimal approximations in Matlab, we would find that k is exactly the value of the coefficient in the exponent of the exponential term in $W(t)$ and $W'(t)$. Thus, if we were to write the function W as $W(t) = Ce^{kt}$, then we could write its derivative as $W'(t) = kCe^{kt} = kW(t)$.

Furthermore, this property holds true for all exponential functions of the form Ce^{kt}. Not all exponential functions are written in this form; some have the form Ca^t. We can rewrite the equation $f(t) = Ca^t$ as $f(t) = Ce^{(\ln a)t}$, and thus, by the property just described,

$$f'(t) = (\ln a)Ce^{(\ln a)t} = (\ln a)Ca^t.$$

To recap, so far we have developed the following derivative rules.

	Function	Derivative
Sum rule	$S(x) = f(x) + g(x)$	$S'(x) = f'(x) + g'(x)$
Constant function	$f(x) = c$	$f'(x) = 0$
Linear function	$f(x) = mx + b$	$f'(x) = m$
Product rule	$P(x) = f(x)g(x)$	$P'(x) = f'(x)g(x) + f(x)g'(x)$
Exponential function	$f(x) = Ce^{kx}$	$f'(x) = kCe^{kx}$
	$f(x) = Ca^x$	$f'(x) = (\ln a)Ca^x \, a > 0$

Example 18.4 (Growing Duckweed in a Growth Medium)

As a control to their experiment, Vermaat and Hanif [74] grew *Lemna minor* in a standard growth medium to determine the growth rate of this species of duckweed under normal conditions. The data from their experiment are shown below, where each data point represents the average value over several experiments.

Time (days)	Plant Fresh Weight (g)
0	0.0416
4	0.1008
8	0.2937
12	0.9588

Find an equation that describes the growth rate of *Lemna minor* over time in a standard growth medium. Does *Lemna minor* grow more quickly in a standard growth medium or in domestic wastewater?

Solution: If we graph the data in the table, we can see that it appears to be approximated by an exponential curve. We can use the methods from Unit 1 to fit an exponential curve to the data and obtain the equation

$$W(t) = 0.0388e^{0.261t},$$

where t is measured in days and W represents the fresh-weight biomass in grams. Using the property of exponential functions,

$$W'(t) = 0.261\left(0.0388e^{0.261t}\right) = 0.0101e^{0.261t}.$$

(Continued)

To determine under which conditions *Lemna minor* is growing more quickly, we can compare the **per capita growth rates** of the plant under each experimental condition. The growth rate of the population in each case is represented by $W'(t)$. If we divide the growth rate at time t by the number of individuals in the population at time t, we obtain the per capita growth rate. Since $W'(t) = kW(t)$, then

$$\frac{W'(t)}{W(t)} = k.$$

Thus, k represents the per capita growth rate.

In the experiment where *Lemna minor* was grown in domestic wastewater, $k = 0.120$. In the experiment where *Lemna minor* was grown in a standard growth medium, $k = 0.261$. Since the value of k (the per capita growth rate) is larger in the experiment when *Lemna minor* was grown in a standard growth medium, we can conclude that this is the condition under which the plant grows more quickly.

Since $W'(t)$ has units of grams/day and $W(t)$ has units of grams, the per capita growth rate has units of 1/day. For the standard growth medium experiment, since $k = 0.261$, we would say that the freshwater biomass of *Lemna minor* is increasing at a rate of 26.1% per day. Note that the value of k often also is called the **intrinsic growth rate**.

Recall that we rewrote

$$A(t) = 18 - 0.2t + e^{-0.0001t} \sin\left(\frac{t}{4.2}\right)$$

as $A(t) = f(t) + g_1(t)g_2(t)$, where $f(t) = 18 - 0.2t$, $g_1(t) = e^{-0.0001t}$, and $g_2(t) = \sin(t/4.2)$. Using the derivative sum and product rules, we determined that

$$A'(t) = f'(t) + g_1'(t)g_2(t) + g_1(t)g_2'(t).$$

Now that we have a rule for finding the derivative function of exponential functions, we know that $g_1'(t) = -0.0001te^{-0.0001t}$.

18.4 Derivatives of Trigonometric Functions

Sinusoidal functions are used to describe a variety of biological processes that are cyclical. As discussed in the beginning of this chapter, sinusoidal functions can be used to describe circadian rhythms. Let us consider the sine term in Equation 18.1, which describes how the rate of photosynthesis of kidney bean leaflets changes over time (in hours). Recall that Equation 18.1 is

$$A(t) = 18 - 0.2t + e^{-0.0001t} \sin\left(\frac{t}{4.2}\right),$$

where t is measured in hours and A is in μmol CO_2 m^{-2} s^{-1}, and we rewrote $A(t)$ as $A(t) = f(t) + g_1(t)g_2(t)$, where $f(t) = 18 - 0.2t$, $g_1(t) = e^{-0.0001t}$, and $g_2(t) = \sin(t/4.2)$. We now find the derivative function of $g_2(t)$.

Table 18.2. Average rates of change of $g_2(t) = \sin(t/4.2)$.

t	$\dfrac{g_2(t + 0.1) - g_2(t - 0.1)}{0.2}$	t	$\dfrac{g_2(t + 0.1) - g_2(t - 0.1)}{0.2}$
0	0.2381	14	−0.2337
2	0.2116	16	−0.1869
4	0.1380	18	−0.1195
6	0.0337	20	0.0118
8	−0.0780	22	0.1195
10	−0.1725	24	0.2006
12	−0.2285		

Recall from trigonometry that a general sinusoidal function can be written as

$$ f(t) = \frac{\alpha}{2} \sin\left(\frac{2\pi}{p}(t - s_h) \right) + s_v, $$

where α is the amplitude, p is the period, s_h is the amount of horizontal shift, and s_v is the amount of vertical shift. Notice that for $g_2(t)$, there is no horizontal or vertical shift, the amplitude is 2, and the period can be found by equating $2\pi/p$ to $1/4.2$ and solving for p, yielding $p \approx 26.4$. If t is measured in hours, then the period represents 26.4 hours (a little more than 1 day).

As we did in Section 18.3, let us approximate the derivative of $g_2(t) = \sin(t/4.2)$ at several points using the slopes of secant lines (see Table 18.2).

Figure 18.4 shows the value of $g_2(t)$ at $t = 0, 2, \ldots, 24$ (as orange dots) and the average rate of change of $g_2(t)$ at the same values of t using $(g_2(t + 0.1) - g_2(t - 0.1))/0.2$ to compute the average rate of change (shown as blue dots). If we compare the two sets of data, we might notice that the set of blue dots appears to form a cosine function with the same period as the sine function in $g_2(t)$, which has been vertically shrunk. Thus, we could express the derivative of $g_2(t)$ as $g_2'(t) = C \cos(t/4.2)$. If we experiment with different values of C, we will find that $C \approx 0.238 \approx 1/4.2$.

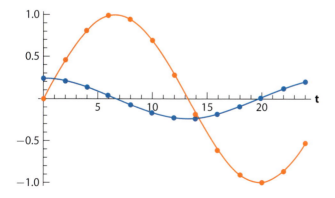

Figure 18.4 Values of $g_2(t) = \sin(t/4.2)$ at selected points (shown as orange dots). Average rate of change of $g_2(t)$ at selected points using $(g_2(t + 0.1) - g_2(t - 0.1))/0.2$ (shown as blue dots). The orange curve is $g_2(t) = \sin(t/4.2)$, and the blue curve is $(1/4.2) \cos(t/4.2)$.

In general, a function of the form $y(t) = \sin(Ct)$ (where C is a constant) will have the derivative function $y'(t) = C\cos(Ct)$.

Recall that Figure 18.1 gives from [31] the rate of photosynthesis in kidney bean leaflets over 100 hours. The data were approximated using the function

$$A(t) = 18 - 0.2t + e^{-0.0001t}\sin\left(\frac{t}{4.2}\right),$$

where t was measured in hours and A represented the rate at which CO_2 was assimilated by the kidney bean leaflets (measured in μmol CO_2 m^{-2} s^{-1}). To find the derivative of $A(t)$ (the rate at which the photosynthetic rate is changing with respect to time), we rewrote $A(t)$ as $A(t) = f(t) + g_1(t)g_2(t)$, where $f(t) = 18 - 0.2t$, $g_1(t) = e^{-0.0001t}$, and $g_2(t) = \sin(t/4.2)$. Using the derivative sum and product rules, we determined that

$$A'(t) = f'(t) + g_1'(t)g_2(t) + g_1(t)g_2'(t).$$

In Section 18.2, we saw that $f'(t) = -0.2$. In Section 18.3, we saw that $g_1'(t) = -0.0001e^{-0.0001t}$. Now in this section, we found that $g_2'(t) = (1/4.2)\cos(t/4.2)$. Putting this all together, we have

$$A'(t) = f'(t) + g_1'(t)g_2(t) + g_1(t)g_2'(t)$$

$$= -0.2 - 0.0001e^{-0.0001t}\sin\left(\frac{t}{4.2}\right) + e^{-0.0001t}\cdot\left(\frac{t}{4.2}\right)\cos\left(\frac{t}{4.2}\right)$$

$$= -0.2 - 0.0001e^{-0.0001t}\sin\left(\frac{t}{4.2}\right) + \left(\frac{te^{-0.0001t}}{4.2}\right)\cos\left(\frac{t}{4.2}\right).$$

To recap, so far we have developed the following derivative rules.

	Function	Derivative
Sum rule	$S(x) = f(x) + g(x)$	$S'(x) = f'(x) + g'(x)$
Constant function	$f(x) = c$	$f'(x) = 0$
Linear function	$f(x) = mx + b$	$f'(x) = m$
Product rule	$P(x) = f(x)g(x)$	$P'(x) = f'(x)g(x) + f(x)g'(x)$
Exponential function	$f(x) = Ce^{kx}$	$f'(x) = kCe^{kx}$
	$f(x) = Ca^x$	$f'(x) = (\ln a)Ca^x, a > 0$
Sine function	$f(x) = \sin(Cx)$	$f'(x) = C\cos(Cx)$

For a more mathematically rigorous derivation of the derivatives of the exponential and sine functions using only the limit definition of a derivative, see Appendix B.

18.5 Derivatives and Continuity

We have now seen examples of derivative functions in the case of several different functions. Do all functions have a derivative at every point? We will see that functions with discontinuities are not differentiable where the discontinuities occur. Is continuity of the function needed for a derivative to exist at a point? It turns out that functions can be continuous everywhere but have points where they are not differentiable.

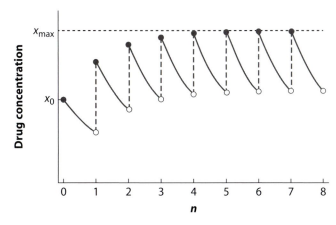

Figure 18.5 Drug does of the same size given every 1 unit of time.

Recall the drug dosing example from Section 5.6. For most drugs, it is assumed that the amount of drug remaining in the body after t hours decays exponentially according to $x(t) = be^{-kt}$, where $k > 0$ is a drug-specific rate-of-decay constant. If an additional dose of size b is given to the patient every τ hours, the amount of drug in the body over time looks like the curve shown in Figure 18.5.

For the curve in Figure 18.5, what is the rate of change exactly at $t = 1$? If we think about the rate of change as the slope of the tangent line at a point, what is the slope of the slope of the tangent line at $t = 1$? Is it defined by the curve to the left of $t = 1$ or to the right of $t = 1$? In order to clarify when the derivative of a function exists, we define the term *differentiable*. If a function is **differentiable** at a point, then the derivative exists at that point. Mathematically, we define differentiability in the following way.

Differentiable

A function f is **differentiable at a point** $x = a$ if

$$\lim_{h \to 0^+} \frac{f(a+h) - f(a)}{h} \quad \text{and} \quad \lim_{h \to 0^-} \frac{f(a+h) - f(a)}{h}$$

exist and

$$\lim_{h \to 0^+} \frac{f(a+h) - f(a)}{h} = \lim_{h \to 0^-} \frac{f(a+h) - f(a)}{h};$$

that is, the right and left limits that define the derivative must exist and be equal. A function f is **differentiable on an open interval** (a, b), (a, ∞), $(-\infty, a)$, or $(-\infty, \infty)$ if f is differentiable at all the points in the interval.

Note that since the rate of change of the curve in Figure 18.5 is different when $t = 1$ is approached from the left than when it is approached from the right, we would say that the curve is not differentiable at $t = 1$. For similar reasons, it is also not differentiable at $t = 2, 3, 4, 5, 6, 7 \ldots$.

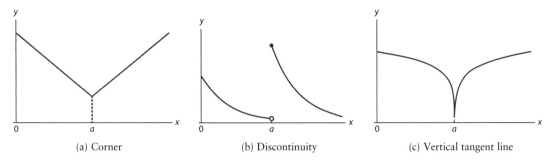

(a) Corner   (b) Discontinuity   (c) Vertical tangent line

Figure 18.6 Graphs demonstrating the three different ways in which a function can be not differentiable at a point.

There are three ways in which a function can be not differentiable at a point.

1. The function can have a corner at $x = a$, where $f(a)$ is continuous but the left and right derivative limits exist and do not agree (see Example 18.5).
2. The function can have a discontinuity at $x = a$. If a function is not continuous at $x = a$, then $f'(a)$ does not exist (see Examples 18.6 and 18.7).
3. The function can have a vertical tangent line at $x = a$ (see Example 18.8). In this case, the limits that define the derivative do not exist.

Differentiability Implies Continuity

If $f'(a)$ exists, then $\lim_{x \to a} f(x) = f(a)$; that is, f is continuous at a.

Note that this says that differentiability at a point implies continuity at that point, but the reverse is not true, as we will see in Example 18.8.

Let us look at examples of functions that are continuous everywhere but not differentiable at a point.

Example 18.5 (Where Differentiable?)

Where is the function $f(x) = |x|$ differentiable?

Solution: When $x > 0$, then $|x| = x$. We know that linear functions are continuous; thus, $f(x) = |x|$ is continuous for $x > 0$. Furthermore, if we choose h small enough, then $x + h > 0$ and $|x + h| = x + h$. Thus, for $x > 0$,

$$f'(x) = \lim_{h \to 0} \frac{|x + h| - |x|}{h} = \lim_{h \to 0} \frac{x + h - x}{h} = \lim_{h \to 0} \frac{h}{h} = 1.$$

Thus, we see that f is differentiable for $x > 0$.

(*Continued*)

Now, when $x < 0$, then $|x| = -x$. Thus, we see that for $x < 0$, f is continuous. If we choose h small enough, then $x + h < 0$ and $|x + h| = -(x + h)$. Therefore, for $x < 0$,

$$f'(x) = \lim_{h \to 0} \frac{|x+h| - |x|}{h} = \lim_{h \to 0} \frac{-(x+h) - (-x)}{h} = \lim_{h \to 0} \frac{-h}{h} = -1.$$

Thus, we see that f is also differentiable for $x < 0$.

For $x = 0$, we examine the right and left limits of

$$\frac{|0 + h| - |0|}{h}.$$

$$\lim_{h \to 0^+} \frac{|0+h| - |0|}{h} = \lim_{h \to 0^+} \frac{0 + h - 0}{h} = \lim_{h \to 0^+} \frac{h}{h} = 1$$

$$\lim_{h \to 0^-} \frac{|0+h| - |0|}{h} = \lim_{h \to 0^-} \frac{-(0+h) - (-0)}{h} = \lim_{h \to 0^-} \frac{-h}{h} = -1.$$

Since

$$\lim_{h \to 0^+} \frac{|0+h| - |0|}{h} \neq \lim_{h \to 0^-} \frac{|0+h| - |0|}{h},$$

the function f is not differentiable at $x = 0$; that is $f'(0)$ does not exist.

The function describing the derivative of f can be written as

$$f'(x) = \begin{cases} 1, & x > 0 \\ -1, & x < 0 \end{cases}.$$

Notice that a value for $f'(0)$ is not given because no such value exists. The function $f(x) = |x|$ is an example of a function that is not differentiable at a point because it has a corner at that point, in this case at $x = 0$.

Example 18.6 (Jump Discontinuity)

Show that the function

$$f(x) = \begin{cases} x^2 + 2x + 1, & x \leq 1 \\ 3 - x, & x > 1 \end{cases}$$

is not differentiable at $x = 1$.

(Continued)

Solution: We consider the left and right limits that define the derivative at $x = 1$. Note that $f(1) = 1^2 + 2(1) + 1 = 4$. However,

$$\lim_{x \to 1^+} f(x) = \lim_{x \to 1^+} 3 - x = 3 - 1 = 2.$$

Thus, the function is not continuous at $x = 1$. Since the function is not continuous at $x = 1$, it is not differentiable at $x = 1$.

Example 18.7 (Not Differentiable at a Removable Discontinuity)

Show that the function

$$f(t) = \begin{cases} 3x + 2, & x \neq 0 \\ 0, & x = 0 \end{cases}$$

is not continuous at $x = 0$ and is not differentiable at $x = 0$.

Solution: To check continuity, we check that the left and right limits of $f(x)$ as $x \to 0$ are equal and equal to $f(0)$. Notice that $\lim_{x \to 0^+} f(x) = \lim_{x \to 0^+} 3x + 2 = 2$ and that $\lim_{x \to 0^-} f(x) = \lim_{x \to 0^-} 3x + 2 = 2$. Thus, $\lim_{x \to 0} f(x) = 2$. However, $f(0) = 0$. Since $f(0) \neq \lim_{x \to 0} f(x)$, f is not continuous at $x = 0$, and hence it is not differentiable at $x = 0$.

Example 18.8 (Vertical Tangent Line)

Show that the function $M(x) = \sqrt[3]{x}$ is continuous but not differentiable at $x = 0$.

Solution: Note that $M(0) = 0$ and that $\lim_{t \to 0^+} \sqrt[3]{x} = 0$ and $\lim_{t \to 0^-} \sqrt[3]{x} = 0$. Thus, the function $M(x)$ is continuous at $x = 0$.

Next we check differentiability. Notice that

$$\lim_{b \to 0^+} \frac{M(0 + b) - M(0)}{b} = \lim_{b \to 0^+} \frac{\sqrt[3]{b} - 0}{b} = \lim_{b \to 0^+} \frac{b^{1/3}}{b} = \lim_{b \to 0^+} b^{-2/3} = \lim_{b \to 0^+} \frac{1}{\left(\sqrt[3]{b}\right)^2}.$$

As $b \to 0^+$, the value of $\frac{M(0+b) - M(0)}{b}$ increases toward infinity. Thus, a real value for this limit does not exist. Since this limit does not exist, the function $M(t)$ is not differentiable at $t = 0$. Note that, it can likewise be shown that as $b \to 0^-$, the value of $\frac{M(0+b) - M(0)}{b}$ increases to ∞.

We can see that the slopes approach a vertical tangent line because the limits of the slopes of the secant lines from the right and left are ∞.

In Sections 18.3 and 18.4, we determined expressions for derivatives of exponential and sine functions. Note that both of these functions are defined for all real numbers. Let us now consider a function that is not defined for all real numbers. In the next section, we determine an expression for the derivative of the natural logarithm function, which is only defined for positive real numbers.

18.6 Derivatives of Logarithmic Functions

Myelin is an electrically insulating membrane that forms a layer, called the myelin sheath, around the axon of a neuron (or nerve cell). At periodic intervals along the axon, there are short segments where there is no myelin sheath and the axon is exposed (see Figure 18.7). These short segments of exposed axon are called the nodes of Ranvier, and they are critical to the proper functioning of the myelin. The length of the myelin sheath between the nodes of Ranvier is called the myelin period and is typically measured in angstroms (1 Å= 1×10^{-10} m).

The study of the formation, growth, and stability of myelin as an individual ages has enabled scientists to better understand how the nervous system functions during critical periods of development. Agrawal et al. [1] measured the total mass of myelin in mouse brains (called myelin yield) in mice from ages 5 to 495 days. The data collected by Agrawal et al. show that the total mass of myelin in a mouse brain increases logarithmically with respect to the mouse's age in days (see Figure 18.8) according to the function

$$M(t) = 23.755 \ln(t) - 56.766,$$

where t is mouse age in days and M is myelin yield measured in milligrams per mouse brain. The function $M(t)$ is shown as the blue curve in Figure 18.8.

The graph in Figure 18.8 shows that the myelin yield increases quickly for young mice and then increases more slowly as a mouse ages. What is the rate of change of myelin yield for a mouse of a given age? The derivative of $M(t)$ is a function that describes the rate of change of myelin yield at age t. To determine the derivative of this logarithmic function, let us approximate the rate of change at several ages using secant lines. For example, the rate of change at age 10 days can be approximated by

$$M'(10) \approx \frac{M(15) - M(5)}{10} = 2.610.$$

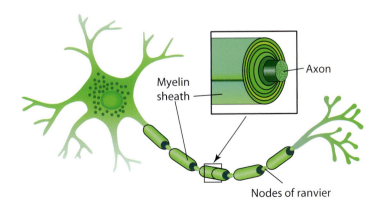

Figure 18.7 Diagram of a neuron.

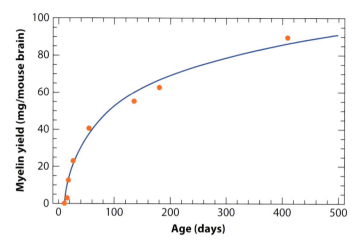

Figure 18.8 Data collected by Agrawal et al. [1] measuring the total mass of myelin in mouse brains in mice from ages 5 to 495 days.

Table 18.3. Average rates of change of myelin yield at selected times. Note that t has units of days and that $(M(t+5) - M(t-5))/10$ has units of grams per day.

t	$\dfrac{M(t+5) - M(t-5)}{10}$
10	2.610
25	0.9632
50	0.4767
100	0.2377
200	0.1188
300	0.07919
400	0.05939

The average rate of change at each age t is shown in Table 18.3.

A plot of the rate-of-change data is shown in Figure 18.9. What type of function might describe the data in Figure 18.9?

You may recognize that the shape of the data in Figure 18.9 looks similar to the graph $f(x) = \frac{1}{x}$ but has been vertically scaled. Try plotting several different curves of the form C/t (where C is a constant) on the same graph as the rate-of-change data points. What value of C causes the data in Figure 18.9 to be most closely approximated?

Figure 18.10 shows the values of the average rate of change at $t = 10, 25, 50, 100, 200, 300$, and 400 days (gray data points) as well as curves for $10/t$, $23.755/t$, and $50/t$ (shown in green, blue, and orange, respectively). The curve that best approximates the average rate of change values is $23.755/t$. Thus, we conclude that

$$M'(t) = \frac{23.755}{t}.$$

Notice that the value of C is the same as the coefficient in front of the natural log term in $M(t)$.

In general, a function of the form $M(t) = C\ln(t) + B$ (where B and C are constants) will have the derivative $M'(t) = C/t$. Note that if $C = 1$, then $M'(t) = 1/t$. Additionally, note that the

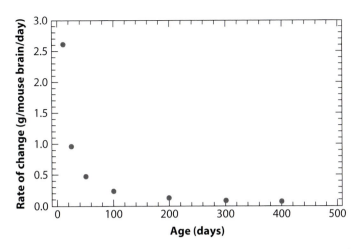

Figure 18.9 Value of the average rate of change of $M(t)$ at $t = 10, 25, 50, 100, 200, 300,$ and 400 days.

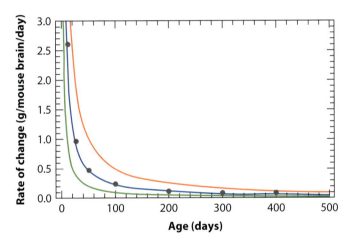

Figure 18.10 The points portray the values of the average rate of change of $M(t)$ at $t = 10, 25, 50, 100, 200, 300,$ and 400 days. The graphs of $10/t$, $23.755/t$, and $50/t$ are shown as the green, blue, and orange curves, respectively.

constant B does not appear in the derivative function. If we think of the function of myelin yield in a mouse brain at age t, $M(t) = 23.755 \ln(t) - 56.766$, as the sum of $f(t) = 23.755 \ln(t)$ and $g(t) = -56.766$, we know from Section 18.2 that $g'(t) = 0$. Since $M'(t) = f'(t) + g'(t)$ and $g'(t) = 0$, then $M'(t) = f'(t)$. Thus, the derivative of $23.755 \ln(t)$ is $23.755/t$.

To recap, so far we have developed the following derivative rules.

	Function	Derivative
Sum rule	$S(x) = f(x) + g(x)$	$S'(x) = f'(x) + g'(x)$
Constant function	$f(x) = c$	$f'(x) = 0$
Linear function	$f(x) = mx + b$	$f'(x) = m$
Product rule	$P(x) = f(x)g(x)$	$P'(x) = f'(x)g(x) + f(x)g'(x)$
Exponential function	$f(x) = Ce^{kx}$	$f'(x) = kCe^{kx}$
	$f(x) = Ca^x$	$f'(x) = (\ln a)Ca^x \quad a > 0$
Sine function	$f(x) = \sin(Cx)$	$f'(x) = C\cos(Cx)$
Logarithmic function	$f(x) = C\ln(x)$	$f'(x) = C/x$

Consider the general logarithmic function of the form $f(x) = C\ln(x) + B$. Where does this function have a discontinuity? What type of discontinuity is it? We know that the function $\ln(x)$ has a vertical assymptote at $x = 0$. Since the constants B and C shift and stretch the function $\ln(x)$ only vertically, $f(x)$ will also have a vertical asymptote at $x = 0$. Since the function $f(x)$ is not defined and not continuous at $x = 0$, the derivative of $f(x)$ should not exist at $x = 0$. Is this true?

We showed that the derivative of $f(x) = C\ln(x) + B$ is $f'(x) = C/x$, which shows that $f'(x)$ is also not defined at 0. Furthermore, note that the slopes of tangent lines to $f(x)$ are always positive (e.g., see Figure 18.8). As we approach $x = 0$ (from the right), the slopes of the tangent lines get steeper, approaching ∞. Now, look at the derivative function $f'(x) = C/x$. Notice that as we approach $x = 0$ (from the right), the value of the derivative approaches infinity (e.g., see Figure 18.10). Finally, notice that though C/x is defined for all real numbers except 0, it does not make sense to define the derivative of $f(x) = C\ln(x) + B$ for values of $x \le 0$ since $f(x)$ does not exist for these values.

Example 18.9 (The Rate of Change of the Myelin Period)

In addition to measuring the total mass of myelin in each mouse's brain, Agrawal et al. [1] also measured the myelin period of myelin along nerves of the central nervous system (CNS) and the peripheral nervous system (PNS). The sciatic nerve was used for measurements along a nerve of the PNS, while the optic nerve was used for measurements along a nerve of the CNS. In this example, we consider the data collected for the sciatic nerve and leave questions relating to the data collected for the optic nerve for investigation in Exercise 18.8.

Agrawal et al. [1] found that the myelin period increased logarithmically with age in mice (data shown below) according to the function

$$P(t) = 0.9572\ln(t) + 170.37,$$

where t is measured in days and P represents the myelin period measured in ansgtroms (Å). Find an equation that describes the growth rate of the myelin period with respect to mouse age. How quickly is the myelin period increasing when a mouse is 1 month old?

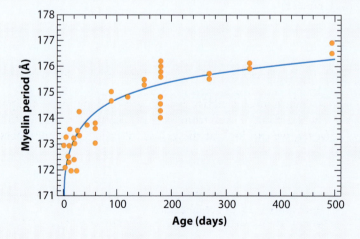

(Continued)

Solution: Use the rule for derivatives of logarithmic functions,

$$P'(t) = 0.9572/t,$$

where P' has units of Å/day. If we approximate 1 month by 30 days, the myelin period will be increasing by

$$P'(30) = 0.9572/30 = 0.3191 \text{Å/day}.$$

18.7 Matlab Skills

In Chapter 15, we used Matlab to numerically approximate limits of functions at a point (see Section 15.3). Here we numerically approximate the derivative of a function at a point.

Recall that the derivative of a function at a point is defined using limits. Thus, the derivative of the function f at the point x is

$$f'(x) = \lim_{h \to 0} \frac{f(x+h) - f(x)}{h}.$$

Suppose, for example, that we want to find the derivative of

$$f(x) = \sqrt[3]{x}$$

at the point $x = 1$. As in Chapter 15, we begin by creating an m-file that defines the function f in which we are interested. Recall that $\sqrt[3]{x} = x^{1/3}$:

—————————— f.m ——————————

```
1   % Creates the function f(x)
2   % Input: x (a real number)
3   % Output: y
4
5   function y = f(x)
6   y = x^(1/3);
```

Next, we use the limit definition of a derivative to write a function that approximates the derivative of f at a particular point. Recall from Section 15.3 that we cannot fully evaluate the limit numerically; however, we can look at the values that the limit approaches from the left and from the right:

—————————— derivative.m ——————————

```
1   % Approximates the derivative of the function f
2   % Make sure the file f.m is contained in the same folder as this file
3
4   % Input: x = value at which you want to find the derivative
5   %        h = how close you want to the limit (h -> 0)
```

```
 6
 7   % Output: LLD = limit of the difference equation from the left side
 8   %          RLD = limit of the difference equation from the right side
 9
10   function [LLD,RLD] = derivative(h,x)
11
12   % left limit
13   LLD = (f(x-h)-f(x))/(-h);
14
15   % right limit
16   RLD = (f(x+h)-f(x))/(h);
```

Now, if we want to approximate the derivative $f'(1)$, we would use the following commands in the command window:

```
————————————————————— Command Window ————————————————————
>> [L,R]=derivative(0.1,1)
L =
    0.3451
R =
    0.3228

>> [L,R]=derivative(0.01,1)
L =
    0.3345
R =
    0.3322

>> [L,R]=derivative(0.001,1)
L =
    0.3334
R =
    0.3332

>> [L,R]=derivative(0.0001,1)
L =
    0.3333
R =
    0.3333
```

We see that as we allow $h \to 0$, both the left and the right limits approach 0.3333. We might guess that the derivative of f at $x = 1$ is $\frac{1}{3}$. In fact, we know from Example 19.1 that this is the case. It is a good idea to check that the function derivative.m works for a function for which we know how to algebraically find the derivative. However, now let us use our newly written function to approximate the derivative of a function for which we do not know how to take the derivative.

Suppose that we want to find the derivative of

$$f(x) = \tan(\sin x).$$

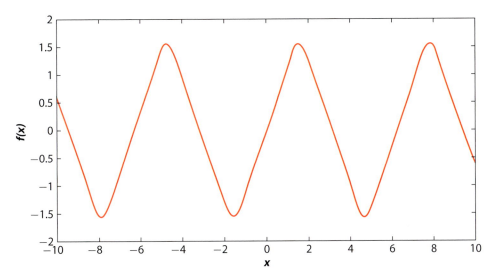

Figure 18.11 Graph of $f(x) = \tan(\sin x)$ for $x \in [-10, 10]$ generated by the graphf.m function.

Before we approximate some derivatives of this function, let us examine a graph of the function to see what we expect. We can utilize the function graphf.m, which we wrote in Section 16.4. First, we modify our f.m file:

```
————————————————— f.m ——————————————————
1   % Creates the function f(x)
2   % Input: x (a real number)
3   % Output: y
4
5   function y = f(x)
6   y = tan(sin(x));
```

Next, to graph the function, we use the graphf.m function that we wrote. Recall that this function has two inputs: xmin (the minimum x-value to graph) and xmax (the maximum x-value to graph). Since we do not know what the function will look like, let us try graphing from $x = -10$ to $x = 10$. Thus, we type the following command in the command window:

```
————————————— Command Window ——————————————
>> graphf(-10,10)
```

The resulting graph is shown in Figure 18.11.

Recall that we formed the limit definition of the derivative by taking the limit of the slopes of secant lines that passed through $(x, f(x))$ and $(x + h, f(x + h))$. The limit of these secant lines is a line that is tangent to the curve at $(x, f(x))$, and the slope of that tangent line is the derivative. Looking at the graph shown in Figure 18.11 we see that the slope of the tangent line at $x = 0$

should be positive and have a value of about 1. Let us utilize the `derivative.m` function to see if this is true:

```
                              ─── Command Window ───
>> [L,R]=derivative(0.1,0)
L =
    1.0017
R =
    1.0017

>> [L,R]=derivative(0.01,0)
L =
    1.0000
R =
    1.0000

>> [L,R]=derivative(0.001,0)
L =
    1.0000
R =
    1.0000
```

It definitely appears that $f'(0) = 1$ based on our numerical estimates. Let us "zoom in" on the graph in Figure 18.11 and focus on the portion of the function between $x = 1$ and $x = 2$. This is where one of the peaks occurs. To produce a graph with x values between 1 and 2, we simply use the `graphf.m` function again:

```
                              ─── Command Window ───
>> graphf(1,2)
```

The resulting graph is shown in Figure 18.12. It appears that the local maximum of this function occurs around $x = 1.5$. Note that the tangent line at this local maximum would have a slope

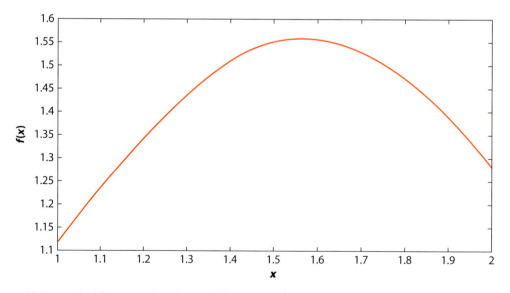

Figure 18.12 Graph of $f(x) = \tan(\sin x)$ for $x \in [1, 2]$ generated by the `graphf.m` function.

of 0. Recall that $\pi/2 = 1.5707....$ If we approximate the derivative at $\pi/2$, we would expect it to be close to 0:

```
────────────────────── Command Window ──────────────────
>> [L,R]=derivative(0.1,pi/2)
L =
     0.1698
R =
    -0.1698

>> [L,R]=derivative(0.01,pi/2)
L =
     0.0171
R =
    -0.0171

>> [L,R]=derivative(0.001,pi/2)
L =
     0.0017
R =
    -0.0017

>> [L,R]=derivative(0.0001,pi/2)
L =
   1.7128e-004
R =
  -1.7128e-004
```

From our numerical approximations, it does appear that $f'(\pi/2) = 0$.

Notice that by using numerical approximations in conjunction with some graphical analysis, we were able to predict the value of a derivative at certain points for a function for which we do not yet know how to take the derivative. This is an illustration of the power and importance of utilizing numerical approximations.

18.8 Exercises

18.1   Find $f'(1)$ for the following functions using the definition of a derivative, that is,

$$\lim_{x \to 1} \frac{f(x) - f(1)}{x - 1}.$$

(a) $f(x) = x^2 - 1$
(b) $f(x) = 3x + 2$
(c) $f(x) = 2x^2 + x$
(d) $f(x) = \dfrac{1}{x}$
(e) $f(x) = \sqrt{x}$

18.2   Find $f'(a)$ for each of the following functions in two ways (1) by computing $\lim_{x \to a} \dfrac{f(x) - f(a)}{x - a}$ and (2) by computing $\lim_{h \to 0} \dfrac{f(a + h) - f(a)}{h}$.

(a) $f(x) = x^2 - 1$

(b) $f(x) = \dfrac{1}{x^2}$

(c) $f(x) = 3x + 2$

(d) $f(x) = \sqrt{x}$

18.3   Use the limit definition of a derivative of a function, that is,

$$\lim_{h \to 0} \frac{f(x + h) - f(x)}{h},$$

to find $f'(x)$ for each of the following functions.

(a) $f(x) = 3x^2$

(b) $f(x) = \dfrac{1}{x^2}$

(c) $f(x) = 3x$

18.4   Find the equation of the tangent line to $f(x) = 3x^2$ at $x = 5$.

18.5   Find the equation of the tangent line to $f(x) = \frac{1}{x^2}$ at $x = 2$.

18.6   If $f'(2) = 5$ and $f(2) = 10$, find the equation of the tangent line to this function at $x = 2$.

18.7   Determine where the following functions are differentiable.

(a) $f(x) = 6x + 2$

(b) $g(x) = 4x^2 - 3x + 7$

(c) $h(x) = 4x + \sqrt{x}$

(d) $s(t) = \sqrt{1 + 2t}$

(e) $m(t) = \dfrac{t^2 - 1}{t + 1}$

(f) $f(t) = \dfrac{1}{ct}$

(g) $g(x) = \begin{cases} 4x - 2 & x < 3 \\ x^2 + 1 & x \geq 3 \end{cases}$

18.8   Agrawal et al. [1] measured the myelin period of myelin along nerves of the central nervous system (CNS) and the peripheral nervous system (PNS). The sciatic nerve was used for measurements along a nerve of the PNS, while the optic nerve was used for measurements along a nerve of the CNS. In Example 18.9, we consider the data collected for the sciatic nerve. The Agrawal et al. data for the optic nerve fit the logarithmic function

$$P(t) = 161.29 - 1.034 \ln(t),$$

representing the length of the myelin period (measured in angstroms) of a mouse at age t.

(a) Graph the function $P(t)$. Is the function increasing or decreasing as mouse age increases?

(b) Find an equation that describes the growth rate of the myelin period with respect to mouse age.

(c) It is biologically reasonable to assume that $t > 0$ (negative age doesn't make sense). When $t > 0$, what is the sign of the derivative? How does this correspond with your answer to part (a)?

18.9 (Adapted from [29]) Suppose that the population of a certain collection of rare Brazilian ants is given by

$$P(t) = (t + 100) \ln(t + 2),$$

where t represents the time in days. Use functions f.m, graphf.m, and derivative.m discussed in Section 18.7 to estimate the values $P'(5)$ and $P(10)$. Note that you will need to modify the function f.m. What might the derivative of a population at a particular time represent biologically?

18.10 (Adapted from [29]) Suppose that a runner's arms swings rhythmically according to the equation

$$y(t) = \frac{\pi}{8} \cos \left[3\pi \left(t - \frac{1}{3} \right) \right],$$

where y denotes the angle between the actual position of the upper arm and the vertical axis and t denotes time in seconds.

(a) Using functions f.m and graphf.m discussed in Section 18.7, graph the first 2 seconds of the arm swinging. Note that you will need to modify the function f.m.

(b) The graph produced in (a) should show three local maxima. These local maxima occur at the times when the arm is swung farthest forward. Estimate the time at which the middle local maximum occurs. What do you expect the derivative of the function y to be at this time point? Explain.

(c) Given the time you estimated in (b), use derivative.m discussed in Section 18.7 to estimate the derivative at this time. Does the numerically estimated derivative concur with your guess from (b)?

(d) What might the derivative of the function y represent biologically?

19
CHAPTER

Computing Derivatives

Biological processes are typically not described by a single function. Organisms are composed of hierarchically structured collections of organelles, cells, and tissues, and numerous interactions occur between these. The normal functioning of organisms requires a complex of interactions driven by genetic control of biochemistry and associated impacts on physiological processes. When this normal functioning goes awry, illness and disease can result. At the scale of ecological systems, webs of interactions between the species that compose the system give rise to properties of the whole system, such as nutrient cycling, that are key to the normal functioning of the ecosystem. When the web of species interactions is perturbed, as can arise when a non native species becomes invasive, the entire system can become degraded and lose its integrity. In all these cases, the biological processes can be analyzed by considering how the building blocks (e.g., organelles, cells, and individuals) respond to changes. This chapter describes how we can use calculus to analyze changes in a standard collection of functions that underlie much of the modeling of biological processes. More important, we develop the rules of calculus that allow us to investigate how more complicated collections of the building blocks of biology respond, accounting for the interactions between them, when there is a change in the biological system.

Just as biological systems are made up of components that interact, we will see how we can build up connections between functions describing different biological components and determine how the entire system responds. These responses can occur naturally, such as when you exercise, leading to changes in hosts of metabolites, such as glycerol, which is involved in the breakdown of fatty tissue. These metabolite changes are correlated with changes in physiological processes, such as heart rate and respiration rate. Physiological responses can also arise from non natural factors, as is the objective when drugs are used for medicinal purposes. Pharmacokinetics seeks to determine how physiological processes respond to drug administration. We've already seen examples of how simple mathematical models for drug metabolism and elimination, assuming exponential decay, can be helpful in describing the effects of the different timing of drug dosages. The calculus developed in this chapter provides the means to describe the dynamics of drug response in more realistic situations.

Table 19.1. Derivative rules for select functions, where C, b, m, and n are constants.

	Function	Derivative
Sum rule	$S(x) = f(x) + g(x)$	$S'(x) = f'(x) + g'(x)$
Constant function	$f(x) = C$	$f'(x) = 0$
Constant multiple	$f(x) = Cg(x)$	$f'(x) = Cg'(x)$
Linear function	$f(x) = mx + b$	$f'(x) = m$
Power rule	$f(x) = x^n$	$f'(x) = nx^{n-1}$
Product rule	$P(x) = f(x)g(x)$	$P'(x) = f'(x)g(x) + f(x)g'(x)$
Exponential function	$f(x) = e^{kx}$	$f'(x) = ke^{kx}$
	$f(x) = a^x$	$f'(x) = (\ln a)a^x$
Sine function	$f(x) = \sin(Cx)$	$f'(x) = C\cos(Cx)$
Logarithmic function	$f(x) = \ln(x)$	$f'(x) = 1/x$

19.1 Derivatives of Frequently Used Functions

By the end of Chapter 18, we had developed several derivatives rules. Here we restate those rules and add to the list the constant multiple rule and power rule. The constant multiple rule can be derived easily from the limit definition of the derivative. Let $f(x) = Cg(x)$; then

$$f'(x) = \lim_{h \to 0} \frac{f(x+h) - f(x)}{h}$$
$$= \lim_{h \to 0} \frac{Cg(x+h) - Cg(x)}{h}$$
$$= \lim_{h \to 0} \frac{C\left(g(x+h) - g(x)\right)}{h}$$
$$= Cg'(x).$$

Note that the restated derivative rules for exponential and logarithmic functions in Table 19.1 can be modified into the derivative rules for exponential and logarithmic functions, as stated in Chapter 18, by using the constant multiple rule.

A derivation of the power rule using the limit definition of a derivative is given in Appendix B.

Example 19.1 (Using the Power Rule)

Find the derivative function of $f(x) = x^4$, $g(x) = 5x^4$, $h(x) = \frac{1}{x^{3/4}}$, and $j(x) = \sqrt[3]{x}$.

Solution: Using the power rule, we have

$$f'(x) = 4x^3.$$

(*Continued*)

Now, with a constant multiplying our power function g, we obtain

$$g'(x) = 5(4x^3) = 20x^3.$$

We can rewrite the function h as $h(x) = x^{-\frac{3}{4}}$. Using the power rule, we get

$$h'(x) = -\frac{3}{4}x^{-\frac{3}{4}-1} = -\frac{3}{4}x^{-\frac{7}{4}}.$$

We can rewrite the function j as $j(x) = x^{\frac{1}{3}}$. Using the power rule, we get

$$f'(x) = \frac{1}{3}x^{-\frac{2}{3}}.$$

Throughout the remainder of this chapter, we will be developing several other derivatives rules. We will start by developing the chain rule.

19.2 The Chain Rule for the Composition of Functions

Much of life science deals with trying to understand how biological processes change in time. This includes biochemical processes, such as enzyme-mediated reaction rates; physiological processes, such as respiration rate; and population processes, such as mortality rates. In each case, the term "rate" means a change with time, and you now know that the instantaneous rate is described mathematically by the derivative of an appropriate function. However, in many cases, the dynamics of the biological process arise due to a change in a separate component. Biochemical reaction rates change due to temperature, which itself may be changing in time; respiration rate changes due to level of exercise, which changes in time; and mortality rates can change due to disease infection, which could change in time, as you are well aware of when flu season arrives. In these cases, to find the rate of change of a biological process, we need first to find how it changes with a factor that affects it and then find how that factor changes in time. Thus, leaf photosynthetic rate changes as a function of temperature, which itself changes through out the course of a day. This is an example of a "chain" of functions: if $P(T)$ gives photosynthetic rate as a function of temperature T and $T(t)$ describes the temperature at time t within a day, then the composite function $P(T(t))$ describes how photosynthesis changes with time. It may actually be easier to use data to find reasonable models for the separate functions $P(T)$ and $T(t)$ than it is to use data to find one for how photosynthesis changes directly as a function of time. This section shows how to find derivatives of composite functions, such as $P(T(t))$.

Let us begin by reviewing notation and some examples of composite functions. Let $y = f(u)$ and $u = g(x)$; then $f \circ g = f(g(x))$ is called the ***composition of f and $g***. Note that, in general, $f \circ g \neq g \circ f$. In describing composite functions, it is often convenient to describe the composite function in terms of its *inner* function and *outer* function. For the composite function $h(x) = f(g(x))$, $g(x)$ is the *inner* function, and $f(g)$ is the *outer* function.

Example 19.2 (The Composition of Two Functions)

Suppose that $f(x) = 2x - 1$ and $g(x) = \sin x$. Compare $f \circ g$ and $g \circ f$.

Solution: First, we find $f \circ g$:

$$(f \circ g)(x) = f(g(x)) = f(\sin x) = 2(\sin x) - 1.$$

Next, we find $g \circ f$:

$$(g \circ f)(x) = g(f(x)) = g(2x - 1) = \sin(2x - 1).$$

Notice that $f \circ g \neq g \circ f$.

Example 19.3 (Photosynthetic Rate)

In Unit 4, we considered examples in which the rate of photosynthesis changes with the time of day and light level. Recall that the photosynthetic rate is the rate at which a unit area of leaf assimilates carbon dioxide (CO_2) or respires oxygen (O_2) per second.

In Example 16.7, we considered data collected on day 164 in the life cycle of a soybean plant [58] and saw that the data could be approximated by the quadratic function

$$P(t) = -0.374t^2 + 9.44t - 41.3,$$

where P is the photosynthetic rate and t is time measured in hours. However, in Section 17.4, we observed that this approximation would be accurate only for times between 5:00 a.m and 8:00 p.m. However, the data collected by Rogers et al. [58] on day 164 in the life cycle of the soybean plant can also be modeled by the function

$$A(t) = 15.17 \sin\left(\frac{2\pi t}{24} - 1.77\right) + 4.00,$$

where t is measured in hours and A is measured in μmol CO_2 m^{-2} s^{-1}. A graph of the collected data and the functions $P(t)$ and $A(t)$ is shown in Figure 19.1.

Describe the function $A(t)$ as a composite function. Identify the outside function and the inside function.

Solution: Let $u = g(t) = \frac{2\pi t}{24} - 1.77$ and $f(u) = 15.17 \sin(u) + 4.00$. Then $A(t) = (f \circ g)(t) = f(g(t))$, where g is the inside function and f is the outside function.

Alternatively, we could view $A(t)$ as the composition of three functions. Let $u = g(t) = \frac{2\pi t}{24} - 1.77$, $v = h(u) = \sin(u)$, and $f(v) = 15.17v + 4.00$. Then $(f \circ h \circ g)(t) = f(h(g(t))) = A(t)$.

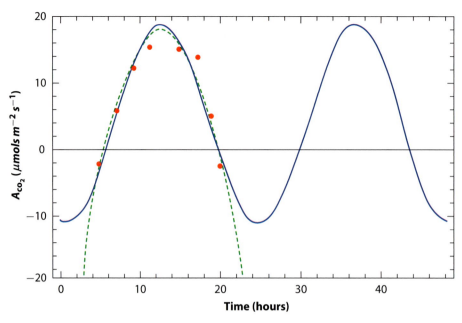

Figure 19.1 Data extracted from figure 2 of [58] showing the photosynthetic rate (measured as the absorption of CO_2 in μmol CO_2 m^{-2} s^{-1}). The points on the graph indicate the actual data (collected over day 164 of the life cycle of a soybean plant). The dashed curve is $P(t) = -0.374t^2 + 9.44t - 41.3$, and the solid curve is $A(t) = 15.17\sin(2\pi t/24 - 1.77) + 4.00$. For each function, t is measured in hours.

How do we find derivatives of composite functions? For composite functions, we have a special derivative rule: the **chain rule**.

Chain Rule

$$\frac{d}{dx}(f \circ g)(x) = f'(g(x)) \cdot g'(x)$$

The chain rule can also be written as

$$\frac{df}{dx} = \frac{df}{dg} \cdot \frac{dg}{dx}.$$

That is, take the derivative of the the outer function with respect to the inner function and multiply by the derivative of the inner function with respect to x.

A derivation of the chain rule is given in Appendix B.

Example 19.4 (Simple Chain Rule Example)

Compute the derivative of $h(x) = \sin(2x - 1)$.

<u>Solution:</u> We can think of this as $h(x) = f(g(x))$, where the outer function is $f(u) = \sin(u)$ and the inner function is $u = g(x) = 2x - 1$:

$$f'(u) = \cos u \quad g'(x) = 2.$$

Putting this together,

$$h'(x) = \cos(2x - 1) \cdot 2 = 2\cos(2x - 1).$$

Example 19.5 (Photosynthetic Rate)

Find an expression for the rate of change of the photosynthetic rate

$$A(t) = 15.17 \sin\left(\frac{2\pi t}{24} - 1.77\right) + 4.00,$$

where t is measured in hours and A is measured in μmol CO_2 m^{-2} s^{-1} (see Example 19.3).

<u>Solution:</u> Let $u = g(t) = \frac{2\pi t}{24} - 1.77$ and $f(u) = 15.17 \sin u + 4.00$, then $A(t) = f(g(t))$. Note that $g'(t) = 2\pi/24 = \pi/12$ and that $f'(u) = 15.17 \cos u$. Then

$$A'(t) = f'(g(t)) \cdot g'(t) = 15.17 \cos\left(\frac{2\pi t}{24} - 1.77\right)\left(\frac{\pi}{12}\right).$$

Example 19.6 (Upwelling)

Upwelling is a process in large bodies of water such as oceans and lakes in which cold water rises from deeper regions toward the surface. The water that rises as a result of upwelling is typically nutrient rich, allowing surface waters to have high biological productivity. In [59], Rykaczewski and Checkley show data on the relationship between zooplankter size and upwelling velocity. They find that the relationship can be modeled by the equation

$$Z(w) = 0.46 \ln(w + 0.71) - 1.20,$$

(Continued)

where w is the upwelling rate measured in m/day and Z is the zooplankter size measured as biomass density in g/mm^2.

Find the rate of change in zooplankter size when upwelling velocity is 0.5 m/day.

Solution: To find the rate of change of zooplankter size with respect to upwelling velocity, we must find $Z'(w)$. Let $u = g(w) = w + 0.71$ and $f(u) = 0.46 \ln(u) - 1.20$; then $Z(w) = f(g(w))$. Note that $g'(w) = 1$ and that $f'(u) = 0.46/u$; thus,

$$Z'(w) = f'(g(w)) \cdot g'(w) = \frac{0.46}{w + 0.71} \cdot 1 = \frac{0.46}{w + 0.71}.$$

When the upwelling velocity is 0.5 m/day, the rate of change in zooplankter size is

$$Z'(0.5) = \frac{0.46}{0.5 + 0.71} = 0.38 \; \frac{\text{g/mm}^2}{\text{m/day}}.$$

So we can estimate that the size of zooplankter is increasing by 0.38 g/mm^2 for each unit increase in upwelling velocity.

Biological processes can involve more complicated chains of interaction than the above cases of a function of a function. The density of a predator species can depend on the density of an herbivore, which is its primary prey; the herbivore density may depend on available plant forage; and plant forage may depend upon time in a year. Thus, finding the rate of change with time of year of predator density requires viewing this as multiple compositions of functions. As this example illustrates, it is not difficult to apply the chain rule several times to find the derivative of a multiple series of compositions of functions.

If we have the composite function $S(x) = f(g(h(x)))$, then the derivative is

$$S'(x) = f'(g(h(x))) \cdot g'(h(x)) \cdot h'(x).$$

In fact, we can create functions that are compositions of as many functions as we would like, and the pattern continues.

Example 19.7 (Applying the Chain Rule Multiple Times)

In Example 19.3, we saw that we could write the function $A(t) = 15.17 \sin \left(\frac{2\pi t}{24} - 1.77 \right) + 4.00$ as a composition of three functions. Let $u = g(t) = \frac{2\pi t}{24} - 1.77$, $v = h(u) = \sin u$, and $f(v) = 15.17v + 4.00$. Then $(f \circ h \circ g)(t) = f(h(g(t))) = A(t)$. Now, $A'(t) = f'(h(g(t)))h'(g(t))g'(t)$. Note that $f'(v) = 15.17$, $h'(u) = \cos u$, and $g'(t) = \pi/12$. Thus,

$$A'(t) = 15.17 \cos \left(\frac{2\pi t}{24} - 1.77 \right) \left(\frac{\pi}{12} \right).$$

Note, this is the same answer we obtained in Example 19.5.

In Chapter 18, we showed that if $f(x) = \sin(Cx)$, then $f'(x) = C\cos(Cx)$. Now, suppose that we want to find the derivative of the cosine function. Recall that $\cos(x) = \sin\left(\frac{\pi}{2} - x\right)$; thus, we can use the chain rule to determine the derivative of the cosine function.

Let $u = g(x) = \frac{\pi}{2} - x$ and $f(u) = \sin(u)$; then $(f \circ g)(x) = \cos(x)$. By the chain rule,

$$\frac{d}{dx}(\cos x) = \cos\left(\frac{\pi}{2} - x\right)(-1) = -\sin x.$$

Example 19.8 (Cycling Hare Population)

In [71], Steury and Murray model the dynamics of a lynx (*Lynx canadensis*) population and its natural food source, snowshoe hares (*Lepus americanus*). They model the cyclical rise and fall of the hare population using the equation

$$h(t) = a + \frac{b - c}{2} \cos\left(\frac{2\pi t}{10}\right),$$

where t is measured in years, a is the average hare density, b is the maximum hare density, and c is the minimum hare density. Find an expression for the rate of change of the hare population at year 5.

Solution: Let $u = g(t) = \frac{2\pi t}{10}$ and $f(u) = a + \frac{b-c}{2}\cos(u)$; then $h(t) = f(g(t))$, and

$$h'(t) = -\frac{b - c}{2} \sin\left(\frac{2\pi t}{10}\right)\left(\frac{2\pi}{10}\right) = -\frac{(b - c)\pi}{10} \sin\left(\frac{2\pi t}{10}\right).$$

At year 5 ($t = 5$),

$$h'(5) = -\frac{(b - c)\pi}{10} \sin\left(\frac{2\pi(5)}{10}\right) = -\frac{(b - c)\pi}{10} \sin \pi = 0.$$

Thus, at year 5, the population is neither increasing nor decreasing.

19.3 Quotient and Reciprocal Rules

In ecology, a functional response is a function that relates the intake rate of a consumer (i.e., a predator, a forager, etc.) to the density of the food item (i.e., prey, plants, etc.). In 1959, the ecologist C. S. Holling proposed three types of functional response curves that are now referred to as Holling's Type I, Type II, and Type III functional responses. The Type I response assumes a linear increase in the intake rate as the density of the food item increases. The Type II response assumes that the consumer is limited by its ability to process/handle its food. This second functional response type is often modeled by the equation

$$C(x) = \frac{ax}{1 + abx},$$

where C is the intake rate, x is the food density, a is the rate at which the consumer encounters food items (per unit of food density), and b is the average amount of time a consumer must spend processing/handling a single unit of food item. The Type III functional response will be discussed in an exercise at the end of this chapter.

Suppose that we want to know the rate at which consumption (C) is changing with respect to a unit increase in food density (x). Note that $C(x)$ is a quotient with linear functions in both the numerator and the denominator. We do not yet have a rule for finding derivatives of quotient functions. However, we can develop such a rule by using the product rule and the chain rule.

First, suppose that we have a function $R(x) = 1/g(x)$; that is, $R(x)$ is the reciprocal of $g(x)$. We could express $R(x)$ as

$$R(x) = [g(x)]^{-1}$$

and think of R as a composite function where the inside function is $g(x)$ and the outside function is raising the inside function to the power of -1. Using the chain rule, we find that

$$R'(x) = (-1)[g(x)]^{-2} \cdot g'(x) = -\frac{g'(x)}{[g(x)]^2}.$$

This is known as the *reciprocal rule*.

Next, suppose that we have a function $Q(x) = f(x)/g(x)$; that is, $Q(x)$ is the quotient of $f(x)$ divided by $g(x)$. We could express $Q(x)$ as

$$Q(x) = f(x) \cdot [g(x)]^{-1}$$

and use the product rule and reciprocal rule to find the derivative of $Q(x)$:

$$Q'(x) = f'(x) \cdot [g(x)]^{-1} + f(x)\left(-\frac{g'(x)}{[g(x)]^2}\right) = \frac{f'(x)g(x) - f(x)g'(x)}{[g(x)]^2}.$$

This is known as the *quotient rule*.

Now, we can find the rate at which consumption (C) is changing with respect to a unit increase in food density (x) by using the quotient rule to find the derivative of $C(x)$. For the function $C(x)$, the function in the numerator is $f(x) = ax$, and the function in the denominator is $g(x) = 1 + abx$. Then

$$C'(x) = \frac{f'(x)g(x) - f(x)g'(x)}{[g(x)]^2}$$

$$= \frac{(a)(1 + abx) - (ax)(ab)}{[1 + abx]^2}$$

$$= \frac{a + a^2bx - a^2bx}{[1 + abx]^2}$$

$$= \frac{a}{[1 + abx]^2}.$$

This is the expression for the rate of change of consumption (C) with respect to food density (x).

We can combine various derivative rules in order to find derivatives of more complicated functions.

Example 19.9 (Multiple Derivative Rules)

Compute the derivative of $m(x) = \ln\left(\sin\left(\frac{2x-1}{3x^2+2x-1}\right)\right)$.

Solution: We can think of this problem as $f(x) = \frac{2x-1}{3x^2+2x-1}$, $g(x) = \sin(x)$, and $h(x) = \ln(x)$, where $m(x) = h(g(f(x)))$. Here we have to use the chain rule twice: once on $g(f(x))$ and then on $h(g(\cdot))$. We can also think of the chain rule as peeling away layers by taking the derivative of each layer. First, we will take the derivative of the outermost layer h with respect to g, then we will take the derivative of the next layer g with respect to f, and then we will take the derivative of the last layer f with respect to x. Thus,

$$h'(g) = \frac{1}{\sin\left(\frac{2x-1}{3x^2+2x-1}\right)}$$

$$g'(f) = \cos\left(\frac{2x-1}{3x^2+2x-1}\right)$$

$$f'(x) = \frac{6x(1-x)}{(3x^2+2x-1)^2}$$

$$m'(x) = \frac{1}{\sin\left(\frac{2x-1}{3x^2+2x-1}\right)} \cdot \cos\left(\frac{2x-1}{3x^2+2x-1}\right) \cdot \frac{6x(1-x)}{(3x^2+2x-1)^2}$$

$$= \cot\left(\frac{2x-1}{3x^2+2x-1}\right) \cdot \frac{6x(1-x)}{(3x^2+2x-1)^2}.$$

The derivatives of the remainder of the trigonometric functions can be derived using the quotient or reciprocal rule and the derivatives of sine and cosine.

The derivation of the derivative of the tangent and cotangent functions uses the facts that $\tan x = \frac{\sin x}{\cos x}$, $\cot x = \frac{\cos x}{\sin x}$, and $\sin^2 x + \cos^2 x = 1$:

$$\frac{d}{dx}(\tan x) = \frac{d}{dx}\left(\frac{\sin x}{\cos x}\right)$$

$$= \frac{\frac{d}{dx}(\sin x) \cdot \cos x - \frac{d}{dx}(\cos x) \cdot \sin x}{\cos^2 x}$$

$$= \frac{(\cos x)(\cos x) - (-\sin x)(\sin x)}{\cos^2 x}$$

$$= \frac{\cos^2 x + \sin^2 x}{\cos^2 x}$$

$$= \frac{1}{\cos^2 x} = \sec^2 x.$$

$$\frac{d}{dx}(\cot x) = \frac{d}{dx}\left(\frac{\cos x}{\sin x}\right)$$

$$= \frac{\frac{d}{dx}(\cos x)\cdot \sin x - \frac{d}{dx}(\sin x)\cdot \cos x}{\sin^2 x}$$

$$= \frac{(-\sin x)(\sin x) - (\cos x)(\cos x)}{\sin^2 x}$$

$$= \frac{-\sin^2 x - \cos^2 x}{\sin^2 x}$$

$$= \frac{-1}{\sin^2 x} = -\csc^2 x$$

The derivations of the derivatives of the secant and cosecant functions use the reciprocal rule and the facts that $\sec x = \frac{1}{\cos x}$, $\csc x = \frac{1}{\sin x}$, $\tan x = \frac{\sin x}{\cos x}$, and $\cot x = \frac{\cos x}{\sin x}$.

$$\frac{d}{dx}(\csc x) = \frac{d}{dx}\left(\frac{1}{\sin x}\right) = -\frac{\cos x}{\sin^2 x} = -\frac{\cos x}{\sin x}\cdot\frac{1}{\sin x} = -\cot x \csc x$$

$$\frac{d}{dx}(\sec x) = \frac{d}{dx}\left(\frac{1}{\cos x}\right) = -\frac{-\sin x}{\cos^2 x} = \frac{\sin x}{\cos x}\cdot\frac{1}{\cos x} = \tan x \sec x$$

19.4 Exponential Models

Exponential functions are used to describe biological and physical phenomenon such as population growth with unlimited resources, radioactive decay, and Newton's Law of Cooling. In each case, an exponential function containing the term $e^{\alpha t}$, where α is some constant and t represents time, is used to mathematically describe a physical or biological quantity that is changing over time. Note that if $a = e^{\alpha}$, then the term could also be written as a^t. Information about the rate of change of these quantities is often relevant and useful information.

In Section 18.3, we showed that if $f(x) = Ce^{kt}$, then $f'(x) = Cke^{kt}$, and that if $f(x) = Ca^x$, then $f'(x) = C(\ln a)a^x$. Note that the derivatives of exponential functions $y = Ce^{kx}$ have a special property:

$$\frac{dy}{dx} = ky;$$

that is, the derivative of the exponential function is directly proportional to the original exponential function.

To show this, onsider $y(x) = Ce^{kx}$. Then

$$y'(x) = Ce^{kx}\cdot k$$
$$= k(Ce^{kx})$$
$$= k\cdot y(x).$$

We call equations like $y' = ky$ *differential equations*. Note that *all* exponential functions of the form $y = Ce^{kx}$ have the derivative $y' = ky$. If we are given the differential equation $y' = ky$, we need additional information to determine the value of C. If we are given information at $x = 0$

(i.e., $y(0) = 5$), this is called an *initial condition*. If we are given information at some other value (i.e., $y(1) = 2$), this is called a *boundary condition*.

Example 19.10 (Derivative of an Exponential Function)

Find the function y that satisfies $y' = 2y$ with $y(0) = 4$.

Solution: Here $k = 2$. So we can write $y = ce^{2x}$. But what is c? Let us use the initial condition to find c:

$$y(0) = 4 = ce^{2 \cdot 0} = c \implies c = 4.$$

So, the equation satisfying the conditions is

$$y(x) = 4e^{2x}.$$

Exponential Growth Model

The exponential growth model describes *growth in an unlimited environment*, meaning that the resources for growth are unlimited.

- $N(t) =$ number or density of individuals in the population at time t
- $N'(t) =$ instantaneous growth rate of the population at time t

---

Exponential Growth

If the rate of growth of a population is directly proportional to the number or density of individuals present in the population, then we can write $N'(t) = kN(t)$, where k is a proportionality constant, which gives

$$N(t) = ce^{kt}.$$

If $N(0) = N_0$, then

$$N(t) = N_0 e^{kt}.$$

Thus, we can conclude that the growth of the population is exponential.

Since $N(t)$ is an exponential function, we know that $N'(t) = kN(t)$ for some constant. Notice what happens if we double the population size, that is, i.e. go from N to $2N$:

$$k[2N(t)] = 2(kN(t)) = 2N'(t).$$

So, if we double the population, we double the instantaneous growth rate because there are twice as many individuals to reproduce. Notice that in this model, there are no effects of crowding.

Space and food are not limited. Death rates do not change with the size of the population. Additionally, an underlying assumption of the model is that doubling the population doubles the number of individuals able to reproduce. We assume that the age structure of the population is not changed by doubling the population. That is, the new individuals added to the population are also able to reproduce (which might not always be the case in species that have an age of sexual maturity).

The value k is often called the **_intrinsic growth rate_** and has units of 1/time. In an exponential growth model, if $k = 0.574$, we would say that the population is growing continuously at a rate of 57.4% per unit time.

For a population that is growing exponentially, we frequently want to know the time it takes for the population to double in size. The **_doubling time_** (denoted T) is the time it takes for N_0 individuals to become $2N_0$ individuals. We can determine a general formula for this by solving $N(t) = 2N_0$ for t:

$$2N_0 = N_0 e^{kt} \;\Rightarrow\; 2 = e^{kt} \;\Rightarrow\; \ln 2 = kt \;\Rightarrow\; T = \frac{\ln 2}{k}.$$

Example 19.11 (Pheasant Population)

A population of 100 pheasants is introduced on an island. Lacking any natural predators, the population flourished to 560 individuals 3 years later. Assume that an exponential growth model is appropriate.

(a) Predict $N(5)$ if time is measured in years.
(b) Find the doubling time; that is, determine how long it takes for N_0 pheasants to increase to $2N_0$.

Solution:

(a) We must find the equation for $N(t)$. Since $N(0) = 100$, we have

$$N(t) = 100 e^{kt},$$

and since $N(3) = 560$, we have

$$N(3) = 560 = 100 e^{3k} \;\Rightarrow\; 5.6 = e^{3k} \;\Rightarrow\; k = \frac{1}{3}\ln 5.6$$

$$\Rightarrow\; N(t) = 100 e^{\frac{t}{3}\ln 5.6} = 100 e^{\ln[(5.6)^{t/3}]} = 100(5.6)^{\frac{t}{3}}$$

$$\Rightarrow\; N(5) = 100(5.6)^{\frac{5}{3}} \approx 1766 \text{ pheasants.}$$

(b) We want to find t for which $N(t) = 2N_0$. Using the formula we derived above,

$$T = \frac{\ln 2}{k} = \frac{\ln 2}{\frac{1}{3}\ln 5.6} = 3\frac{\ln 2}{\ln 5.6} \approx 1.2 \text{ years.}$$

Model for Radioactive Decay

The differential equation $y' = ky$ can also be used to model radioactive decay. Radioactive substances decay due to the emission of α-particles.

1. $A(t) =$ the number of grams of the radioactive substance at time that is decreasing
2. $A'(t) =$ the rate of decay (this is negative)

Radioactive Decay

If the rate of decay is directly proportional to the amount present, then we can write $A'(t) = -kA(t)$, where k is the positive proportionality constant. Notice that the negative sign indicates that the rate of change of the radioactive substance is negative; that is, the amount is decreasing. Similar to exponential growth, if $A(0) = A_0$, we can find that

$$A(t) = A_0 e^{-kt}.$$

Thus, we conclude that the decay of the radioactive substance is exponential.

Just as we looked at the doubling time of a population, we are often interested in the **half-life** (denoted $t_{1/2}$) of a decaying material, that is, the amount of time that it takes for the material to decay to half of its initial amount, that is, $A(t) = \frac{1}{2}A_0$. It can easily be shown that

$$t_{1/2} = \frac{\ln 2}{k}.$$

Here k is called the **decay rate**. Again, the units of k are 1/time.

Example 19.12 (Carbon Dating)

All living things contain small amounts of radioactive C^{14}. The ratio of C^{14} to the stable C^{12} in the atmosphere is constant. Once death occurs, C^{14} is no longer taken in, and the amount of C^{14} in the body begins to decay. The half-life of C^{14} is 5760 years, which gives $k = 0.000120$. Let A_0 be the amount of C^{14} in a living organism and A_1 the present amount in the object to be dated. To determine the age of the object, we solve $A_1 = A_0 e^{-kt}$ for t:

$$e^{-kt} = \frac{A_1}{A_0} \quad \Rightarrow \quad -kt = \ln\left(\frac{A_1}{A_0}\right) \quad \Rightarrow \quad t = \frac{1}{k}\ln\left(\frac{A_0}{A_1}\right) \approx 8333\ln\left(\frac{A_0}{A_1}\right).$$

Newton's Law of Cooling

If an object with temperature T_1 is placed in a room of constant temperature T_0, $T_1 > T_0$ initially, the object will begin to cool and will eventually reach the temperature of the room T_0. We take $T(t)$ to be the temperature of the object at time t and $T'(t)$ to be the rate at which the temperature of the object is changing.

Rate of Cooling

Newton's Law of Cooling states that the rate of cooling is directly proportional to the difference between the object temperature and the room temperature. Then we can write

$$T'(t) = -k(T(t) - T_0),$$

where $k > 0$ is the proportionality constant. To solve this differential equation, set $y = T - T_0$; then

$$y' = T' = -k(T - T_0) = -ky.$$

We know that the solution to $y' = -ky$ is $y = ce^{-kt}$, so

$$T(t) - T_0 = ce^{-kt} \quad \Rightarrow \quad T(t) = T_0 + ce^{-kt}.$$

Using the initial condition $T(0) = T_1 = T_0 + c$, we get $c = T_1 - T_0$. Thus, we can conclude that the temperature at time t is given by

$$T(t) = T_0 + (T_1 - T_0)e^{-kt},$$

which is an exponential function. Note that if $T_1 = T_0$, then $T(t) = T_0$ for all t. Thus, if the object is at the same temperature as the room at time $t = 0$, then it will remain at that temperature.

Let us consider an example from forensic anthropology using the rate of cooling.

Example 19.13 (Determining the Time of Death [16])

A dead body is discovered in the basement of a building. At the time the dead body is discovered, the room temperature is $70°$ and the body temperature $85°$. One hour later, the body temperature is $80°$. Determine the time of death. What assumptions do you have to make in order to apply the model?

Solution: In order to solve this problem, we must make two assumptions:

1. The body temperature at death was the standard $98.6°$.
2. The room temperature has been constant since the time of death.

(*Continued*)

Let $t = 0$ be the time of death. The time line is as follows:

Using $T_0 = 70$ and $T_1 - T_0 = 98.6 - 70 = 28.6$, we obtain the function for the temperature of the body over time:

$$T(t) = 70 + 28.6e^{-kt}.$$

What information do we have?

- $T(t_1) = 85$
- $T(t_1 + 1) = 80$

$$85 = 70 + 28.6e^{-kt_1} \quad \Rightarrow \quad \frac{15}{28.6} = e^{-kt_1}$$

$$80 = 70 + 28.6e^{-k(t_1+1)} \quad \Rightarrow \quad \frac{10}{28.6} = e^{-kt_1 - k} = e^{-kt_1}e^{-k} \quad \Rightarrow \quad \frac{10}{28.6}e^k = e^{-kt_1}$$

Set $e^{-kt_1} = e^{-kt_1}$ to get

$$\frac{15}{28.6} = \frac{10}{28.6}e^k \quad \Rightarrow \quad \frac{3}{2} = e^k \quad \Rightarrow \quad k = \ln\frac{3}{2}.$$

Substitute the value of k back into the first equation and solve for t:

$$\frac{15}{28.6} = e^{-t_1 \ln\frac{3}{2}} \quad \Rightarrow \quad \ln\frac{15}{28.6} = -t_1 \ln\frac{3}{2} \quad \Rightarrow \quad t_1 = \frac{\ln\frac{15}{28.6}}{\ln\frac{2}{3}} \approx 1.6 \text{ hours}$$

So the death occurred approximately 1.6 hours prior to the discovery of the body.

Example 19.14 (Wound Healing Example)

To assess how quickly a wound was healing, its area was estimated every 4 days. The results are shown in the table.

time t (days)	0	4	8	12	16	20	24	28
area A (cm^2)	107	88	75	62	51	42	34	27

(a) Use a linear regression to find an equation that best fits the data.
(b) Using the equation from part (a), determine the average rate of healing from day 2 to day 10.
(c) Using the equation from part (a), what was the instantaneous rate of healing on day 6?

(Continued)

(d) Using just the data, estimate the number of days it takes for the wound to heal to half of its original size.

(e) Using the equation from part (a), estimate the number of days it takes for the wound to heal to half of its original size.

Solution:

(a) Suppose that the area is decreasing exponentially; then we would expect a curve like $y = ce^{-kt}$. Notice that

$$A(t) = ce^{-kt}$$

$$\ln A(t) = \ln[ce^{-kt}]$$

$$= \ln c + \ln e^{-kt}$$

$$= \ln c - kt.$$

Let $Y = \ln A(t)$; then we have the equation for a line $Y = -kt + \ln c$, and we fit the line to the data.

t	0	4	8	12	16	20	24	28
Y	4.67	4.48	4.32	4.13	3.93	3.74	3.53	3.295

Using the linear regression methods from Chapter 3, we can obtain an equation for the line best fitting these data. Using Matlab to accomplish this, the equation for the linear regression is

$$Y = -0.048597t + ln(108.98).$$

When we transform this back to an equation for $A(t)$, we obtain the equation

$$A(t) = (108.98)e^{-0.048597t}.$$

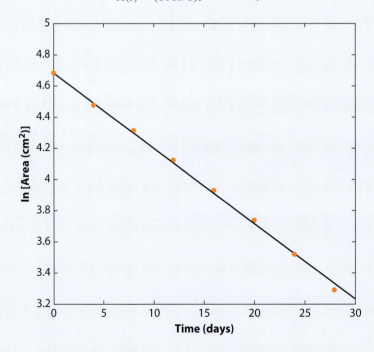

(*Continued*)

(b) The average rate of healing from day 2 to day 10 is

$$\frac{A(10) - A(2)}{10 - 2} = \frac{(108.98)e^{-0.048597(10)} - (108.98)e^{-0.048597(2)}}{10 - 2}$$

$$= \frac{67.034 - 98.886}{10 - 2}$$

$$= \frac{-31.852}{8} \approx -3.98 \frac{\text{cm}^2}{\text{day}}.$$

The negative sign indicates that the area is decreasing. Thus, the wound healed by an average area of 3.98 cm^2 per day.

(c) To find the instantaneous rate of change at day 6, we need the derivative, $A'(6)$:

$$A'(t) = (108.98)(-0.048597)e^{-0.048597t} = -5.2961e^{-0.048597t}$$

$$A'(6) = -5.2961e^{-0.048597(6)} = -3.96 \frac{\text{cm}^2}{\text{day}}.$$

Thus, at day 6, the area of the wound was decreasing by 3.96 cm^2 per day.

(d) The data indicate that the original size of the wound was 107 cm^2. Half that size would be 53.5 cm^2. From the data, we see that at day 16 the wound is at 51 cm^2. Thus, we might guess that the wound would decrease to half of its original area by around day 14 or 15.

(e) Using the equation we formed in part (a), the original size of the wound is $A(0) = 108.98$. We will solve $(108.98)e^{-0.048597t} = \frac{1}{2}A(0)$ for t to determine the time at which the wound is half of its original area:

$$(108.98)e^{-0.048597t} = 0.5(108.98)$$

$$e^{-0.048597t} = 0.5$$

$$-0.048597t = \ln 0.5$$

$$t = -\frac{\ln 0.5}{0.048597} \approx 14.263.$$

So we see that between days 14 and 15, the wound decreases to half of its original area, which is the same result that we estimated directly from the data.

Notice that the graph above is a semi log plot, which was discussed in Chapter 4.

19.5 Higher Derivatives

We have seen many examples in which biological processes have a rate of change that can be calculated as a derivative. A cell has an individual growth rate of its volume that can be described by the derivative of cell volume with units mm^3 per second or of its mass with units micrograms per second. The population size of cells in a culture has a growth rate describing how the number of cells changes, so the units of the derivative are cells per unit time. In both cases, we may be interested in whether the growth rate itself is changing. Is the cell growing more rapidly or less rapidly in volume as it ages? Is the cell culture growing less rapidly as the cells utilize more and more of the nutrients available in a petri dish? If we wish to understand how the growth rate is changing, we can determine this by finding the derivative of the growth rate. But the growth rate is itself a derivative, so now we are taking the derivative of a derivative: this is called the second derivative. It tells us how rapidly the rate of change of something is changing. If the second derivative is positive, it means that the derivative is increasing. If the second derivative is negative, it means that the derivative is decreasing. We expect in a cell culture that as the culture grows and utilizes more and more of the nutrients, the growth rate of the culture measured in cells per second being added in the culture will decrease. So if $C(t)$ gives the number of cells in the culture at time t, the derivative $C'(t)$ will be positive since the culture is still adding cells, but the derivative of $C'(t)$, which we denote $C''(t)$, will become negative, as there is more and more competition between the cells for the available nutrients.

You have a lot of direct experience with second derivatives if you have ever driven an automobile. When you accelerate by pushing on the gas pedal, you are causing the car to speed up. The derivative of the car's location is its speed, measured on the car's speedometer. There is typically no gauge in the car to measure the car's acceleration, which would be the derivative of its velocity, but you can observe this directly by watching the rate at which the speed is increasing (as you accelerate) or decreasing (as you brake).

If a function f is differentiable, then its derivative f' is a function. If we take the derivative of the function f', we say that we are taking the second derivative, which we denote by $f''(x)$ or $\frac{d^2f}{dx^2}$. In a similar fashion, if f'' is differentiable, then we can take the derivative of the function f'', which will result in the third derivative, denoted by $f'''(x)$ or $\frac{d^3f}{dx^3}$. In general,

$$\frac{d^nf}{dx^n}$$

denotes the nth derivative of $f(x)$. Note that $f'(x)$ is referred to as the first derivative. Higher derivatives provide further information about the nature of changes in a process. For the cell culture, the second derivative $C''(t)$ tells us how the growth rate is changing. The third derivative $C'''(t)$ tells us whether the rate at which the rate of growth of the culture is itself changing. So, if $C''(t)$ is positive and $C'''(t)$ is negative, it means that the growth rate of the cell culture is increasing but that the rate at which it is increasing is decreasing at time t. If $C''(t)$ is negative and $C'''(t)$ is negative, it means that the growth rate of the cell culture is decreasing and that the rate at which it is decreasing is decreasing at time t.

Example 19.15 (Decelerating Consumption)

Show that for a Holling Type II functional response given by

$$C(x) = \frac{ax}{1 + abx},$$

consumption decelerates as food density increases. That is, show that $C''(x) < 0$ for all $x > 0$.

(*Continued*)

Solution: Recall that we have already shown that

$$C'(x) = \frac{a}{[1+abx]^2}.$$

Using the quotient rule and the chain rule, we obtain

$$C''(x) = \frac{(0)\left[1+abx\right]^2 - (a)2\left[1+abx\right](ab)}{\left[1+abx\right]^4} = -\frac{2a^2b(1+abx)}{\left[1+abx\right]^4} = -\frac{2a^2b}{\left[1+abx\right]^3}.$$

Recall that a is the rate at which the consumer encounters food items (per unit of food density) and that b is the average amount of time a consumer must spend processing/handling a single unit of food item. By biological constraints, a and b must be positive, and thus $C''(x) < 0$ for all $x > 0$.

Example 19.16 (Higher Derivatives)

Let $f(x) = x \cos x$. Find $f'''(x)$.

Solution: We start by taking the first derivative using the product rule:

$$f'(x) = (1)\cos x + x(-\sin x) = \cos x - x \sin x.$$

To find the second derivative, we take the derivative of the first derivative using the sum rule and the product rule:

$$f''(x) = -\sin x - [(1)\sin x + x \cos x] = -\sin x - \sin x - x \cos x = -2\sin x - x \cos x.$$

To find the third derivative, we take the derivative of the second derivative using the sum rule and product rule:

$$\begin{aligned}
f'''(x) &= -2\cos x - [(1)\cos x + x(-\sin x)] \\
&= -2\cos x - \cos x + x \sin x \\
&= -3\cos x + x \sin x.
\end{aligned}$$

Example 19.17 (Patterns in Higher Derivatives of Cosine)

If $y = \cos t$, show that $\frac{d^4y}{dt^4} = y$.

Solution: We start by taking the first derivative:

$$\frac{dy}{dt} = -\sin t.$$

Next, we take the derivative of the first derivative to get the second derivative:

$$\frac{d^2y}{dt^2} = -\cos t.$$

Next, we take the derivative of the second derivative to get the third derivative:

$$\frac{d^3y}{dt^3} = -(-\sin t) = \sin t.$$

We take the derivative of the third derivative to get the fourth derivative,

$$\frac{d^4y}{dt^4} = \cos t.$$

Since the fourth derivative is equal to the original function y, we can write

$$\frac{d^4y}{dt^4} = y.$$

19.6 Exercises

19.1   Find the derivatives of the following functions.

(a) $f(x) = 3x^{10} + 2x^5 + x - \sin x + e^x - \ln x$

(b) $f(x) = e^x \sin x$

(c) $f(x) = (x^2 - 1)\cos x$

(d) $f(x) = \dfrac{x^2 - 2x + 5}{\tan(x)}$

(e) $f(x) = \dfrac{1}{1 - \sqrt{x} + x^2}$

(f) $f(x) = (x^2 + 4x + 6)^5$

(g) $f(x) = \sin(5x)$

(h) $f(x) = \cos(\tan x)$

(i) $f(x) = \dfrac{1 + \sin 2x}{1 - \sin 2x}$

(j) $f(x) = x \sin\left(\frac{1}{x}\right)$

(k) $f(x) = \tan^2(x^3)$

(l) $f(x) = \ln(\sin x)$

(m) $f(x) = \ln\left(\sin^2 x\right)$

(n) $f(x) = \ln\left(\sin^2(x^2 - 1)\right)$

(o) $f(x) = e^{x^2}$

19.2 Find $f'(1)$, the derivative at $x = 1$, of the following functions.

(a) $f(x) = 3x^{10} + 2x^5 + x + e^x$

(b) $f(x) = \dfrac{x^2 - 2x + 5}{1 + x}$

(c) $f(x) = \dfrac{1}{1 - \sqrt{x} + x^2}$

(d) $f(x) = (x^2 + 4x + 6)^5$

19.3 Find the equation of the tangent line at $x = 1$ for the following functions.

(a) $f(x) = 3x^{10} + 2x^5 + x$

(b) $f(x) = \dfrac{x^2 - 2x + 5}{1 + x}$

(c) $f(x) = \dfrac{1}{1 - \sqrt{x} + x^2}$

(d) $f(x) = (x^2 + 4x + 6)^5$

19.4 Find the equation of the tangent line at $x = 0$ for the following functions.

(a) $f(x) = \dfrac{1}{1 - \sqrt{x} + x^2}$

(b) $f(x) = (x^2 + 4x + 2)^5$

19.5 Use the power rule to find the first and second derivatives of the following functions.

(a) $f(x) = 3x^2$

(b) $g(x) = \dfrac{1}{x^2}$

(c) $h(x) = 3x$

(d) $s(t) = \sqrt{t}$

(e) $M(t) = \dfrac{2}{3} t^{3/2}$

19.6 Find $f''(0)$ for the following functions.

(a) $f(x) = \sin(3x)$
(b) $f(x) = (x^2 + 4x + 2)^5$

19.7 (From [34]) If t is the number of years since 1993, the population P of China, in billions, can be approximated by the function

$$P = f(t) = 1.15(1.014)^t.$$

Estimate $f(6)$ and $f'(6)$, giving units. What do these two numbers tell you about the population of China?

19.8 Using the derivative rules, show that if $y = \sin x + \cos x$, then $y'' = -y$.

19.9 Using the derivative rules, show that if $f(x) = e^x \sin x$, then $\frac{d^4 f}{dx^4} = -4f(x)$.

19.10 Using the derivative rules, show that if $m(t) = ke^{-rt}$, where r and k are positive constants, then $m''(t) = r^2 m(t)$.

19.11 Using the derivative rules, show that if $s(t) = A\cos(\omega t + \delta)$, where A, ω, and δ are constants, then $s''(t) = -\omega^2 s(t)$.

19.12 (From [34]) For some painkillers, the size of the dose, D, given depends on the weight of the patient, W. Thus, $D = f(W)$, where D is in milligrams and W is in pounds.

(a) Interpret the statements $f(140) = 120$ and $f'(140) = 3$ in terms of this painkiller.
(b) Use the information from part (a) to estimate $f(145)$. Explain how you estimated $f(145)$.

19.13 (From [34]) If you invest P dollars in a bank account at an annual interest rate of $r\%$, then after t years, you will have B dollars, where

$$B = P\left(1 + \frac{r}{100}\right)^t.$$

(a) Find $\frac{dB}{dt}$ assuming that P and r are constant. In terms of money, what does $\frac{dB}{dt}$ represent?
(b) Find $\frac{dB}{dr}$ assuming that P and t are constant. In terms of money, what does $\frac{dB}{dr}$ represent?

19.14 Carbon dating was used on the following objects. Estimate their age.

(a) An artifact with 79% to 81% loss in C^{14} content.
(b) An artifact with 98.2% to 98.5% loss in C^{14} content.

19.15 (From [34]) A yam is put into a hot oven, maintained at a constant temperature of 200°C. At time $t = 30$ minutes, the temperature T of the yam is 120° and is increasing at an instantaneous rate of 2°/min. Newton's Law of Cooling (or, in our case, warming) implies that the temperature at time t is given by

$$T(t) = 200 - ae^{-bt}.$$

Find a and b.

19.16 In Example 19.13, we assumed that the temperature of the victim was normal at the time of death. How would the time of death change if it were known that the victim had been running a fever of 101° at the time of death?

19.17 A Holling's Type III response is similar to Type II since it assumes that the consumer is limited by its ability to process/handle food. However, Type III additionally assumes that at low food densities, consumption accelerates as food density increases. A Holling's Type III functional response can be modeled by the equation

$$C(x) = \frac{ax^2}{1 + abx^2}.$$

(a) Show that $C(x)$ is increasing for all $x > 0$; that is, show that $C'(x) > 0$ for all $x > 0$.

(b) Show that $C(x)$ is accelerating for low food densities and decelerating for large food densities. To do this, find a value $x = \alpha$ such that $C''(\alpha) = 0$, $C''(x) > 0$ for $0 < x < \alpha$, and $C''(x) < 0$ for $x > \alpha$.

Using Derivatives to Find Maxima and Minima

In the previous chapter, we noted that many biological processes are periodic, with a regular changing pattern sometimes arising due to environmental factors that are diurnal and seasonal. In addition, many biological processes are maintained close to some particular value, with only small variations from this. An example is human body temperature, which stays quite constant except for small variations due to day-night cycles and illness, which can induce a larger change from the normal temperature. We gave many examples in the matrix section of this text of biological systems that move toward an equilibrium value over time and, if perturbed from the equilibrium, return to it. So we saw that landscapes can approach an equilibrium structure, given by an eigenvector, which describes the fraction of the area across the landscape in each vegetation class.

This chapter provides the basic mathematical tools that are needed to describe equilibrium properties of biological processes, and makes use of many properties of derivatives. The basic idea is that for a biological process to be at an equilibrium, it must be neither increasing nor decreasing. So a population that is at equilibrium has either no net increase or decrease in members of that population. Note that we used the term "net" here, as there could be new births in the population, but to achieve equilibrium, these births must be balanced by an equivalent number of deaths. We've already seen that a derivative of a function is positive when the function is increasing and that the function is decreasing when its derivative is negative. Logically, therefore, if a function is neither increasing nor decreasing, then its derivative will be 0, and wherever a function has a 0 derivative, it has an equilibrium. An equilibrium for which the derivative of a function is 0 can arise in several ways, however: (i) the function can be at a peak or maximum value, (ii) the function can be at a trough or minimum value, or (iii) the function can be at a point where its graph is flat. The following sections will give us a way to test which of these three cases hold when a derivative is 0. In general, (i) arises when a function changes from increasing, to decreasing, (ii) arises when a function changes from decreasing to increasing, and (iii) arises when the function remains increasing or decreasing but has a change in the sign of its second derivative.

20.1 Maxima and Minima

In order to find where functions obtain their maximum and minimum values, we must define what we mean by maximum and minimum. If we graph a function f on small open intervals of the domain of the function, we can find the values of x in those intervals for which $f(x)$ has a maximum or minimum. We say that the function $f(x)$ has a *local maximum* at $x = c_1$ if $f(c_1) \geq f(x)$ for x near c_1 (for the x in an open interval that contains the point c_1). Similarly, we say that the function $f(x)$ has a *local minimum* at $x = c_2$ if $f(c_2) \leq f(x)$ for x near c_2. If we graph the function f over its entire domain, we can find the global maximum and minimum.

We say that the function $f(x)$ has an *absolute maximum* or *global maximum* at $x = k_1$ if $f(k_1) \geq f(x)$ for all x in the domain of f. Likewise, we say that the function $f(x)$ has an *absolute minimum* or *global minimum* at $x = k_2$ if $f(k_2) \leq f(x)$ for all x in the domain of f. Note that *maxima* and *minima* are the plural versions of the words "maximum" and "minimum", respectively.

In addition to defining the maxima and minima of a function, we must also define what it means for a function to be increasing or decreasing. The function f is *increasing* on some interval (a, b) if for all x and y in (a, b) such that $x \leq y$, then $f(x) \leq f(y)$. We say that the function f is *decreasing* on the interval (a, b) if for all x and y in (a, b) such that $x \leq y$, then $f(x) \geq f(y)$.

20.2 First Derivative Test

We now make some further connections between functions and their derivatives. Consider the two graphs in Figure 20.1.

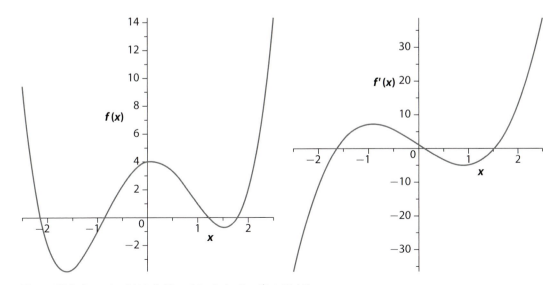

Figure 20.1 A graph of $f(x)$ (left) and its derivative $f'(x)$ (right).

What observations can we make about the relation of the graph of the function to the graph of the derivative?

Observation 1: If f is increasing on some interval $(a, b) \subset \mathbb{R}$, then $f' > 0$ on (a, b).

Observation 2: If f is decreasing on some interval $(a, b) \subset \mathbb{R}$, then $f' < 0$ on (a, b).

Observation 3: The point(s) at which f has a local maximum or minimum satisfy $f' = 0$. That is, if $f(c)$ is a local maximum or minimum, then $f'(c) = 0$. Note that when $f'(c) = 0$, the line tangent to the function f at $x = c$ is horizontal (i.e., the slope is 0).

Observation 4: When $f(c)$ is a maximum, $f' > 0$ to the left of c and $f' < 0$ to the right of c. When $f(c)$ is a minimum, $f' < 0$ to the left of c and $f' > 0$ to the right of c.

We use a ***sign chart*** like

to mean that $f'(x) > 0$ in an interval to the left of but not including $x = c$, that $f'(c) = 0$, and that $f'(x) < 0$ in an interval to right of but not including $x = c$.

This leads to the first derivative test.

First Derivative Test

Suppose that $f'(x)$ exists in the vicinity of $x = c$ and that $f'(c) = 0$. Then

1. <image: sign chart with + 0 − , f', point at c> implies that a local maximum occurs at $x = c$.

2. <image: sign chart with − 0 + , f', point at c> implies that a local minimum occurs at $x = c$.

3. <image: sign chart with + 0 + , f', point at c> or <image: sign chart with − 0 − , f', point at c> implies

that neither a local maximum nor a local minimum occurs at $x = c$.

We can use this information to help us sketch the graph of a function. Below is an outline of the steps used to sketch a function's graph, followed by a few examples. The idea here is in part to help us determine under what circumstances a function describing a biological process might have a stable equilibrium and how this equilibrium might be affected by changes in the parameters (constants) in the function. An example we will see later in this chapter is a function that describes how a population's growth rate depends on the population's size. From this, we will be able to determine whether there is a population size at which the growth rate is 0, so that the population has an equilibrium population size at that value. We can then determine how that equilibrium population size might shift as we change parameters in the function, such as a parameter that determines how much food or other resources are available to the population.

Curve Sketching Using the First Derivative Test

1. Determine the domain of the function.
2. Find all **critical numbers**. A critical number is a value of x such that $f'(x) = 0$ or $f'(x)$ does not exist. If x_0 is a critical number, then the pair $(x_0, y(x_0))$ is called a **critical point**. Notice that all local maxima and minima are critical points; however, not all critical points are a maximum or a minimum.
3. Evaluate the function $f(x)$ at each critical point. Plot these points. Also, take note of any points on the function that are undefined.
4. Construct a derivative sign chart.

Example:

$$\begin{array}{ccccc} & + & 0 & - & f' \\ \longleftarrow & - - - - - - - - & \bullet & - - - - - - - - - & \longrightarrow \\ & & c & & x \end{array}$$

5. Use the first derivative test to determine whether the critical points are local maxima, minima, or neither.
6. Look for any asymptotic behavior; that is, examine the limits of $f(x)$ as $x \to \pm\infty$.
7. Connect the points you have thus far plotted using the information from the first derivative test.

Now let us try sketching a few functions.

Example 20.1 (Sketching a Polynomial Function)

Classify the critical points of the function $f(x) = x^4 - 4x^2 + 3$ and sketch the graph.

Solution:

1. The domain is $(-\infty, \infty)$.

2. $f'(x) = 4x^3 - 8x = 0 \;\Rightarrow\; 4x(x^2 - 2) = 0 \;\Rightarrow\; x = -\sqrt{2}, 0, \sqrt{2}$. Thus, the possible locations for the local maxima/minima values are $x = -\sqrt{2}, 0, \sqrt{2}$.

3. $f(0) = 3$ and $f(-\sqrt{2}) = -1$ and $f(\sqrt{2}) = -1$.

4.
$$\begin{array}{ccccccccc} & - & 0 & + & 0 & - & 0 & + & f'(x) \\ \longleftarrow & - - - - - - & \bullet & - - - - - - - & \bullet & - - - - - - - & \bullet & - - - - - - & \longrightarrow \\ & & -\sqrt{2} & & 0 & & \sqrt{2} & & x \end{array}$$

5. Local minimum at $x = -\sqrt{2}$ and $x = \sqrt{2}$, and local maximum at $x = 0$.

6. $\displaystyle\lim_{x \to \infty} f(x) = \infty$ and $\displaystyle\lim_{x \to -\infty} f(x) = \infty$.

(Continued)

7. Plotting the graph, we get

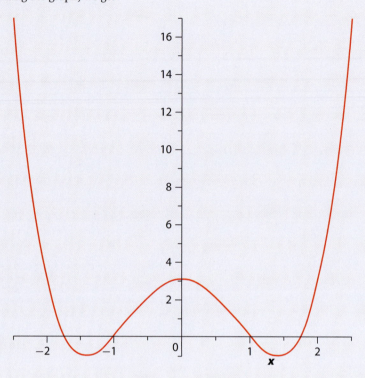

Example 20.2 (Sketching a Function with Asymptotes)

Classify the critical points of the function $f(x) = \frac{x}{2} + \frac{2}{x^2}$ and sketch the graph.

Solution:

1. The domain is $(-\infty, 0) \cup (0, \infty)$.

2. $f'(x) = \frac{1}{2} - 4x^{-3} = 0 \;\Rightarrow\; x^3 = 8 \;\Rightarrow\; x = 2.$

3. $f(2) = \frac{3}{2}$. Notice also that $f(0)$ is undefined.

4.
$$
\begin{array}{ccccccc}
+ & & dne & & - & & 0 & & + & & f'(x) \\
\end{array}
$$

<----------- • ----------- • ----------->

          0   2   x

5. Local minimum at $x = 2$ and a vertical asymptote at $x = 0$ since $\lim\limits_{x \to 0^+} f(x) = \infty$ and $\lim\limits_{x \to 0^-} f(x) = \infty$.

6. Notice that for large x^2, the term $2/x^2 \approx 0$. Thus, as $x \to \pm\infty$, the function can be approximated by $y = x/2$. This is called a ***slant asymptote***.

(*Continued*)

7. Plotting the graph, we obtain

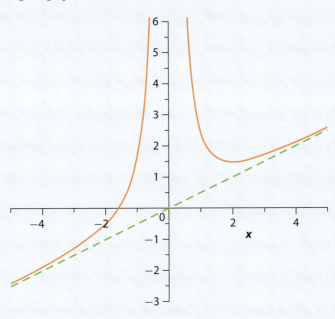

Example 20.3 (Sketching Another Function with Asymptotes)

Classify the critical points of the function $f(x) = \frac{x^2+1}{x^2-1}$ and sketch the graph.

Solution:

1. The domain is $(-\infty, -1) \cup (-1, 1) \cup (1, \infty)$.

2. $f'(x) = -\frac{4x}{(x^2-1)^2} = 0 \Rightarrow x = 0$.

3. $f(0) = 0$. Notice also that $f(\pm 1)$ is undefined.

4.
$$\begin{array}{ccccccccc}
 & + & & dne & & + & & 0 & & - & & dne & & - & & f'(x) \\
\leftarrow & \text{- - - - - - -} & \bullet & \text{- - - - - - - -} & \bullet & \text{- - - - - - -} & \bullet & \text{- - - - - - - -} & > \\
 & & -1 & & & 0 & & & 1 & & & x
\end{array}$$

5. Local maximum at $x = 0$ and vertical asymptotes at $x = -1$ (since $\lim\limits_{x\to -1^+} f(x) = -\infty$ and $\lim\limits_{x\to -1^-} f(x) = \infty$) and at $x = 1$ (since $\lim\limits_{x\to 1^+} f(x) = \infty$ and $\lim\limits_{x\to 1^-} f(x) = -\infty$).

6. $\lim\limits_{x\to\infty} f(x) = 1$ and $\lim\limits_{x\to -\infty} f(x) = 1$.

(*Continued*)

7. Plotting the graph, we get

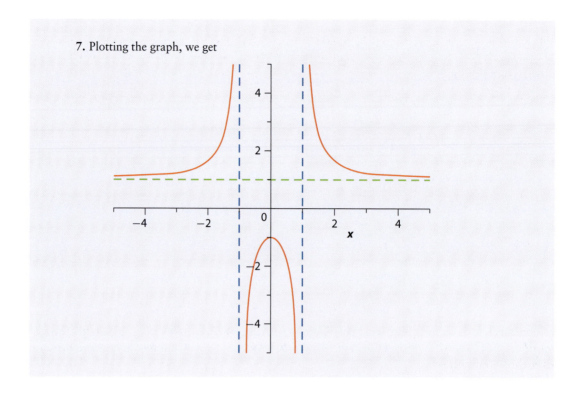

20.3 Mean Value Theorem

Most functions describing biological processes do not have "jumps"; that is, they are continuous, and they are also smooth enough that they do not have "corners" where there is a sudden change in the direction of the function (which you now know means that there is a sudden change in the derivative of the function). In these cases, the following theorem allows us to says something about how the instantaneous rate of change of a biological process over some time period relates to the overall, average rate of change of a process over the time period. Suppose that we measure a biological process at only two times, a and b with $a < b$, and the measured values are $f(a)$ and $f(b)$. We also suppose that the function describing the biological process is "nice enough" (it does not have sudden jumps or corners). Then this theorem guarantees that somewhere in the time period between a and b, the process had an instantaneous rate of change that is identical to the average rate of change of the process over the time period from a to b.

Mean Value Theorem

If $f(x)$ is continuous on $[a, b]$ and differentiable on (a, b), then there is at least one $c \in (a, b)$ such that

$$f'(c) = \frac{f(b) - f(a)}{b - a}.$$

The mean value theorem says that *there is a tangent line at a point on the graph between the points a and b that has the same slope as the line connecting the point (a, f(a)) and the point (b, f(b)).*

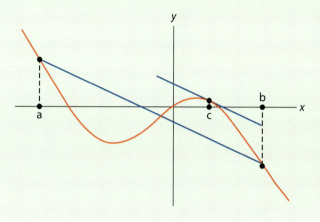

Suppose that we think about this theorem in terms of growth rates of a population. If we monitored the size of a population over 6 months, we could compute the average growth rate over those 6 months as

$$\frac{\Delta \text{ population size}}{\Delta \text{ time}} = \frac{\text{change in population size}}{6 \text{ months}}.$$

The mean value theorem says that at some point during those 6 months, the instantaneous rate of growth is equal to the average rate of growth.

Let us consider another example.

Example 20.4 (Mean Rate of Change of the Photosynthetic Rate of a Soybean Plant)

Figure 16.4 shows the photosynthetic rate (measured as the absorption of CO_2 in μmol of CO_2 per m^2 of leaf area per second) over the course of 1 day (24 hours) at day 164 in the life cycle of soybean plant [58]. The data can be approximated by the quadratic function

$$P(t) = -0.374t^2 + 9.44t - 41.3,$$

where P is the photosynthetic rate measured in μmol m^{-2} s^{-1} and t is measured in hours. The formula for the quadratic can be found using a non linear least-squares regression with a second-order polynomial.

(a) Find the average rate of change of the photosynthetic rate P over the interval $[8, 18]$.

(b) Use the mean value theorem to show that there exists a time $t = c$ between $t = 8$ and $t = 18$ such that the instantaneous rate of change of P at c is equal to the average rate of change over $[8, 18]$.

(c) Estimate the value of c.

(Continued)

Solution:

(a) $\dfrac{\Delta P}{\Delta t} = \dfrac{P(18) - P(8)}{18 - 8} = \dfrac{7.52 - 10.30}{10} = \dfrac{-2.78}{10} = -0.278 \; \mu\text{mol m}^2 \text{ s}^{-2}$

(b) Before applying the mean value theorem, let us make sure that the conditions of the theorem are satisfied. We have estimated the change in the photosynthetic rate of the soybean plant with a continuous curve over the interval $[0, 24]$ (1 day); thus, the function is continuous over the interval $[8, 18]$ (between 8 am and 6 pm). Additionally, since the change in the photosynthetic rate of the soybean plant is modeled using a quadratic curve, the function is differentiable on $(8, 18)$. Thus, by the mean value theorem there exists a value $c \in (8, 18)$ such that $P'(c) = -0.278$.

(c) Note that $P'(t) = -0.748t + 9.44$. To find the value of c, we can set $P(t)$ equal to -0.278 and solve for t:

$$-0.748t + 9.44 = -0.278 \quad \Rightarrow \quad t = 12.6 \text{ hours (i.e., 12:36 pm)}$$

The mean value theorem can also be used to find upper and lower bounds on quantities if we have data only on the rates of change of those quantities. In Example 20.4, we considered the data in Figure 16.4, which shows the photosynthetic rate (measured as the absorption of CO_2 in μmol of CO_2 per m^2 of leaf area per second) over the course of 1 day (24 hours) at day 164 in the life cycle of soybean plant. Note that these data give the *rate of change* of absorbed CO_2. From Figure 16.4, the photosynthetic rate of change is never greater than 20 μmol m^{-2} s^{-1}.

Since the photosynthetic rate has time units of s^{-1} but t is in hours, we convert the equation P to have its independent variable in terms of seconds. Let s be measured in seconds; then $t = \frac{s}{3600}$ (i.e., there are 3600 seconds in 1 hour). Thus,

$$P(t) = P\left(\frac{s}{3600}\right).$$

Now, let $A(s)$ represent the total amount of CO_2 absorbed by the soybean plant where s is measured in seconds. Then $A'(s)$ is the rate of change of the amount of CO_2 absorbed per m^2 per second, which is by definition the photosynthetic rate, and thus $A'(s) = P\left(\frac{s}{3600}\right)$. Since the photosynthetic rate is never greater than 20 μmol m^{-2} s^{-1}, $A'(s) = P\left(\frac{s}{3600}\right) < 20$ for $t \in [0, 24]$ (over the course of 1 day).

If we assume that the function $A(s)$ is continuous on $[0, 24 \times 3600]$ and differentiable on $(0, 24 \times 3600)$ and that at the start of the day ($s = 0$) the net CO_2 absorbed by the soybean plant is 0 (i.e., $A(0) = 0$), then we can find an upper bound on $A(\tau)$ for any value $\tau \in (0, 24 \times 3600)$.

By the mean value theorem, there exists a value $c \in (0, \tau)$ such that

$$A'(c) = \frac{A(\tau) - A(0)}{\tau - 0}.$$

Since $c \in (0, \tau)$ and $A'(s) < 20$ for all $t \in [0, 24 \times 3600]$, then $A'(c) < 20$. Thus,

$$A'(c) < 20$$

$$\frac{A(\tau) - A(0)}{\tau - 0} < 20$$

$$A(\tau) - A(0) < 20\tau$$

$$A(\tau) < 20\tau,$$

since $A(0) = 0$. Thus, even though we do not have an equation for $A(s)$, we are able to obtain an upper bound on the total amount of CO_2 that the soybean plant would be able to absorb by time τ (in seconds) on day 164 in its life cycle. A few examples follow:

- At 8 hours into the day, or $8 \times 3600 = 28{,}800$ seconds, the soybean plant could absorb no more than

$$A(28{,}800) < 20 \times 28{,}800 = 560{,}000 \, \mu\text{mol of } CO_2 \text{ per m}^2 \text{ of leaf.}$$

- At 16 hours into the day, or $16 \times 3600 = 57{,}600$ seconds, the soybean plant could absorb no more than

$$A(57{,}600) < 20 \times 57600 = 1{,}152{,}000 \, \mu\text{mol of } CO_2 \text{ per m}^2 \text{ of leaf.}$$

In reality, the total CO_2 assimilated by the plant will be considerably lower than these values since it is dark early and late in the day and the plant is respiring and releasing CO_2 during periods of darkness.

20.4 Concavity

We have used the first derivative to determine the location of local maxima and minima and find where a function is increasing or decreasing. Using this information, we were able to sketch a graph of the function. Now we will explore another aspect of the graph of a function: concavity. Consider the graphs of x^2 and $\ln x$ (see Figure 20.2).

The x^2 graph has a "cupped-up" shape; more precisely, the slope of the tangent line increases as x increases. Mathematically, we refer to this graph as *concave up*.

The $\ln x$ graph has a "cupped-down" shape; more precisely, the value of the slope of the tangent line decreases as x increases. Mathematically, we refer to this graph as *concave down*.

For these functions, it is simple to determine the concavity (up or down) because the entire graph has the same concavity. But what about a graph that switches concavity?

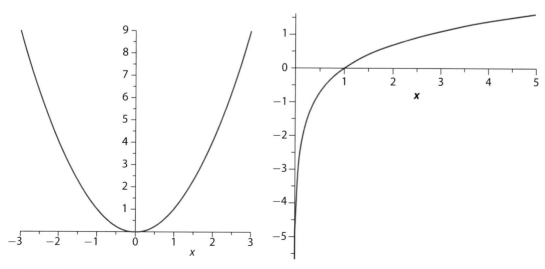

Figure 20.2 The graph of $y = x^2$ (left) is concave up, while the graph of $y = \ln x$ (right) is concave down.

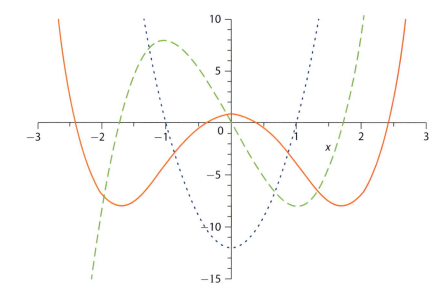

Figure 20.3 Graphs of the function $f(x) = x^4 - 6x^2 + 1$ (orange) and its first and second derivatives, $f'(x)$ (green) and $f''(x)$ (blue).

Consider the function graphed in Figure 20.3. Using the techniques we have learned so far, we can sketch a graph of this function. Using information about the first derivative, we can determine where the maxima and minima are located and also where the graph is increasing and decreasing.

Now we would like to know for what values of x the graph of $f(x)$ is concave up and concave down. We can see in Figure 20.2 that from $x = 0$ to about $x = 2$, the graph is decreasing. Over the interval $(0, 2)$, where does the graph switch from being concave down to concave up? It turns out we can use information about the second derivative to determine the exact point at which a graph switches concavity. Just as we used the roots of the first derivative to determine maxima and minima, we will use the roots of the second derivative to determine *points of inflection*.

We say that the graph of a function f is **concave up** on the open interval (a, b) if $f''(x) > 0$ on (a, b). The graph of f is **concave down** on (a, b) if $f''(x) < 0$ on (a, b). We say that an **inflection point** occurs at $x = c$ if $f''(c) = 0$ and the concavity of the graph changes as x passes through c. The points where $f''(x) = 0$ are the only possible points of inflection. It may be that $f''(c) = 0$ but that the concavity of f does not change as x passes through c.

Looking back at the graph of f in Figure 20.3 and its second derivative (the dotted blue line), can you now tell where the concavity of the graph switches?

Now we can use the first derivative and the second derivative of a function in sketching graphs.

Curve Sketching Using the First Derivative and the Second Derivative

1. Determine the domain of the function.
2. Solve the equation $f'(x) = 0$. This will yield a set of points, $x = c$, called *critical points*.
3. Take note of any values of x for which the function is undefined.

4. Construct a first derivative sign chart.

   Example:

5. Use the first derivative test to determine whether the critical points are local maxima, minima, or neither.
6. Evaluate the function $f(x)$ at each local maxima or minima.
7. Look for any asymptotic behavior; that is, examine the limits of $f(x)$ as $x \to \pm\infty$.
8. Solve the equation $f''(x) = 0$ for possible inflection points, $x = a$.
9. Construct a second derivative sign chart.

   Example:

10. The inflection points occur where the second derivative changes sign. Evaluate the function $f(x)$ at each inflection point.
11. Plot the local maxima and minima and the inflection points. Connect the points to form the graph using increasing and decreasing and concave-up and concave-down properties.

Note that in steps 4 and 9, you should also include any values found in step 3 for which the function is not defined as possible locations where the derivatives change sign in the sign chart.

Example 20.5 (Determining Concavity)

Using the function $f(x) = x^4 - 6x^2 + 1$, where are the local maxima and minima, and where are the points of inflection? Where is the function concave up? Concave down?

Solution: Note that the domain of the function is $(-\infty, \infty)$. To find local minima and maxima, we use the first derivative test:

$$f'(x) = 4x^3 - 12x = 0 \quad \Rightarrow \quad x = -\sqrt{3}, 0, \sqrt{3}$$

Thus, the local minima occur at $x = \pm\sqrt{3}$, and the local maximum occurs at $x = 0$.
   To find the possible points of inflection, we use the roots of the second derivative:

$$f''(x) = 12x^2 - 12 = 0 \quad \Rightarrow \quad x = -1, 1$$

Thus, the points of inflection occur at -1 and 1, and the graph is concave up on $(-\infty, -1) \cup (1, \infty)$ and concave down on $(-1, 1)$. See the graph of f in Figure 20.2.

What do points of inflection and concavity represent biologically?

Recall the equation for the logistic growth curve,

$$N(t) = \frac{KN_0 e^{rt}}{K + N_0(e^{rt} - 1)},$$

and assume that $K \neq N_0$ and $N_0 > 0$. If we take the derivative of $N(t)$, we obtain

$$N'(t) = \frac{rKN_0 e^{rt}\left[K + N_0(e^{rt} - 1)\right] - KN_0 e^{rt}\left[rN_0 e^{rt}\right]}{[K + N_0(e^{rt} - 1)]^2}$$

$$= r\left[\frac{K^2 N_0 e^{rt} + KN_0^2 e^{rt}(e^{rt} - 1) - KN_0^2 e^{2rt}}{[K + N_0(e^{rt} - 1)]^2}\right] \tag{20.1}$$

$$= rKN_0\left[\frac{Ke^{rt} + N_0 e^{2rt} - N_0 e^{rt} - N_0 e^{2rt}}{[K + N_0(e^{rt} - 1)]^2}\right]$$

$$= rKN_0\left[\frac{Ke^{rt} - N_0 e^{rt}}{[K + N_0(e^{rt} - 1)]^2}\right]$$

$$= rKN_0\left[\frac{e^{rt}(K - N_0)}{[K + N_0(e^{rt} - 1)]^2}\right]. \tag{20.2}$$

We can see from Equation 20.2 that there exists no time t for which $N'(t) = 0$ since for all real numbers t, $e^{rt} > 0$. This means that the function $N(t)$ has no critical points (no local minima and maxima). At $t = 0$, $N(0) = N_0$, and

$$\lim_{t \to \infty} N(t) = \lim_{t \to \infty} \frac{KN_0 e^{rt}}{K + N_0(e^{rt} - 1)} \cdot \frac{e^{-rt}}{e^{-rt}}$$

$$= \lim_{t \to \infty} \frac{KN_0}{Ke^{-rt} + N_0(1 - e^{-rt})}$$

$$= \frac{KN_0}{K(0) + N_0(1 - 0)} = \frac{KN_0}{N_0} = K.$$

If $0 < N_0 < K$, then $N'(t) > 0$ for all t, and thus the population is increasing over the interval $[0, \infty)$. Since $\lim_{t \to \infty} N(t) = K$, if $0 < N_0 < K$, then the population increases toward the horizontal asymptote $N = K$. Biologically, if the population starts below the carrying capacity, it will increase toward the carrying capacity. However, if $N_0 > K > 0$, then $N'(t) < 0$ for all t, and thus the population is decreasing over the interval $[0, \infty)$. Again, since $\lim_{t \to \infty} N(t) = K$, if $N_0 > K > 0$, the population decreases toward the horizontal asymptote $N = K$. Biologically, if the population starts above the carrying capacity, it will decrease toward the carrying capacity.

Continuing from Equation 20.1, we obtain

$$N'(t) = r\left[\frac{KN_0 e^{rt}}{K + N_0(e^{rt} - 1)}\right]\left[\frac{K + N_0(e^{rt} - 1) - N_0 e^{rt}}{K + N_0(e^{rt} - 1)}\right]$$

$$= r\left[\frac{KN_0 e^{rt}}{K + N_0(e^{rt} - 1)}\right]\left[\frac{K + N_0(e^{rt} - 1)}{K + N_0(e^{rt} - 1)} - \frac{N_0 e^{rt}}{K + N_0(e^{rt} - 1)}\right]$$

$$= r[N(t)]\left[1 - \frac{N(t)}{K}\right]. \tag{20.3}$$

Notice from Equation 20.3 that we can now see that $N'(t) = 0$ if either $N(t) = 0$ or $N(t) = K$. Notice that if $N_0 = 0$, then

$$N(t) = \frac{KN_0 e^{rt}}{K + N_0(e^{rt} - 1)} = \frac{K(0)e^{rt}}{K + (0)(e^{rt} - 1)} = \frac{0}{K} = 0.$$

Thus, if the population starts with no individuals ($N_0 = 0$), then it will continue to have no individuals through time. Now notice that if $N_0 = K$, then

$$N(t) = \frac{KN_0 e^{rt}}{K + N_0(e^{rt} - 1)} = \frac{K(K)e^{rt}}{K + K(e^{rt} - 1)} = \frac{K^2 e^{rt}}{K + Ke^{rt} - K} = \frac{K^2 e^{rt}}{Ke^{rt})} = K.$$

Thus, if the population starts with exactly K individuals ($N_0 = K$), then it will continue to have exactly K individuals through time. In each case, $N_0 = 0$ and $N_0 = K$, the population remains constant over time. We define a population whose size is not changing over time to be at **equilibrium**. When $N(t) = 0$, the population is extinct. When $N(t) = K$, the population is at carrying capacity. For $N(0) = 0$ and $N(0) = K$, $N'(t) = 0$ for all times t.

If we take the second derivative of $N(t)$ (using the chain rule), we obtain

$$N''(t) = rN'(t) - \frac{2rN(t)N'(t)}{K} = rN'(t)\left(1 - \frac{2N(t)}{K}\right).$$

Setting the second derivative equal to 0 implies that $N'(t) = 0$ or $N(t) = \frac{K}{2}$. For $K \neq N_0, N_0 > 0$, we have already observed that $N'(t) = 0$ does not provide any critical points. When $N(t) = \frac{K}{2}$, the population is at half of its carrying capacity. For what time does $N = \frac{K}{2}$? Setting $N(t) = \frac{K}{2}$ in the original logistic equation shows that $t^* = \frac{1}{r}\ln\left(\frac{K-N_0}{N_0}\right)$. This is the time at which there is a point of inflection. Biologically, this is when the rate of change (the instantaneous population growth rate) stops increasing and starts declining. So, for the time up to t^*, the population is accelerating, and for the time after t^*, the population is decelerating. See Figure 20.4 for graphs of the population size over time (black), the growth rate of the population over time (green),

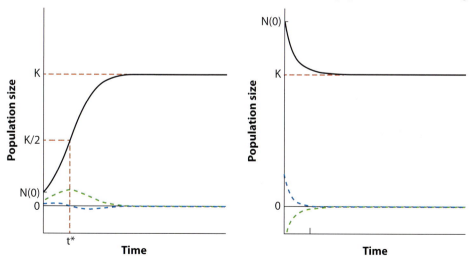

Figure 20.4 The figure on the left shows the population size $N(t)$ (black), the growth rate $N'(t)$ (green), and the population acceleration $N''(t)$ (blue) when $0 < N_0 < K$. The figure on the right shows the population size $N(t)$ (black), the growth rate $N'(t)$ (green), and the population acceleration $N''(t)$ (blue) when $0 < K < N_0$.

and the population acceleration over time (blue) for the cases when $0 < N(0) < K$ (left) and $N(0) > K$ (right).

Before we proceed with a few more examples, we condense the general procedure for graphing $y = f(x)$.

General Procedure for Graphing $y = f(x)$

1. Determine the domain of definition of $f(x)$. Vertical asymptotes may occur at points not in the domain.
2. Set $f'(x) = 0$ and determine the critical points.
3. Construct a sign chart for $f'(x)$. Determine the local maxima and minima.
4. Determine whether horizontal asymptotes exist by examining $\lim\limits_{x \to -\infty} f(x)$ and $\lim\limits_{x \to \infty} f(x)$.
5. Compute $f''(x)$ and construct a sign chart for $f''(x)$. Determine inflection points.
6. Plot local maxima, local minima, and inflection points and draw the graph, incorporating concavity features and vertical, horizontal, and/or slant asymptotes.

Now let us try an example.

Example 20.6 (Curve Sketching)

Sketch the function $f(x) = \dfrac{x}{x^4 + 2x^2 + 1}$.

Solution:

1. Notice that the function can be rewritten as

$$f(x) = \frac{x}{(x^2 + 1)^2}.$$

Thus, the denominator is always positive and can never equal 0. Therefore, the domain is $(-\infty, \infty)$, and there are no vertical asymptotes.

2. Taking the first derivative, we obtain

$$f'(x) = \frac{(1)(x^2 + 1)^2 - (x)[2(x^2 + 1)(2x)]}{(x^2 + 1)^4} = \frac{1 - 3x^2}{(x^2 + 1)^3},$$

and this gives critical points $x = -\dfrac{\sqrt{3}}{3}, \dfrac{\sqrt{3}}{3}$.

3.
$$
\begin{array}{ccccccc}
 & - & 0 & + & 0 & - & f'(x) \\
\longleftarrow\!-\!-\!-\!-\!-\!-\!- & \bullet & -\!-\!-\!-\!-\!-\!- & \bullet & -\!-\!-\!-\!-\!-\!-\!\longrightarrow \\
 & & -\frac{\sqrt{3}}{3} & & \frac{\sqrt{3}}{3} & & x
\end{array}
$$

Thus, $f\left(-\dfrac{\sqrt{3}}{3}\right) = -\dfrac{3\sqrt{3}}{16}$ is a minimum, and $f\left(\dfrac{\sqrt{3}}{3}\right) = \dfrac{3\sqrt{3}}{16}$ is a maximum.

(Continued)

4. $\lim\limits_{x \to -\infty} f(x) = 0$ and $\lim\limits_{x \to \infty} f(x) = 0$.

5. $f''(x) = \frac{12x(x^2-1)}{(x^4+2x^2+1)^2} = 0$ gives the possible points of inflection occuring at $x = -1, 0, 1$.

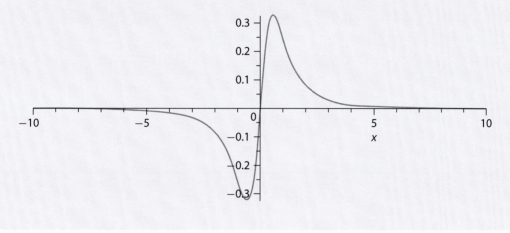

Checking the changes in concavity gives inflection points at $x = -1, 0, 1$.

6. Plotting the graph, we get

There is another method for classifying maxima and minima. If $f''(x)$ exists nearby a local maximum or minimum point, then that minimum will always be on a concave-up portion of the graph, and that maximum will always be on a concave-down portion of the graph. Thus, we get the second derivative test:

Second Derivative Test

Assume that $f''(x)$ is continuous in the vicinity of c and that c is a critical point of $f(x)$. Then the following hold:

1. If $f''(c) > 0$, then a local minimum occurs at $x = c$.
2. If $f''(c) < 0$, then a local maximum occurs at $x = c$.

So, to find the maximum of a given function, you could use this test.

1. Take the first derivative, set it equal to 0, and find the critical points.
2. Take the second derivative and substitute in the critical points to see which ones evaluate to a negative second derivative value. If $f''(c) = 0$, this test cannot be used.

Figure 20.5 Smiley face up illustrates a concave-up function with positive second derivatives on both sides of the critical point, and smiley face down illustrates a concave-down function with negative second derivatives on both sides of the critical point.

An easy way to remember the second derivative test is to use the "smiley" test, which is illustrated in Figure 20.5. As an exercise, you should think about how these "smileys" might change for other changes in concavity at a critical point than the two illustrated.

Example 20.7 (Finding Local Minima)

Find all the local minima of $f(x) = 1 + x^2 - \frac{x^6}{3}$.

Solution:
$$f'(x) = 2x - 2x^5 = 2x(1 - x^4) = 0 \;\Rightarrow x = -1, 0, 1.$$
Thus, the critical points are at $x = -1, 0, 1$.

$$f''(x) = 2 - 10x^4$$
$$f''(-1) = 2 - 10 = -8 \;\Rightarrow\; \text{maximum at } x = -1$$
$$f''(0) = 2 \;\Rightarrow\; \text{minimum at } x = 0$$
$$f''(1) = 2 - 10 = -8 \;\Rightarrow\; \text{maximum at } x = 1$$

The only local minimum is at $x = 0$, and the local minimum value of the function is $f(0) = 1$.

Example 20.8 (Finding Local Minima)

Find all the local minima of $f(x) = 3x^4 - 4x^3$.

Solution:
$$f'(x) = 12x^3 - 12x^2 = 12x^2(x - 1) = 0 \;\Rightarrow x = 0, 1.$$

Thus, the critical points are at $x = 0, 1$. But notice that $f''(x) = 36x^2 - 24x = 12x(3x - 2)$ and that

$$f''(0) = 0,$$

(Continued)

which means that the second derivative test will not give information about a local maximum or minimum occuring at $x = 0$.

When we draw a first derivative sign chart, we see that f' does not change sign at $x = 0$ because of the x^2 factor in f'.

Notice that $f''(1) = 12$, which means that a local minimum occurs at $x = 1$.

We could continue to examine this function, and we would find that there are inflection points at $x = 0$ and $x = 2/3$ by examining the factors in its second derivative.

The only local minimum is at $x = 1$, and the local minimum value of the function is $f(1) = -1$.

What is the largest value that a function $f(x)$ attains over some closed interval $[a, b]$? What is the smallest value? In Example 24.7, we saw that a local minimum occurred at $x = 0$, where $f(0) = 1$. However, this is by no means the smallest value that $f(x)$ attains over its whole domain. For example, $f(5) = -15,599$. So, how do we find the absolute maxima and minima over an interval?

Finding the Absolute Maximum and Minimum over a Closed Interval

1. For $[a, b]$, calculate $f(a)$ and $f(b)$. These are the function values at the end points.
2. Find all critical points of f; that is, set $f'(x) = 0$ and solve for x to get values $c_1, c_2, \ldots$.
3. For each critical point c in the open interval (a, b), evaluate $f(c)$.
4. You should now have $f(a), f(b), f(c_1), f(c_2), \ldots$. Select the largest value (this is your maximum over the interval) and the smallest value (this is your minimum over the interval).

Note: For this method to work, $f'(x)$ must exist at all $x \in (a, b)$.

Example 20.9 (Find the Minimum and Maximum of a Function)

Find the largest and smallest values of $f(x) = x - x^3$ for $0 \le x \le 1$.

<u>Solution:</u> Using the method to find the absolute maximum and minimum, we find the following.

1. $f(0) = 0$ and $f(1) = 1 - 1 = 0$.
2. $f'(x) = 1 - 3x^2 = 0 \Rightarrow x = \pm\frac{1}{\sqrt{3}}$.
3. Notice that $-\frac{1}{\sqrt{3}}$ is not in our closed interval, so we need not consider it.

   However, $\frac{1}{\sqrt{3}}$ is in our interval. $f(\frac{1}{\sqrt{3}}) \approx 0.385$.
4. The minimum value of the function over the interval $[0,1]$ is 0, and the maximum value of the function over the interval $[0,1]$ is 0.385.

Example 20.10 (Maxima and Minima of the Sine Function)

Find the local and absolute maxima and minima of $f(x) = \sin x$ over the closed interval $[0, 4\pi]$.

Solution: We start by taking the first derivative of $f(x) = \sin x$ to get

$$f'(x) = \cos x.$$

Notice that $\cos x = 0$ when $x = (2n-1)\frac{\pi}{2}$ for $n = \pm 1, \pm 2, \pm 3, \ldots$, that is, when $x = \pm\frac{\pi}{2}, \pm\frac{3\pi}{2}, \ldots$. Over the interval $[0, 4\pi]$, this corresponds to $x = \frac{\pi}{2}, \frac{3\pi}{2}, \frac{5\pi}{2}, \frac{7\pi}{2}$ (these are the critical points of f). Taking the second derivative, we find

$$f''(x) = -\sin x.$$

Since

$$-\sin\left(\frac{\pi}{2}\right) = -(1) = -1 < 0,$$

$$-\sin\left(\frac{3\pi}{2}\right) = -(-1) = 1 > 0,$$

$$-\sin\left(\frac{5\pi}{2}\right) = -(1) = -1 < 0, \text{ and}$$

$$-\sin\left(\frac{7\pi}{2}\right) = -(-1) = 1 > 0,$$

by the second derivative test, we see that the local maxima occur at $x = \frac{\pi}{2}$ and $x = \frac{5\pi}{2}$ and that the local minima occur at $x = \frac{3\pi}{2}$ and $x = \frac{7\pi}{2}$. Checking the end points of the interval, we find that $f(0) = \sin 0 = 0$ and $f(4\pi) = \sin(4\pi) = 0$. Thus, we find that this function attains its absolute maximum value of 1 at $x = \frac{\pi}{2}$ and $x = \frac{5\pi}{2}$ and its absolute minimum value of -1 at $x = \frac{3\pi}{2}$ and $x = \frac{7\pi}{2}$.

20.5 Optimization Problems

Calculus is a very useful tool in many biological applications since it deals with rates of change that arise in many areas of biology. Beyond this, any biological process that can be described by a function will depend on some parameters, and calculus allows us to address a wide array of questions about how the biological process depends on these parameters. As we will see in this section, a large group of applications involve determining when some aspect of a biological process is optimized, that is, maximized or minimized subject to some constraints. Examples include determining the best way to harvest a resource, such as a forest to maximize long-term profit; choosing an optimum drug dosage to maximize therapeutic benefit while minimizing potential side effects; and evaluating conditions under which we expect to see certain morphological features in organisms (e.g., body shape and size) based on environmental constraints on energy gain and heat loss. Sometimes finding a solution is as simple as finding the derivative of a function representing the quantity to be optimized and using one of the tests for maxima or minima described in this chapter. However, biological processes often involve many different functions that depend on many variables. The examples we develop here are particularly easy, but the same basic concepts apply when dealing with much more complicated biological processes. Although the techniques for dealing with these more complicated cases are beyond the scope of this text, after completing this course you will have the skills to go on to read about optimizing features in more complicated models.

Deciding what quantity is to be optimized can sometimes be difficult; the quantity to be optimized is called the "objective function," a type of underlying goal. These goals can be very practical, such as maximizing net profit from harvesting a forest, where you take account of the cost of harvesting the timber, the sales price of the logs, and constraints on how many logs can be harvested over the time period. The goals can also be of a more general nature, such as determining the optimal shapes and sizes of leaves across environmental gradients that vary in temperature and moisture conditions. In the latter case, the objective is to determine an objective function that might explain the tendency for plants to have small leaves under very dry, desert conditions as compared to larger leaves in tropical conditions. In this leaf size case, the objective function is chosen to indicate why smaller leaves are more adaptive (e.g., lead to higher overall survival and reproduction of the individual plant) in some environmental conditions than in others. This is an example of a common method in evolutionary biology in which an analysis is carried out to determine what circumstances might lead to certain organismal traits and characteristics arising because of natural selection; this is called the adaptationist approach.

Before we continue, let us look at some helpful suggestions for successfully solving optimization problems, which are frequently called "max/min" problems.

Suggestions for Solving Max/Min Problems

1. Gather the data for your problem.
2. Define key variables.
3. Draw a picture, giving the relationship between the variables, if possible.
4. Determine the *objective function* by answering the question, "What is the quantity Q to be maximized or minimized?"
5. Write Q in terms of the variables.
6. Look for *constraints*; that is, look for bounds on the variables or determine the relationships among the variables.
7. Write the objective function Q as a function of *one variable only*, using the relationships found in step 4.
8. Apply the appropriate calculus procedures to find the maximum or minimum of Q. If appropriate, remember to check whether the maximum or minimum occurs at the end points of the domain of the variables.
9. Check that your answer is biologically/physically reasonable. If it is not, then you may need to rethink your objective function or constraints.

Now let us consider several problems.

Example 20.11 (Grazing Land Example [16])

Consider a farmer who wishes to fence in a grazing area next to a river. The farmer has only 3000 feet of fencing available. Given that the farmer constructs a rectangular grazing area, what are the height and width that maximize the grazing area?

Solution: First, we draw a picture of the situation to get a better idea of what is going on.

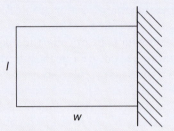

Let l = length of the fence parallel to the river and w = the length of the other two sides. Next, we write down the function that we are trying to maximize. Since we want to maximize the area of a rectangle, we use the formula for the area of a rectangle:

$$\text{the objective function} \quad A = lw.$$

Then we write down a function that describes the constraints we have. There is only 3000 feet of fence, and that fence must be used for the perimeter of three sides (river side not included):

$$\text{the constraint function} \quad 2w + l = 3000.$$

We need our objective function to be a function of only one variable. We can solve the second equation for l and substitute that into our equation for the area so that A depends only on w:

$$l = 3000 - 2w$$
$$A = (l)(w) = (3000 - 2w)w$$
$$= 3000w - 2w^2 \quad 0 \le w \le 1500.$$

Now, how do we find the maximum area? Take the first derivative, find the critical points, and find the critical points that give a maximum using either the first or the second derivative test:

$$A'(w) = 3000 - 4w = 0, \ 0 < w < 1500 \quad \Rightarrow \quad w = 750 \text{ feet.}$$

Using the second derivative test,

$$A''(w) = -4 \text{ for all } w \ \Rightarrow w = 750 \text{ is a local maximum.}$$

We need not check the end points $w = 0, 1500$ since both of these give areas of 0. Thus, $w = 750$ and $l = 1500$ give the maximum area of grazing land given the fencing constraint.

Example 20.12 (Crop Yield [49])

Suppose that

$$Y(N) = \frac{kN}{1 + N^2} \quad \text{for} \quad N \geq 0$$

models the yield of agricultural crop as a function of nitrogen levels measured in the appropriate units and that k is a constant that converts to crop yield units. Take $k = 1$. Find the amount of nitrogen that maximizes the crop yield.

Solution: The domain of the function is $[0, \infty)$ since the denominator of the function is always positive (no discontinuities) and since a negative nitrogen concentration does not make biological sense. We already have the objective function in terms of one variable. Differentiating, we get

$$Y'(N) = \frac{1 - N^2}{(1 + N^2)^2}, \quad N > 0.$$

$$Y'(N) = 0 \implies N = \pm 1$$

Notice that $N = -1$ is not in the domain. Using the first derivative test,

```
        0        +        0        –        f'(x)
        •─ ─ ─ ─ ─ ─ ─ ─ •─ ─ ─ ─ ─ ─ ─ ─>
        0                 1                 x
```

Since $N = 1$ is the only critical point in the interval $(0, \infty)$, looking at the sign chart gives a local maximum at $N = 1$. Note that $Y(0) = 0$, and by taking the limit

$$\lim_{N \to \infty} \frac{N}{1 + N^2} = 0,$$

we see that as the nitrogen levels get large, the crop yield approaches 0. This makes biological sense because crops need soil with a balance of nitrogen, phosphorus, and potassium. No concentration or a very large concentration of nitrogen is indicative that the balance between these three elements is lost. Thus, crop yield is maximized when $N = 1$.

Example 20.13 (Bacilli Surface Area)

Bacteria come in a variety of shapes, including spherical, cylindrical, and spiral. Bacteria with a cylindrical shape are often referred to as *rod-shaped* bacteria, or *bacilli*. Bacilli are not strictly cylindrical but can be approximated as a cylinder with rounded ends. We could model the shape of a bacillus using a cylinder with hemispheres on either end. Let the cylinder have height h and radius r; then the radius of each hemisphere is also r.

(Continued)

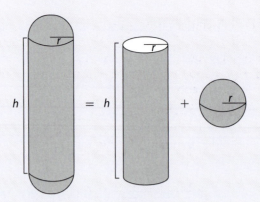

Given this geometric configuration, we can see that the volume of the cell is the volume of the cylinder plus the two hemispheres (i.e., one whole sphere),

$$V = \underbrace{\pi r^2 h}_{\text{volume of cylinder}} + \underbrace{\frac{4}{3}\pi r^3}_{\text{volume of sphere}} ,$$

and that the surface area of the cell is the surface area of the side of the cylinder plus the surface area of the two hemispheres,

$$S = \underbrace{2\pi r h}_{\text{S.A. from cylinder}} + \underbrace{4\pi r^2}_{\text{S.A. from hemispheres}} .$$

Suppose that the volume of the bacillus is fixed to be $V = \frac{4}{3}\pi \alpha^3$, where α depends on the specific species of bacilli. If there is loss through the cell surface (e.g., it is a semipermeable membrane) of some internal compound that is costly for the cell to produce, then we might be interested in considering a cell shape that minimizes the surface area. At what radius does the bacilli have a minimum surface area? What constraints on h and r must be considered when trying to find the radius at which the surface area of the bacilli is minimized? Finally, what happens to the values of h and S as $r \to 0^+$? What can we infer about the morphology (i.e., form and structure) of bacteria cells from these results?

Solution: To find the radius at which the surface area is minimized, we need to find the derivative of S with respect to r. However, the equation for S depends on r and h. To write an equation for S strictly in terms of r, we use the fact that the volume is constant and solve the volume equation for h:

$$V = \pi r^2 h + \frac{4}{3}\pi r^3 = \frac{4}{3}\pi \alpha^3$$

$$\pi r^2 h = \frac{4}{3}\pi \alpha^3 - \frac{4}{3}\pi r^3$$

$$h = \frac{4\pi(\alpha^3 - r^3)}{3\pi r^2}$$

$$h = \frac{4(\alpha^3 - r^3)}{3r^2}. \tag{20.4}$$

(*Continued*)

Biologically, the value of h should never be negative. If $h = 0$, the cell is simply a sphere, and for any $h > 0$, the cell is a bacillus. If $h \geq 0$, then, by Equation (20.4), $r \leq \alpha$. Thus, realistic biological constraints require that $0 < r \leq \alpha$.

Next, we substitute the expression for h into the equation for the surface area of the cell to derive a formula for S as a function of r:

$$S = 4\pi r^2 + 2\pi r h$$

$$= 4\pi r^2 + 2\pi r \left(\frac{4(\alpha^3 - r^3)}{3r^2} \right)$$

$$= \frac{12\pi r^3 + 8\pi (\alpha^3 - r^3)}{3r}$$

$$= \frac{12\pi r^3 + 8\pi \alpha^3 - 8\pi r^3}{3r}$$

$$= \frac{4\pi r^3 + 8\pi \alpha^3}{3r}.$$

Now that the equation for surface area is in terms of r, we can determine any local maxima or minima by taking the derivative of S with respect to r using the quotient rule:

$$\frac{dS}{dr} = \frac{12\pi r^2 (3r) - \left(4\pi r^3 + 8\pi \alpha^3\right)(3)}{9r^2}$$

$$= \frac{36\pi r^3 - 12\pi r^3 - 24\pi \alpha^3}{9r^2}$$

$$= \frac{24\pi r^3 - 24\pi \alpha^3}{9r^2}$$

$$= \frac{8\pi \left(r^3 - \alpha^3\right)}{3r^2}.$$

The only critical number is $r = \alpha$, which occurs at the right end point of the domain. Note that S is decreasing over the entire domain $(0, \alpha]$, and thus the minimum surface area occurs at the end point $r = \alpha$. Note that when $r = \alpha$, then $h = 0$, and thus the minimum surface area that is obtained with the cell is a sphere, which is actually not a bacilli bacteria but rather a cocci bacteria.

Finally, we consider the limits of h and S as $r \to 0^+$:

$$\lim_{r \to 0^+} h = \lim_{r \to 0^+} \frac{4(\alpha^3 - r^3)}{3r^2} = \infty$$

$$\lim_{r \to 0^+} S = \lim_{r \to 0^+} \frac{4\pi r^3 + 8\pi \alpha^3}{3r} = \infty$$

since in each case the numerator approaches a positive constant, and the denominator approaches zero from the right.

(Continued)

A wide variety of constraints on sizes of bacteria are not taken into account in our simple model. A cell may gain or lose compounds through its surface so that a very high surface area may be adaptive if the objective is to gain compounds through the surface, while a very small surface area may be adaptive if the objective is to reduce the loss of compounds. Whether a particular size of bacteria is adaptive depends on multiple factors. For example, the transport of material within a cell through the process of diffusion will be very slow if the cell has a very elongated rod shape. One of the remarkable aspects of life is the fact that cell sizes have remained quite small and larger organisms have arisen by building up large numbers of small cells. It has been argued that the development of multicellular organisms rather than organisms made up of a "giant" cell occurs because of the inherent limitations of diffusion of material within cells. This is illustrated by the average size of bacilli, though this varies somewhat among different bacteria species, which have radius r from 0.25 to 0.5 μm with a length h in the range of 1.0 to 4.0 μm. Understanding why these sizes are the ones observed for bacilli would require modifying our simple model to incorporate other factors. See the wonderful book *Life's Devices* by Steven Vogel [75] for discussions about the limitations of diffusion and many other aspects of the physical basis for constraints on organism form.

Example 20.14 (Optimal Clutch Size)

Smith and Fretwell [65] first proposed the following model in 1974 to consider the trade-off between clutch size and clutch fitness. Let N be the size of a bird's clutch (i.e., the number of eggs laid) and R be the total amount of resources allocated to producing that clutch (a positive constant). Smith and Fretwell assumed that the amount of resources allocated to each offspring x was equally divided to each individual in the clutch, and thus $x = \frac{R}{N}$. Next, they assumed that the probability of survival of the clutch (referred to as clutch fitness) depended on the number of offspring in the clutch (N) and the fitness of each offspring. Here we assume that the fitness of each offspring is a function that depends on the amount of resources allocated to that offspring; thus, we define $f(x)$ as the probability of survival of an individual offspring (or offspring fitness). Since f is a probability the range of the function f must be between 0 and 1, with $f(0) = 0$ (when no resources are allocated, the probability of survival of an offspring is 0) and $\lim_{x \to \infty} f(x) = 1$ (as larger amounts of resources are devoted to the clutch, the probability of survival of an individual offspring approaches 1). Clutch fitness is then given by the equation

$$w(x) = [\# \text{ offspring}] \times [\text{survival probability of offspring}]$$

$$= Nf(x) = \frac{R}{x}f(x), \text{ for all } x > 0.$$

(Continued)

We wish to find the optimal clutch size, that is, the clutch size N that maximizes fitness of the clutch $w(x)$. We will do this by finding the value of x that maximizes the $w(x)$.

Solution:

$$\frac{dw}{dx} = R\frac{d}{dx}\left(\frac{f(x)}{x}\right) = R\left(\frac{f'(x)x - f(x)}{x^2}\right) = \frac{R}{x}\left(f'(x) - \frac{f(x)}{x}\right), \text{ for all } x > 0.$$

Setting $w'(x) = 0$ yields

$$f'(x) = \frac{f(x)}{x}.$$

So, our critical point $x = \hat{x}$ satisfies

$$\hat{x} = \frac{f(\hat{x})}{f'(\hat{x})}.$$

Taking the second derivative so that we can apply the second derivative test to our critical point,

$$\frac{d^2w}{dx^2} = -\frac{R}{x^2}\left(f'(x) - \frac{f(x)}{x}\right) + \frac{R}{x}\left(f''(x) - \frac{d}{dx}\frac{f(x)}{x}\right) \quad \Rightarrow \quad \frac{d^2w}{dx^2}\bigg|_{x=\hat{x}} = \frac{R}{\hat{x}}f''(\hat{x}).$$

Thus, $w(x)$ has a local maximum at $x = \hat{x}$ if $f(x)$ is concave down at $\hat{x}$ or $f''(\hat{x}) < 0$.

Suppose that we determine from data that the chance of survival for an individual offspring, given that x is the amount of resources allocated to each of the offspring, is determined by the function $f(x) = \frac{x^2}{1+x^2}$. Taking the first derivative, we obtain

$$f'(x) = \frac{2x}{(1+x^2)^2} \quad \Rightarrow \quad \hat{x} = \frac{\hat{x}^2}{1+\hat{x}^2} \cdot \frac{(1+\hat{x}^2)^2}{2\hat{x}}$$

$$\Rightarrow \quad 2 = 1 + \hat{x}^2 \quad \Rightarrow \quad x = -1, 0, 1.$$

Since -1 and 0 are not in the range we are considering, we look at $x = 1$:

$$f''(x) = \frac{2 - 6x^2}{(1+x^2)^3} \quad \Rightarrow \quad f''(1) = -\frac{1}{2} < 0.$$

What is the optimal clutch size?

$$N = \frac{R}{x} = \frac{R}{1} = R,$$

with $w(1) = \frac{R}{2}$. Note that $N = R$ is the optimal clutch size since $\lim_{x\to 0^+} w(x) = 0$ and $\lim_{x\to\infty} w(x) = 0$.

These results could be drastically different for a different survival function $f(x)$.

20.6 Matlab Skills

There are functions for which we cannot algebraically solve for the local maximum or minimum values. For these functions, we turn to Matlab for assistance.

Consider the function

$$f(x) = \frac{\ln(x)}{x+1}.$$

If we take the derivative of this function, we have

$$f'(x) = \frac{1 + \frac{1}{x} - \ln(x)}{(x+1)^2}.$$

The critical points occur where the numerator of $f'(x)$ equals 0 and at any points in the domain of $f(x)$ where the function $f'(x)$ is undefined. Notice that the domain of $f(x)$ is $x > 0$, so there are no points in the domain of $f(x)$ where $f'(x)$ is undefined. Thus, to find any critical points, we set $f'(x) = 0$, which gives us the equation

$$1 + \frac{1}{x} - \ln(x) = 0. \tag{20.5}$$

How can we solve this equation for x?

It turns out that it cannot be done algebraically. However, we can explore the graph of the left-hand side of Equation 20.5 and find where that graph is equal to 0.

Let us plot the function $g(x) = 1 + \frac{1}{x} - \ln(x)$ over a large portion of its domain and see where the function crosses the x-axis. To do this, we will use the `plot` command in Matlab. Since we do not know where the function might cross the x-axis, we will start by plotting function from 0.1 to 100, plotting points at every 0.1 increment:

Command Window

```
>> x = 0.1:0.1:100;
>> g = 1 + (1./x) - log(x);
>> plot(x,g)
>> hold on
>> plot(x,0)
>> hold off
```

Notice that when we construct the vector of function values f we have to use the ./ operator to divide by the vector x. Additionally, notice that after we plotted just the vectors x and y, we added to the plot the line $y = 0$ so that we could easily see where the function g was passing through the x-axis. The resulting plot is shown in Figure 20.6(a).

Where does $g(x) = 1 + \frac{1}{x} - \ln(x)$ appear to cross the x-axis? Since $g(x)$ crosses the x-axis somewhere between 0 and 10, let us zoom in on that region and see if we can get a better estimate:

Command Window

```
>> x = 0.1:0.01:10;
>> g = 1 + (1./x) - log(x);
>> plot(x,g)
>> hold on
>> plot(x,0)
>> hold off
```

The plot produced is shown in Figure 20.6(b).

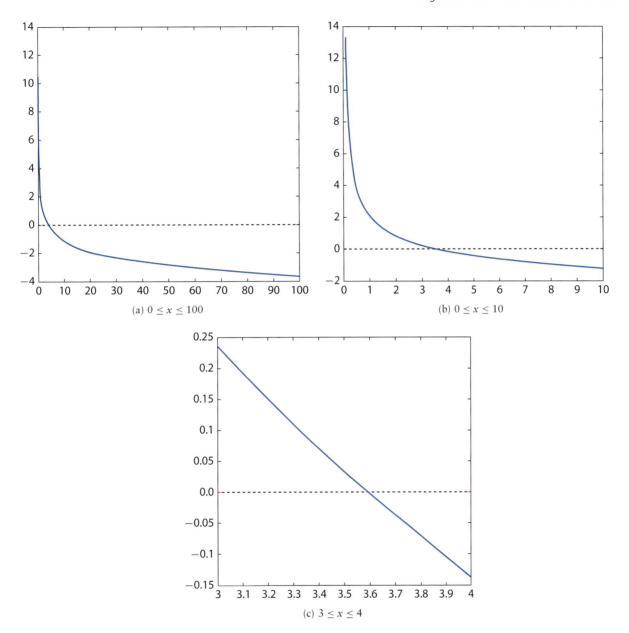

(a) $0 \leq x \leq 100$

(b) $0 \leq x \leq 10$

(c) $3 \leq x \leq 4$

Figure 20.6 Plot of $g(x) = 1 + \frac{1}{x} - \ln(x)$.

Where does $g(x) = 1 + \frac{1}{x} - \ln(x)$ appear to cross the x-axis now? We can see that $g(x)$ crosses the x-axis somewhere between 3 and 4; let us zoom in one more time so we can get a better estimate:

```
───────────────── Command Window ─────────────────
>> x = 3:0.001:4;
>> g = 1 + (1./x) - log(x);
>> plot(x,g)
>> hold on
>> plot(x,0)
>> hold off
```

The plot produced is shown in Figure 20.6(c).

With this last plot, we can estimate that the function $g(x)$ crosses the x-axis at approximately 3.6. We could, of course, continue to zoom in to increase the accuracy of our estimate. However, we will stop here.

20.7 Exercises

20.1   Find all the critical points for each function f. Classify the critical points using the first derivative test and draw the sign chart for f'. Determine where the function is concave up and concave down, locate all inflection points, and draw a sign chart for f''. Sketch the graph of f.

(a) $f(x) = 4x - x^2$

(b) $f(x) = x^3 - 6x - 4$

(c) $f(x) = 3x^4 - 4x^3 + 1$

(d) $f(x) = (x^2 - 4)^2$

(e) $f(x) = (x - 1)(x - 2)(x - 3)$

(f) $f(x) = \dfrac{(x + 2)^2}{x}$

(g) $f(x) = x + \dfrac{1}{x}$

(h) $f(x) = \dfrac{x}{x^2 - 1}$

(i) $f(x) = \dfrac{9}{x} + x + 1$

(j) $f(x) = x^3 + x$

20.2   Given the sign charts for f', determine the general shape of the graph of f. Assume that a vertical asymptote occurs at any points where f and f' do not exist (dne).

(a)

$$\begin{array}{ccccccccc} & + & & 0 & & + & & 0 & & - & & 0 & & - & & f'(x)\\ \leftarrow & - - - - - - - & \bullet & - - - - - - - & \bullet & - - - - - - - & \bullet & - - - - - & \rightarrow \\ & & & -4 & & & & 0 & & & & 3 & & & & x \end{array}$$

(b)

$$\begin{array}{ccccccccccc} & - & & 0 & & + & & 0 & & + & & 0 & & - & & 0 & & + & & f'(x)\\ \leftarrow & - - - - - & \bullet & - - - - & \bullet & - - - - - - & \bullet & - - - - - - & \bullet & - - - - - - & \rightarrow \\ & & & -3 & & & & -1 & & & & 0 & & & & 3 & & & & x \end{array}$$

(c)

$$\begin{array}{ccccccc} & - & & 0 & & - & & dne & & + & & f'(x)\\ \leftarrow & - - - - - - & \bullet & - - - - - & \bullet & - - - - - & \rightarrow \\ & & & -1 & & & & 1 & & & & x \end{array}$$

(d)

$$\begin{array}{ccccccc} & + & & dne & & + & & dne & & - & & f'(x)\\ \leftarrow & - - - - - - & \bullet & - - - - - - & \bullet & - - - - - - & \rightarrow \\ & & & -1 & & & & 1 & & & & x \end{array}$$

20.3   For each of the following functions f, classify the critical points of f using the second derivative test. If the test fails $[f''(c) = 0]$, then use the first derivative test.

(a) $f(x) = 3x^2 - 8x + 4$

(b) $f(x) = x + \dfrac{9}{x}$

(c) $f(x) = x^4 + 8x^3 + 18x^2 + 5$

(d) $f(x) = \dfrac{x}{x^2 + 1}$

(e) $f(x) = 3x - \ln(6x)$

(f) $f(x) = x^3 e^{-x}$

20.4   Find the absolute maximum and minimum of these functions on $[-1, 4]$.

(a) $f(x) = x^2 - 1$

(b) $f(x) = x^3 + x$

20.5   In [56], Reis et al. used slow-motion videos to collect time-series data on the vertical position of a cat's tongue as it drank water in an effort to better understand the kinematics (i.e., motion) of a cat's tongue while it laps up water. Reis et al. collected time-series data of the vertical position of the tongue through 11 laps of the tongue. The graph below shows the average vertical position of the cat's tongue over one lap of the tongue (i.e., the average position of the tongue at each time over the 11 laps of the tongue).

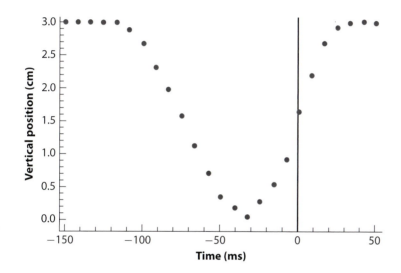

(a) Over what interval is the vertical position increasing?

(b) Over what interval is the vertical position decreasing?

(c) Sketch a graph of the velocity of the cat's tongue over time.

(d) Over what intervals is the vertical position curve concave downward?

(e) Over what intervals is the vertical position curve concave upward?

(f) Sketch a graph of the acceleration of the cat's tongue over time.

20.6   The population of a flock of geese is modeled by

$$G(t) = 2000 + 150 \sin\left(\frac{\pi}{6}t\right) + 75 \sin\left(\frac{\pi}{3}t\right),$$

where t is measured in months from January 1.

(a) Find any critical points and any inflection points of G. State the location of any local maxima, minima, and inflection points. Use this information to sketch a graph of $G(t)$ over a 2-year period.

(b) Interpret any local maxima, minima, and inflection points over the first 2 years in biological terms.

(c) How fast is the flock growing on May 1? On May 1, is the *rate* at which the flock is growing increasing or decreasing?

20.7   A hedgehog population on a small island off the coast of Ireland is approximated by

$$P(t) = \frac{1647}{1 + e^{4.5 - 0.3t}},$$

where t is measured in years since 1900.

(a) What happens to the population as $t \to \infty$? Interpret this result in biological terms.

(b) Estimate when the hedgehog population was growing most rapidly; that is, find when $P'(t)$ is maximized. How large was the population at this time?

(c) Use your answers from (a) and (b) along with information about the first and second derivative to sketch P. What natural causes may explain the shape of the graph of P?

20.8   Suppose that a drug is administered to a patient by injection. Let the proportion of the administered dose of the drug that is in the bloodstream t hours later be given by

$$D(t) = \frac{6t}{2t^2 + 1}.$$

On what time intervals is the concentration of drug increasing, and on what intervals it is decreasing? What happens to the proportion of the administered dose as $t \to \infty$? Does this make biological sense? Explain.

20.9   During an epidemic, the number of infected people $I(t)$ is approximated by

$$I(t) = \frac{a \ln(bt + 1)}{bt + 1},$$

where t is measured in days since the initial outbreak.

(a) Find an expression for the time when the number of infected people starts to decline.

(b) Suppose that you introduce a treatment for infected individuals. When treatment is added to the model, the value of b is increased. What effect does this have on the time at which the number of infected people starts to decline? Given your answer, describe (in biological terms) the population level effect of your treatment.

(c) Consider your answers from (a) and (b). Is there a biological/physical limitation to how large the value of b can be? (*Hint*: Try calculating the time at which the number of infected people starts to decline for a *very large* value of b.)

20.10   In [46], Mezzadra et al. estimate the average individual daily milk consumption for Charolais, Angus, and Hereford calves to be the function

$$M(t) = 6.281 t^{0.242} e^{-0.025t}, \quad 1 \le t \le 26,$$

where $M(t)$ is the milk consumption (in kg) and t is the age of the calf (in weeks). Find the age of a calf at which maximum daily consumption occurs. How much milk is consumed on this day? Would you expect this value to be exactly the same for all calves? Explain.

20.11 In [37], Kulesa et al. present a mathematical model that estimates the number of teeth $N(t)$ on day t in the lower half jaw of *Alligator mississippiensis*,

$$N(t) = 71.8e^{-8.96e^{-0.0685t}},$$

where t is measured in days of incubation prior to hatching.
(a) Show that the number of teeth is always increasing.
(b) Does the number of teeth approach a constant value as the number of incubation days increases? If so, what is this value?
(c) The average incubation period for a *Alligator mississippiensis* is 65 days. Is it biologically reasonable to consider what occurs when $t \to \infty$? Why or why not?
(d) Find the inflection point of $N(t)$. Give a biological interpretation of the inflection point.

20.12 In [67], Speer et al. propose the following model to approximate the growth of breast cancer tumors:

$$N(t) = e^{\frac{A_0}{\alpha}(1-e^{-\alpha t})},$$

where $N(t)$ is the number of cancer cells t days after the start of tumor growth, A_0 is the initial specific growth rate, and α is the proportional rate of decay of A_0. For a particular set of data, it was estimated that $A_0 = 0.3$ and $\alpha = 0.011$.

(a) Evaluate $\lim_{t \to \infty} N(t)$. Interpret your answer biologically.
(b) Use the first derivative to show that the number of cancer cells is always increasing but that as $t \to \infty$, the rate of change of the number of cancer cells goes to 0.
(c) Find the inflection point of $N(t)$. What does the inflection point indicate in biological terms?
(d) Use Matlab to generate a graph of $N(t)$, $N'(t)$, and $N''(t)$ for $t \in [0, 200]$.

20.13 In [58], Rogers et al. collected data on the photosynthetic rate of soybean leaves over the course of 1 day. On day 215 in the soybean plant life cycle, the photosynthetic rate of the upper leaves of the plant can be approximated with the equation

$$P(t) = (-1.83 \times 10^8)t^2 + 0.00139t - 6.58,$$

where t is measured in seconds and $P(t)$ has units $\mu\text{mol m}^{-2}\text{ s}^{-1}$.

(a) Find the maximum photosynthetic rate (in $\mu\text{mol m}^{-2}\text{ s}^{-1}$) over the course of 1 day.
(b) Assume that the net amount of CO_2 absorbed by the soybean plant at the beginning of the day $t = 0$ is $0 \ \mu\text{mol m}^{-2}$. Given the maximum photosynthetic rate (found in (a)), find an upper bound on the total amount of CO_2 absorbed by the plant at the following times:

(i) $t = 18,000$ (5 hours)
(ii) $t = 36,000$ (10 hours)
(iii) $t = 82,800$ (23 hours)

20.14 Find the rectangle with the largest area that can be placed inside the region bounded between the curve $y = 4 - x^2$ and the horizontal axis $y = 0$.

20.15 (From [16]) For infants less than 9 months old, the relationship between the rate of growth R (in pounds/month) and the present weight W (in pounds) is approximated by

$$R = cW(21 - W)$$

for some constant c. At what weight is the growth rate the largest?

20.16 (From [16]) Orville is trying to decide when to sell the turkey farm that his great uncle left him. All of that gobbling is driving him nuts. Right now, he has 1000 turkeys ready for market that he can sell at \$5 per turkey. If he waits t years, the number of marketable turkeys will grow to

$$N(t) = 1000 + 300t,$$

but the actual buying power derived from the sale of a single turkey diminishes according to

$$B(t) = 5 - 1.2t.$$

When should Orville sell the turkeys?

20.17 (From [16]) In a simple model of territory, a single animal defends a circular territory of radius x (miles). The following assumptions are made:

(a) The energy L spent per day in looking for food and defending the territory is directly proportional to the area of the region.
(b) The energy G gained per day is directly proportional to the radius x.

Suppose that $L = 3000$ calories and $G = 3500$ calories when $x = 1$. Find the territorial size that will result in maximum benefit to the animal, that is, maximize the net energy $G - L$.

20.18 (From [29]) Imagine that you have just been given a vacation home situated on an island 10 miles off the coast. The nearest town is 60 miles up the coast. Every month, you must travel into town to restock your supplies. Naturally, you have an amphibious auto. It travels 30 mph in water and 55 mph along the coastal highway. Find the route that is the quickest (shortest time) to the nearest town.

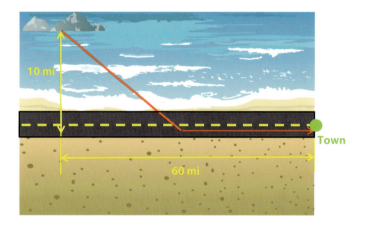

20.19 (From [29]) Epidemiologists have found a new communicable disease running rampant in Knoxville, Tennessee. They estimate that t days after the disease is first observed in the community, the percentage of the population infected by the disease is approximated by

$$P(t) = \frac{20t^3 - t^4}{1000},$$

for $0 \leq t \leq 20$.

(a) After how many days is the percentage of the population infected at a maximum?

(b) What is the maximum percentage of the population infected?

20.20   (From [34]) A smokestack deposits soot on the ground with a concentration inversely proportional to the square of the distance from the stack. With two stacks 20 miles apart, the concentration of the combined deposits on the line joining them, at a distance x miles from one stack, is given by

$$S = \frac{k_1}{x^2} + \frac{k_2}{(20 - x)^2},$$

where k_1 and k_2 are positive constants that depend on the quantity of smoke that each stack is emitting. If $k_1 = 7k_2$, find the point on the line joining the stacks where the concentration of the deposit is a minimum. (You might need to solve the equation for critical points graphically.)

20.21   (From [29, 40]) The distance that a shot-putter can throw a shot depends on the height, angle, initial velocity, and release angle of the shot. For some shot-putters, the distance that a shot will travel can be determined by

$$d(\theta) = 5.1 \sin(2\theta)\left[1 + \sqrt{1 + \frac{0.41}{\sin^2 \theta}}\right],$$

where θ is the release angle (in degrees) and $d(\theta)$ is the distance the shot travels (in meters). Determine the release angle that maximizes the distance traveled by the shot and the corresponding maximum distance.

Unit 5 Student Projects

Optimal Foraging Theory

This project requires you to use the ideas of Chapter 20 regarding maximization of a function. Here we assume that evolution has acted to generate highly efficient foragers. By *highly efficient*, we mean that foraging animals that are able to more rapidly obtain a high food intake rate (food eaten per unit time) will be more likely to survive and reproduce. Thus, if the characteristics that lead to high food intake efficiency (which may depend on speed, visual or hearing skills, size, etc.) are heritable, then the characteristics that lead to higher efficiency will become more prevalent (increase in frequency) in the population. This area of science is called optimal foraging theory. For a readable description of this theory, see David Stephens and John Krebs' *Foraging Theory* [69].

One of the main areas of foraging theory deals with animals that move from "patch" to "patch" across a landscape, depleting the food in each patch the longer they stay there. Patches could be individual plants with nectar in the plant's flowers that is eaten by bees or individual plants with seeds that are eaten by birds. One question in foraging theory involves deducing the rules for when an individual will leave one patch to search for a new patch. Numerous complications arise in this, but we will consider only a very simple situation in which we have the following assumptions:

1. There are no interactions with other foragers, so depletion of food in a patch is due only to the single forager under consideration.
2. There is no randomness in the environment, so each patch is identical in terms of how much resource it has before foraging starts.
3. Once the patch is depleted, it stays depleted.
4. Many patches are available.
5. The travel time of the forager between patches is constant.
6. The food available in the patch decreases exponentially as the forager spends time there.

Let t be the time spent in a patch before leaving, M be the time it takes to move between two patches, and K be the amount of food in a patch before any food has been eaten. Then the total amount of food eaten in a patch is

$$K(1 - e^{-ct}), \tag{20.6}$$

where c is the decay rate of available food in a patch once foraging starts. In Equation 20.6, e^{-ct} represents the proportion of food left in the patch after the forager has eaten from the patch over time t. Thus, $1 - e^{-ct}$ represents the proportion of food eaten from the patch over time t. When we multiply this term by K (the total amount of food in the patch before any is eaten), we obtain the amount of food eaten from the patch after the foraging is complete.

We define one *foraging bout* as the time spent moving to the patch plus the time spent foraging in that patch. Thus, the total time of one foraging bout is

$$M + t.$$

The food intake rate is the amount of food consumed divided by the time spent consuming the food, that is,

$$\frac{K\left(1 - e^{-ct}\right)}{t + M}. \tag{20.7}$$

Following the example from Section 20.6, complete the following.

1. Think of Equation 20.7 as a function of the time spent in a patch before leaving, t. Take the first derivative of this function with respect to t. Notice that when the derivative is set equal to 0, you cannot explicitly solve for t. Write an equation, in terms of t, c, and M that can be graphed and used to find the critical points.
2. Use Matlab to estimate the critical point of the derivative of Equation 20.7, using $c = 4$ and $M = 10$. Based on the graph you produced, find the value of the critical point correct to two decimal places. Explain why you know that a local maximum occurs at this point.
3. The critical point found in part 2 is the *optimal time* that a forager should spend foraging in one particular patch. Explain why it makes sense biologically that this point is where there is a local maximum.
4. Repeat part 2 but use $M = 1, 5, 15$, and 20. Does the graph in each case indicate that the critical point is still a maximum? How do the optimal foraging times in these cases compare to the optimal foraging time you found in part 2?
5. Repeat part 2 but use $c = 0, 2, 4, 6, 8$, and 10. Does the graph in each case indicate that the critical point is still a maximum? How do the optimal foraging times in these cases compare to the optimal foraging time you found in part 2?
6. In Matlab, create the following m-file:

```
────────────────── OptimalForaging.m ──────────────────
1    M = [1 5 10 15 20];
2    c = [0 2 4 6 8 10];
3    z =
4
5    surface(c,M,z);
6    xlabel('c');
7    ylabel('M');
```

For the matrix z in this m-file, construct a matrix where each column corresponds to an M value, (1, 5, 10, 15, 20); each row corresponds to a c value, (0, 2, 4, 6, 8, 10); and the value of each entry in the matrix is the corresponding optimal foraging time. For example, the optimal foraging time found in part 2 would be the entry in row 3, column 3. You already have several of the values from parts 2, 4, and 5. You will need to use Matlab to determine the rest of the values. The resulting matrix z should be a 6×5 matrix.

Once you have constructed matrix z, run the m-file `OptimalForaging.m`. A plot will be generated. Use the button to rotate the graph to look at the surface plot from different angles. Describe what this three-dimensional surface plot is showing you. What are the advantages and disadvantages of viewing a three-dimensional plot such as this? ■

Integration

As we have seen in numerous examples, derivatives are useful in describing the rates of change of various biological processes. We have considered many ways to calculate derivatives of functions using different "rules." These rules make it simple to determine how rapidly a biological process (photosythesis, species diversity, population size, etc.) changes as some factor that affects it varies (light level for photosynthesis, area for species diversity, time for population size, etc.). Information about derivatives could then be used to determine whether a measurement (e.g., population size) was increasing, decreasing, or stable and to determine the shape of the graph of the function (concave up or down).

This unit deals with the reverse of finding derivatives. That is, given the rate of change of some biological process, how do measurements of that process change through time? So, if we are given an equation that expresses the derivative of the population size, how can we find the population size at any particular time? This procedure of going from the derivative of some measurement to the measurement itself is called "antidifferentiation" or "integration." The idea is that if we are given a formula for $\frac{dN}{dt} = f(t)$, where $N(t)$ is the population size at time t and $f(t)$ is a function of time that gives the population growth rate at time t, then the process of antidifferentiation can be used to find $N(t)$.

Several examples were developed earlier in which a differential equation was used to find the function that satisfied the equation, which we call the solution of the differential equation. In the case of exponential growth, we saw that if $N(t)$ is population size and $\frac{dN}{dt} = kN$ for some positive constant k, then the solution is exponential growth, that is, $N(t) = N(0)e^{kt}$. Similarly, we saw that Newton's Law of Cooling, which expresses the derivative of temperature, T of a body as proportional to the difference between the ambient temperature and the body's current temperature, has a solution that gives $T(t)$. In both of these cases, we did not show how to find the solution $N(t)$ or $T(t)$ but instead demonstrated that the formula given was a solution. In this unit, we will learn some rules to help us find solutions to differential equations such as these.

Estimating the Area under a Curve

One way to describe changes in biological processes across time and space is by specifying the rates of change of these processes. These can be instantaneous rates (giving derivatives such as $\frac{dN}{dt}$ for rate of change of the population density or $\frac{dC}{dt}$ for rate of change of drug concentration in the blood). Alternatively, for processes that change in a discrete manner or that we can follow only over discrete time periods, these are described by difference equations (e.g., giving $C_{n+1} = kC_n + D$, where C_n is the blood concentration of a drug just after the nth dose D of the drug is given and k is the fraction of drug remaining in the system between doses). Although the biological processes may be described by their rates, we often want to know the actual values of the variables describing the process, not simply their rates. We wish to know the actual population densities and drug concentrations in order to determine whether the population level is too low to be maintained or the drug concentration level is too low to be effective or too high so that it is toxic. How we proceed from information about the rates of change of specific variables to estimate the total change of those variables is the subject of this chapter.

Perhaps the simplest way to use information about rates of change to find the total change of a biological variable is to sum up the changes of the process that occur over different time periods. If you have the average, not instantaneous, rate of change of some process over a time period, this is particularly easy. If you know that the average elimination rate of a drug is 50 mg per hour (which occurs because the drug is filtered out by the kidneys or is being metabolized) for a person of a given body composition (e.g., weight, height, gender, etc.), then over 4 hours the person would have eliminated 200 mg of the drug. Clearly, this procedure cannot be used over long time periods since the amount of drug eliminated cannot be more than the amount taken. The average elimination rate of a drug declines as you consider longer time periods. So we need a method to account for how the rate of change varies through time. For a drug with a loading dose of 500 mg given at time 0, we might have an average elimination rate of 100 mg per hour over the first 2 hours after administration, 50 mg per hour over hours 2 through 3, 25 mg per hour over hours 4 through 5, and so on. Then the amount of drug remaining at the start of the sixth hour after administration is 150 mg since 350 mg have been eliminated.

If the rate of change is constant on a time interval, then to calculate the total change over that time interval, we use

$$\text{total change} = (\text{rate of change}) \times (\text{length of time interval}).$$

When the rate of change is not constant over the whole time period involved, the above approach in which we break down the rates of change of the biological process over smaller time periods will provide a better approximation for the total change over the whole time period. We multiply each individual rate of change by the length of its corresponding time subinterval and then sum up those products:

$$\text{total change} = \sum_{i=1}^{n} (r_i)(\text{length}_i),$$

where r_i is the rate of change on the ith subinterval (with length denoted by length_i).

This summation idea is the main concept for what we call "integration" in calculus. As the time periods become smaller and smaller, the corresponding rates of change get closer and closer to the instantaneous rate of change given by the derivative ($\frac{dC}{dt}$ for drug concentration).

This unit develops the mathematics needed to determine the total change in a biological process across time when the rate of change of the process is specified. We illustrate this geometrically by first considering areas under curves. For problems involving the calculation of the total change, this can be found by estimating the area under a curve that represents the rate of change function. Estimating the area under a rate of change function from time a to b gives the total change of the quantity over $[a, b]$. For example, the total CO_2 assimilated by a plant leaf over a time period can be obtained by summing up the photosynthetic rate over parts of the time period, accounting for light variations and their impact on CO_2 uptake over the time period. This allows us to estimate much more accurately the total CO_2 assimilated than the very coarse estimates for an upper bound on this that we obtained in Chapter 20 using the maximum photosythetic rate.

21.1   The Area under a Curve

Suppose that for a function, $y = f(x)$ we want to find the area bounded by the graph of this function and the horizontal axis over some interval $[a, b]$. How might we estimate the area?

Let us start with the simple case of $f(x) = 5$. How would you estimate the area bounded by the horizontal line that is the graph of this function and the horizontal axis over the interval $[0, 1]$? Since this area forms a rectangle, we can use the formula for a rectangle, *area = height × width*. The width (along the horizontal axis) is 1, and the height (along the vertical axis) is 5. Thus, the area is 5 units^2 (see Figure 21.1(a)).

Next, consider the case of $f(x) = x + 2$. How would you estimate the area bounded by the graph of this line and the horizontal axis over the interval $[0, 1]$? Since this area forms a trapezoid, we can use the formula for a trapezoid, *area = altitude × (base$_1$ + base$_2$)/2*. In this case, the altitude is the width along the horizontal axis, 1. The length of the two bases are the vertical heights on either end of the trapezoid: $f(0) = 2$ and $f(1) = 3$. Thus, the area is $1 \times (2+3)/2 = 2.5$ units^2 (see Figure 21.1(b)).

What if the graph of $f(x)$ is not a line but rather that of a curve, such as x^3, $\sin(x)$, or e^{3x^2}? How will we estimate the area bounded between the curve and the horizontal axis? A logical

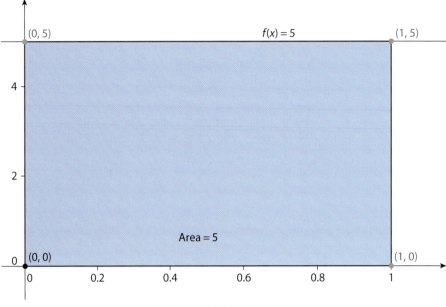

(a) Area under a horizontal line.

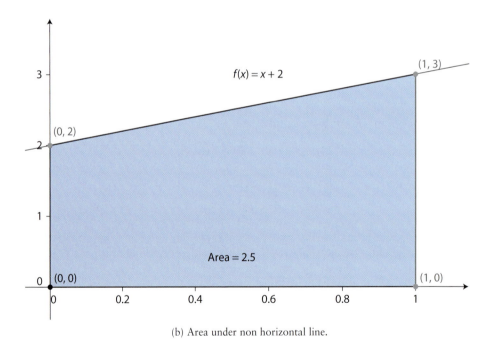

(b) Area under non horizontal line.

Figure 21.1 Area under lines.

method is to partition the area under the curve into many rectangles or trapezoids (whose area can be found easily) and then sum the areas of all the rectangles or trapezoids (see Figure 21.2). In this chapter, we will use the phrase "area under the curve" to mean the area bounded between the curve and the horizontal axis.

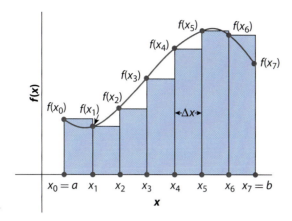

Figure 21.2 The area under the curve $f(x)$ has been partitioned into seven rectangles. The height of the rectangles is determined by using the left end point of each of the seven intervals.

Using Rectangles to Estimate the Area under a Curve

First, we develop the method that uses many small rectangles to estimate the area under a curve. For now, we will consider only functions with non negative values, that is, $f(x) \geq 0$ for all x in the interval $[a, b]$, where $[a, b]$ is the interval over which we are estimating the area under the curve. At the end of this chapter, the method will be extended to the case of functions that take on negative values.

Method to Estimate the Area under a Curve Using Rectangles

1. Divide the interval $[a, b]$ into n intervals, each of length $\Delta x = (b - a)/n$.
2. Let $x_0 = a, x_1 = a + \Delta x, x_2 = a + 2\Delta x, \ldots, x_n = a + n\Delta x = a + n(b - a)/n = b$.
3. Form the sum

$$\sum_{i=0}^{n-1} f(x_i)\Delta x.$$

This sum gives the approximate area under the curve.

Note that the left end point of each of the n intervals was used to determine the height of the rectangle. One could also use the right end point of each of the n intervals to estimate the area under the curve. In this case, the sum would be

$$\sum_{i=1}^{n} f(x_i)\Delta x.$$

Now, let us consider an example for which we estimate the area under a curve using both the left end points and the right end points.

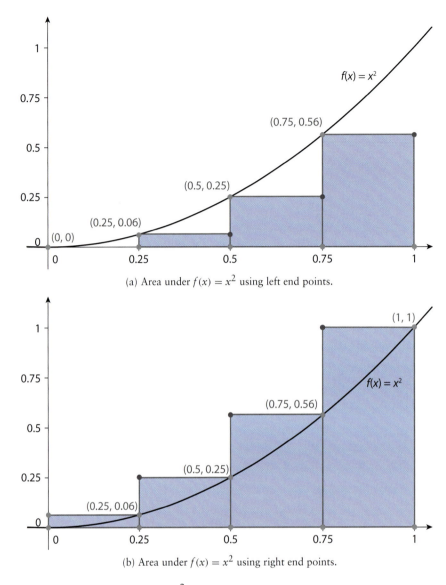

(a) Area under $f(x) = x^2$ using left end points.

(b) Area under $f(x) = x^2$ using right end points.

Figure 21.3 Estimating the area under $f(x) = x^2$.

Example 21.1 (Estimating the Area under a Curve)

Use rectangles to estimate the area under the parabola $f(x) = x^2$ from 0 to 1. Use $n = 4$. Make the estimate first using the left end points of the four intervals and then using the right end points.

(*Continued*)

Solution: If we partition the interval $[0, 1]$ into four equal intervals, each will have length

$$\Delta x = (1 - 0)/4 = 0.25.$$

Thus, the intervals are $[0, 0.25]$, $[0.25, 0.50]$, $[0.50, 0.75]$, and $[0.75, 1.00]$. We can compute the value of f at each of the end points:

$$f(0) = 0, \; f(0.25) = 0.0625, \; f(0.5) = 0.25, \; f(0.75) = 0.5625, \; f(1) = 1.$$

To estimate the area under the curve using the left end points (see Figure 21.3(a)), we have

$$\sum_{i=0}^{3} f(x_i)\Delta x = [0 + 0.0625 + 0.25 + 0.5625](0.25) = 0.21875.$$

To estimate the area under the curve using the right end points (see Figure 21.3(b)), we have

$$\sum_{i=1}^{4} f(x_i)\Delta x = [0.0625 + 0.25 + 0.5625 + 1](0.25) = 0.46875.$$

Notice that the first sum underestimates the area under the curve (see Figure 21.3(a)) and that the second sum overestimates the area under the curve (see Figure 21.3(b)). Thus, for this specific function, we have a lower and upper bound for the exact area A:

$$0.21875 < A < 0.46875.$$

If we had chosen to use more rectangles by using, for example, $n = 8$, how do you think this will affect the upper and lower bounds on the estimated area? Will this improve our estimate of the area or not?

Example 21.2 (Estimating the Area under a Curve)

Use rectangles to estimate the area under the parabola $f(x) = 1 - x^2$ from -1 to 1. Use $n = 10$. Make the estimate first using the left end points of the 10 intervals and then using the right end points.

Solution: If we partition the interval $[-1, 1]$ into 10 equal intervals, each will have length

$$\Delta x = (1 - (-1))/10 = 0.2.$$

(Continued)

Thus, the intervals are $[-1.0, -0.8]$, $[-0.8, -0.6]$, $[-0.6, -0.4]$, $[-0.4, -0.2]$, $[-0.2, 0]$, $[0, 0.2]$, $[0.2, 0.4]$, $[0.4, 0.6]$, $[0.6, 0.8]$, and $[0.8, 1.0]$. We can compute the value of f at each of the end points.

x	$f(x)$	x	$f(x)$
-1.0	0.00	0.2	0.96
-0.8	0.36	0.4	0.84
-0.6	0.64	0.6	0.64
-0.4	0.84	0.8	0.36
-0.2	0.96	1.0	0.00
0	1.00		

To estimate the area under the curve using the left end points (see Figure 21.4(a)), we have

$$\text{Area} \approx \sum_{i=0}^{9} f(x_i)\Delta x$$
$$= [0 + 0.36 + 0.64 + 0.84 + 0.96 + 1 + 0.96 + 0.84 + 0.64 + 0.36] \, (0.2)$$
$$= 1.32 \text{ units}^2.$$

To estimate the area under the curve using the right end points (see Figure 21.4(b)), we have

$$\text{Area} \approx \sum_{i=1}^{10} f(x_i)\Delta x$$
$$= [0.36 + 0.64 + 0.84 + 0.96 + 1 + 0.96 + 0.84 + 0.64 + 0.36 + 0] \, (0.2)$$
$$= 1.32 \text{ units}^2.$$

Notice that whether we use the left or the right end points, some of the rectangles overestimate the area under their portion of the curve, while others underestimate it. Thus, we cannot use this method to obtain bounds on the exact area A. Additionally, notice that for this function over this particular interval, we obtain the same estimate using the left end points as we do with the right end points. Although we have approximated the area under the function $f(x) = 1 - x^2$ over the interval $[-1, 1]$ as 1.32 units^2, we really cannot have too much confidence that this is highly accurate.

Example 21.3 (Estimating the Area under a Curve)

Use rectangles to estimate the area under the function $f(x) = (x - 1)^3 + 2$ from 0 to 4. Use $n = 8$. Make the estimate first using the left end points and then using the right end points.

(*Continued*)

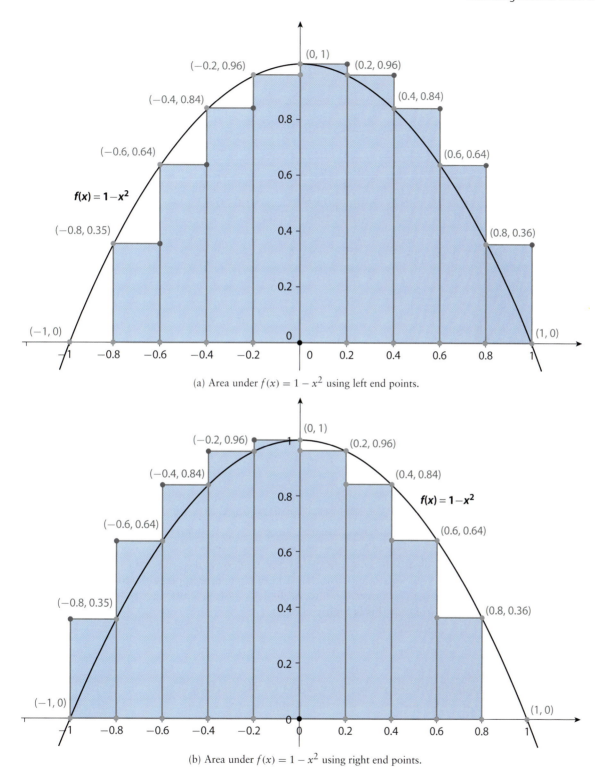

(a) Area under $f(x) = 1 - x^2$ using left end points.

(b) Area under $f(x) = 1 - x^2$ using right end points.

Figure 21.4 Estimating the area under $f(x) = 1 - x^2$.

Solution: If we partition the interval $[0, 4]$ into eight equal intervals, each will have length

$$\Delta x = (4 - 0)/8 = 0.5.$$

We can compute the value of f at each of the end points.

x	$f(x)$
0.0	1.0
0.5	1.875
1.0	2.0
1.5	2.125
2.0	3.0
2.5	5.375
3.0	10.0
3.5	17.625
4.0	29.0

To estimate the area under the curve using the left end points, we have

$$\sum_{i=0}^{7} f(x_i)\Delta x = [1 + 1.875 + 2 + 2.125 + 3.0 + 5.375 + 10 + 17.625] (0.5) = 21.5.$$

To estimate the area under the curve using the right end points, we have

$$\sum_{i=1}^{8} f(x_i)\Delta x = [1.875 + 2 + 2.125 + 3.0 + 5.375 + 10 + 17.625 + 29] (0.5) = 33.5.$$

If you were to draw the graph of this function, you would see that some of the rectangles are completely below the graph and that the tops of some rectangles are above the graph. Thus, the estimates of the area obtained will not be clearly underestimates or overestimates.

The next example involves estimating the area under a function that is a rate of change. The graph shows the photosynthetic rate, and the objective is to estimate the total CO_2 assimilated over the whole time interval.

Example 21.4 (Photosynthetic Rates)

Fay and Knapp [22] measured the amount of CO_2 absorbed by the lower leaves of soybean plants over a 30-minute period when the plant was grown under normal and then under drought conditions. During the 30-minute experiment, the soybean plant is initially in full sunlight. After 8 to 10 minutes, the plant is shaded for approximately 8 to 10 minutes, after which it is returned to full sunlight. Below is a graph showing the uptake rate of CO_2 in μmol of CO_2 per m^2 of leaf area per second over the full experiment for the soybean

(Continued)

plant under drought conditions. The CO_2 assimilation rate was sampled at roughly 1-minute intervals.

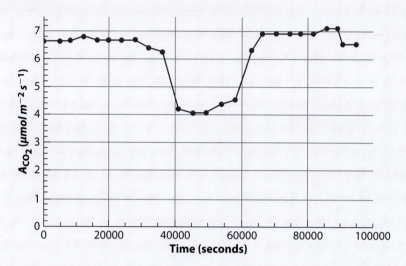

Use a combination of rectangles to estimate the area under the curve over the duration of the experiment. What does the total area represent?

<u>**Solution:**</u> In previous examples, when we had an algebraic expression for the function describing the curve, we used evenly spaced intervals. Here, however, we have only the data shown in the graph. Rather than using evenly spaced intervals, we will construct the rectangles with widths that best capture certain features of the data. The graph shown below divides the length of the experiment into four regions:

$$[0, 38,000], \quad [38,000, 60,000], \quad [60,000, 90,000], \quad \text{and} \quad [90,000, 95,000].$$

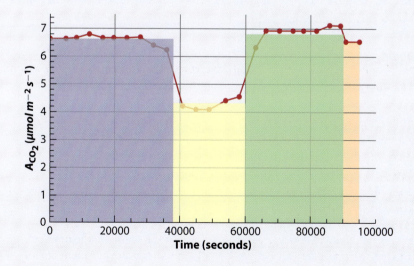

(*Continued*)

The height of the blue rectangle is 6.6 μmol CO_2 m^{-2} s^{-1}, so the area of the rectangle is

$$(38{,}000 - 0) \text{ s} \times 6.6 \; \frac{\mu\text{mol } CO_2}{\text{m}^2 \text{ s}} = 237{,}600 \; \frac{\mu\text{mol } CO_2}{\text{m}^2}.$$

Notice that the resulting units do not include time since the seconds have canceled out. The height of the yellow rectangle is 4.3 μmol m^{-2} s^{-1}, so the area of the rectangle is

$$(60{,}000 - 38{,}000) \text{ s} \times 4.3 \; \frac{\mu\text{mol } CO_2}{\text{m}^2 \text{ s}} = 94{,}600 \; \frac{\mu\text{mol } CO_2}{\text{m}^2}.$$

The height of the green rectangle is 6.8 μmol m^{-2} s^{-1}, so the area of the rectangle is

$$(90{,}000 - 60{,}000) \text{ s} \times 6.8 \; \frac{\mu\text{mol } CO_2}{\text{m}^2 \text{ s}} = 204{,}000 \; \frac{\mu\text{mol } CO_2}{\text{m}^2}.$$

The height of the orange rectangle is 6.5 μmol m^{-2} s^{-1}, so the area of the rectangle is

$$(95{,}000 - 90{,}000) \text{ s} \times 6.5 \; \frac{\mu\text{mol } CO_2}{\text{m}^2 \text{ s}} = 32{,}500 \; \frac{\mu\text{mol } CO_2}{\text{m}^2}.$$

Thus, the total area is

$$237{,}600 + 94{,}600 + 204{,}000 + 32{,}500 = 568{,}700 \; \frac{\mu\text{mol } CO_2}{\text{m}^2}.$$

The area represents an estimate of the total amount of CO_2 assimilated by the lower soybean leaves over the entire duration of the experiment. Notice that the units of total area are not a rate per unit time, while the original units of the data were; the total area under the rate curve gives the total change in CO_2 assimilated.

Using Trapezoids to Estimate the Area under a Curve

Next, we consider estimating the area under a curve using trapezoids.

Method to Estimate the Area under a Curve Using Trapezoids

1. Divide the interval $[a, b]$ into n intervals, each of length $\Delta x = (b - a)/n$.
2. Let $x_1 = a + \Delta x, x_2 = a + 2\Delta x, \ldots, x_n = a + n\Delta x = a + n(b - a)/n = b$.

3. Form the sum

$$\sum_{i=0}^{n-1} \Delta x \left(\frac{f(x_i) + f(x_{i+1})}{2} \right) = \frac{\Delta x}{2} \sum_{i=0}^{n-1} (f(x_i) + f(x_{i+1}))$$

$$= \frac{\Delta x}{2} \left[f(x_0) + \left(\sum_{i=1}^{n-1} 2f(x_i) \right) + f(x_n) \right].$$

This is the approximate area under the curve.

In Example 21.1, we estimated the area under the function $f(x) = x^2$ over the interval $[0, 1]$ using rectangles, and we now reconsider this example using trapezoids.

Example 21.5 (Estimating the Area under a Curve)

Use trapezoids to estimate the area under the parabola $f(x) = x^2$ from 0 to 1. Use $n = 4$.

Solution: If we partition the interval $[0, 1]$ into four equal intervals, each will have length

$$\Delta x = (1 - 0)/4 = 0.25.$$

Thus, the intervals are $[0, 0.25]$, $[0.25, 0.50]$, $[0.50, 0.75]$, and $[0.75, 1.00]$. We can compute the value of f at each of the end points:

$$f(0) = 0, \ f(0.25) = 0.0625, \ f(0.5) = 0.25, \ f(0.75) = 0.5625, \ f(1) = 1.$$

To estimate the area under the curve using trapezoids (see Figure 21.5), we find that

$$\text{Area} \approx \frac{0.25}{2} \left[f(0) + 2f(0.25) + 2f(0.5) + 2f(0.75) + f(1) \right]$$

$$= \frac{0.25}{2} \left[0 + 2(0.0625) + 2(0.25) + 2(0.5625) + 1 \right]$$

$$= 0.34375.$$

Thus, the area under the curve of $f(x) = x^2$ from 0 to 1 is estimated to be 0.34375, and this value is between the two estimates we obtained earlier using rectangles.

In Example 21.2, we estimated the area under the function $f(x) = 1 - x^2$ over the interval $[-1, 1]$ using rectangles, which we now reconsider using trapezoids.

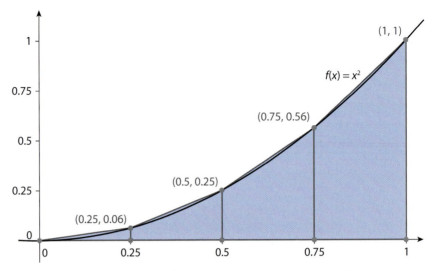

Figure 21.5 Estimating the area under $f(x) = x^2$ using trapezoids.

Example 21.6 (Estimate the Area under a Curve)

Use trapezoids to estimate the area under the parabola $f(x) = 1 - x^2$ from -1 to 1. Use $n = 10$.

Solution: If we partition the interval $[-1, 1]$ into 10 equal intervals, each will have length

$$\Delta x = (1 - (-1))/10 = 0.2.$$

Example 21.2 shows the resulting intervals. Now, to estimate the area under the curve using trapezoids (see Figure 21.6), we find that

$$\text{Area} \approx \frac{0.2}{2} \left[f(-1) + 2f(-0.8) + 2f(-0.6) + \cdots + 2f(0.6) + 2f(0.8) + f(1) \right]$$

$$= \frac{0.2}{2} \left[0 + 2(0.36) + 2(0.64) + \cdots + 2(0.64) + 2(0.36) + 0 \right]$$

$$= 1.32 \text{ units}^2.$$

Thus, the area under the curve is approximately 1.32 units^2. Notice that this is the same area we computed using rectangles. As the previous example illustrated, typically the estimates using these two methods will not be the same. See Exercise 21.10 to explore why this happens in this particular example.

21.2 Increasing the Accuracy of the Area Estimation

Example 21.1 used four rectangles ($n = 4$) to estimate the area under $f(x) = x^2$, using both left end points and right end points. Visually from Figure 21.3(a) it was clear that using left end points would produce an underestimation of the area and from Figure 21.3(b) that using right end points would produce an overestimation of the area. Thus, we were able to conclude that

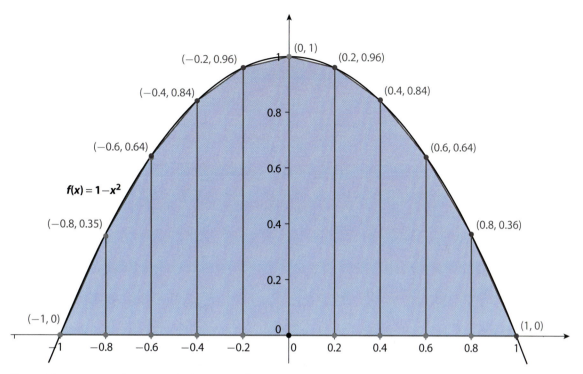

Figure 21.6 Estimating the area under $f(x) = 1 - x^2$ using trapezoids.

the exact area, A, under $f(x) = x^2$ for $x \in [0, 1]$ is $0.21876 < A < 0.46875$. Is it possible to make this estimate more exact? What happens if we repeat Example 26.1 using larger values of n, such as $n = 10$?

Example 21.7 (Estimating the Area under a Curve ($n = 10$))

Use rectangles to estimate the area under the parabola $f(x) = x^2$ from 0 to 1. Use $n = 10$. Make the estimate first using the left end points of the intervals and then using the right end points.

<u>Solution:</u> If we partition the interval $[0, 1]$ into 10 equal intervals, each will have length

$$\Delta x = (1 - 0)/10 = 0.1.$$

We can compute the value of f at each of the end points of the intervals.

x	$f(x)$	x	$f(x)$
0.0	0.00	0.6	0.36
0.1	0.01	0.7	0.49
0.2	0.04	0.8	0.64
0.3	0.09	0.9	0.81
0.4	0.16	1.0	1.00
0.5	0.25		

(*Continued*)

To estimate the area under the curve using the left end points (see Figure 21.3(a)), we have

$$\sum_{i=0}^{9} f(x_i)\Delta x = [0 + 0.01 + 0.04 + \cdots + 0.64 + 0.81](0.1)$$
$$= 0.285.$$

To estimate the area under the curve using the right end points (see Figure 21.3(b)), we have

$$\sum_{i=1}^{10} f(x_i)\Delta x = [0.01 + 0.04 + 0.09 + \cdots + 0.81 + 1](0.1)$$
$$= 0.385.$$

Notice that first sum still underestimates the area under the curve and that the second sum still overestimates the area under the curve. Thus, we have a lower and upper bound for the exact area A:

$$0.285 < A < 0.385.$$

This estimate gives an smaller interval for the exact area, when compared to Example 21.1, so increasing n gave us a more accurate estimate of the range of values for A.

Later in this chapter, we will use Matlab to obtain an estimate of .33 for the area A by using $n = 100$ intervals. Later in the text, we will show that the exact value for the area is $\frac{1}{3}$, so the estimate using 100 intervals is quite accurate.

Example 21.8 (Photosynthetic Rates Revisited)

We will now revisit Example 21.4 using both rectangles and trapezoids to estimate the area under the curve obtained from data on photosynthetic rates.

Solution: We will divide the area under the curve into six regions.

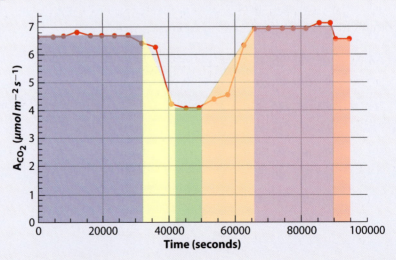

(*Continued*)

The blue region is a rectangle with height 6.7 μmol CO_2 m^{-2} s^{-1} and width 32,000 seconds. The yellow region is a trapezoid with heights of 6.7 and 4.1 μmol CO_2 m^{-2} s^{-1} and width 42,000 − 32,000 = 10,000 seconds. The green region is a rectangle with height 4.1 μmol CO_2 m^{-2} s^{-1} and width 50,000 − 42,000 = 8000 seconds. The orange region is a trapezoid with heights 4.1 and 7.0 μmol CO_2 m^{-2} s^{-1} and width 66,000 − 50,000 = 16,000 seconds. The purple region is a rectangle with height 7.0 μmol CO_2 m^{-2} s^{-1} and width 90,000 − 66,000 = 24,000 seconds. Finally, the pink region is a rectangle with height 6.5 μmol CO_2 m^{-2} s^{-1} and width 95,000 − 90,000 = 5000 seconds.

The total estimated area under the curve represented by the data is

$$\text{Area} \approx (32{,}000 \times 6.7) + \left(10{,}000 \times \frac{1}{2}(6.7 + 4.1)\right) + (8000 \times 4.1)$$

$$+ \left(16{,}000 \times \frac{1}{2}(4.1 + 7.0)\right) + (24{,}000 \times 7.0) + (5000 \times 6.5)$$

$$= 580{,}900 \ \frac{\mu\text{mol CO}_2}{\text{m}^2}.$$

This is slightly larger than the area we computed in Example 21.4.

Example 21.9 (Pesticides [45])

Several insect species are important agricultural pests, and some pesticides have their greatest effects at particular stages of the insect's development. Timing of the application of the pesticide can be very significant. However, the maturation of insects is often dependent on temperature rather than age. Thus, it can be important to track the cumulative temperature an insect has experienced, in units of degrees × time unit, rather than its age. Cumulative temperature T_c (in °C × hr) is found by estimating the area under the function for temperature $T(t)$ over a period of time. Suppose that the following data are found on the temperatures for part of a day (between noon and 7 p.m.).

Time	12:00	13:00	14:00	15:00	16:00	17:00	18:00	19:00
Temp°C	33	34	36	35	32	30	26	24

Use the trapezoid method and the data in the table above to approximate the cumulative temperature from noon to 7 p.m.

Solution: Since the length of time between the temperature measurements is 1 hour, we choose $\Delta t = 1$. The trapezoid method for approximating the area gives

$$T_c \approx \frac{\Delta t}{2}[T(12) + 2\,(T(13) + T(14) + T(15) + T(16) + T(17) + T(18)) + T(19)]$$

$$= \frac{1}{2}[33 + 68 + 72 + 70 + 64 + 60 + 52 + 24]$$

$$= 221.5°\text{C} \times \text{hr}.$$

Example 21.10 (World Population [34])

The rate of change of the world's human population in units of millions of people per year is given in this table.

Year	1950	1960	1970	1980	1990
Rate of change	37	41	78	77	86

Use the trapezoid method and the data in the table to approximate the total change in world population from 1950 to 1990. If the world's population was about 2555 million people in 1950, estimate the population at the end of 1990 using the total change estimate.

Solution: Since the length of time between the data is 10 years, we choose $\Delta t = 10$. The trapezoid method for approximating the area gives

$$\text{Total change} \approx \frac{10}{2} [37 + 2(41 + 78 + 77) + 86]$$
$$= 2575.$$

The estimate for the population at the end of 1990 is obtained by adding the total change over the 40 years to the population number in year 1950 to obtain $2555 + 2575 = 5130$ million people.

21.3 Area below the Horizontal Axis

Example 19.3 approximated data from [58] on the photosynthetic rate (measured as the absorption of CO_2 in μmol CO_2 m^{-2} s^{-1}) of leaves on a soybean plant using the function $A(t) = 15.17 \sin(2\pi t/24 - 1.77) + 4.00$. The data were collected on day 164 of the life cycle of the soybean plant. Figure 19.1 shows the data and the curve $A(t)$. Examples 21.4 and 21.8 illustrated the fact that the area under the photosynthetic rate curve (and above the horizontal axis) provides an estimate for the total amount of CO_2 assimilated by the soybean leaves over a time period. Note, however, that $A(t) \not\geq 0$ for all t. In Figure 21.7, the shaded region shows the area between the curve and the horizontal axis for $0 \leq t \leq 24$. The subregions R_1 and R_3 are below the horizontal axis, while the subregion R_2 is above the horizontal axis. How can we interpret the area of the subregions below the horizontal axis?

A negative photosynthetic rate ($A(t) < 0$) indicates that the leaf is producing (respiring) more carbon dioxide than it is consuming (assimilating); thus, the overall amount of CO_2 assimilated is decreasing. If we wish to estimate the total CO_2 assimilated by the soybean leaves over the course of 1 day, $0 \leq t \leq 24$, the regions above the horizontal axis will increase the total CO_2 assimilated, while the regions below the horizontal axis will decrease the total CO_2 assimilated. Specifically,

$$\text{Total } CO_2 \text{ assimilated} = \text{Area of } R_2 - \text{Area of } R_1 - \text{Area of } R_3.$$

To estimate the area of each subregion (and subsequently the total assimilated CO_2), we can still use the method of summing to approximate the area between the curve and the horizontal axis. The approximate area is represented by

$$\sum_{i=1}^{n} f(x_i) \Delta x.$$

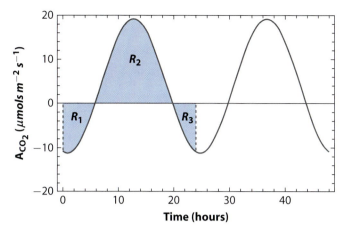

Figure 21.7 The photosynthetic rate (measured as the uptake of CO_2 in μmol CO_2 m^{-2} s^{-1}) of soybean leaves over a 48-hour period. The curve shown approximates data collected by Rogers et al. [58] on the 164th day of the life cycle of the soybean plant. The shaded region shows the area between the curve and the horizontal axis. Note that the subregions R_1 and R_3 are below the horizontal axis, while the subregion R_2 is above the horizontal axis.

Notice now that if $f(x_i) < 0$ (when the function is below the horizontal axis), then $f(x_i)\Delta x < 0$. This does not mean that the area of the rectangle is negative (since that does not make physical sense). The area of the rectangle is positive, but the negative sign indicates that the area is below the horizontal axis. Thus, when we form the sum, the rectangles that are above the horizontal axis will add to the total area, while the rectangles that are below the horizontal axis will subtract from the total area. Furthermore, if we find that

$$\sum_{i=1}^{n} f(x_i)\Delta x > 0,$$

this indicates that there is more area above the horizontal axis than below it. Conversely, if we find that

$$\sum_{i=1}^{n} f(x_i)\Delta x < 0,$$

this means that there is more area below the horizontal axis than above it.

Before we estimate the total assimilated CO_2, notice that the units of the function $A(t)$ are

$$\frac{\mu\text{mol } CO_2}{\text{m}^2 \text{ s}},$$

while t is measured in hours. Thus, the units of $A(t_i)\Delta t$ would be

$$\frac{\mu\text{mol } CO_2 \text{ h}}{\text{m}^2 \text{ s}}.$$

To match the time units for the dependent variable A with the underlying time units in hours, we let x have units of seconds, so $x = 3600t$. To express A as a function of x, note that $t = x/3600$; thus,

$$A(x) = 15.17 \sin\left(\frac{2\pi x}{(24)(3600)} - 1.77\right) + 4.00.$$

Figure 21.8 shows $A(x)$ over a 24-hour period (86,400 seconds) and 12 rectangles each of width 2 hours (7200 seconds) with height determined by the value of $A(x)$ at the right end point of each subinterval. Notice that five of the rectangles lie below the horizontal axis and thus decrease the total assimilated CO_2 over the 24-hour period. The area of each subrectangle is given in Table 21.1 with areas below the horizontal axis being denoted with a negative sign. From this table, the estimate of the total carbon dioxide assimilated over 1 day is

$$\text{Total assimilated } CO_2 \approx \sum_{i=1}^{12} A(x_i)\Delta x = 34.59 \times 10^4 \ \mu\text{mol } CO_2 \ \text{m}^{-2}$$

$$= 0.3459 \ \text{mol } CO_2 \ \text{m}^{-2}.$$

Note that $1 \ \mu\text{mol} = 10^{-6} \ \text{mol}$.

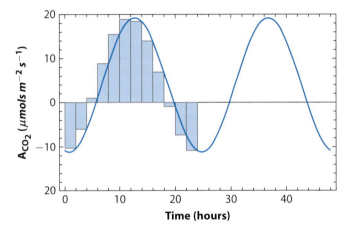

Figure 21.8 The photosynthetic rate (measured as the uptake of CO_2 in $\mu\text{mol } CO_2 \ \text{m}^{-2} \ \text{s}^{-1}$) of soybean leaves over a 48-hour period. The curve shown estimates data collected by Rogers et al. [58] on the 164th day of the life cycle of the soybean plant. The total CO_2 assimilated over $[0, 24]$ (1 day) is estimated to be $34.59 \times 10^4 \ \mu\text{mol } CO_2 \ \text{m}^{-2}$ (or $0.3459 \ \text{mol } CO_2 \ \text{m}^{-2}$) using 12 rectangles with heights defined by the right end points of 12 subintervals each of length 2 hours.

Table 21.1. Values of $A(x_i)\Delta x$ using 12 rectangles.

t_i hours	x_i seconds	$A(x_i)\Delta x \ \dfrac{\mu\text{mol } CO_2}{\text{m}^2}$
2	7200	-7.83×10^4
4	14,400	-7.45×10^4
6	21,600	-4.31×10^4
8	28,800	0.77×10^4
10	36,000	6.41×10^4
12	43,200	11.10×10^4
14	50,400	13.59×10^4
16	57,600	13.22×10^4
18	64,800	10.07×10^4
20	72,000	5.00×10^4
22	79,200	-0.64×10^4
24	86,400	-5.34×10^4

If we increase the number of rectangles from 12 to 24, then the estimate of the total carbon dioxide assimilated over 1 day is

$$\text{Total assimilated CO}_2 \approx \sum_{i=1}^{24} A(x_i)\Delta x = 34.57 \times 10^4 \ \mu\text{mol CO}_2 \ \text{m}^{-2}$$

$$= 0.3457 \ \text{mol CO}_2 \ \text{m}^{-2}.$$

21.4 Matlab Skills

Example 21.2 used a total of 10 rectangles to estimate the area bounded between the curve $f(x) = 1 - x^2$ and the horizontal axis. If greater accuracy is desired, we would need to use smaller intervals and a greater number of rectangles. As the number of rectangles increases, the calculations of the total area quickly become tedious to do by hand, so it is helpful to do these calculations in Matlab.

To estimate the area bounded between the curve $f(x) = 1 - x^2$ and the horizontal axis using 20 rectangles, start by editing the function f.m, which we wrote and used in Sections 15.3, 16.4, and 18.7:

f.m

```
1   % Creates the function f(x)
2   % Input: x (a real number)
3   % Output: y
4
5   function y = f(x)
6   y = 1-x^2;
```

Next, write a function in Matlab that will compute the total area by summing the areas of all of the rectangles. To do this, we can use a loop. We first write a function that utilizes the left end point of each subinterval:

Lrect.m

```
1    % Approximates the area under the function f over a given interval
2    % Make sure the file f.m is contained in the same folder as this file
3
4    % Input: xmin = left end of interval
5    %        xmax = right end of interval
6    %        n = how many rectangles to use
7
8    % Output: sum = approx of the area under f over a given interval
9
10   function sum = Lrect(xmin,xmax,n)
11
12   % Initially set sum to zero
13   sum = 0;
14
15   % Calculate the width of the rectangles
16   deltax = (xmax-xmin)/n;
17
18   % One pass thru the for loop for each rectangle
19   for i = 0:n-1
20       % Add onto the current sum the area of the current rectangle
21       sum = sum + f(xmin + i*deltax) * deltax;
22   end
```

We can check that this function works correctly (doing such a check is always a good idea when you are coding something new) by testing it using the case $n = 10$, since we found the answer for this case in Example 21.2. Following this, we can use the same code to determine how the result changes using $n = 20$, 50, and 100:

```
───────── Command Window ─────────
>> A = Lrect(-1,1,10)
A =
    1.3200

>> A = Lrect(-1,1,20)
A =
    1.3300

>> A = Lrect(-1,1,50)
A =
    1.3328

>> A = Lrect(-1,1,100)
A =
    1.3332
```

First, for $n = 10$, the total area of the rectangles is 1.32 units^2, which is the same answer obtained in Example 21.2. Next, as the number of rectangles used increases, (i.e., increasing the value of n), the total area approaches 1.3333... units^2, or $\frac{4}{3}$ units^2.

Next, consider a function similar to `Lrect.m` that uses the right end point of each subinterval:

```
──────────── Rrect.m ────────────
1   % Approximates the area under the function f over a given interval
2   % Make sure the file f.m is contained in the same folder as this file
3
4   % Input: xmin = left end of interval
5   %        xmax = right end of interval
6   %        n = how many rectangles to use
7
8   % Output: sum = approx of the area under f over a given interval
9
10  function sum = Rrect(xmin,xmax,n)
11
12  % Initially set sum to zero
13  sum = 0;
14
15  % Calculate the width of the rectangles
16  deltax = (xmax-xmin)/n;
17
18  % One pass thru the for loop for each rectangle
19  for i = 1:n
20      % Add onto the current sum the area of the current rectangle
21      sum = sum + f(xmin + i*deltax) * deltax;
22  end
```

Notice that the only difference between `Lrect.m` and `Rrect.m` is on line 19 of the code, where we start the *for* loop. For the left end points of the subintervals, we use `i = 0:n-1`, whereas for the right end points of the subintervals we use `i = 1:n`.

We can additionally write the code for a method that uses trapezoids instead of rectangles:

```
─────────────────────── Trap.m ───────────────────────
% Approximates the area under the function f over a given interval
% Make sure the file f.m is contained in the same folder as this file

% Input: xmin = left end of interval
%        xmax = right end of interval
%        n = how many trapezoids to use

% Output: sum = approx of the area under f over a given interval

function sum = Trap(xmin,xmax,n)

% Initially set sum to zero
sum = 0;

% Calculate the width of the rectangles
deltax = (xmax-xmin)/n;

% One pass thru the for loop for each trapezoid
for i = 0:n-1
    % Add onto the current sum the area of the current trapezoid
    sum = sum + ...
        (f(xmin + i*deltax) + f(xmin + (i+1)*deltax))/2 * deltax;
end
```

Example 21.7 gave estimates for the exact area A bounded between $f(x) = x^2$ and the horizontal axis over the interval $[0, 1]$ as

$$0.285 < A < 0.410.$$

The estimate was obtained by first using the left end points of the subintervals with $n = 10$ and then using the right end points of the subintervals with $n = 10$. Now use the functions Lrect.m, Rrect.m, and Trap.m to obtain a better estimate:

```
─────────────────────── Command Window ───────────────────────
>> % Use n = 10
>> A = Lrect(0,1,10)
A =
    0.2850

>> A = Rrect(0,1,10)
A =
    0.3850

>> A = Trap(0,1,10)
A =
    0.3350

>> % Use n = 50
>> A = Lrect(0,1,50)
A =
    0.3234

>> A = Rrect(0,1,50)
A =
```

```
      0.3434

>> A = Trap(0,1,50)
A =
      0.3334

>> % Use n = 100
>> A = Lrect(0,1,100)
A =
      0.3284

>> A = Rrect(0,1,100)
A =
      0.3384

>> A = Trap(0,1,100)
A =
      0.3333
```

For $n = 10$, the estimate was

$$0.285 < A < 0.410$$

and, from the trapezoid method,

$$A \approx 0.3350.$$

For $n = 50$, the estimate is

$$0.3234 < A < 0.3434$$

and, from the trapezoid method,

$$A \approx 0.3334.$$

For $n = 100$, the estimate is

$$0.3284 < A < 0.3384$$

and, from the trapezoid method,

$$A \approx 0.3333.$$

It would appear that the estimates using the rectangles with left and right end points indicate that the exact area is $\frac{1}{3}$. This is supported by the estimates using the trapezoid method that also show the estimated area is approaching $\frac{1}{3}$. Thus, we might hypothesize that the exact area bounded between $f(x) = x^2$ and the horizontal axis over the interval $[0, 1]$ is $A = \frac{1}{3}$. Using the techniques in the next chapter, we will be able to show that this is correct.

21.5 Exercises

21.1   For each of the following functions, find the area bounded between the curve and the horizontal axis over the given interval. In each case, use $n = 4$ intervals, the rectangle method, and the left end points of the intervals.

(a) $f(x) = 2x + 1$ over $[0, 2]$

(b) $f(x) = \dfrac{1}{4 - x}$ over $[-1, 3]$

(c) $f(x) = -x^2 + 2x + 1$ over $[0, 2]$

21.2 For each of the following functions, find the area bounded between the curve and the horizontal axis over the given interval. In each case, use $n = 4$ intervals, the rectangle method, and the right end points of the intervals.

(a) $f(x) = -x^2 + 2x + 1$ over $[0, 2]$

(b) $f(x) = \dfrac{1}{x + 1}$ over $[1, 5]$

21.3 For each of the following functions, find the area bounded between the curve and the horizontal axis over the given interval. In each case, use $n = 4$ intervals and the trapezoid method.

(a) $f(x) = 2x + 1$ over $[0, 2]$

(b) $f(x) = \dfrac{1}{4 - x}$ over $[-1, 3]$

(c) $f(x) = -x^2 + 2x + 1$ over $[0, 2]$

21.4 The growth rate (in inches per year) of the diameter at breast height of a certain tree is given by

$$y = 1.2e^{-t^2/2},$$

where t is time in years. Find the total growth over the time interval from $t = 1$ to $t = 4$ using $n = 6$ intervals and the trapezoid method.

21.5 Suppose that flow rate data (in m^3/day) for the spatial spread of an oil spill are as given below. Estimate the total spatial spread of oil over the 21 days. Comment on the accuracy of your estimate.

Day	0	7	14	21
Flow Rate	8400	8200	8000	7900

21.6 Suppose that a car travels 30 mph for 1 hour, 40 mph for 2 hours, and then 50 mph for 3 hours. Sketch a graph of the velocity function from time 0 to 6 hours. Find the area under this curve and relate it to the total distance traveled.

21.7 Suppose that a town's population grows at the rate of 1000 people/year for 5 years and then grows at the rate of 2000 people/year for 3 years. What is the change in this population during these 8 years?

21.8 (From [29, 46]) The daily milk consumption (in kilograms) for calves can be approximated by the function

$$y = b_0 w^{b_1} e^{b_2 w},$$

where w is the age of the calf in weeks and b_0, b_1, and b_2 are constants.

(a) The age in days is given by $t = 7w$. Use this fact to convert the function above to a function in terms of t days (instead of w weeks).

(b) For a group of Angus calves, $b_0 = 5.955$, $b_1 = 0.233$, and $b_2 = 0.027$. Use trapezoids to find the total amount of milk consumed by one Angus calf over the first 25 weeks of life (modify the function f.m appropriately and use function Trap.m). Use $n = 10, 25, 175$, and 350. Does the answer seem to approach a specific value as n increases? If so, what is that value?

(c) For a group of Nelore calves, $b_0 = 8.409$, $b_1 = 0.143$, and $b_2 = 0.037$. Use trapezoids to find the total amount of milk consumed by one Nelore calf over the first 25 weeks of life (modify the function f.m appropriately and use function Trap.m). Use $n = 10, 25, 175$, and 350. Does the answer seem to approach a specific value as n increases? If so, what is that value?

(d) Which group of calves, Angus or Nelore, consumes more milk over the first 25 weeks of life?

Use the function Trap.m developed in Section 21.4 to answer (b) and (c). Remember to change the function f.m appropriately.

21.9   (From [45]) The toxicity of a drug is affected by the amount of drug in the blood times the length of time it remains at that level. This cumulative effect is found by estimating the area under the curve presenting the amount of drug over the time that the dose is effective. Suppose that the amount of a drug $A(t)$ (mg) is measured over a period of time after taking a pill and that its quantity in the relevant body organ is found to be as follows:

Hour	0	1	2	3	4	5	6	7	8	9	10
$A(t)$	0.05	0.46	0.87	0.54	0.43	0.36	0.28	0.21	0.16	0.12	0.09

The cumulative dose effect is given by the area under $A(t)$ from $t = 0$ to 10. Using trapezoids with $n = 10$, estimate the cumulative dose effect from the data in the table.

21.10   Show that for $f(x) = 1 - x^2$, $x_0 = -1$, $x_{10} = 1$, and $\Delta x = 0.2$,

$$\sum_{i=0}^{9} f(x_i)\Delta x = \frac{\Delta x}{2} \sum_{i=0}^{9} \left(f(x_i) + f(x_{i+1})\right) = \sum_{i=1}^{10} f(x_i)\Delta x.$$

Hint: Utilize the fact that $f(x_0) = f(x_{10})$, $f(x_1) = f(x_9)$, and so on to show that each of the sums above is equal to

$$\Delta x \left[f(x_0) + 2f(x_1) + 2f(x_2) + 2f(x_3) + 2f(x_4) + f(x_5)\right].$$

21.11   Fay and Knapp [22] measured the amount of CO_2 assimilated by the upper leaves of soybean plants over a 30-minute period when the plant was grown under normal and then under drought conditions. During the 30-minute experiment, the soybean plant is initially in full sunlight. After 8 to 10 minutes, the plant is shaded for approximately 8 to 10 minutes, after which it is returned to full sunlight. Below are graphs showing the assimilation rate of CO_2 in μmol per m^2 of leaf area per second over the full experiment for the soybean plant under normal conditions (shown in blue) and drought conditions (shown in red). The assimilation rate was sampled roughly at 1-minute intervals.

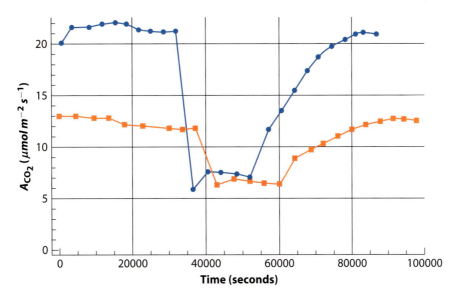

(a) Use a combination of rectangles and trapezoids to estimate the area under each curve for the first 25 minutes of the experiment.
(b) What does the total area represent?
(c) When the areas under each curve are compared, what conclusions can be drawn from the two experiments?

Antiderivatives and the Fundamental Theorem of Calculus

Rates of change are a central feature of essentially every biological process. The basic metabolic processes of photosynthesis and respiration arise from changes in many chemical reactions in cells. Circulatory systems are described by the rates at which blood and other fluids move. Physiology is affected directly by rates of diffusion of oxygen out of the bloodstream and into tissues. The rate at which carbon dioxide diffuses through the stomates of a leaf to the site at which the chemical reactions of photosynthesis occur within the chloroplast. At biological levels of organization above the individual, we describe rates of disease spread, rates at which organisms move throughout their environment, and rates at which species composition within a habitat change over time (the process of ecological succession).

As we noted in the previous chapter, there is a direct way to connect the rates of biological processes with measurements of the underlying variables that are changing (e.g., the total amount of carbon fixed by photosynthesis, the total oxygen consumption through respiration, the numbers of individuals in a population infected with a disease, etc.). We illustrated this using the area under a curve and the cumulative temperature to which an insect has been exposed since insect development is a function of cumulative temperature. In this case, we showed how to sum up changes described by a rate of change over small time intervals to get a cumulative total.

One of the marvelous facts of calculus is that there is a much easier way to find cumulative totals of biological processes from rates of change than taking lots of sums. This is the main idea of this chapter, and it is called the "fundamental theorem of calculus" because it connects the two main ideas of calculus: derivatives for instantaneous rates of change and integrals, which are a quick way to find sums similar to area and cumulative temperature. The first idea we will discuss is how to make a very accurate calculation of cumulative totals, such as cumulative temperature exposure of an insect, by considering larger and larger numbers of pieces in the sums, using the idea of limits to expand the sums in the previous chapter. The second idea is that of reversing the process of taking a derivative of a function: the antiderivative is a way of finding a function whose derivative gives back a particular function of interest. Thinking back to the cumulative temperature example, the temperature at time t, $T(t)$, is an antiderivative of $T'(t)$, the instantaneous rate of change of temperature at time t. The final idea is that we can use antiderivatives to find cumulative totals, or integrals: the fundamental theorem of calculus tells us how to do this.

In the beginning of Chapter 21, we used rate of change of drug concentration as an example of how we can estimate the amount of drug remaining in the body after some period of elimination or metabolism of the drug. We did this by summing up the change of the drug over short time periods. The fundamental theorem of calculus provides a way to very accurately estimate the amount of drug remaining without having to do lots of sums once we have a formula for the rate at which a drug is being eliminated. It provides a way to go from drug elimination rate, which can be estimated directly from data on concentration in the blood and is often found to be proportional to the drug concentration (this is the basic assumption of pharmacokinetics), to determine the amount of drug left in the body at any time after drug administration. This is just one example, though: the fundamental theorem is a critical concept that is applied in virtually every area of biology, and the methods we describe here have been used to solve a large number of biological problems.

22.1 Definition of an Integral

In the previous chapter, we used sums of areas of rectangles or trapezoids to estimate the area under a curve over an interval. This method of approximation partitioned the interval into smaller intervals of width Δx. As we saw, the smaller Δx became (i.e., the more intervals we used), the more accurate the estimation of the area became.

Recall that in defining the derivative, we calculated first the slopes of secant lines (i.e., average rates of change) and then took the limit of these slopes to find the slope of the tangent line at a point (i.e., the derivative or instantaneous rate of change). We will use the same idea here, taking the limit of the sum of the area as we increase the number of intervals we use in the partition.

More formally, given a function $f(x)$ defined over an interval $[a, b]$, we define the ***integral from a to b*** as

$$\int_a^b f(x)\, dx = \lim_{n \to \infty} \sum_{i=1}^{n} f(x_i) \Delta x,$$

where $\Delta x = (b - a)/n$ and $x_i = a + i\Delta x$. The word "integral" is derived from the Latin verb "integro," which means "to make whole." The integral takes very small parts of a physical quantity, measures them, and then adds up all the measurements, making a whole.

The standard symbol for the integral is

$$\int_a^b f(x)\, dx.$$

The symbol $\int$ is an elongated S, standing for "sum." Thus, we are summing up the terms $f(x)\, dx$ from $x = a$ to $x = b$. The function $f(x)$ is the function over which we are integrating and is referred to as the ***integrand*** of the integral. The values a and b are called the ***limits of integration***.

Example 22.1 (Relation between Derivatives and Integrals)

Consider calculating this integral:

$$\int_0^x t \, dt.$$

We are finding the area under the curve $y = t$ from 0 to x. Looking at the graph of $y = t$, we see that area is the area of a triangle with base length x and height x, which is (base)(height)/2. Thus,

$$\int_0^x t \, dt = \frac{x^2}{2}.$$

We can define a function $F(x)$ as

$$F(x) = \int_0^x t \, dt = \frac{x^2}{2}.$$

This function $F(x)$, built from integrating $f(t) = t$, is differentiable, and

$$F'(x) = x.$$

We see that $F'(x) = f(x)$. Putting this together gives

$$\frac{d}{dx}\left[\int_0^x f(t) \, dt\right] = f(x).$$

We are illustrating in this case that integration and differentiation are inverse operations. We start with $f(t)$ and integrate from 0 to x and then differentiate that function, and our result is $f(x)$.

22.2 Antiderivatives

In the derivatives unit, we concentrated on the rules for computing $f'(x)$ for a given function $f(x)$. Now, we will develop rules for the reverse process: given $f'(x)$, we would like to find the original function $f(x)$.

Definition of Antiderivative

If a function $F(x)$ satisfies $F'(x) = f(x)$, then we say that $F(x)$ is an **antiderivative** of $f(x)$.

Example 22.2 (Example of an Antiderivative)

Suppose that $f'(x) = 2x$. What is $f(x)$?

Solution: We know that $\frac{d}{dx}\left[x^2\right] = 2x$, so $f(x) = x^2$. We say that x^2 is an antiderivative of $2x$.

However, notice that $\frac{d}{dx}\left[x^2 + 1\right] = 2x$, and thus $g(x) = x^2 + 1$ is also an antiderivative of $2x$. In fact, we could show that $h(x) = x^2 + c$, where c is any constant, is an antiderivative of $2x$.

From the previous examples, we discover an important property of antiderivatives, namely, that for a given function, there is a not a single, unique antiderivative. In fact, since the derivative of a constant is 0, if $F(x)$ satisfies $\frac{d}{dx}[F(x)] = f(x)$, then $F(x) + c$, where c is a constant, satisfies $\frac{d}{dx}[F(x) + c] = f(x)$. If we want to denote all of the antiderivatives of a function, $f(x)$, which we will call the *family of antiderivatives*, we denote this by adding an arbitrary constant c. Thus, we write $\frac{d}{dx}[F(x) + c] = f(x)$ to denote that $F(x) + c$ is the family of antiderivatives of $f(x)$.

Antiderivatives can often be discovered by thinking about reversing the process of taking derivatives.

Example 22.3 (Derivatives in Reverse)

Find the family of antiderivatives of the following functions:

(a) $f(x) = x^4$
(b) $g(x) = e^{3x}$

Solution: Let us try a few things.

(a) Notice that $\frac{d}{dx}\left[x^5\right] = 5x^4$. Thus, $\frac{d}{dx}\left[x^5\right] = 5f(x)$. This is close, but we have a constant of 5 multiplied by the function $f(x)$. We can get rid of it if we multiply by $\frac{1}{5}$:

$$\frac{1}{5}\frac{d}{dx}\left[x^5\right] = \frac{d}{dx}\left[\frac{1}{5}x^5\right] = \frac{1}{5} \cdot 5x^4 = x^4.$$

Thus, $F(x) = \frac{1}{5}x^5$ is an antiderivative of $f(x) = x^4$. The family of antiderivatives of $f(x)$ vary by a constant c, and thus $\frac{1}{5}x^5 + c$ is the family of antiderivatives of $f(x) = x^4$.

(b) First, we know that e^x is its own derivative. Now, notice that $\frac{d}{dx}\left[e^{3x}\right] = 3e^{3x}$. To get rid of the 3, we multiply by $\frac{1}{3}$:

$$\frac{1}{3}\frac{d}{dx}\left[e^{3x}\right] = \frac{d}{dx}\left[\frac{1}{3}e^{3x}\right] = \frac{1}{3} \cdot 3e^{3x} = e^{3x}.$$

Thus, $G(x) = \frac{1}{3}e^{3x}$ is an antiderivative of $g(x) = e^{3x}$. Thus, $\frac{1}{3}e^{3x} + c$ is the family of antiderivatives of $g(x) = e^{3x}$.

Every formula for a derivative given in Chapter 19 has a corresponding formula for computing the antiderivative. In fact, we can use the table given in Chapter 19 in reverse to get antiderivatives of many functions.

22.3 Fundamental Theorem of Calculus

We now make the idea of integration and differentiation as inverse operations precise in the most important theorem in calculus.

Fundamental Theorem of Calculus

If f is continuous on the interval $[a, b]$, then the function F defined by

$$F(x) = \int_a^x f(t)\, dt \quad \text{for } a \le x \le b$$

is continuous on $[a, b]$ and differentiable on (a, b), then $F'(x) = f(x)$; that is, F is an antiderivative of f. Furthermore, if $F(x)$ is any antiderivative of f, then

$$\int_a^b f(x)\, dx = F(b) - F(a).$$

Note that we will use the notation $F(b) - F(a) = F(x)\big|_a^b$.

Let us consider a few examples where we use the fundamental theorem of calculus.

Example 22.4 (Illustrating This Inverse Relation)

Using geometry, we can find the area under $f(t) = t + 3$ from $t = 0$ to $t = x$. Using the area of a trapezoid, we obtain

$$F(x) = \int_0^x (t + 3)\, dt = \frac{(3 + x + 3)(x)}{2},$$

with the sum of the bases, 3 and $3 + x$, and height x. Now take the derivative of $F(x) = 3x + \frac{x^2}{2}$ to give $F'(x) = x + 3 = f(x)$, which verifies the fundamental theorem of calculus for this example.

Example 22.5 (Using the Fundamental Theorem of Calculus)

Find the derivative of the function $g(x) = \int_0^x \sin(1 + t^2) \, dt$.

Solution: Since $f(t) = \sin(1 + t^2)$ is continuous, the fundamental theorem of calculus gives $g'(x) = \sin(1 + x^2)$.

Example 22.6 (Gompertz Survival Function [28])

Although a formula of the form $g(x) = \int_a^x f(t) \, dt$ may seem like a strange way to define a function, many biological processes are defined this way. For example, suppose that $f(t)$ denotes the instantaneous death rate for members of a population at age t. Then the number of individuals who do not survive to age T is given by

$$F(T) = \int_0^T f(t) \, dt.$$

In 1825, the biologist Benjamin Gompertz proposed that $f(t) = kb^t$. Given Gompertz's proposed function for the instantaneous death rate, what is the rate of change of individuals who do not survive to age T?

Solution: Using the fundamental theorem of calculus, the rate of change of individuals who do not survive to age t is

$$F'(t) = kb^t.$$

Thus, Gompertz had proposed that the rate of change of individuals who do not survive to age T is exponential, or equivalently that the mortality rate is an exponential function of age.

Example 22.7 (Chain Rule)

Find the derivative of

$$H(x) = \int_0^{x^2} \sin(t) \, dt.$$

Solution: Note that H is a composition of functions, $H(x) = F(G(x))$, where $G(x) = x^2$ and $F(x) = \int_0^x \sin(t) \, dt$. Using the chain rule, $H'(x) = F'(G(x))G'(x)$, giving

$$H'(x) = \sin(x^2)2x.$$

22.4 Antiderivatives and Integrals

Using the fundamental theorem of calculus, we can connect the concepts of antiderivatives and integrals. The fundamental theorem of calculus tells us that if f is continuous on the interval $[a, b]$, then the function

$$F(x) = \int_a^x f(t)\, dt$$

is continuous on $[a, b]$, and that $F'(x) = f(x)$. Thus, $F(x)$ is an antiderivative of $f(x)$. The fundamental theorem of calculus is written in terms of what is known as a **definite integral**, which is an integral over a defined interval. Definite integrals help us determine the area under a curve over a specific interval. However, we might be interested in the integral as a function and the antiderivative as a function, not necessarily over a specific interval (see Example 22.6 for a biological example). In this case, we write

$$\int f(x)\, dx = F(x) + c,$$

where $\int f(x)\, dx$ is called an **indefinite integral** and $F(x) + c$ is the family of antiderivatives of $f(x)$, where c is a constant.

   Using the indefinite integral notation, we can rewrite the antiderivatives in Table 22.1 as integrals (see Table 22.2).

Table 22.1. Families of antiderivatives.

Function $f(x)$	$F(x)$, where $F'(x) = f(x)$
0	c (any constant)
m	$mx + c$
x^n	$\frac{1}{n+1}x^{n+1} + c, \ n \neq -1$
$\frac{1}{x}$	$\ln(x) + c, \ x > 0$
$\sin x$	$-\cos(x) + c$
$\cos x$	$\sin(x) + c$
$e^{\alpha x}$	$\frac{1}{\alpha}e^{\alpha x} + c, \ \alpha \neq 0$

Table 22.2. Families of antiderivatives.

(1) $\int 0\, dx = 0$	(2) $\int k\, dx = kx + c \ (k \neq 0)$
(3) $\int x^n\, dx = \dfrac{1}{n+1}x^{n+1} + c \ (n \neq -1)$	(4) $\int \dfrac{1}{x}\, dx = \ln x + c \ (x > 0)$
(5) $\int \sin(x)\, dx = -\cos(x) + c$	(6) $\int \cos(x)\, dx = \sin(x) + c$
(7) $\int \sec^2(x)\, dx = \tan(x) + c$	(8) $\int \csc^2(x)\, dx = -\cot(x) + c$
(9) $\int \sec(x)\tan(x)\, dx = \sec(x) + c$	(10) $\int \csc(x)\cot(x)\, dx = -\csc(x) + c$
(11) $\int e^{\alpha x}\, dx = \dfrac{1}{\alpha}e^{\alpha x} + c \ \alpha \neq 0$	(12) $\int a^x\, dx = \dfrac{a^x}{\ln(a)} + c \ (a > 0)$

We have the following properties for integrals.

> ### Properties of Integrals
>
> Given functions $f(x)$ and $g(x)$ and constants $a < b < c$ and k,
>
> (1) $\displaystyle\int_a^b kf(x)\,dx = k\int_a^b f(x)\,dx$
>
> (2) $\displaystyle\int_a^b \big(f(x) + g(x)\big)\,dx = \int_a^b f(x)\,dx + \int_a^b g(x)\,dx$
>
> (3) $\displaystyle\int_a^b f(x)\,dx + \int_b^c f(x)\,dx = \int_a^c f(x)\,dx$
>
> (4) $\displaystyle\int_a^b f(x)\,dx = -\int_b^a f(x)\,dx$

Using the properties of integrals and the fundamental theorem of calculus, we can find the definite and indefinite integrals of many functions. Let us consider a few examples.

Example 22.8 (Indefinite Integrals)

Find each of the following.

(a) $\displaystyle\int (3x - 5)\,dx$

(b) $\displaystyle\int \big(2^x + 3\sin(x)\big)\,dx$

Solution:

(a) First, we use the properties of integrals to write the integral as the sum of two integrals:
$$\int (3x - 5)\,dx = 3\int x\,dx - \int 5\,dx.$$

Using antiderivatives rules (2) and (3) in Table 22.2, we see that
$$3\int x\,dx = 3\left[\frac{1}{2}x^2\right] + c = \frac{3}{2}x^2 + c \ \text{ and } \ \int 5\,dx = 5x + c.$$

Thus, we have
$$\int (3x - 5)\,dx = \frac{3}{2}x^2 + 5x + c.$$

Notice that since c is an unknown constant for each integral in the sum, we reuse the same letter c as the integration constant for the integral of $3x - 5$.

(Continued)

(b) First, we use the properties of integrals to write the integral as the sum of two integrals:

$$\int \left(2^x + 3\sin(x)\right) \, dx = \int 2^x \, dx + 3 \int \sin(x) \, dx.$$

Using antiderivative rules (5) and (12) in Table 22.2, we see that

$$\int 2^x \, dx = \frac{2^x}{\ln(2)} + c \quad \text{and} \quad 3 \int \sin(x) \, dx = -3\cos(x) + c.$$

Thus, we have

$$\int \left(2^x + 3\sin(x)\right) \, dx = \frac{2^x}{\ln(2)} - 3\cos(x) + c.$$

Example 22.9 (Definite Integrals)

Find each of the following.

(a) $\int_0^3 (x - 1) \, dx$

(b) $\int_0^\pi \left(2^x + 3\sin(x)\right) \, dx$

Solution:

(a) From Example 22.8(a), we know that an antiderivative of $f(x) = x - 1$ is $F(x) = \frac{1}{2}x^2 - x$. Using the fundamental theorem of calculus, we know that

$$\int_0^3 (x - 1) \, dx = F(3) - F(0)$$

$$= \left[\frac{1}{2}3^2 - 3\right] - \left[\frac{1}{2}(0)^2 - 0\right]$$

$$= \left[\frac{9}{2} - 3\right] - [0 - 0] = \frac{3}{2}.$$

Looking at the graph of this function, we see that the area of triangle above the horizontal axis from $x = 1$ to $x = 3$ is 2, while the area of the triangle below

(*Continued*)

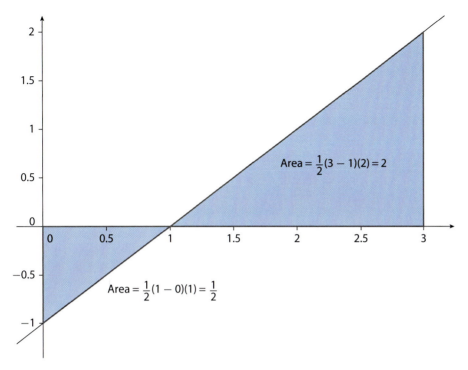

Figure 22.1 Note the triangle is above the horizontal axis from $x = 1$ to 3 and the triangle is below the horizontal axis from $x = 0$ to 1.

the horizontal axis from $x = 0$ to $x = 1$ is $\frac{1}{2}$. So the integral gives the area under the triangle above the horizontal axis minus the area under the horizontal axis, $2 - \frac{1}{2} = \frac{3}{2}$. So the area under the horizontal axis is subtracted (see Figure 22.1). Notice that

$$\int_0^1 (x - 1)\, dx = -\frac{1}{2}.$$

You should calculate this integral yourself and check it.

(b) From Example 22.8(b), we know that an antiderivative of $g(x) = 2^x + 3\sin(x)$ is $G(x) = \frac{2^x}{\ln(2)} - 3\cos(x)$. Using the fundamental theorem of calculus, we know that

$$\int_0^\pi \left(2^x + 3\sin(x)\right)\, dx = G(\pi) - G(0)$$

$$= \left[\frac{2^\pi}{\ln(2)} - 3\cos(\pi)\right] - \left[\frac{2^0}{\ln(2)} - 3\cos(0)\right]$$

$$= \left[\frac{2^\pi}{\ln(2)} - 3(-1)\right] - \left[\frac{1}{\ln(2)} - 3(1)\right]$$

$$= \frac{2^\pi}{\ln(2)} + 3 - \frac{1}{\ln(2)} + 3 = \frac{(2^\pi - 1)}{\ln(2)} + 6.$$

Thus, we would say that the area bounded between the function $2^x + 3\sin(x)$ and the horizontal axis on the interval $[0, \pi]$ is $\frac{(2^\pi - 1)}{\ln(2)} + 6$ units^2.

22.5 Average Values

This section relates to the basic ideas of descriptive statistics of Chapter 1, in which we discussed ways to assess the central tendency and dispersion of data sets. We pointed out that the average value of a data set can sometimes be a useful way to provide a rapid summary of the data. This is particularly useful in comparing observations giving rise to different data sets (you may remember that we used it as a way to compare average heights of different groups of individuals as well as whether heights changed over time). We can use an integral to find an average value of some function that describes how some biological phenomenon changes over time or space, such as whether the average population size of a species changes with latitude or whether the average hourly body temperature changes throughout a day. A second example considers integrals that measure average differences between two measurements, which we will see is related to finding the area between two curves. If over the course of 1 day we measured the net photosynthetic rate of two crop fields, one irrigated and the other not, we can use integrals to compute the difference of total net carbon gain in the two fields over the day. If we have a way to convert carbon gain to yield for the crop, we would then have a way to estimate the potential benefits of irrigation and, once we account for the costs, determine what might be a most appropriate irrigation strategy.

The ***average value of a function*** $f(x)$ over the interval $[a, b]$ is given by

$$\bar{f} = \frac{1}{b-a} \int_a^b f(x) \, dx.$$

We use the notation $\bar{f}$ to indicate the average value of the function f over the interval. This notation is similar to the notation we used in Chapter 1 to denote the arithmetic mean (or average) of a set of data. It is one of the most commonly used measures of central tendency, in this case for a biological property that can take on a continuum of values, such as an individual's life span, the amount of radiation an individual is exposed to, or the amount of mercury or some other heavy metal in an organism's diet.

Let us consider two biological applications of the average value of a function.

Example 22.10 (Population Growth)

Suppose that the size of a population N at time t is given by $N(t) = 1000e^{2t}$, where t is measured in years. Find the average population size from year 0 to year 2.

Solution: If $N(t) = 1000e^{2t}$ over $[0, 2]$, then

$$\bar{N} = \frac{1}{2-0} \int_0^2 1000e^{2t} \, dx = \frac{1000}{2} \int_0^2 e^{2t} \, dx = \frac{500}{2} \cdot e^{2t} \Big|_0^2 = 250 \left(e^4 - 1 \right) \approx 13{,}400.$$

If $N(t)$ is the population size at time t, then $\bar{N}$ is the average population size over time $0 \le t \le 2$. Thus, we say that the average size of the population from year 0 to year 2 was approximately 13,400 individuals.

Example 22.11 (Tendon Strain [6, 44])

The human gastrocnemius tendon is located in the calf. The force on the gastrocnemius tendon as it is stretched can be approximated by the function

$$f(x) = 71.3x - 4.15x^2 + 0.434x^3,$$

where x is the tendon elongation in millimeters and $f(x)$ is the force exerted by the tendon measured in newtons (N). Compute the average force on the tendon as it is elongated from 2 mm to 11 mm.

Solution: If $f(x) = 71.3x - 4.15x^2 + 0.434x^3$ over $[2, 11]$, then

$$\bar{f} = \frac{1}{11 - 2} \int_2^{11} 71.3x - 4.15x^2 + 0.434x^3 \ dx$$

$$= \frac{1}{9} \left(\frac{71.3}{2}x^2 - \frac{4.15}{3}x^3 + \frac{0.434}{4}x^4 \right) \Big|_2^{11}$$

$$\approx 436.4 \text{ N}.$$

Note that 436.4 N of force is approximately 98.1 lbs of force.

Recall that we approximated areas under curves using the sums of areas of rectangles, but we were really approximating definite integrals.

Example 22.12 (Approximate a Definite Integral)

Approximate $\int_0^2 (1 - x^2)dx$. Use $n = 4$ intervals and compute this first using the left end points of the intervals and then using the right end points of the intervals.

Solution: If we partition the interval $[0, 2]$ into four equal intervals, each will have length

$$\Delta x = (2 - 0)/4 = 0.5.$$

Thus, the intervals are $[0, 0.5]$, $[0.5, 1.0]$, $[1.0, 1.5]$, and $[1.5, 2.0]$. We can compute the value of f at each of the end points.

x	$f(x)$
0	1.00
0.5	0.75
1.0	0
1.5	−1.25
2.0	−3.00

(Continued)

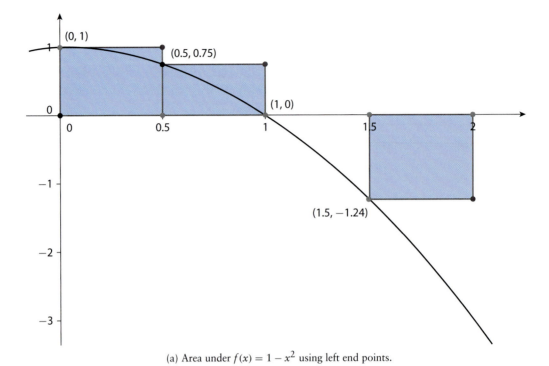

(a) Area under $f(x) = 1 - x^2$ using left end points.

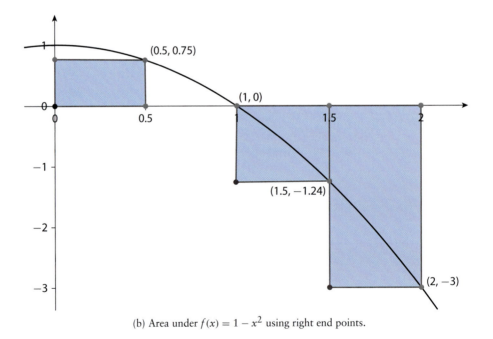

(b) Area under $f(x) = 1 - x^2$ using right end points.

Figure 22.2 Estimating the integral of $f(x) = 1 - x^2$.

Now, if we do the estimate using the left end points (see Figure 22.2(a)), we have

$$\sum_{i=0}^{3} f(x_i)\Delta x = [1.00 + 0.75 + 0 + (-1.25)](.5) = 0.25 \text{ units}^2.$$

Using the right end points (see Figure 22.2(b)), we have

$$\sum_{i=1}^{4} f(x_i)\Delta x = [0.75 + 0 + (-1.25) + (-3.00)](.5) = -1.75 \text{ units}^2.$$

Notice that when we use the right end points, we compute an approximate integral of -1.75 units^2 due to the function values. Notice that rectangles below the horizontal axis contribute negative values to the calculation due to their $f(x_i)$ values being negative. To get a more accurate approximation, one should use a much larger number of rectangles.

22.6 Matlab Skills

As we saw in Example 22.12, we can approximate a definite integral using areas of rectangles bounded between the integrand function and the horizontal axis over the interval defined by the limits of integration. We could have also used trapezoids. In Section 21.4, we developed a set of Matlab functions to help us approximate the area under a curve. Since those Matlab functions do not depend on the curve's being nonnegative, we can use those Matlab functions to approximate the values of definite integrals.

Let us use the functions f.m, Lrect.m, Rrect.m, and Trap.m to estimate the area bound between the function $f(x) = 1 - x^2$ and the horizontal axis over the interval $[0, 2]$. If we use four intervals (i.e., $n = 4$), we should get the same answer we found in Example 22.12. As we increase the number of intervals, we approach the exact value of the integral:

$$\int_0^2 \left(1 - x^2\right) \, dx.$$

First, we must modify the function f.m appropriately:

```
                                          f.m
1   % Creates the function f(x)
2   % Input: x (a real number)
3   % Output: y
4
5   function y = f(x)
6   y = 1-x^2;
```

Next, using `Lrect.m`, let us estimate the value of the integral using $n = 4, 10, 25, 100, 500$, and 1000:

```
                            Command Window
>> A = Lrect(0,2,4)
A =
      0.2500

>> A = Lrect(0,2,10)
A =
    -0.2800

>> A = Lrect(0,2,25)
A =
    -0.5088

>> A = Lrect(0,2,100)
A =
    -0.6268

>> A = Lrect(0,2,500)
A =
    -0.6587

>> A = Lrect(0,2,1000)
A =
    -0.6627
```

Notice that the initial approximation, using $n = 4$, produces a very different estimation than when we increase the value of n. Now, using `Rrect.m`, let us estimate the value of the integral using $n = 4, 10, 25, 100, 500$, and 1000. Do you expect these approximations to be greater or smaller than the estimates made using the left end points of the intervals?

```
                            Command Window
>> A = Rrect(0,2,4)
A =
    -1.7500

>> A = Rrect(0,2,10)
A =
    -1.0800

>> A = Rrect(0,2,25)
A =
    -0.8288

>> A = Rrect(0,2,100)
A =
    -0.7068

>> A = Rrect(0,2,500)
A =
    -0.6747

>> A = Rrect(0,2,1000)
A =
    -0.6707
```

Let $A = \int_0^2 \left(1 - x^2\right) \, dx$; thus, A is the exact value of the integral. Using the estimates using Lrect.m and Rrect.m, when $n = 4$, we estimate

$$-1.7 < A < 0.25;$$

when $n = 10$, we estimate

$$-1.08 < A < -0.28;$$

when $n = 25$, we estimate

$$-0.8288 < A < -0.5088;$$

when $n = 100$, we estimate

$$-0.7068 < A < -0.6268;$$

when $n = 500$, we estimate

$$-0.6747 < A < -0.6587;$$

and when $n = 1000$, we estimate

$$-0.6707 < A < -0.6627.$$

Notice that as n increases, the interval we use to estimate A becomes smaller and smaller. Thus, we see that we are approaching the exact value of A.

Finally, let us use Trap.m to estimate the value of the integral using $n = 4, 10, 25, 100, 500$, and 1000. How do you expect these estimates to compare to the estimates made using Lrect.m and Rrect.m?

```
———————————————————————— Command Window ————————————————————————
>> A = Trap(0,2,4)
A =
   -0.7500

>> A = Trap(0,2,10)
A =
   -0.6800

>> A = Trap(0,2,25)
A =
   -0.6688

>> A = Trap(0,2,100)
A =
   -0.6668

>> A = Trap(0,2,500)
A =
   -0.6667

>> A = Trap(0,2,1000)
A =
   -0.6667
```

What value do the estimates seem to be approaching? It appears that as we increase the size of n, the estimates approach the value -0.6667. In fact, if we allowed Matlab to show us a great number of digits in the calculation, we would see that the estimates approach the value $-0.666666\ldots = -\frac{2}{3}$. If we solve $A = \int_0^2 \left(1 - x^2\right) \, dx$ algebraically, we would find that, indeed, $A = -\frac{2}{3}$.

22.7 Exercises

22.1   Answer the following.

(a) How is the antiderivative of a function related to the function?
(b) What must be true of $F(x)$ and $G(x)$ if both are antiderivatives of $f(x)$?

22.2   If

$$F(x) = \int_0^x e^{s^2}\, ds,$$

what is $F'(x)$?

22.3   Find each of the following.

(a) $\displaystyle\int 6\, dx$

(b) $\displaystyle\int (2x + 3)\, dx$

(c) $\displaystyle\int (x^2 - 4x + 5)\, dx$

(d) $\displaystyle\int (15\sqrt{x} + 2x^2)\, dx$

(e) $\displaystyle\int (5\sec^2(x) - 8\sin(x))\, dx$

(f) $\displaystyle\int 6e^{2x}\, dx$

(g) $\displaystyle\int (e^{3x} - \csc^2(x))\, dx$

(h) $\displaystyle\int (2^x + 3^x + 4^x + 5^x)\, dx$

22.4   Find each of the following.

(a) $\displaystyle\int_{-1}^{2} (x^2 - 15)\, dx$

(b) $\displaystyle\int_{1}^{4} \sqrt{x}\, dx$

(c) $\displaystyle\int_{-2}^{3} (2e^x + 1)\, dx$

(d) $\displaystyle\int_{0}^{1} e^{2x}\, dx$

(e) $\displaystyle\int_{1}^{2} \frac{1}{x}\, dx$

22.5   Find the area bounded between $f(x)$ and the horizontal axis for the following functions over the interval $[0, 1]$.

(a) $f(x) = 6$
(b) $f(x) = 2x + 3$

(c) $f(x) = x^2 - 4x + 5$

(d) $f(x) = \sqrt{x}$

(e) $f(x) = 15\sqrt{x} + 2x^2$

(f) $f(x) = e^{-x}$

22.6   Find the average values of these functions on the indicated interval.

(a) $f(x) = x^2$ on $[0, 3]$

(b) $f(x) = \sqrt{x+1}$ on $[-1, 8]$

22.7   (From [29]) If the rate of excretion of a biochemical compound is given by

$$f'(t) = 0.01e^{-0.01t},$$

the total amount excreted by time t (in minutes) is $f(t)$.

(a) Find an expression for $f(t)$.

(b) If 0 units are excreted at time $t = 0$, how many units are excreted in 10 minutes?

22.8   (From [29, 35]) The average annual increment in the horn length (in cm) of bighorn rams born since 1986 can be approximated by

$$y = 0.1762x^2 - 3.986x + 22.68,$$

where x is the ram's age in years for x between 3 and 9. Integrate to find the total increase in the length of a ram's horn during the age period from 3 to 9 years.

22.9   In [56], Reis et al. collected time-series data of the velocity of the tongue through 11 laps of a cat's tongue. The graph below shows the average velocity of the cat's tongue over one lap of the tongue (i.e., the velocity at each time averaged over the 11 laps of the tongue).

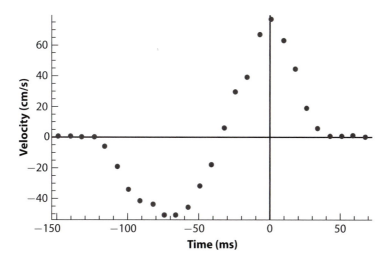

(a) Sketch a curve through the velocity data.

(b) Use trapezoids to estimate the area between the velocity curve and the horizontal axis.

(c) Interpret your answer from (b) in biological and physical terms.

22.10 Suppose that a chemical reaction produces compound A at a rate of

$$f'(t) = 0.01e^{-0.01t} \text{ mg/min,}$$

where t is in minutes. Then $f(t)$ is the total amount of compound A produced (in mg) by time t; that is,

$$f(t) = \int_0^t 0.01e^{-0.01s} \, ds,$$

by the fundamental theorem of calculus.

(a) Using `Trap.m` (see Section 21.4), estimate how many milligrams of compound A are produced in 5 minutes, using $n = 10, 50, 100,$ and 500. Remember to change `f.m` appropriately.

(b) Suppose that a similar chemical reaction produces compound B at a rate of

$$g'(t) = 0.02e^{-0.02t} \text{ mg/min,}$$

where t is in minutes. Using `Trap.m`, estimate how many milligrams of compound B are produced in 5 minutes, using $n = 10, 50, 100,$ and 500. Remember to change `f.m` appropriately.

(c) After 10 minutes, is more of compound A or compound B produced?

22.11 The growth rate of a certain population is given by

$$x'(t) = \frac{1 - e^{2-t}}{1 + e^{2-t}},$$

where t is measured in months. By the fundamental theorem of calculus, the change in the population size after t months is

$$x(t) = \int_0^t \frac{1 - e^{2-s}}{1 + e^{2-s}} \, ds.$$

Thus, if $x(t) > 0$, then the population has increased over the interval $[0, t]$; if $x(t) < 0$, then the population has decreased over the interval $[0, t]$; and if $x(t) = 0$, then the population size has not changed since the intitial time.

(a) Use Matlab to graph $x'(t)$. Utilize the function `graphf.m` that we constructed in Section 15.3. Don't forget to change `f.m` appropriately. Look at the graph of $x'(t)$ over several different ranges. Over what interval is $x'(t) > 0$? Over what interval is $x'(t) < 0$?

(b) Use the function `Trap.m` (see Section 21.4) to estimate the change in the population size after 2 months. Use $n = 100, 500,$ and 1000. Has the population increased or decreased?

(c) Use the function `Trap.m` (see Section 21.4) to estimate the change in the population size after 12 months. Use $n = 100, 500,$ and 1000. Has the population increased or decreased?

(d) After how many months will the change in population size be 0; that is, for what t will $x(t) = 0$? Explain.

Methods of Integration

If you have ever done laboratory experiments or taken measurements in the field, you had to learn or develop the techniques appropriate for the project (e.g., pipetted a liquid, used a microscope to count cells, laid out a meter-squared quadrant and made counts of plant stems). In all cases, some time and effort were required to learn the technique and become suitably proficient to complete the experiment. Along the way, you also developed some intuition about the system that you were studying. Thus, you developed a feasible way to estimate what would happen, such as approximately how many cells are a reasonable number to expect to see in a microscope slide at different levels of magnification.

Developing a similar level of intuition about calculus requires effort as well as an appreciation for the techniques used to solve problems that apply calculus. It is the objective of this chapter to illustrate two of the techniques that are used to find integrals and antiderivatives. This goes beyond the relatively simple antiderivatives discussed in Chapter 22 that were found by inverting the formulas for derivatives. If you want to integrate the photosynthetic rate from Chapter 18,

$$A(t) = 18 - 0.2t + e^{-0.0001t} \sin\left(\frac{t}{4.2}\right),$$

you would need to integrate functions that involve the composition and products of functions. In this chapter, you will have the opportunity to learn two of the most useful techniques for finding antiderivatives of many functions built using composition and products. Our hope is that as you work through the problems, you will develop some intuition about the process because these methods are used in a wide variety of forms to address more realistic biological problems than the ones we have already covered.

23.1 Substitution Method

In Chapter 19, we approximated the rate of photosynthesis in soybean leaves by the sinusoidal function

$$A(t) = 15.17 \sin\left(\frac{2\pi t}{24} - 1.77\right) + 4.00,$$

where t is in hours and A is in μmols CO_2 m^{-2} s^{-1}. Recall that $A(t)$ is a function that represents the rate of carbon dioxide (CO_2) assimilation with respect to time. If we want to calculate the total change in CO_2 over a specific time interval, we need to integrate the rate function $A(t)$ over that time interval. To get the time units on the dependent variable A to match with the underlying time units in hours, we let $x = \frac{t}{3600}$, giving x units of seconds, and our new function (with time units in seconds) can be written as

$$A(x) = 15.17 \sin\left(\frac{2\pi x}{(24)(3600)} - 1.77\right) + 4.00.$$

Notice that the function $A(x)$ is a composition of function with

$$\frac{2\pi x}{(24)(3600)} - 1.77$$

as the "inside" function. The substitution method can be used to integrate such composition of functions, and requires deciding what is the "inside" function.

The substitution method is based on the chain rule, which involves differentiation of a composition of functions. Recall that the chain rule says that if we want to differentiate $f(g(x))$, where $g(x)$ is the "inside" function in the composition, then

$$\frac{d}{dx} f(g(x)) = f'(g(x))g'(x).$$

As we saw in Chapter 22, we can obtain antiderivative formulas from corresponding derivative formulas. This says that the general antiderivative of $f'(g(x))g'(x)$ is

$$f(g(x)) + c,$$

where c is an arbitrary constant. This result gives rise to the antiderivative formula

$$\int f'(g(x))g'(x)\, dx = f(g(x)) + c.$$

Note that f is the antiderivative of f'.

To use the substitution method on

$$\int f'(g(x))g'(x)\, dx,$$

we change variables and let $u = g(x)$ and to simplify the integral in terms of $f'(u)$.

Substitution Method

1. Change variables by letting $u = g(x)$, where $g(x)$ is usually the "inside" function in a composition of functions.
2. Then $\frac{du}{dx} = g'(x)$. Imagine $\frac{du}{dx}$ to be a fraction and solve for $du = g'(x)\, dx$.

3. In the definite integral, replace every expression in terms of x by the corresponding expression in terms of u to obtain

$$\int f'(g(x))g'(x)\,dx = \int f'(u)\,du.$$

Note that in the first integral, the differential dx indicates the variable to be integrated is x, while in the second integral, the differential du indicates that u is the variable of integration.

4. Find $f(u)$ as an antiderivative of $f'(u)$ and then, going back to the original variable, replace u by $g(x)$ to find the general antiderivative in terms of the original variable x, that is,

$$\int f'(u)\,du = f(u) + c = f(g(x)) + c.$$

Let us try a few examples, starting with a simple example leading into the case of photosynthetic rate described above.

Example 23.1 (Substitution Method for a Simple Sine Function)

Find

$$\int \sin\left(\frac{2x}{3}\right)\,dx.$$

Solution: Let $u = \frac{2x}{3}$; then $du = \frac{2}{3}\,dx$. Solving for dx gives

$$\frac{3}{2}\,du = dx.$$

Converting to an integral in terms of u,

$$\int \sin\left(\frac{2x}{3}\right)\,dx$$
$$= \int \sin u \frac{3}{2}\,du$$
$$= -\frac{3}{2}\cos u + c$$
$$= -\frac{3}{2}\cos\left(\frac{2x}{3}\right) + c.$$

Later, we can use this general antiderivative (indefinite integral) to calculate the change in CO_2 over a time interval, given the initial amount.

Example 23.2 (Substitution Method for Photosynthetic Rate)

In Section 21.3, we estimated the total amount of assimilated carbon dioxide by soybean leaves over the course of 1 day when the rate of carbon dioxide assimilation is given by $A(x) = 15.17 \sin \left(\frac{2\pi x}{(24)(3600)} - 1.77 \right) + 4.00$, where x is measured in seconds and A is measured in μmol CO_2 m^{-2} s^{-1}. Find $\int A(x) \, dx$. Assume that at time $x = 0$, no carbon dioxide has yet been assimilated for that day. What does this function represent?

Solution: Let $u = \frac{2\pi x}{(24)(3600)} - 1.77$; then $du = \frac{2\pi}{(24)(3600)} \, dx$. Solving for dx gives

$$\frac{(24)(3600)}{2\pi} \, du = dx.$$

Converting to an integral in terms of u,

$$A_{CO_2}(x) = \int \left[15.17 \sin \left(\frac{2\pi x}{(24)(3600)} - 1.77 \right) + 4.00 \right] dx$$

$$= \int [15.17 \sin u + 4.00] \frac{(24)(3600)}{2\pi} \, du$$

$$= -\frac{(24)(3600)}{2\pi} 15.17 \cos u + 4.00u + c$$

$$= -\frac{(24)(3600)}{2\pi} 15.17 \cos \left(\frac{2\pi x}{(24)(3600)} - 1.77 \right) + 4.00 \left[\frac{2\pi x}{(24)(3600)} - 1.77 \right] + c.$$

The function $A_{CO_2}(x)$ represents the total amount of CO_2 assimilated up to x seconds, where $x = 0$ represents the beginning of a day, that is, 12:00 a.m. Since $A_{CO_2}(0) = 0$, then

$$-\frac{(24)(3600)}{2\pi} 15.17 \cos (0 - 1.77) + 4.00 [0 - 1.77] + c = 0,$$

and solving for c gives

$$c = \frac{(24)(3600)}{2\pi} 15.17 \cos (-1.77) + 4.00(1.77) \approx -41,287.$$

Later, we will show how to use this general antiderivative (indefinite integral) to calculate the change in CO_2 over a time interval, given the initial amount.

Example 23.3 (Substitution Method with a Power Function)

Find

$$\int \left(1+x^3\right)^3 x^2 \ dx.$$

Solution: Let $u = 1 + x^3$; then $du = 3x^2 \ dx$. Dividing both sides of the du equation by 3 gives $\frac{1}{3} \ du = x^2 \ dx$. Notice that $x^2 \ dx$ appears in the integral of the integral that we wish to solve. Thus,

$$\int \underbrace{\left(1+x^3\right)^3}_{u^3} \underbrace{x^2 \ dx}_{\frac{1}{3}du} = \frac{1}{3}\int u^3 \ du = \frac{1}{12}u^4 + c = \frac{1}{12}\left(1+x^3\right)^4 + c.$$

Example 23.4 (Substitution Method with a Rational Function)

Find

$$\int \frac{1}{(4-3x)^2} \ dx.$$

Solution: Let $u = 4 - 3x$; then $du = -3 \ dx$. Dividing both sides of the du equation by -3 gives $-\frac{1}{3} \ du = dx$. Thus,

$$\int \underbrace{\frac{1}{(4-3x)^2}}_{\frac{1}{u^2}} \underbrace{dx}_{-\frac{1}{3}du} = -\frac{1}{3}\int \frac{1}{u^2} \ du = -\frac{1}{3}\int u^{-2} \ du$$

$$= -\frac{1}{3}\left(-u^{-1}\right) + c = \frac{1}{3u} + c = \frac{1}{3(4-3x)} + c.$$

So far, our examples have been indefinite integrals. Let us now consider some definite integrals. When we change variables in a definite integral, we must change the bounds of integration of the original variable to the corresponding bounds of integration for the new variable. Then we can simply continue and finish the integration calculation with the new variable, and there is no need to go back to the original variable. Thus,

$$\int_a^b f'(g(x))g'(x) \ dx = \int_{g(a)}^{g(b)} f'(u) \ du,$$

where, when $x = a$, we have $u = g(a)$, and when $x = b$, we have $u = g(b)$. Therefore, our new bounds of integration are $g(a)$ and $g(b)$.

Example 23.5 (Substitution Method with a Definite Integral)

Compute the definite integral

$$\int_1^2 \frac{x \ln(1 + x^2)}{1 + x^2} \, dx.$$

Solution: Let $u = \ln(1 + x^2)$; then $du = \frac{2x}{1+x^2} \, dx$, or $\frac{1}{2} \, du = \frac{x}{1+x^2} \, dx$. Additionally, now our limits of integration will change. When $x = 1$, $u = \ln 2$, and when $x = 2$, $u = \ln 5$. So we have

$$\int_1^2 \frac{x \ln(1 + x^2)}{1 + x^2} \, dx = \frac{1}{2} \int_{\ln 2}^{\ln 5} u \, du = \frac{u^2}{4} \Big|_{\ln 2}^{\ln 5} = \frac{(\ln 5)^2 - (\ln 2)^2}{4}.$$

Example 23.6 (Substitution Method Involving a Trigonometric Function)

Compute the definite integral

$$\int_0^\pi x \sin x^2 \, dx.$$

Solution: Let $u = x^2$; then $du = 2x \, dx$, or $\frac{1}{2} \, du = x \, dx$. When $x = 0$, $u = 0$, and when $x = \pi$, $u = \pi^2$. Then

$$\int_0^\pi x \sin x^2 \, dx = \frac{1}{2} \int_0^{\pi^2} \sin u \, du = -\frac{1}{2} \left(\cos \pi^2 - \cos 0 \right) = \frac{1}{2} \left(1 - \cos \pi^2 \right).$$

Example 23.7 (Definite Integral for Photosynthetic Rate)

Find

$$\int_0^{7200} \left[15.56 \sin \left(\frac{2\pi x}{(24)(3600)} - 1.787 \right) + 2.763 \right] \, dx.$$

Solution: Since we already calculated the general antiderivative for this rate, we can use that result and change bounds accordingly. Recall that $u = \frac{2\pi x}{(24)(3600)} - 1.787$; then

(Continued)

$du = \frac{2\pi}{(24)(3600)} dx$. The new bounds are -1.787 and $\frac{\pi}{6} - 1.787$ after some simplication on the upper bound (when substituting $x = 7200$):

$$\int_0^{7200} \left[15.56 \sin \left(\frac{2\pi x}{(24)(3600)} - 1.787 \right) + 2.763 \right] dx$$

$$= \left[\frac{(24)(3600)}{2\pi} 15.56 \cos u + 2.763u \right] \Big|_{-1.787}^{\frac{\pi}{6} - 1.787}.$$

23.2 Integration by Parts

Recall the product rule:

$$\frac{d}{dx} f(x)g(x) = f'(x)g(x) + f(x)g'(x).$$

This gives rise to

$$\int \left[f'(x)g(x) + f(x)g'(x) \right] dx = f(x)g(x) + c.$$

We rearrange this equation to become the ***integration by parts*** formula:

$$\int f(x)g'(x) dx = f(x)g(x) - \int f'(x)g(x) dx.$$

In order to recognize how to "throw the derivative" from $g(x)$ in one integral to $f(x)$ in the other integral, we use the following notation. In

$$\int f(x)g'(x) dx,$$

let $u = f(x)$ and $\frac{dv}{dx} = g'(x)$. Then we obtain

$$v = g(x) \quad \text{and} \quad dv = g'(x) dx.$$

Similarly,

$$u = f(x) \quad \text{and} \quad du = f'(x) dx.$$

Thus, our integration by parts formula can be rewritten as

$$\int u \, dv = uv - \int v \, du.$$

The wording "by parts" reminds us that we need to choose which part of the integrand will be u and which part will be dv.

How do we use this new method for integration in practice? Let us consider a few examples.

Example 23.8 (Integration by Parts)

Find
$$\int xe^x \, dx.$$

Solution: We have the product xe^x. We need to choose which factor in the product will be u and which will be dv. The general rule of thumb is that the derivative of u should be a "simpler" function than u and that the integral of dv should not be any more complicated a function than dv. This will not always be the case, but it is a good rule of thumb.

So, we pick $u = x$ and $dv = e^x dx$. Then $du = dx$ and $v = e^x$. We will take care of adding the constant at the end. So, by integration by parts,

$$\int xe^x \, dx = xe^x - \int e^x \, dx = xe^x - e^x + c.$$

Example 23.9 (Integration by Parts with a Logarithm Function)

Evaluate
$$\int \ln(x) \, dx.$$

Solution: Let $dv = dx$ and $u = \ln(x)$. Then $v = x$ and $du = \frac{1}{x} \, dx$. Thus, we obtain

$$\int \ln(x) \, dx = (\ln(x))(x) - \int x \left(\frac{1}{x} \right) \, dx = x \ln(x) - x + c.$$

For a definite integral, we rewrite the integration by parts rule.

Integration by Parts for a Definite Integral

$$\int_a^b u \, dv = uv \Big|_a^b - \int_a^b v \, du$$

Let us consider a few examples of definite integrals that require integration by parts.

Example 23.10 (Integration by Parts for a Definite Integral)

Find $\int_1^2 x^2 \ln x \, dx$.

Solution:

$$u = \ln x \;\; \Rightarrow \;\; du = \frac{1}{x} dx$$

$$dv = x^2 \, dx \;\; \Rightarrow \;\; v = \frac{1}{3} x^3$$

Note that to find v from $dv = x^2 \, dx$, we are doing taking an antiderivative of x^2:

$$\int_1^2 x^2 \ln x \, dx = \frac{1}{3} x^3 \ln x \Big|_1^2 - \int_1^2 \frac{1}{3} x^2 \, dx$$

$$= \left(\frac{8}{3} \ln 2 - \frac{1}{3} \ln 1 \right) - \frac{1}{9} x^3 \Big|_1^2$$

$$= \frac{8}{3} \ln 2 - \left(\frac{8}{9} - \frac{1}{9} \right)$$

$$= \frac{8}{3} \ln 2 - \frac{7}{9}.$$

Example 23.11 (Integrating a Rate by Parts)

Suppose that new cases of the flu are reported at the rate $R(t) = 100te^{-0.2t}$ persons per day. Assuming that at time $t = 0$ there were no reported cases yet, what is the total number of reported cases when $t = 10$ days?

Solution: Since $R(t)$ is the rate of new infections, we integrate $R(t)$ over $[0, 10]$ to find the total number of reported cases in the first 10 days. We pick $u = t$ and $dv = e^{-0.2t} dt$. Then $du = dt$ and $v = -e^{-0.2t}/0.2$:

$$\int_0^{10} 100te^{-0.2t} \, dt = -\frac{1}{0.2} te^{-0.2t} \Big|_0^{10} - \int_0^{10} \frac{1}{0.2} e^{-0.2t} \, dt$$

$$= \left(-\frac{1}{0.2} 10e^{-0.2(10)} + 0 \right) - \frac{1}{0.04} e^{-.2t} \Big|_0^{10}$$

$$\approx 2444.$$

Note that the answer has been rounded up to 2444 cases so that we are not counting fractional cases.

Next, we consider an example for which substitution is used together with integration by parts.

Example 23.12 (Substitution and Integration by Parts)

Find $\int e^{\sqrt{x}}\, dx$.

__Solution:__ First, let $u = \sqrt{x}$. Then

$$x = u^2 \quad \Rightarrow \quad \frac{dx}{du} = 2u \quad \Rightarrow \quad dx = 2u\, du.$$

Now we can write

$$\int e^{\sqrt{x}}\, dx = \int 2u e^u\, du = 2 \int u e^u\, du = 2(u e^u - e^u) + c$$

using the solution from Example 23.8 which used integration by parts to find $\int x e^x\, dx$. Now, we substitute back in $u = \sqrt{x}$:

$$\int e^{\sqrt{x}}\, dx = 2\left(\sqrt{x} e^{\sqrt{x}} - e^{\sqrt{x}} \right) + c.$$

Next, we consider an example for which it is necessary to apply integration by parts twice.

Example 23.13 (Using Integration by Parts Twice)

Find $\int e^x \sin x\, dx$.

__Solution:__

$$u = e^x \quad \Rightarrow \quad du = e^x\, dx$$

$$dv = \sin x\, dx \quad \Rightarrow \quad v = -\cos x$$

$$\int e^x \sin x\, dx = -e^x \cos x + \int e^x \cos x\, dx$$

Now, we will need to apply integration by parts to $\int e^x \cos x\, dx$:

$$u = e^x \quad \Rightarrow \quad du = e^x\, dx$$

$$dv = \cos x\, dx \quad \Rightarrow \quad v = \sin x$$

$$\int e^x \cos x\, dx = e^x \sin x - \int e^x \sin x\, dx$$

(Continued)

Putting this together, we get

$$\int e^x \sin x \, dx = -e^x \cos x + e^x \sin x - \int e^x \sin x \, dx$$

$$2\int e^x \sin x \, dx = -e^x \cos x + e^x \sin x$$

$$\int e^x \sin x \, dx = \frac{1}{2}e^x (\sin x - \cos x) + c$$

There are many computer programs and Web-based tools that will automatically help you find antiderivatives (as well as derivatives and many other mathematical objects). One of our objectives is to provide you with some of the background to understand how these tools obtain their answers. Remember that just because a Website provides you with an answer does not mean that the answer is correct. The intuition that you develop in working through this chapter should help you develop your own ability to "check" whether the answer from a software product or Webtool is appropriate.

23.3 Exercises

23.1   Evaluate the following indefinite integrals using substitution.

(a) $\int 2x\sqrt{x^2 + 1} \, dx$

(b) $\int \cos(2x + 1) \, dx$

(c) $\int \tan x \, dx$

(d) $\int \cos x \sin^4 x \, dx$

23.2   Evaluate the following indefinite integrals using integration by parts.

(a) $\int xe^{-x} \, dx$

(b) $\int x^2 e^x \, dx$

(c) $\int x \cos(3x) \, dx$

(d) $\int x \cos x \, dx$

23.3   Evaluate the following integrals.

(a) $\int \frac{1}{x \ln(x)} \, dx$

(b) $\int_2^3 \frac{1}{x \ln(x)} \, dx$

23.4   A specific drug dose was taken orally at $t = 0$. The total amount of this drug dose removed from the body at time T (in hours) is

$$\int_0^T 0.5te^{-.1t} \, dt,$$

in units of milligrams. Find the amount of drug removed after 2 hours.

23.5   Suppose that during a specific part of its life cycle, the biomass M of a small tree frog grows with this relation:

$$\frac{dM}{dt} = (1 + t^2)e^{-3t}$$

with t in years. Measuring time from when this relationship holds, $M(0) = 3\,$g. What is $M(1)$?

23.6   For a particular town, new cases of flu are reported at the rate

$$R(t) = 4te^{-.05t},$$

where R is people per day and t is time in days. If 500 flu cases have been reported at $t = 0$, what is the estimated number of cases after 10 days?

23.7   If the population P of bacteria (in units of 1000) is declining because of a toxin, its rate of change is

$$P'(t) = -0.01t\sqrt{1 + t^2},$$

where t is in hours. If $P(0) = 6000$, find the population after 2 hours.

23.8   Evaluate these definite integrals.

(a) $\int_1^3 x\sqrt{x^2 + 1}\ dx$

(b) $\int_0^3 xe^{-x}\ dx$

(c) $\int_0^{\frac{\pi}{2}} x^2 \sin(2x)\ dx$

(d) $\int_0^{\frac{\pi}{4}} \tan x \sec^2 x\ dx$

(e) $\int_0^1 r\sqrt{1 - r^2}\ dr$

(f) $\int_0^3 \sqrt{y + 1}\ dy$

(g) $\int_{-1}^1 (2x^3 + 1)^3 x^2\ dx$

(h) $\int_2^{10} \frac{3}{\sqrt{5x-1}}\ dx$

23.9   Find the area under the curve of f from 0 to 3.

(a) $f(x) = \sqrt{x + 4}$

(b) $f(x) = x(x^2 + 1)^3$

23.10   (From [6]) The volume of a tree between heights a and b for a tree with total height H may be approximated by

$$\int_a^b K(H - x)^{3/2}\ dx,$$

where K is a constant. Use this to approximate the total volume of a tree, and note that your answer will include the constants H and K.

Applications of Integrals to Area and Volume

Since rates of change occur so often in describing biological systems, it is reasonable to expect that the process by which we sum up these rates of change over many small time periods to obtain total change in a biological process, through the process of integration, arises whenever we wish to measure the processes that are changing. Our objective in this chapter is to point out a few general examples of how integrals arise in biology, but realize that these are only some of the possible examples. Essentially every field in modern biology relies on integration to solve problems.

One example considers integrals that measure average differences between two measurements, which is related to the problem of finding the area between two curves. If over the course of 1 day we measured the net photosynthetic rate of two crop fields, one irrigated and the other not, we can use integrals to compute the difference of total net carbon gain in the two fields over the day. If we have a way to convert carbon gain to yield for the crop, we would then have a way to estimate the potential benefits of irrigation and, once we account for the costs, determine what might be the most appropriate irrigation strategy. Expanding this to three dimensions, we will illustrate the use of integration for problems in which we need to calculate volumes of objects or how some quantity varies throughout a region in three-dimensional space.

We have already seen applications of the area under a curve in integrating rates to calculate total change in specific quanitites. We will look at the area between curves and then, in three dimensions, illustrate the use of integration for problems in which we need to calculate volumes of objects or how some quantity varies throughout a region in three-dimensional space.

Another example of the use of integrals to analyze biological problems that vary in space is in oceanography, for which integration can be used to summarize total amounts of some biological process that changes with depth, such as plankton density.

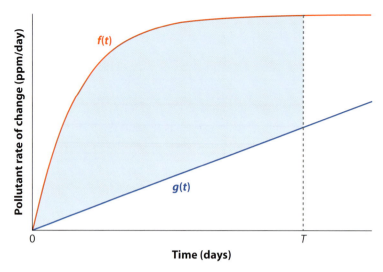

Figure 24.1 The shaded region is the area bounded between the functions $f(x)$ and $g(x)$ over the interval $[0, T]$.

24.1 The Area between Two Curves

Suppose that you are studying pollution entering a lake through a contaminated stream and are able to determine that the rate at which a particular pollutant enters the lake is represented by the function $f(t)$ measured in parts per million (ppm) per day. Then

$$\int_0^T f(t)\, dt$$

represents the amount of the pollutant in the lake (in ppm) after T days, assuming that there is no removal of the pollutant. However, if the pollutant is being removed from the lake at a rate $g(t)$, measured in ppm, then $f(t) - g(t)$ represents the total rate of change of the pollutant in the lake, and

$$\int_0^T \left[f(t) - g(t)\right]\, dt$$

represents the total amount of the pollutant in the lake (in ppm) after T hours. Moreover, geometrically, $\int_0^T \left[f(t) - g(t)\right]\, dt$ represents the area between $f(t)$ and $g(t)$ over the interval $[0, T]$.

Note that this model of the amount of pollutant in the lake makes sense only when

$$\int_0^T \left[f(t) - g(t)\right]\, dt > 0$$

since negative parts per million of a pollutant does not make physical sense. If $f(t) \geq g(t)$ over the interval $[0, T]$, then we know that $\int_0^T \left[f(t) - g(t)\right] dt > 0$.

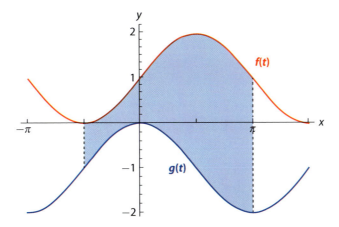

Figure 24.2 The shaded region is the area bounded between $f(x) = \sin(x) + 1$ and $g(x) = \cos(x) - 1$ over the interval $[-\frac{\pi}{2}, \pi]$.

Area between Two Curves

For the general interval $[a, b]$, if $f(x) \geq g(x)$ for $a \leq x \leq b$, then the area bounded by the two graphs and the lines $x = a$ and $x = b$ is given by

$$A = \int_a^b \left[f(x) - g(x) \right] \, dx.$$

Let us try an example.

Example 24.1 (Area between Two Curves)

Find the area bounded between the curves $f(x) = \sin(x) + 1$ and $g(x) = \cos(x) - 1$ over the interval $[-\pi/2, \pi]$.

Solution: To find the area bounded between $f(x)$ and $g(x)$ over $[-\pi/2, \pi]$, we first must identify which function is greater over the interval $[-\pi/2, \pi]$. One way to determine this is to examine the sketch of the graphs of the two functions (see Figure 24.2).

$$\int_{-\pi/2}^{\pi} (\sin(x) + 1) - (\cos(x) - 1) \, dx$$

$$= \int_{-\pi/2}^{\pi} \sin(x) - \cos(x) + 2 \, dx$$

$$= (-\cos(x) - \sin(x) + 2x)|_{-\pi/2}^{\pi}$$

$$= (-\cos(\pi) - \sin(\pi) + 2\pi) - (-\cos(-\pi/2) - \sin(-\pi/2) + 2(-\pi/2))$$

$$= (1 - 0 + 2\pi) - (0 + 1 - \pi) = 3\pi \text{ units}^2.$$

Thus, the area bounded between $\sin(x) + 1$ and $\cos(x) - 1$ over the interval $[-\pi/2, \pi]$ is 3π units^2.

Example 24.2 (Area between Two Curves)

Find the area bounded between the curves $f(x) = 2 - x^2$ and $g(x) = -x$.

<u>Solution:</u> Notice that we are not told over what interval to find the area bounded between $f(x)$ and $g(x)$. If we look at the graphs of $f(x)$ and $g(x)$, we see that the two functions intersect in such a way as to form a closed-off, bounded area (see Figure 24.3). This is the region for which we need to find the area. In order to find the area between the two curves, we first need to know where the functions intersect so that we know what x values to integrate over. First, we find the points of intersection:

$$2 - x^2 = -x \implies 0 = x^2 - x - 2 = (x - 2)(x + 1) \implies x = -1, 2.$$

Thus, the graphs of the functions $f(x)$ and $g(x)$ intersect at $x = -1$ and $x = 2$. Therefore, we will find the area bounded between $f(x)$ and $g(x)$ over the interval $[-1, 2]$.

Next, we determine which function is greater over the interval $[-1, 2]$. We can either examine the sketch (see Figure 24.3) or evaluate the each function at a point between $x = -1$ and $x = 2$ and see which is greater. If we do the latter, and choose $x = 0$, we see that $f(0) = 2 > 0 = g(0)$.

Finally, we use integration to find the area bounded between the curves.

$$A = \int_{-1}^{2} 2 - x^2 - (-x)\, dx = \int_{-1}^{2} 2 - x^2 + x\, dx$$

$$= 2x - \frac{x^3}{3} + \frac{x^2}{2}\, dx \bigg|_{-1}^{2} = \left(4 - \frac{8}{3} + \frac{4}{2}\right) - \left(-2 + \frac{1}{3} + \frac{1}{2}\right) = 4.5 \text{ units}^2$$

Thus, the area bounded between the functions $f(x)$ and $g(x)$ is 4.5 units^2.

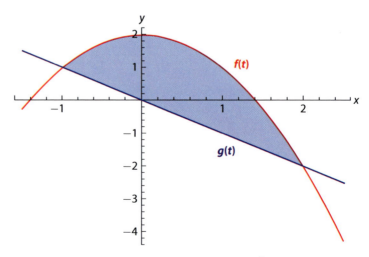

Figure 24.3 The shaded region is the area bounded between $f(x) = 2 - x^2$ and $g(x) = -x$.

Example 24.3 (Find area using two subregions)

Find the area of the region in the first quadrant bounded by the curves, $f(x) = \sqrt{x}$, $y = 0$ and $g(x) = x - 2$.

<u>Solution:</u> If we graph the curves and look at the area, we see that the curve, $f(x) = \sqrt{x}$, forms the top boundary of the area (see Figure 24.4). However, the bottom boundary is formed by $y = 0$ from $x = 0$ to $x = 2$ and then by $g(x) = x - 2$ from $x = 2$ to $x = 4$. We calculate the area by breaking our area in two subregions: Region 1 has $y = 0$ on the bottom and Region 2 has $g(x) = x - 2$ on the bottom. We use integration to find the area bounded between the curves.

$$A = \int_0^2 \sqrt{x}\, dx + \int_2^4 \left(\sqrt{x} - (x - 2) \right)\, dx$$

$$= \frac{2}{3}x^{\frac{3}{2}} \Big|_0^2 + \left(\frac{2}{3}x^{\frac{3}{2}} - \frac{x^2}{2} + 2x \right) \Big|_2^4$$

$$= \frac{10}{3} \text{ units}^2$$

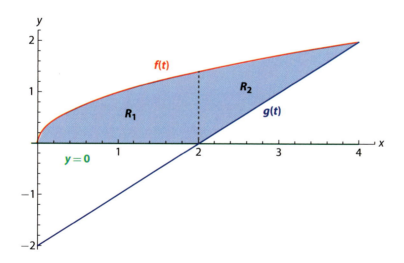

Figure 24.4 The shaded region is the area bounded between $f(x) = \sqrt{x}$, $y = 0$, and $g(x) = x - 2$. The region is divided into two subregions R_1 and R_2.

Now, let us consider a biological example.

Example 24.4 (Lake Pollution)

A small paper plant was discharging waste containing carbon tetrachloride into a lake. The chemical was entering at 16 cubic yards per year. After 3 years, filters were installed and begin to remove the chemical from the discharged waste at a rate of $16 - (t-7)^2$ cubic yards per year. Find the time when the inflow and outflow rates of this chemical are equal. (when the net flow rate is 0). Under this situation, how much of this pollutant entered the lake up until the time when the net rate of flow was 0?

Solution: If $f(t) = 16$ is the rate at which pollution is entering the lake, then

$$\int_0^3 16 \, dt$$

represents the amount of this type of pollution in the lake (in cubic yards) after 3 years. After 3 years, removal of the chemical is taking place at a rate of $g(t) = 16 - (t-7)^2$ cubic yards per year. Thus, after 3 years, $f(t) - g(t)$ represents the net rate at which this chemical enters and exits the lake. Solving for t in

$$16 = 16 - (t-7)^2$$

gives $t = 7$ years when the net flow is 0.
  The total amount of this chemical entering the lake over 7 years is

$$\int_0^3 16 \, dt + \int_3^7 (16 - (16 - (t-7)^2)) \, dt = 48 + \frac{(t-7)^3}{3}\bigg|_3^7 = 48 + 64/3 = 59\frac{1}{3} \text{ cubic yards.}$$

Example 24.5 (Population Change)

Suppose that the population of a certain town decreases at the rate of $g(t) = 6000e^{-0.01t}$ persons per year. Suppose that 900 persons move into the town per year. Given that the current population is 100,000, what would the population be in 10 years if these rates were valid for 10 years?

Solution: The entering rate is $f(t) = 900$. The net change is given by

$$\int_0^{10} (900 - 6000e^{-0.1t}) \, dt = 9000 - 6000\frac{e^{-0.01t}}{-0.01}\bigg|_0^{10}$$

$$= 9000 + 600,000(e^{-0.1} - 1) = 9000 - 57,098 = -46,098.$$

Add this (negative) net change to the initial population, $100,000 - 46,098 = 53,902$, and the population would be estimated to be 53,902 individuals.

24.2 The Volume of a Solid of Revolution

Poiseuille's Law

Around 1846, the French physician Jean Poiseuille was interested in the forces that affect the flow of blood in small blood vessels. He derived formulas to describe the flow of a liquid through cylindrical tubing. When the velocities of the fluid are not too large, the flow is *laminar*, meaning that the fluid flows in parallel layers with no disruption between layers. This means that the paths, or stream lines, of flow are parallel to the walls of the tube. See Figure 24.5 for an illustration of laminar flow.

However, the walls of the tube create friction against the fluid and slow down the fluid that is closer to the walls. The fluid is not moving at $r = R$, where R is the radius of the tube. Let r be the distance from a point in the tube to the center axis of the tube and $v(r)$ be the velocity of the fluid flowing through the tube at distance r from the center of the tube. Specifically, if r is small (close to the center), then $v(r)$ will be larger than for radii closer to R.

Through experiments, Poiseuille found the relationship between the v and r, specifically,

$$v(r) = k(R^2 - r^2),$$

where k is a constant determined by the length of the tube, the pressure differences and the viscosity of the fluid itself. We could use this to calculate the amount of blood flowing through a specific length of an artery.

More specifically, we would like to compute the amount of liquid that passes through a circular cross section of the tube during some time period. We can use a technique in calculus to do this, called solids of revolution.

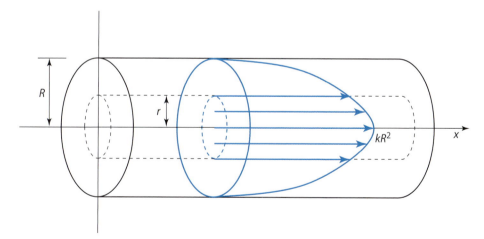

Figure 24.5 A tube of radius R is shown with a velocity profile of fluid flowing through the tube as a function of r, where r is the distance from a point in the tube to the center axis of the tube.

Generating Solids of Revolution

We have seen how to find areas between curves, but now we would like to add another dimension (literally) and find the volume of solids. We will form solids by revolving the area under a curve

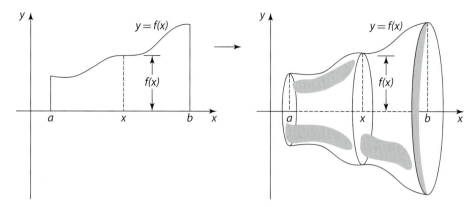

Figure 24.6 A solid may be formed from the graph of a function $y = f(x)$, $a \leq x \leq b$, by rotating the bounded area about the horizontal axis. The cross section at x is a circle with area $\pi[f(x)]^2$.

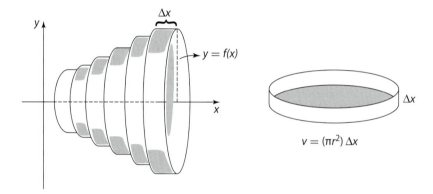

Figure 24.7 For a solid formed by rotating around the horizontal axis, we approximate the volume of the solid with tiny disks of width Δx. Each disk has volume $\pi r^2 \Delta x = \pi [f(x)]^2 \Delta x$. The volume of the solid can be approximated by summing the volume of all the disks.

around an axis. A solid may be formed from the area under the graph of a function $y = f(x)$, $a \leq x \leq b$, by rotating the bounded area about the horizontal axis (see Figure 24.6).

We can form many solids in this fashion: spheres, cones, cylinders, paraboliods, and so on. Notice that the cross section of any such solid is a circle and thus has area $\pi r^2 = \pi [f(x)]^2$ (see Figure 24.6). We will exploit this fact. To find the volume of any solid formed in this fashion, we will approximate the solid with tiny disks of width Δx, and thus volume of each disk is $V = \pi r^2 \Delta x = \pi (f(x))^2 \Delta x$ (see Figure 24.7). The volume of the solid can be approximated by summing the volume of all the disks. We sum up the volumes of all disks and then take the limit as $\Delta x \to 0$ to find that the volume of the solid is

$$V = \int_a^b \pi [f(x)]^2 \, dx.$$

Let us try an example.

Example 24.6 (Paraboloid)

Rotate the area under the curve, $y = \sqrt{x}$, $0 \leq x \leq 4$, about the horizontal axis and find the volume of the resulting solid.

<u>Solution:</u> Since $f(x) = \sqrt{x}$, we have $\pi[f(x)]^2 = \pi x$. Thus,

$$V = \int_0^4 \pi x \, dx = \frac{\pi x^2}{2}\bigg|_0^4 = \frac{16\pi}{2} = 8\pi.$$

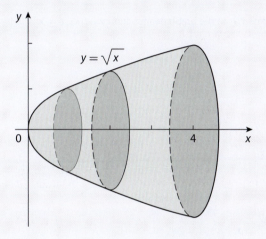

We return to our motivating problem of finding the amount of liquid that passes through a circular cross section per unit time. Suppose that the velocity v is measured in cm/sec. After 1 second, a particle traveling along a line that is radius r away from the center axis will have traveled a distance of $x = k(R^2 - r^2)$ cm. Particles closer to $r = 0$ will travel farther than those closer to $r = R$. This is what is meant by *laminar* flow. Notice that when $r = 0$, $x = kR^2$ (see Figure 24.5).

  Notice that $x(r) = k(R^2 - r^2)$ forms a parabola opening to the left with the x-axis intercept at $x = kR^2$ and the r-axis intercept at $r = R$ (see Figure 24.8). If we revolve this curve about the x-axis, we form a paraboloid, the volume of which represents the amount of liquid that passes

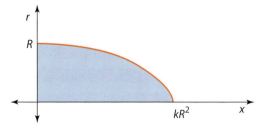

Figure 24.8 The top half of a parabola opening to the left, $r(x) = \sqrt{(kR^2 - x)/k}$. Notice that the x-intercept of the parabola is at $x = kR^2$ and that the r-intercept of the parabola is $r = R$.

through the cylinder of radius R in a given unit of time. To find the volume of this paraboloid, we can imagine slicing it into disks of width dx. Thus, we will need to integrate with respect to x. So we solve the equation $x = k(R^2 - r^2)$ for r as a function of x:

$$r(x) = \sqrt{\frac{kR^2 - x}{k}}.$$

If we revolve the area under the curve r from $x = 0$ to $x = kR^2$, we can find the volume of the resulting paraboloid:

$$V = \int_0^{kR^2} \pi \left[\sqrt{\frac{kR^2 - x}{k}} \right]^2 dx = \frac{\pi}{k} \int_0^{kR^2} [kR^2 - x] \, dx.$$

Using the power rule for integration, we find

$$V = \frac{\pi}{k} \int_0^{kR^2} [kR^2 - x] \, dx$$

$$= \frac{\pi}{k} \left[kR^2 x - \frac{1}{2} x^2 \right] \Big|_0^{kR^2}$$

$$= \frac{\pi}{k} \left[(kR^2)(kR^2) - \frac{1}{2}(kR^2)^2 \right] = \frac{1}{2} \frac{\pi}{k} k^2 R^4 = \frac{\pi k}{2} R^4.$$

Let F denote the flow rate (in cm^3/sec). Thus, if we have a tube of radius R and the fluid that passes through the tube has laminar flow, then the flow rate of the fluid in the tube is given by

$$F = \frac{k\pi}{2} R^4. \tag{24.1}$$

Equation (24.1) is often referred to as **Poiseuille's Law for the Rate of Flow**. We can think of F as the volume of fluid that passes through a slice of the tube in 1 second.

Let us consider an example in which we compare the rate of flow in different-sized tubes.

Example 24.7 (Comparing Flow Rates)

Compare the flow rates of water through a $\frac{1}{2}''$ diameter pipe versus a more standard $\frac{5}{8}''$ diameter pipe.

Solution: The radius of the first pipe is $R_1 = \frac{1}{4}$, and the radius of the second pipe is $R_2 = \frac{5}{16}$. Using Poiseuille's Law for the Rate of Flow, we see that the flows for the two pipes are

$$F_1 = \frac{k\pi}{2} \left(\frac{1}{4} \right)^4 \quad \text{and} \quad F_2 = \frac{k\pi}{2} \left(\frac{5}{16} \right)^4,$$

(*Continued*)

respectively. To compare the two rates, we take the ratio F_1/F_2:

$$\frac{F_1}{F_2} = \frac{\frac{k\pi}{2}\left(\frac{1}{4}\right)^4}{\frac{k\pi}{2}\left(\frac{5}{16}\right)^4} = \frac{\left(\frac{1}{4}\right)^4}{\left(\frac{5}{16}\right)^4} = \left(\frac{4}{5}\right)^4 = \frac{256}{625} = 0.4096$$

We can rewrite this as $F_1 = 0.4096 F_2$. Thus, the flow through the first pipe (of radius $\frac{1}{2}''$) is less than half of the flow through the second pipe (of radius $\frac{5}{8}''$). Thus, by decreasing the radius of the pipe by $\frac{1}{16}''$, we decrease the flow by more than half.

In general, if we want to compare the flow between two different-size tubes, where the flow in the first tube is given by

$$F_1 = \frac{k\pi}{2} R_1^4$$

and the flow in the second tube is given by

$$F_2 = \frac{k\pi}{2} R_2^4,$$

then relative rate of flow is given by

$$\frac{F_1}{F_2} = \left(\frac{R_1}{R_2}\right)^4.$$

Example 24.8 (Hypertension)

Hypertension (high blood pressure) is caused by constriction of the arteries. To increase the flow rate, the heart is forced to pump much harder, increasing the pressure in the cardiovascular system. In Equation 24.1, the constant k is directly proportional to the pressure P on the liquid; that is, $k = cP$ where c is a constant of proportionality. Thus, we can rewrite Poiseuille's Law for the Rate of Flow as

$$F = \frac{c\pi P}{2} R^4.$$

Suppose that a person with hypertension has arteries that are constricted to 80% of their normal size. How much more pressure is needed to maintain the same flow rate?

__Solution:__ Suppose that R_1 represents the normal radius of the arteries in the person with hypertension and that R_2 represents the radius of the constricted arteries. Then the flow through the normal arteries is given by

$$F_1 = \frac{c\pi P_1}{2} R_1^4,$$

(Continued)

and the flow through the constricted arteries is given by

$$F_2 = \frac{c\pi P_2}{2} R_2^4.$$

We want to find the relation P_2/P_1. If we consider the ratio F_2/F_1, we have the equation

$$\frac{F_2}{F_1} = \frac{P_2}{P_1}\left(\frac{R_2}{R_1}\right)^4.$$

Since the arteries are constricted to 80% of their normal size, $R_2 = 0.80R_1$, or

$$\frac{R_2}{R_1} = 0.80.$$

Since we want to maintain the same flow, let $F_2 = F_1$. Then our equation becomes

$$1 = \frac{P_2}{P_1}(0.80)^4.$$

If we solve for P_2/P_1, we get

$$\frac{P_2}{P_1} = \frac{1}{(0.80)^4} = (1.25)^4 \approx 2.44.$$

Thus, we see that by constricting the arteries by 20%, more than twice the pressure is needed for the cardiovascular system to maintain the same rate of flow through the constricted arteries.

24.3 Density Functions

The oceans are about 96% water. Within the oceans are also salts, nitrates, plankton, and fish. If we looked down a square column of water, we would find that the density of various elements or life forms in the water changes density with the depth of the water. For example, plankton have higher densities near the surface of the water, whereas bottom-feeding fish, such as halibut, have higher densities nearer to the ocean floor.

Let x be the distance from the ocean's surface (usually in meters), $A(x)$ be the total amount of a given solute (i.e., salt in grams) or organism (i.e. biomass of plankton in kilograms) in a unit column of water from the ocean surface to the depth x, and $\rho(x)$ be the average concentration or density of the solute at depth x.

Consider a column with 1 m^2 cross section of ocean water (see Figure 24.9). If we look from depth x to depth $x + \Delta x$, then we have a volume of

$$(x + \Delta x - x) \text{ m} \times 1 \text{ m}^2 = \Delta x \text{ m}^3.$$

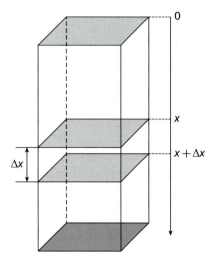

Figure 24.9 A water column with cross-sectional area of 1 m². The value x is the distance from the surface of the water. The volume from depth x to depth $x + \Delta x$ is $(x + \Delta x - x)$ m × 1 m² $= \Delta x$ m³.

Furthermore, we can define the average concentration at $x + \Delta x$ as

$$\rho(x + \Delta x) = \frac{A(x + \Delta x) - A(x)}{\Delta x} \; \frac{\text{g}}{\text{m}^3}.$$

If we take the limit on both sides as $\Delta x \to 0$, we get

$$\rho(x) = \lim_{\Delta x \to 0} \frac{A(x + \Delta x) - A(x)}{\Delta x} = A'(x).$$

Typically, there is no practical way to measure $A(x)$; however, a function $\rho(x)$ can be estimated from sampling. Using the fundamental theorem of calculus, we can determine the function $A(x)$ from the function for $\rho(x)$, using $A(0) = 0$:

$$\rho(x) = A'(x) \quad \Rightarrow \quad A(x) = A(x) - A(0) = \int_0^x \rho(y) \, dy.$$

We will use this formula in some examples.

Example 24.9 (Fishing for Sardines [16])

Suppose that the density of sardines is given by

$$\rho(x) = .005x(75 - x) \text{ fish/m}^3, \quad 0 \le x \le 75.$$

(Continued)

(a) Determine the number of sardines in the total length of the water column (using a standard 1 m^2 cross section).

(b) Find the depth where the density of sardines is greatest.

(c) Suppose that a net has a 10 m × 10 m square opening and is lowered so that the center of the net is at the depth where the density is greatest. If the boat pulling the net is trawling at 20 m/min, how many sardines could be caught in 15 minutes?

Solution:

(a) The total length of the water column is from the surface $x = 0$ to a depth of $x = 75$ m, thus the total number of sardines in the water column is:

$$A(75) = \int_0^{75} .005x(75 - x)\, dx = .005\left(37.5x^2 - \frac{1}{3}x^3\right)\Big|_0^{75} = 351 \text{ sardines.}$$

(b) To find the maximum of the density function, we take the derivative, set it equal to 0 and solve for x:

$$\rho'(x) = .005(75 - 2x) = 0 \quad \Rightarrow \quad 75 = 2x \quad \Rightarrow \quad x = 37.5 \text{ m}$$

(c) If the center of the net is at 37.5 m below the surface, then the net goes from a depth of 32.5 m to 42.5 m. The number of sardines in 1 m^2 of water between 32.5 m and 42.5 m is

$$\int_{32.5}^{42.5} .005x(75 - x)\, dx = 0.005\left(37.5x^2 - \frac{1}{3}x^3\right)\Big|_{32.5}^{42.5} \approx 69.9 \text{ sardines.}$$

This result is for 1 m^2 of water between those depths, but the net is also 10 m wide, so as the net moves through 1 m of water, it can capture

$$10 \times 69.9 = 699 \text{ sardines.}$$

If the boat trawls for 15 minutes at 20 m per minute, then the boats moves

$$15 \text{ min} \times 20\ \frac{\text{m}}{\text{min}} = 300 \text{ m.}$$

So, in theory, the net could capture

$$300 \times 699 \approx 209{,}700 \text{ sardines.}$$

However, the escape rate for sardines is quite high, and thus the actual yield would be much smaller.

Example 24.10 (Salinity [16])

The density of salt (*salinity*) in the central Pacific Ocean (in October) at depth x is given by

$$s(x) = 34.7 \left[1 - .0176 e^{-.05x}(1 + .04x) \right],$$

where s is measured in grams of salt per kilogram of seawater. Given the fact that 1 m^3 seawater has mass 1000 kg, compute the total number of kilograms of salt in a square meter water column from $x = 0$ to $x = 100$.

Solution: The concentration of the salt in the water (in kg/m^3) is given by

$$\rho(x) = s(x) \, \frac{\text{g salt}}{\text{kg seawater}} \times \frac{0.001 \text{ kg}}{\text{g}} \times 1000 \, \frac{\text{kg seawater}}{\text{m}^3}$$

$$= 34.7 - .611 e^{-.05x} - .0244 x e^{-.05x} \, \frac{\text{kg}}{\text{m}^3}$$

In computing $\int_0^{100} \rho(x) \, dx$, we will use the integral formulas

$$\int e^{ax} \, dx = \frac{e^{ax}}{a} \quad \text{and} \quad \int x e^{ax} \, dx = \frac{e^{ax}}{a} \left(x - \frac{1}{a} \right),$$

which can be found using the substitution and integration by parts methods. Hence,

$$\int_0^{100} \rho(x) \, dx = 34.7x + 12.2144 e^{-.05x} + .488 e^{-.05x}(x + 20) \Big|_0^{100}$$

$$= 3470.477 - 21.974 = 3448.50 \text{ kg.}$$

24.4 Exercises

24.1   Find the area between the curves.

(a) $f(x) = x^2$ and $g(x) = \sqrt{x}$
(b) $f(x) = e^x$ and $g(x) = x$

24.2   Find the area bounded by the graphs of $y = 6 + x$, $y = x^3$, and $y = -\frac{1}{2}$ by breaking the region into two subregions.

24.3   Find the volume of the solid generated by revolving the region under the curve, $f(x) = x^2 + 1$, from $x = -1$ to $x = 1$, about the x-axis.

24.4   Find the volume of the solid generated by revolving the region bounded by the curves, $f(x) = \frac{1}{x}$, $x = 1$, $x = 3$, and the x-axis, about the x-axis.

24.5   Find the volume of the solid generated by revolving the region bounded the curves, $y = \sqrt{x}$, $x = 4$, and the x-axis, about the x-axis.

24.6   Find the volume of the solid when the region between the graph of $y = x\sqrt{\sin x}$ and the x-axis from $x = 0$ to $x = \frac{\pi}{2}$ is revolved about the x-axis.

24.7   (From [3, 6]) The concentration of phenylbutazone in the plasma of a calf injected with this anti-inflammatory agent is given approximately by

$$C(t) = 42.03e^{-0.01050t},$$

where C is the concentration in μg/ml and $0 \le t \le 120$ is the number of hours after the injection.

(a) What is the concentration of the initial dose given?
(b) What is the average amount of phenylbutazone in the calf's body for the time between 10 and 120 hours?

24.8   The average temperature each month for Knoxville, Tennessee, is given in the table below (data from http://www.worldclimate.com).

Month	Jan	Feb	Mar	Apr	May	Jun	Jul	Aug	Sep	Oct	Nov	Dec
°C	2.8	4.8	10.3	15.2	19.5	23.5	25.4	25.1	22.0	15.5	10.3	5.2

(a) Using only the data, determine the average (i.e., arithmetic mean) temperature in Knoxville over the course of 1 year.
(b) Let $t = 1$ denote the month January, $t = 2$ denote the month February, and so on. Using Matlab, plot the data and the function

$$C(t) = 12 \sin\left(\frac{\pi}{6}(t - 4)\right) + 14$$

for $t = 1, 2, \ldots, 12$. Does the function $C(t)$ give a reasonable approximation of the temperature cycle in Knoxville over the course of 1 year?
(c) Determine the average value of the temperature function $C(t)$ in Knoxville over the course of 1 year. Does this value differ from the average you calculated in (a)? If so, explain why this difference occurs.

24.9   (From [6, 44]) The forces on the human gastrocnemius tendon during traction and during recoil can be approximated by the functions

$$f(x) = 71.3x - 4.15x^2 + 0.434x^3 \text{ and}$$
$$g(x) = 71.0x - 10.3x^2 + 0.986x^3,$$

respectively, where $0 \le x \le 11$ is the tendon elongation in millimeters. The forces are measured in newtons (N). Integrate from $x = 0$ to $x = 11$ to compute the *elastic strain energy*, given by the area bounded by the graph of the two functions. Your units will be N·mm, which is equivalent to millijoules.

24.10   (From [6]) In a certain memory experiment, subject A is able to memorize words at the rate given by

$$m_1(t) = -0.009t^2 + 0.2t \text{ words/min}.$$

In the same memory experiment, subject B is able to memorize at the rate given by

$$m_2(t) = -0.003t^2 + 0.2t \ \text{words/min.}$$

(a) Which subject has the higher rate of memorization?
(b) How many more words does that subject memorize during the first 10 minutes of the memory experiment?

24.11 (From [16]) Suppose that a person with hypertension has arteries that are constricted to 60% of their normal size. Use the equations derived in Example 24.8 to determine the increase in pressure that is needed to maintain the same blood flow in the cardiovascular system.

24.12 (From [16]) Poiseuille discovered that the velocity function for laminar fluid flow is parabolic, with maximum velocity in the center of the tube. Thus, $v(r) = ar^2 + br + c$ and $v'(0) = 0$. Show that if we assume that $v(R) = 0$ (i.e., the velocity at the edge of the tube is 0), then v can be written in the form

$$v = k(R^2 - r^2), \ \ 0 \le r \le R.$$

24.13 Show that the volume obtained by rotating the graph of the power function $y = \sqrt[3]{x}$ over the interval $[0, 1]$ is given by

$$V = \pi \frac{3}{5}.$$

24.14 (From [16]) The velocity profile of fluid flowing through a tube is not always strictly parabolic but sometimes has a much flatter shape around the vertex and is given by the function

$$v = k(R^n - r^n),$$

where n is some integer power. Use the volume formula obtained in Exercise 24.13 to derive the generalized flow rate formula

$$F = \pi \frac{n}{n+2} k R^{n+2}.$$

24.15 (From [16]) The total number $A(x)$ of organisms from the ocean's surface to depth x m into the water column is given below. Find the corresponding density function $\rho(x)$ in each situation.

(a) $A(x) = 5x$
(b) $A(x) = x^2 + x$
(c) $A(x) = 75x^2 - x^3, \ 0 \le x \le 50$
(d) $A(x) = 50 \left(1 - e^{-2x}\right)$
(e) $A(x) = -10xe^{-0.1x} - 100e^{-0.1x} + 100$

24.16 (From [16]) The water column measures 50 m in a certain part of the sea. Given $\rho(x)$, the hypothetical density function (number of cod per cubic meter) below, find (1) the total number of cod in the water column, (2) the depth where the density is the largest,

(3) the number of cod between 20 and 30 m, and (4) the percentage of the total number of cod in the water column that are between 20 and 30 m.

(a) $\rho(x) = 10$

(b) $\rho(x) = \frac{1}{10}x(50 - x)$

(c) $\rho(x) = 25 - \frac{1}{2}x$

(d) $\rho(x) = 150x(x^2 + 900)^{-1}$

(e) $\rho(x) = xe^{-0.025x}$

(f) $\rho(x) = 5 \sin\left(\frac{\pi x}{50}\right)$

24.17 (From [16]) The blue whale feeds on krill, a small shrimp-like animal of length 2 to 5 cm. Suppose that the density of krill (the number per cubic meter) is given by $\rho(x) = 0.70xe^{-0.001x}$, where x is the distance (in meters) from the Antarctic coast. The distribution in this problem is *horizontal* rather than vertical, but this does not change the fact that $\int_a^b \rho(x)\,dx$ gives the total number of krill from distance a to distance b. Suppose that the blue whale acts as a strainer with a cross-sectional area of 1 square meter.

(a) Find the total number of krill the whale can catch in a run from the coast of Antarctica to 1500 m off the coast.

(b) Find the distance from the coast at which the density of krill is the largest. What is the density of krill at this distance?

(c) Suppose that a blue whale starts feeding 500 m off the coast of Antarctica and continues swimming away from the coast at a speed of 2.5 km per hour for 20 minutes. Find the total number of krill the blue whale catches over this feeding run.

Probability in a Continuous Context

Many problems in biology involve uncertainty. A common feature of biological observations is variation in measurements since exactly the same measurement made on seemingly similar entities can vary. Some of that variation may arise from measurement error, as the instrument being used may produce different results when a measurement is repeated. For example, typical estimates of human blood pressure made with a sphygmomanometer can easily vary by several mm Hg when repeated on the same individual. Some of the variation may arise because of actual differences between the entities being measured, just as human heart rate measurements generally differ when taken just before and just after a period of exertion. While we might expect such variation in measurements to arise in the case of a manipulative change, such as heart rate measurements before and after exercise, in many cases variation arises because of intrinsic character differences.

We can use the concepts of probability to describe the variation that occurs in any set of biological observations by specifying the probability that an individual drawn at random from the group under consideration has a particular value of the measurement. This is used to describe how cholesterol levels vary across a group of humans, how nitrogen levels vary across different soil samples, and how wingspread varies across a collection of birds of the same species. In each situation, we might construct a histogram for the observations and use this to estimate the fraction of the measured individuals or samples that take on a certain range of values. We would then use this as a way to estimate the probability that a random individual has a certain value. For example, for U.S. males between 65 and 74 years of age, 27.4% were found to have high total cholesterol (over 240 mg/dL) in the period 1998–2004. So, if you were to choose a random male in this age range during that time period, there would be a 0.726 probability that the person would not have high cholesterol.

Basic concepts in probability were discussed in Unit 3; however, in this chapter we will see how we can first describe variation among measurements that can take on a continuum of values using functions and then use integration to analyze questions about the observations being made. Suppose that we are taking wingspan measurements from a sample of birds within a population or measuring the amount of nitrogen in a collection of soil samples. We are using the term *sample* here to mean that we are not going to measure every bird in the population or every possible part of the soil; rather, we have measurements on only some of these. Each data point in these measurements could have any value within a reasonable interval of real numbers. So the wingspread of a peregrine falcon can vary from about 40 cm for a fledgling to about 120 cm for an adult. We refer to such observations that can take on a range of continuous values and that have potential variation between different observations as ***continuous random variables***. Although each data point could be any real number within the interval, measurements may well be taken to only a certain level of accuracy (say, to the nearest centimeter for wingspan or to the nearest milligram of nitrogen per kilogram of soil for nitrogen concentration).

After we have amassed our data, we might wish to know the probability that a peregrine falcon we just saw flying by would have a wingspan greater than 80 cm. Or we might wish to know the probability that a randomly selected soil sample from the collection will be within 225 to 240 mg/kg (the optimal nitrogen concentration for growing corn and soy). Suppose that X represents the concentration of nitrogen within a soil sample. Uppercase letters are typically used to represent a random variable, so think of X as taking on a range of all possible values for soil nitrogen concentration. Then we denote the probability of the concentration of a randomly selected soil sample from the region of soil under consideration being within 225 and 240 mg/kg by

$$P(225 \leq X \leq 240).$$

Just as we often thought of probabilities as proportions in Unit 3, here $P(225 \leq X \leq 240)$ represents the proportion of all possible soil samples in the region of soil under consideration with nitrogen concentration falling within the range of 225 to 240 mg/kg. And, just as we saw in Unit 3, the proportion and therefore the probability must be between 0 and 1.

For any continuous random variable X, we can construct a ***probability density function*** f, where the area under the function f over the interval $[a, b]$ represents the probability that the random variable X lies within $[a, b]$, that is,

$$P(a \leq X \leq b) = \int_a^b f(x) \, dx,$$

where the continuous variable x represents possible values for the random variable X. Note that a probability density function must satisfy

$$f(x) \geq 0$$

for all x values of the random variable since negative probabilities make no sense. If X represents the concentration (in mg/kg) of nitrogen in a soil sample, then

$$P(225 \leq X \leq 240) = \int_{225}^{240} f(x) \, dx$$

gives the probability that a randomly selected soil sample will have a concentration of 225 to 240 mg/kg, where $f(x)$ gives the chance that the random variable X has the value x. Here, if we knew the function $f(x)$, we could determine the probability $P(225 \leq X \leq 240)$.

In Unit 3, we discussed dart tosses as one way of considering probabilities. Imagine a line segment going from 0 on the left to 500 mg/kg on the right representing possible nitrogen concentrations in soil samples. Now imagine tossing darts that randomly hit this line segment in different locations, and the point where the dart hits giving the measurement of nitrogen concentration in that soil sample. If there were a magnetic field underlying the line (and steel-tipped darts), then darts would be more attracted to certain parts of the line segment than to others. This is one way of thinking about the probability density function: the magnetic field is stronger where there is a higher probability of obtaining that nitrogen concentration, and thus it is more likely that a dart will be attracted to that part of the line segment.

Example 25.1 (Nitrogen Concentration in Soil)

Suppose that we can represent the probability density function for the concentration of nitrogen x (in g/kg) in a particular soil sample among a database of soil samples as $f(x) = 192x \left(\frac{1}{2} - x \right)^2$ when $0 \le x \le 0.5$ and $f(x) = 0$ otherwise. Notice that this means that x ranges from concentrations of 0 to 0.5 g/kg (or 500 mg/kg). What is the probability that a randomly selected soil sample will have a concentration within the range of 225 to 240 mg/kg?

Solution: Since 225 and 240 are measured in mg/kg and the variable x in the probability density function $f(x)$ has units g/kg, we must convert to 0.225 and 0.240 g/kg. Then

$$P(0.225 \le X \le 0.240) = \int_{0.225}^{0.240} 192x \left(\frac{1}{2} - x \right)^2 dx$$

$$= 192 \int_{0.225}^{0.240} \left(\frac{1}{4}x - x^2 + x^3 \right) dx$$

$$= 192 \left[\frac{1}{8}x^2 - \frac{1}{3}x^3 + \frac{1}{4}x^4 \right]_{0.225}^{0.240} \approx 0.0479.$$

This means that if we randomly picked a soil sample within the database, there would be a probability of 0.0479 of selecting one with a nitrogen concentration within 225 to 240 mg/kg, or, equivalently, about 4.79% of the soil samples have nitrogen concentration in this range.

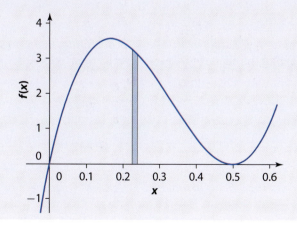

Note that because the area under the probability density function represents a probability, the integral over the entire domain of f should be 1. For many biological applications, measured values are always nonnegative, so it does not make sense for the underlying random variable to have negative values. In this case, we can represent integration over the entire domain of possible observations as

$$\int_0^\infty f(x)\,dx = \lim_{b\to\infty} \left(\int_0^b f(x)\,dx \right) = 1.$$

If the random variable X can take on negative values, then we can represent integration over the entire domain as

$$\int_{-\infty}^\infty f(x)\,dx = \lim_{b\to\infty} \left(\lim_{a\to-\infty} \left(\int_a^b f(x)\,dx \right) \right) = 1.$$

Another function associated with any continuous random variable X is its **probability distribution function**, which is defined to be

$$F(x) = P(X \le x) = \int_{-\infty}^x f(y)\,dy.$$

The probability distribution function for a random variable, which is often called simply the distribution function, is an antiderivative of the probability density function, or, equivalently, the density function is the derivative of the distribution function. Because the area under the probability density function is 1, there is only one antiderivative $F(x)$ with the property that $\lim_{x\to-\infty} F(x) = 0$ and $\lim_{x\to\infty} F(x) = 1$. This implies that there is a unique probability distribution function for any probability density function and vice versa.

Example 25.2 (Nitrogen Concentration in Soil, Part 2)

Show that the total area under the probability density function in Example 25.1 is 1.

Solution: The probability density function is $f(x) = 192x(0.5 - x)^2$ for $0 \le x \le 0.5$ and $f(x) = 0$ otherwise. Thus, if we integrate over the entire domain,

$$\int_{-\infty}^\infty f(x)\,dx = \int_{-\infty}^0 0\,dx + \int_0^{1/2} 192x(0.5 - x)^2\,dx + \int_{1/2}^\infty 0\,dx$$

$$= 0 + 192 \int_0^{1/2} \left(\frac{1}{4}x - x^2 + x^3 \right) dx + 0$$

$$= 192 \left[\frac{1}{8}x^2 - \frac{1}{3}x^3 + \frac{1}{4}x^4 \right]_0^{1/2} = 1.$$

The simplest continuous probability distribution is one in which it is equally likely to obtain any value in some range of possible values. You can think of this as tossing a dart at a line segment, where it is equally likely that the dart could hit any point on the line segment (e.g., you are not aiming for the middle or one of the segments when you toss the dart). When it is equally likely that any particular value is selected or sampled from a finite range of values, we call the the probability distribution function the ***uniform distribution***. The density function for a uniform random variable on the interval from a to b is

$$f(t) = \begin{cases} 0, & \text{if } t < a \\ 1/(b-a), & \text{if } a \le t \le b \\ 0, & \text{if } t > b. \end{cases}$$

Example 25.3 (Uniform Distribution)

Suppose that a random variable has a uniform distribution on the interval $[0, 50]$. What is the probability that a sample from this distribution gives a value between 10 and 20?

Solution: The probability density function is $f(x) = \frac{1}{50}$ for $0 \le x \le 50$ and $f(x) = 0$ otherwise. So the probability that a sample has a value between 10 and 20 is

$$\int_{10}^{20} \frac{1}{50} \, dx = \frac{1}{50}(20 - 10)$$
$$= 0.2.$$

25.1 Expected Value and Median Value

Recall that a data set could be summarized in part by finding its mean, which we obtained by adding up all the values in the data set and dividing by the number of data points. The mean was a measure of central tendency in that it provided a sort of "average value" of the data set. Similarly, we now would like to find a measure of central tendency for a measurement that can be described by a continuous random variable X. This could be the average wingspan for peregrine falcons or the average soil nitrogen concentration for a crop field. We define the mean value of the probability density function $f(x)$ to be

$$\mu = \int_{-\infty}^{\infty} xf(x) \, dx.$$

Here, we can think of the mean value as the average value obtained when you measure many random samples all of which have the same probability density function. If you think of $f(x) \, dx$ as giving the probability that a particular random sample a measured value between x and $x + dx$, then this integral multiplies the value x times the probability that the random variable takes on that value and then sums up over all possible values of the random variable. The mean is also often called the ***expected value*** of the random variable X, where X has probability density function $f(x)$, and is denoted $E[X]$.

The *median value* of a probability density function $f(x)$ is the value α such that

$$\int_{\alpha}^{\infty} f(x)\, dx = \frac{1}{2}.$$

The median value indicates the point at which exactly half of the area under $f(x)$ lies to the right of α and the other half lies to the left of α. So you can think of the median as giving the value where, if you made many observations of a random variable with this probability density function, about half of the observed values would be above the median α and half would be below it.

For the uniform distribution on $[a, b]$, the mean is

$$\mu = \int_a^b x \frac{1}{b-a}\, dx = \frac{1}{b-a}\left[\frac{x^2}{2}\right]_a^b = \frac{b^2 - a^2}{2(b-a)} = \frac{a+b}{2}.$$

The median value for this distrbution would be also be $\frac{a+b}{2}$ since

$$\int_{\alpha}^b \frac{1}{b-a}\, dx = \frac{b-\alpha}{b-a} = \frac{1}{2}$$

when $\alpha = \frac{a+b}{2}$.

Example 25.4 (Nitrogen Concentration in Soil, Part 3)

Find the mean and median of the of probability density function from Example 25.1.

Solution: First we will find the mean:

$$\int_{-\infty}^{\infty} xf(x)\, dx = \int_{-\infty}^0 0\, dx + \int_0^{1/2} 192x^2(0.5 - x)^2\, dx + \int_{1/2}^{\infty} 0\, dx$$

$$= 0 + 192\int_0^{1/2}\left(\frac{1}{4}x^2 - x^3 + x^4\right) dx + 0$$

$$= 192\left[\frac{1}{12}x^3 - \frac{1}{4}x^4 + \frac{1}{5}x^5\right]_0^{1/2} = \frac{1151}{8450} \approx 0.136.$$

Thus, if we sampled the soil database many times, we would expect the average value of the nitrogen concentration of the collection of samples to be close to 0.136 g/kg. The more samples we took, the closer the average should get to 0.136. Notice that the mean value of the soil sample database is outside the optimal nitrogen concentration range for growing corn and soy, 0.225 to 0.240 g of nitrogen per kg of soil.

(Continued)

Next, we find the median value. Since the probability function has values of 0 outside the interval $[0, 0.5]$, we can assume that $0 \leq \alpha \leq 0.5$:

$$\frac{1}{2} = \int_\alpha^\infty f(x)\, dx = \int_\alpha^{1/2} 192x(0.5 - x)^2\, dx$$

$$= 192 \int_\alpha^{1/2} \left(\frac{1}{4}x - x^2 + x^3 \right)\, dx$$

$$= 192 \left[\frac{1}{8}x^2 - \frac{1}{3}x^3 + \frac{1}{4}x^4 \right]_\alpha^{1/2}$$

$$= 192 \left[\frac{1}{8} \left(\frac{1}{4} - \alpha^2 \right) - \frac{1}{3} \left(\frac{1}{8} - \alpha^3 \right) + \frac{1}{4} \left(\frac{1}{16} - \alpha^4 \right) \right]$$

$$= 192 \left[\frac{1}{192} - \frac{1}{8}\alpha^2 + \frac{1}{3}\alpha^3 - \frac{1}{4}\alpha^4 \right]$$

$$= 1 - 24\alpha^2 + 64\alpha^3 - 48\alpha^4$$

$$0 = \frac{1}{2} - 24\alpha^2 + 64\alpha^3 - 48\alpha^4.$$

Using a polynomial root finder on your calculator or Matlab (see Appendix A.2) indicates that there are four roots: two real and two imaginary. Of the two real roots, one is negative (and thus outside of the interval $[0, 0.5]$), and the other is $\alpha \approx 0.193$. This means that if we sampled the soil many times, we would expect roughly half of the samples to have nitrogen concentration below 193 mg/kg and roughly half to be above 193 mg/kg. So the mean and median for the distribution of soil nitrogen are not the same, which often is the case in observations.

25.2 Normal Distribution

As we noted above, measurement error can arise anytime an observation is made. Such errors may arise if there is a bias in the measuring device, such as a thermometer that is miscalibrated so that it regularly measures temperatures as being higher than the correct value. More typical than bias errors is what is usually called "noise," meaning that many small factors are affecting the measurement, each of which produces very small effects and does not tend to cause the measurement to be particularly higher or lower. When these many small factors are added up, they can produce a sum effect that causes the measurement to be different from the correct value, which may be larger or smaller. Because this noise is made up of the sum of many small factors that do not interact, a commonly observed probability density function describes them. This density function is observed to occur in many biological situations, not only in cases of measurement error. The probability density function that describes this is the ***normal probability density***, which is a family of functions described by

$$f(x) = \frac{1}{\sigma \sqrt{2\pi}} e^{-(x-\mu)^2/(2\sigma^2)}, \tag{25.1}$$

where μ is the mean value and σ is the standard deviation. For given μ and σ values, the graph of the normal density can be shown to be symmetric around the mean value, just as we would expect measurement errors to be when there is no bias.

Recall from Chapter 1 that the variance (and its square root, the standard deviation) is a measure of dispersion of a data set. A measure of the dispersion for a continuous random variable is the *variance*, defined for a random variable with density function $f(x)$ and mean μ to be

$$\sigma^2 = \int_{-\infty}^{\infty} (x - \mu)^2 f(x) \, dx.$$

The variance determines how dispersed, measured as the squared distance away from the mean value, the distribution is, and the standard deviation σ is the square root of the variance. Note that the units of the standard deviation are the same as those of the measurements, while the variance has units that are the square of that of the measurement.

For the normal distribution, the value of σ indicates how spread out we should expect data to be when sampled from a random variable with an underlying normal density. Notice in Figure 25.1 that smaller values of sigma correspond to a sharper peak, indicating that sample values will be more clustered about the mean value. Additionally, notice that the height of the peak of the normal density decreases as the standard deviation increases. This change in heights of the peaks ensures that the total area under the curve (for $-\infty \leq x \leq \infty$) will always be 1, that is,

$$\int_{-\infty}^{\infty} \frac{1}{\sigma\sqrt{2\pi}} e^{-(x-\mu)^2/(2\sigma^2)} \, dx = 1$$

for all μ and $\sigma > 0$.

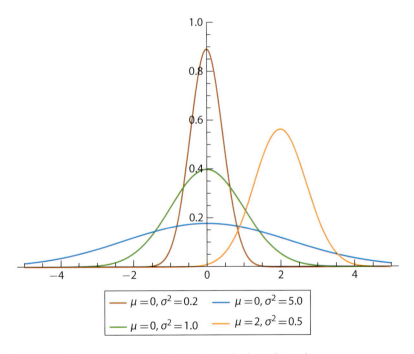

Figure 25.1 The normal distribution (Equation 25.1) for several values of μ and σ.

One reason that the normal distribution arises in so many situations is because this distribution occurs for the sum of a collection of random variables all taken from the same distribution. As we noted, this is the case for measurement errors, which can occur because any measurement involves a potentially large number of separate factors, each of which has a small effect. However, it also arises when there are many factors (genetic, developmental, and environmental) that affect some characteristic of an organism and act independently. The central limit theorem guarantees that the normal distribution arises in the limit from sums of many separate factors, all having the same distribution. However, you should keep in mind that although "bell-shaped curves" often occur, they may not have a normal distribution exactly and that other distributions are unimodal (e.g., they have one peak in the graph of their density functions) and not normal.

Example 25.5 (Length of Human Pregnancies [5])

A study of pregnancies in Swedish women found the mean duration of pregnancy from last menstrual cycle to vaginal birth (pregnancies ending in a cesarean section were not included in this study) to be 281 days with a standard deviation of 13 days. A full-term pregnancy is considered to be between 38 and 42 weeks. If the distribution is normal, what is the probability that a Swedish woman has a full-term pregnancy (ending in a vaginal birth)? Babies born after 42 weeks of gestation are considered postterm. Within the study, 10% of pregnancies ended postterm. What is the probability that a Swedish woman will have a postterm pregnancy?

<u>Solution:</u> According to the study, $\mu = 281$ and $\sigma = 13$. Thus,

$$f(x) = \frac{1}{13\sqrt{2\pi}} e^{-(x-281)^2/\left(2\cdot(13)^2\right)}.$$

Therefore, the probability that a Swedish woman will have a full-term pregnancy between 38 and 42 weeks (or between 266 and 294 days) is

$$P(266 \leq X \leq 294) = \int_{266}^{294} \frac{1}{13\sqrt{2\pi}} e^{-(x-281)^2/388} \, dx.$$

Now, it turns out that the function $y = e^{x^2}$ does not have an elementary antiderivative that we can express as an algebraic function. However, we can approximate the value of this integral using the techniques we developed in Chapter 21 to approximate the area under $f(x)$ over the interval $[266, 294]$. Using the trapezoid rule with 1000 subintervals,

$$P(266 \leq X \leq 294) \approx 0.7171.$$

This means that roughly 72% of Swedish women who have vaginal births will have full-term pregnancies.

(Continued)

Next, we would like to calculated the probability that a Swedish woman who has vaginal birth has a postterm birth, that is,

$$P(X \geq 294) = \int_{294}^{\infty} \frac{1}{13\sqrt{2\pi}} e^{-(x-281)^2/388} \, dx.$$

However, if we wish to numerically estimate the value of this integral, we need a finite value for the upper limit of integration. Since pregnancies lasting longer than 48 weeks (336 days) are very rare, a good approximation of $P(X \geq 294)$ would be $P(294 \leq X \leq 400)$, that is, the probability that the Swedish woman had a vaginal birth between 294 and 400 days. Using the trapezoid rule with 1000 subintervals,

$$P(X \geq 294) \approx P(294 \leq X \leq 400) \approx 0.1587.$$

If you are unconvinced that $P(294 \leq X \leq 400)$ is a good approximation, using the trapezoid rule with 1000 subintervals, $P(294 \leq X \leq 1000) \approx 0.1587$. Therefore, the area under the curve for $294 \leq x \leq 400$ is essentially equal to (at least to an accuracy of 0.0001) the area under the curve for $294 \leq x \leq 1000$. Thus, the probability that a Swedish woman will have has a postterm vaginal birth is approximately 16%, which is a bit higher than was shown in the data used to generate the normal distribution.

25.3 Waiting Times

All of us have been in circumstances that require us to wait, be it waiting for a cashier in a store, a service representative on the phone, or a bus or a subway car. Biological situations that involve waiting include spiders waiting for a suitable prey to hit their web (this is a general problem for sit-and-wait predators), a chemical reaction in a cell that is slowed because of the lack of a necessary enzyme, and a nerve that does not respond until sufficient action potential arrives for it to fire. Each of these involves uncertainty, and answers to questions, such as how long a spider must wait for prey to arrive therefore involve probability. We can address questions such as the probability that a prey item does not arrive in 10 minutes and find the expected time until the prey arrives. In these questions, we are essentially asking how long we should expect to wait for something to happen; that is, what should we expect the waiting time to be? The waiting time is a random variable, and a commonly used probability density function for this random variable is exponential with the form

$$f(t) = \begin{cases} 0, & \text{if } t < 0 \\ ae^{-bt}, & \text{if } t \geq 0. \end{cases}$$

Since it would be unreasonable for us to expect something to happen before we enter a service line or before a spider establishes a web, the probability for $t < 0$ is 0.

Recall that the area below the probability density function over the entire domain of the function must be equal to 1. Thus,

$$1 = \int_{-\infty}^{\infty} f(t)\, dt = \int_{-\infty}^{0} 0\, dt + \int_{0}^{\infty} ae^{-bt}\, dt$$

$$= \int_{0}^{\infty} ae^{-bt}\, dt = \lim_{x \to \infty} \int_{0}^{x} ae^{-bt}\, dt$$

$$= \lim_{x \to \infty} -\frac{a}{b} e^{-bt} \Big|_{0}^{x}$$

$$= \lim_{x \to \infty} -\frac{a}{b} \left(e^{-bx} - e^{0} \right)$$

$$= -\frac{a}{b}(0 - 1) = \frac{a}{b}.$$

Since $1 = a/b$, then $a = b$, and we can rewrite the probability density function as

$$f(t) = \begin{cases} 0, & \text{if } t < 0 \\ ae^{-at}, & \text{if } t \geq 0. \end{cases}$$

To find the mean waiting time, we evaluate $\int_{-\infty}^{\infty} tf(t)\, dt$. For this calculation, we will have to use integration by parts:

$$\mu = \int_{-\infty}^{\infty} tf(t)\, dt = \int_{-\infty}^{0} 0\, dt + \int_{0}^{\infty} ate^{-at}\, dt$$

$$= \lim_{x \to \infty} \int_{0}^{x} ate^{-at}\, dt \quad u = t, \ \ dv = ae^{-at}\, dt$$

$$= \lim_{x \to \infty} \left[-te^{-at} \Big|_{0}^{x} + \int_{0}^{x} e^{-at}\, dt \right]$$

$$= \lim_{x \to \infty} \left[-xe^{ax} - \frac{e^{-at}}{a} \Big|_{0}^{x} \right]$$

$$= \lim_{x \to \infty} \left[-xe^{-ax} - \frac{e^{-ax}}{a} + \frac{1}{a} \right] = \frac{1}{a}.$$

Note that you can verify graphically that $\lim_{x \to \infty} -xe^{-ax} = 0$. Thus, the mean waiting time is $\mu = 1/a$. Therefore, $a = 1/\mu$, and we can rewrite the probability density function in terms of the mean waiting time:

$$f(t) = \begin{cases} 0, & \text{if } t < 0 \\ \frac{1}{\mu} e^{-t/\mu}, & \text{if } t \geq 0. \end{cases}$$

This is called the *exponential density function*, and it is one of the most commonly used distributions for waiting times until the next arrival. Queuing theory deals with lengths of waiting lines in restaurants or at service stations and for packet arrivals on computer networks. The exponential distribution has the "memoryless property" since the probability that the next arrival, say, a prey item at a spider web, occurs in the next 10 minutes does not depend on whether the last prey just arrived or whether it arrived 20 minutes ago. The memoryless property says that the distribution for the waiting time starts over anew at any time you start to observe the process. This may seem unusual, but it has been found to be a reasonable first assumption for many cases of waiting times. However, other distributions may be more appropriate to use if, for example, you have a lower bound on the time to the next arrival or there is a most likely waiting time. For example, buses that run on a fixed schedule have some variability in the exact times between arrivals at a station but not very much if they stay mostly on schedule.

25.4 Matlab Skills

Waiting Times: Birders and Bad Luck

Birding is one of the most popular outdoor activities in the world, and many birders keep a "life list" of species of birds they have observed. It is common for the local "birder network" to inform each other when a particularly unusual or rare species has been spotted in an area. Suppose that you wanted to add the broad-winged hawk (*Buteo platypterus*) to your life list and someone told you that two of these hawks were observed in an hour at a nearby vantage point (perhaps during their migration south to Florida for the winter). If you then rushed over to the viewing site, how long do you suppose you would have to wait to see a hawk? Since it appears that the mean time between bird sightings is 30 minutes (there were two birds seen in an hour) and you might think of yourself as randomly arriving at the vantage point between two bird sightings, you would expect that on average you'd have to wait about 15 minutes to see a hawk and add that species to your life list. This is thinking of yourselfs, on average, arriving in the middle of two sightings of hawks. We will see below whether this intuition is correct, but first we will calculate the probability of seeing a hawk in some time period.

If we assume an exponential distribution of times of arrivals for hawks passing this vantage point, what is the probability that you will see at least one hawk in 30 minutes? If we let T be a random variable that has the exponential distribution with mean waiting time of $\mu = 0.5$, where time is measured in hours, then the probability that a waiting time is less than 30 minutes is

$$P(T \le 0.5) = \int_0^{0.5} \frac{1}{0.5} e^{-t/0.5}\, dt = e^{-t/0.5}\Big|_0^{0.5} = 1 - e^{-1} = 0.632.$$

This would seem to imply that you would have a reasonable chance (63%) of seeing at least one hawk if you spent 30 minutes at the vantage point.

Now let us determine if your intuition is correct about the mean time you would need to wait to see a hawk. We will use Matlab to randomly generate 10 waiting times. Essentially, we will "randomly sample" the exponential probability distribution 10 times.

The exponential probability distribution with a mean waiting time 30 minutes (or 0.5 hours) is

$$P(T \le t) = \int_0^t \frac{1}{0.5} e^{-u/0.5}\, du = 1 - e^{-2t}.$$

Let p be the probability that your waiting time T is less than or equal to t; then

$$p = 1 - e^{-2t} \quad \Rightarrow \quad e^{-2t} = 1 - p \quad \Rightarrow \quad t = -0.5 \ln(1 - p).$$

Since p is a probability, it has a value between 0 and 1. In Matlab, we can use the built-in function `rand` to generate a random number between 0 and 1 (see Matlab Skills Section 10.5). Using `rand(1,10)` or `rand(10,1)`, we can generate 10 random numbers between 0 and 1. In the command window, we can generate 10 waiting times from the exponential probability distribution:

```
———————————————— Command Window ————————————————
>> t = -0.5 * log(1-rand(10,1))

t =
    0.9587
    1.0354
    0.1576
    0.1169
    0.4162
    0.5113
    0.2698
    0.1153
    1.4776
    0.0428
```

This is one way to simulate a random variable if you can both find the distribution function and take its inverse (which is what we have done by solving for t in terms of p). The mean time between arrivals for the sample of waiting times is obtained by simply taking their arithmetic mean, `mean(t)`, to get 0.5102 hours, or 30.6 minutes.

This particular set of times between arrivals for a hawk includes some times that are very short compared to the mean of 0.5 hour and some that are very long compared to it. Note that the values in the array t are in hours. We can use Matlab to convert those values to hours and minutes using the `floor` and `mod` functions. For a real number a, the command `floor(a)` finds the largest integer less than or equal to a. For real numbers a and b, the command `mod(a,b)` finds the remainder of $a \div b$. Typically, we use integer values for b. Once we have the array t, we can find convert those values to hours and minutes in the command window:

```
———————————————— Command Window ————————————————
>>W = [floor(t) round(mod(t*60,60))]

W =
    0    58
    1     2
    0     9
    0     7
    0    25
    0    31
    0    16
    0     7
    1    29
    0     3
```

The command `floor(t)` determines the number of hours for each waiting time. Note that the floor of any value in the array t with a value less than 1 will be 0. The command `mod(t*60,60)` multiplies each value in the array t by 60 (converting it to minutes) and then divides that number by 60 and finds the remainder. Note that we round the resulting value to the nearest minute (the nearest whole number).

Suppose that someone was at the vantage point, stayed there all day, saw the first hawk at 5:00 a.m., and observed 10 more hawks such that the time between sightings (in fractions of an hour) is given by the sampled values in t. Then the list of times at which hawks were observed for this particular set of interarrival times is

05:00, 05:58, 07:00, 07:09, 07:16, 07:41, 08:12, 08:28, 08:35, 10:04, and 10:06.

We can use Matlab to calculate these values as well. For this calculation, we will use the Matlab function cumsum, which calculates the cumulative sum of an array. For example, if want to find the cumulative sum of the array

$$[0 \quad 1 \quad 2 \quad 3 \quad 4 \quad 5],$$

then the function cumsum will produce the array

$$[0 \quad 1 \quad 3 \quad 6 \quad 10 \quad 15],$$

where each value x_i in the latter array is the sum of the first i terms in the original array. In the command window, we compute the time at which each of the 10 hawks was sighted:

```
────────────────── Command Window ──────────────────
>> S = cumsum(t)+5

S =
    5.9587
    6.9941
    7.1517
    7.2686
    7.6848
    8.1961
    8.4659
    8.5812
   10.0588
   10.1016

>> W = [floor(S) round(mod(S*60,60))]

W =
    5    58
    6    60
    7     9
    7    16
    7    41
    8    12
    8    28
    8    35
   10     4
   10     6
```

Note that we add 5 to the cumulative sum since the initial hawk sighting occurred at 5:00 a.m. After we calculate the time in hours of each sighting, we again converted those value to hours and minutes to obtain the list of times of the sightings. Note that for this type of conversion, we would compute the times in military time, so any time occurring after 12:00 p.m. would have hour values greater than 12.

The first sighting occurred at 5:00 a.m., and the 11th sighting occurred at 10:06 a.m. Imagine a birder going to the vantage point sometime during this period. Lacking additional information, we will assume that the birder arrives at some "random" time between 5:00 a.m. and 10:06 a.m. The definition of "random" here is that it is equally likely that the birder will arrive at any time in this time period. When it is equally likely that any particular value is selected from an interval, we use the the probability distribution function: the *uniform distribution*. Recall that the density function for a uniform random variable on the interval from a to b is

$$f(t) = \begin{cases} 0, & \text{if } t < a \\ 1/(b-a), & \text{if } a < t \le b \\ 0, & \text{if } t > b. \end{cases}$$

We have already seen how to sample from a uniform probability distribution in Matlab Skills Section 10.5, though we did not call it a uniform probability distribution at that point. To use rand to sample from a uniform distribution over $[a, b]$, we use the command

```
rand(1)*(b-a)+a
```

We will use Matlab to simulate 10 birders arriving at random times between 5:00 a.m. and 10:06 a.m. Recall that we saved the time values in the array S. Thus, we want to sample from the uniform distribution over $[5, max(S)]$. In the command window, we sample 10 values from the uniform distribution and convert those values to time of day:

```
———————————— Command Window ————————————
>> R = rand(10,1)*(max(S)-5)+5

R =
    8.3453
    5.1822
    9.3319
    9.7649
    8.4626
    8.8657
    8.7912
    7.0010
    8.3440
    5.8733

>> [floor(R) round(mod(R*60,60))]

ans =
    8    21
    5    11
    9    20
    9    46
    8    28
    8    52
    8    47
    7     0
    8    21
    5    52
```

Thus, the times at which the birders arrive are

$$8:21, 5:11, 9:20, 9:46, 8:28, 8:52, 8:47, 7:00, 8:21, \text{ and } 5:52.$$

How long does each birder have to wait to see a hawk? Consider the birder who shows up at 8:21 a.m. The next hawk passes the vantage point at 8:28. So this birder will have to wait only 7 minutes. However, for the birder who shows up at the vantage point at 5:11, the next hawk will not pass by until 5:58. This birder will have to wait 47 minutes. Let us use Matlab to compute the time each birder has to wait to see the next hawk and find the average time that a randomly arriving birder will have to wait for a sighting.

The array S contains the times at which each hawk passes the vantage point. In Matlab, the command S(S>R(1)) finds all the values in S that are greater than R(1) (the time that the first sample birder arrives). Thus, the command returns the times of all the sightings that the birder who arrived at time R(1) would be able to see. If we want the time of the first such sighting, we can take the minimum of the resulting array. Thus, min(S(S>R(1))) is the minimum value of S, which is greater than R(1). If we then subtract the time at which the birder arrived, we have calculated how long the birder had to wait for the first sighting. We can construct a for-loop to find the waiting time for each birder (looping through the length of R). Note that when you construct a for-loop in the command window, pressing the "Enter" or "Return" key will not execute the loop until after you have typed end:

```
─────────────────── Command Window ───────────────────
>> for i=1:length(R)
D(i) = min(S(S>R(i))) - R(i);
end

>>  [floor(D') round(mod(D'*60,60))]

ans =
     0      7
     0     47
     0     44
     0     18
     0      0
     1     12
     1     16
     0      9
     0      7
     0      5

>>  meanD = mean(D*60)

meanD =
    28.4395
```

After we computed the waiting times, we converted those values to hours and minutes. Note that we had to use the transpose of D to have the data print out in two columns. It is interesting to observe that while some birders had to wait fewer than 10 minutes for a sighting, others had to wait over an hour. When we compute the mean waiting time of the 10 birders (in minutes), we found that the average time a birder had to wait for a sighting was roughly 28 minutes.

Recall that we hypothesized that a randomly arriving birder would have to wait on average 15 minutes. However, our random simulation shows the average waiting time to be closer to double that amount of time. Why is this? It is possible that this is an artifact of the particular arrival times of the hawks and birders we obtained. To test this idea, we could try rerunning the "experiment" of simulating bird and birder arrival times and computing the average waiting time of each birder. The Matlab script file HawkWaitingTimes.m simulates the appearance of 10 hawks at the vantage point after 5:00 a.m., along with the random arrival of 10 birders and the amount of time they have to wait to see the next hawk. Run the HawkWaitingTimes.m script file multiple times. What are the average birder waiting times?

────── HawkWaitingTimes.m ──────

```
% Hawk Waiting Times

% Generate a sample of 10 waiting times
T = -0.5 * log(1-rand(1,10));

% Calculate times of sightings
S = 5 + cumsum(T);

% Display waiting times in terms of hours and minutes
fprintf('\nWaiting Times  //  Sighting Times\n')
fprintf('                //   5:00  \n')
for i=1:length(T)
  fprintf('%2.f h  %2.f min',floor(T(i)),round(mod(T(i)*60,60)))
  fprintf('   //  ')
  fprintf('%2.f:%02.f\n',floor(S(i)),round(mod(S(i)*60,60)))
end

% Calculate and print mean waiting time in minutes
meanT = mean(T*60);
fprintf('\nMean waiting time is %4.1f minutes\n\n',meanT)

% Generate a random time at which the birder shows up
R = rand(1,10)*(S(length(S))-5)+5;

% For each birder, find the time of the next hawk sighting
for i=1:length(R)
  v = min(S(S>R(i))); %time of next sighting
  D(i) = v - R(i);    %time of next sighting - time of birder arrival
end

% Display arrival and waiting times
fprintf('10 Birders randomly show up\n')
fprintf('\nArrival Times  //  Waiting Times\n')
for i=1:length(T)
  fprintf('  %02.f:%02.f',floor(R(i)),round(mod(R(i)*60,60)))
  fprintf('       //  ')
  fprintf('%3.f min\n',D(i)*60)
end

% Calculate and print mean waiting time in minutes of the 10 birders
meanD = mean(D*60);
fprintf('\nMean waiting time of 10 random birders is %4.1f minutes\n\n',meanD)
```

When we run the `HawkWaitingTimes.m` script two times, we generate the following output in the command window:

```
───────────────────────── Command Window ─────────────────────────
>> HawkWaitingTime

Waiting Times  //  Sighting Times
//    5:00
0 h   37 min  //    5:37
0 h    1 min  //    5:38
0 h   10 min  //    5:47
0 h    1 min  //    5:49
0 h    3 min  //    5:52
0 h   52 min  //    6:44
0 h   36 min  //    7:20
0 h   11 min  //    7:31
1 h   30 min  //    9:01
0 h    1 min  //    9:02


Mean waiting time is 24.2 minutes

10 Birders randomly show up

Arrival Times  //  Waiting Times
06:46       //    33 min
06:32       //    12 min
08:05       //    56 min
08:12       //    49 min
05:45       //     2 min
06:59       //    21 min
06:48       //    32 min
07:36       //    85 min
07:52       //    69 min
08:03       //    58 min

Mean waiting time of 10 random birders is 41.6 minutes

>> HawkWaitingTime

Waiting Times  //  Sighting Times
//    5:00
0 h   10 min  //    5:10
0 h   34 min  //    5:44
0 h   32 min  //    6:16
0 h    5 min  //    6:21
0 h    4 min  //    6:25
0 h   21 min  //    6:46
1 h   36 min  //    8:22
0 h   12 min  //    8:34
0 h   26 min  //    9:01
0 h    8 min  //    9:08


Mean waiting time is 24.8 minutes

10 Birders randomly show up

Arrival Times  //  Waiting Times
08:07       //    15 min
06:03       //    12 min
```

```
07:06        //      76 min
07:54        //      28 min
08:41        //      20 min
08:58        //       3 min
07:16        //      66 min
05:34        //       9 min
05:37        //       7 min
06:04        //      12 min

Mean waiting time of 10 random birders is 24.8 minutes
```

If you run the HawkWaitingTimes.m script file many times, you will observe that the mean waiting time of the birders is rarely as low as 15 minutes. The reason for this is a bit subtle. You can visualize what happens by thinking of a time line starting at 5:00 a.m. and ending at 10:06 a.m. and marking on it the arrival times of the hawks. Now imagine tossing a dart randomly at this time line. Where is the dart most likely to hit? It is much more likely that it will land in a time segment that has a long interarrival time (such as the time between 8:35 a.m. and 10:04 a.m.) than in one of the shorter interarrival times (such as the one from 10:04 a.m. To 10:06 a.m.). Thus, it is more likely that a randomly arriving birder will have a longer wait than the wait time that we intuitively thought of as one-half of the average waiting time between birds that the same birder would observe if he or she had stayed at the vantage point all day. This is commonly called "bad luck." The official name for this is "length-biased sampling," which means that you are more likely to arrive in a longer interarrival time than in a shorter one.

25.5 Exercises

25.1 The probability density function of a random variable X is given by

$$f(x) = \frac{1}{300}e^{-\frac{x}{300}},$$

for $x \geq 0$ and $f(x) = 0$ for $x < 0$. Find $P(X \leq 600)$.

25.2 The probability density function of a random variable X on $[2,5]$ is given by

$$f(x) = 1 - \frac{1}{\sqrt{x-1}}.$$

Find $P(3 \leq X \leq 5)$.

25.3 Consider the function $f(x) = \beta - \frac{x}{4}$ for $1 \leq x \leq 2$ and $f(x) = 0$ otherwise. Determine the value of β so that this function is a probability density function on the interval $[1, 2]$.

25.4 Consider the function $g(x) = \alpha - \left(x - \frac{1}{2}\right)^2$ for $0 \leq x \leq 1$ and $g(x) = 0$ otherwise.

(a) Determine the value of α so that this function is a probability density function on the interval $[0, 1]$.

(b) Graph the function $g(x)$ for your value of α and from this graph make a guess as to the mean and median for this probability density function.

(c) Calculate the mean and median of the probability density function and compare these to your guesses.

25.5 For the uniform probability density function on the interval $[a, b]$, do the following.

(a) Find the variance and standard deviation
(b) Find the ratio of the standard deviation to the mean and describe what happens to this ratio in the special case for the interval being $[0, b]$

25.6 A traffic light remains red for 60 seconds at a time before it changes to green. You arrive at this light and find it red. Assuming that you arrive at a uniformly distributed time during the period the light is red, the probability density function for how long you wait until the light changes would be $f(x) = \frac{1}{60}$ for $0 \le x \le 60$ and $f(x) = 0$ otherwise. What is the probability that you have to wait at least 20 seconds for the light to turn green?

25.7 Consider times at which a parent provisions nestlings, that is, the times between arrivals of a parent carrying food for the nestlings. Suppose that the waiting time between arrivals of a parent bird follows the exponential density function $f(t) = .6e^{-0.6t}$ for $t \ge 0$ and $f(t) = 0$ for $t < 0$, where the waiting times are in minutes.

(a) What is the mean waiting time?
(b) Find the probability that the wait will be between 2 and 5 minutes.
(c) Find the probability that the wait will be at least 3 minutes.

25.8 For the previous exercise, suppose that a different pair of parent birds are one-half as efficient at finding food. This means that the mean times between arrivals of a provisioning parent are twice as long as in the previous exercise.

(a) What is the probability density function for waiting times between arrivals of a parent in this situation?
(b) Compare the less efficient parent's ability to provision the nest to the more efficient parent's by finding the probabilities in the two cases of parents that the wait will be at least 1 minute, 2 minutes, and 4 minutes.

25.9 The vertical distribution of some marine invertebrates follows a probability distribution with depth $\rho(x) = 2\beta x e^{-\beta x^2}$, where $\beta > 0$. This means that the probability that any particular individual invertebrate is found between depths x and $x + \triangle x$ m is approximately $\rho(x) \triangle x$.

(a) Show that the depth at which the probability density function is maximized (e.g., the depth at which there is the highest probability of finding an invertebrate) is $x = (2\beta)^{-\frac{1}{2}}$.
(b) Multiplying the density function by a constant α gives a function that provides the organism density $S(x) = \alpha \rho(x)$ at each depth (the units for α will be individuals per cubic meter). Find α and β so that the maximum density of 115 organisms per cubic meter occurs at a depth of 35 m.
(c) Using the values of α and β found above, find the total number of invertebrates between $x = 0$ m and $x = 50$ m. Hint: Take the integral of $S(x)$ to find this.

25.10 For an exponential distribution of waiting times with probability density function given by $f(t) = \frac{1}{\mu} e^{-t/\mu}$ for $t \ge 0$ and $f(t) = 0$ for $t < 0$, do the following.

(a) Show that the variance of the waiting time distribution is μ^2. Hint: Use integration by parts.
(b) Compute the standard deviation for the data in the two cases of the birder waiting times produced by the HawkWaitingTimes.m script and find the ratio of the standard

deviation to the mean for these two cases. What do you expect these ratios to be in the two cases, and how do the data compare to these expectations?

25.11 Susan Riechert, arachnologist and behavioral ecologist extraordinaire, has spent many years observing the foraging behavior of a funnel-web-building spider, *Agelenopsis aperta*, in the southwestern U.S. desert. The web of this spider is not sticky: the web acts as a sensory mechanism by which the spider can ascertain that a potential prey item is encountered, and the spider can then quickly attack it. The spider eats a range of insect prey, including flies, homopterans, beetles, grasshoppers, lepidopterans, wasps, bees, and hemipterans. Overall, Susan has found that at one of her study sites, a suitable prey item that the spider can include in its diet arrives every 96 minutes on average over the course of 24 hours. The arrival distribution of prey is exponential based on her observations.

(a) What is the probability that a spider encounters no prey item that can be included in its diet over a 4-hour time period?

(b) For optimal growth, this spider requires about 22 mg of dry weight of insect prey a day. A single grasshopper can supply all of the spider's daily food needs, but on average a grasshopper arrives only every 2857 minutes. What is the probability that a spider's energy needs are met in a single day by grasshoppers? Hint: This is equivalent to asking whether at least one grasshopper arrives at the spider's web over a day.

Unit 6 Student Projects

Trapezoid Rule

We have investigated integrals in two ways: by explicitly using antiderivatives and by approximation using the sum of rectangles. For example, we can calculate

$$\int_0^1 (x^2 + 4)dx = \left(\frac{x^3}{3} + 4x\right)\Big|_0^1 = \frac{13}{3}.$$

This integral is easy to calculate since the antiderivative of its integrand, $x^2 + 4$, can be found exactly. But some important functions do not have explicit antiderivatives. Consider this function, which is used in representing normal distributions:

$$e^{-(x^2)}.$$

No explicit antiderivative exists for this function, so we must approximate its integral numerically instead of using antiderivatives and the fundamental theorem of calculus.

We will use Matlab m-files to calculate integrals numerically with the trapezoid rule.

1. Using the trapezoid rule, estimate the area under the curve of $f(x) = 3x^2 + 4$ on the interval $[1, 5]$ using four subintervals.
2. Now calculate the approximate integrals for the cases in which the number of intervals is 40 and then 100.
3. Calculate the exact value of this integral using antiderivatives.
4. Estimate using the trapezoid rule the area under the curve of $g(x) = e^{-(x^2)}$ over the interval $[-2, 2]$ using the number of subintervals, $n = 4, 40, 100,$ and 1000.

Write a one-page report, describing your procedures and results. Attach copies of your m-files on separate pages. Discuss the accuracy of your answers. ■

Lake Pollution

Pollution enters a lake starting at time $t = 0$ at a rate (in gallons per hour) given by the formula

$$f(t) = 12\left(1 - e^{-0.08t}\right),$$

where t is the time in hours. While there is still pollution in the lake, a filter begins to remove it at the rate of

$$g(t) = 0.35t.$$

(a) How much pollution is in the lake after 6 hours?

(b) Use Matlab to graph $f(t)$ and $g(t)$. From the graph, estimate where the rate of pollution entering the lake is equal to the rate of pollution being removed from the lake, that is, estimate the t for which $f(t) = g(t)$.

(c) Find the amount of pollution in the lake at the time found in part (b).

(d) Estimate the time T when all the pollution from the lake has been removed. To do this, construct an equation

$$H(T) = \int_0^T f(t) - g(t) \, dt.$$

Use the fundamental theorem of calculus to obtain a formula for $H(T)$ by using the specific antiderivatives of $f(t)$ and $g(t)$ and integrating from 0 to T. Then use Matlab to plot $H(T)$ as a function of T. Then, using your graph, determine the T for which $H(T) = 0$. ■

Introduction to Differential Equations

The living world is intimately affected by changes in time and space of many processes that are mostly external to living systems, including environmental conditions, such as light, temperature, and humidity, as well as processes that are internal to these living systems, such as biochemical changes in cells, respiration, and the circulatory transport of nutrients. One of the most fascinating aspects of living systems is the ability of these systems to modify environmental conditions. Lizards move into and out of sun and shade to maintain body temperature, the branching structures of a tree affects light levels in the lower portions of the tree, and complex systems of genetic networks regulate many cellular-level biochemical processes. The general term used to describe the interplay between actions and processes within and between organisms that impact their environment is "feedback." The mechanisms that cause this feedback can be behavioral, such as the movement of individual lizards between different environmental conditions; they can be physiological, such as changes in an individual's breathing that occur under exertion to maintain oxygen levels; and they can be genetic, such as the evolution of characteristics that allow individuals to survive and reproduce under novel conditions. An example of genetic changes occurs in plants growing on mine tailings, which have very high levels of toxic compounds. Because of natural selection in these toxic environments, certain populations have evolved the capacity to tolerate high levels of heavy metals that would kill individuals from populations not adapted to such soil conditions.

Differential equations make up a mathematical tool that allows us to describe and predict the impact of feedbacks between and within living systems and their environment. We have already seen several examples of such equations in situations without any explicit feedback. Recall that we analyzed exponential growth and decay in Chapter 19 using the equation

$$\frac{dy}{dt} = ky$$

and pointed out that this has the solution

$$y(t) = Ce^{kt},$$

where C is a constant. This equation is perhaps the simplest and most important differential equation for describing how some biologically relevant variable changes in time. A **differential equation** is any equation containing derivatives of one or more unknown functions (also called dependent variables) depending on one or more unknown variables. In this unit, we will study only **ordinary differential equations**, in which the unknown functions depend on a single independent variable. In the above exponential model, the function $y(t)$ is the unknown function, where t is the sole independent variable. In many biological systems, it is useful to describe how some variable changes in both time and space. For example, infectious diseases, such as the flu, change locally within a region and between cities or regions, so describing a flu epidemic often requires understanding how the rate of change of the number of infected individuals changes in time and across locations. This requires the use of partial differential equations, which are outside the scope of this text; however, the background on differential equations that we present in this unit will help you understand these more complicated types of models for biological systems as well.

There are many types of ordinary differential equations that arise in biological applications, but we will focus in this unit only on those that are rather similar to the exponential differential equation above. This is an example of a first-order differential equation since the only derivative it includes is the first derivative of the variable of interest. The order of a differential equation refers to the highest derivative in the equation. A **first-order ordinary differential equation** is an ordinary differential equation in which the highest derivative of the dependent variables is a first derivative. We will use the terminology "first-order differential equation" to mean first-order ordinary differential equation. The following are examples of first-order differential equations:

- $\dfrac{dy}{dt} + 10y - e^t = 0$
- $\left(\dfrac{dy}{dt}\right)^2 = y(1 - y)$

Note that in the second differential equation, even though the first derivative of y is squared, it is still only a first derivative.

Differential equations that are important in some biological examples can include higher-order derivatives, or they can include systems of several differential equations (e.g., more than one differential equation) that are connected. For example, a standard model in epidemiology is to have a differential equation for the number of individuals who are susceptible $S(t)$ to the disease of interest and another equation for those who are infected by the disease $I(t)$. The equations for susceptible and infected individuals are linked because the rate of change of infected individuals, $\frac{dI}{dt}$, depends on both $I(t)$ and $S(t)$. Another example would be the situation of a predator-prey system in which the rate of change of the predator density depends on the density of both the predator and the prey and, similarly, the rate of change of the prey density depends on both the prey and the predator densities.

While systems of differential equations arise in many areas of biology, they are beyond the scope of this text, and we encourage you to consult more advanced books on modeling in biology. The topics in this unit are designed to provide you with the necessary background to read the literature in areas of biology of interest to you and provide some intuition about how differential equations are used to describe living systems. We also take this as an opportunity to discuss one of the most important notions in analyzing biological models, that of an equilibrium and whether it is stable. This a mathematical means to describe homeostatic mechanisms in biology.

Separation of Variables

Describing biological systems and predicting the response of these systems to changing conditions is often done using differential equations. In many cases, the interactions between biological processes are considered not through an equation relating the variables, as we will discuss in detail in Chapter 28, but by developing equations for the processes that explicitly incorporate derivatives. These differential equations are often not simple enough to express directly as a function of the independent variable. In most cases, the process we are trying to describe has dynamics that depend on itself. The first-order differential equation that describes exponential growth and decay is an example of this. If $N(t)$ is population size at time t, then the differential equation for exponential growth $N'(t) = rN(t)$ has the growth rate as a linear function of the current population size. It is rather rare in biology to have the case of a derivative depending only on the independent variable (time, or t in this case). If indeed we could express $N'(t)$ as a function of just t, then we need only find the antiderivative of $N'(t)$ to be able to describe how the population size $N(t)$ changed through time.

In many realistic biological situations, however, the equation for $N'(t)$ also includes $N(t)$. This is true not only in population biology but also in pharmacokinetics, in which the rate of decay of a drug in the body depends on the current concentration of that drug. Most biological systems described by differential equations are like this because the feedback in biological systems, mentioned in the introduction to this unit, that distinguishes living systems from most examples in physics. It turns out that even in cases in which there is this feedback present, so that the equation for $N'(t)$ depends on $N(t)$, we can often still find a formula for $N(t)$, but we can't do this by simply taking an antiderivative. It requires the use of a method that allows us to "separate out" the $N(t)$ parts of the differential equation from all other parts. We start with some simple examples of this method. Then, in the next chapter, we will use this method for the differential equation with logistic growth:

$$\frac{dN}{dt} = rN\left(1 - \frac{N}{K}\right).$$

Another theme of this chapter concerns how we account for some basic data in describing the response of a biological system. We already saw some examples of this in the case of exponential growth and decay, in which we considered some basic data about the variables. For example, in the case of describing a pheasant population on an island, we noted the number of pheasants that were initially introduced onto the island and how many were present 3 years later. In the morbid case from forensic anthropology, we saw that Newton's Law of Cooling could be applied to find the time of death but required information about the body's temperature at the time the body was found and the room temperature. In both cases, we needed to use observations of the variables of interest (e.g., population size or body temperature) in order to find the solution of the differential equation. We will see in this chapter how to use data, which are called initial or boundary conditions, to determine what value the biological variable of interest has at any time in the past or the future.

If we have a differential equation of the form

$$\frac{dy}{dt} = f(t)$$

and we know that the antiderivative of $f(t)$ is $F(t)$ (meaning $F'(t) = f(t)$), then we can easily find all the solutions of this differential equation:

$$y(t) = F(t) + C.$$

Example 26.1 (Simple Differential Equation)

The first AIDS case in the United States was identified in mid-1981. During the early phases of the AIDS epidemic, it can be estimated that the total growth in number of cases among males follows a differential equation of the form $\frac{dy}{dt} = 642t^2$, where t is in years after 1981 (i.e., $t = 0$ corresponds to 1981). This assumes a very small number of initial cases. Find all solutions of this differential equation.

Solution: All of the solutions of this differential equation can be found by taking an antiderivative to obtain

$$y = 642\frac{t^3}{3} + C.$$

Notice that since C can be any real number, there are infinitely many solutions that satisfy the differential equation. In the example above,

$$y = 642\frac{t^3}{3} + C$$

represents a *family of solutions*, while the equation

$$y = 642\frac{t^3}{3} + 27$$

represents one specific solution.

If we have a family of solutions, we can specify a particular solution by indicating the value of the solution at a particular point. For example, if we know that at $t = 0$ the function value is $y = 27$, we write this as $y(0) = 27$. This is based on an assumption that the initial number of U.S. AIDS cases near the beginning of 1981 was 27 males.

In biological applications, we can often determine a differential equation that defines the rate of change of some quantity and know the initial state of the quantity or some information about what occurs at some point in the domain of the independent variable. For example, if we are modeling a population, we might have a differential equation defining the rate of change of the population $\frac{dy}{dt} = f(y)$ and the size of the population at an initial time $y(0) = y_0$. The extra information given by $y(0) = y_0$ is called the *initial condition*. If $y(10) = 1000$ is given, that extra information is called a *boundary condition*. As an example in which the independent variable is not time, it is possible to model the changes in chlorophyll content C of some leaves as depending on the current chlorophyll concentration and daily light level L. In this case, the rate of change of chlorophyll content, $C'(L)$, depends on L and C and so might be written as a differential equation $\frac{dC}{dL} = h(C, L)$ for some function h, which depends on both C and L. The solution to the differential equation, that is, the function $C(L)$, is a function of average daily light levels, and thus the average daily light level is the independent variable. In this case, if we know both the function h and the chlorophyll concentration for a particular average daily light level, we can determine one particular solution to the differential equation and be able to specify the chlorophyll content of the leaf at any light level.

Before treating the separation of variables method, we illustrate solving a simple differential equation with an associated boundary condition.

Example 26.2 (Simple Differential Equation with a Boundary Condition)

Find the solution to the differential equation $\frac{dy}{dt} = 642t^2$ with boundary condition $y(6) = 46{,}251$. Note that this corresponds to the total number of AIDS cases reported in U.S. males by 1987 being 46,251, which is an estimate based on data from the Centers for Disease Control.

<u>Solution:</u> The general solution of this differential equation is

$$y = 642\frac{t^3}{3} + C.$$

Now we need to find the constant C so that $y(6) = 46{,}251$. Using $t = 6$ and $y = 46{,}251$ together, we have $46{,}251 = \frac{642(6^3)}{3} + C$, which gives $C = 27$. This is consistent with the assumption that there were a small number of initial cases. Note that this approximation of the growth of reported AIDS cases is reasonable only over the initial phases of the AIDS epidemic, for years t that are fairly soon after the start of the epidemic, since by the late 1980s the epidemic growth rate had been thankfully reduced. The number of total cases continued to grow but not as rapidly as during the earliest phases of the epidemic.

26.1 Separation of Variables Method

It is possible that the derivative $\frac{dy}{dt}$ is represented in terms of t and y. For example,

$$\frac{dy}{dt} = ty.$$

In this case, we consider another method for solving the differential equation. The idea is to separate the y terms on one side of the equation and t terms on the other side. Assuming that y does take the value 0 on our domain, we would do the separation as

$$\frac{1}{y}\frac{dy}{dt} = t$$

and, continuing to separate,

$$\frac{dy}{y} = tdt.$$

We are able to pull apart the "fraction"

$$\frac{dy}{dt}$$

by multiplying across by dt, as we can think about the differentials, dy and dt, as changes in y and t, respectively.

One of the most commonly used methods for solving a differential equation $y' = f(t, y)$ is the technique known as **separation of variables**. We would write the derivative y' as $\frac{dy}{dt}$. In this method, we attempt to get the terms with y on the left-hand side of the equation and the terms with t on the right-hand side of the equation. The process for solving a differential equation by separation of variables is as follows.

The Separation of Variables Method

For a differential equation of the form $\frac{dy}{dt} = h(y)g(t)$:

(a) Rearrange the differential equation to be in the form

$$h(y)\frac{dy}{dt} = g(t).$$

(b) Moving the dt to the right-hand side, we obtain

$$h(y)\, dy = g(t)\, dt.$$

(c) Using indefinite integrals on both sides of the equation, we obtain antiderivatives giving our relationship between y and t:

$$\int h(y)\, dy = \int g(t)\, dt \quad \Rightarrow \quad H(y) = G(t) + C,$$

where $H(y)$ is an antiderivative of $h(y)$, $G(t)$ is an antiderivative of $g(t)$, and C is an integration constant.

(d) Solve for y in terms of t.

If you have additional information about the specific y-values at $x = 0$ or another x-value, you can solve for the constant C in equation for y as a function of t.

Let us consider an example.

Example 26.3 (Solving a Differential Equation with a Boundary Condition)

Solve the differential equation $\frac{dy}{dt} = \frac{t}{y}$ subject to the boundary condition $y(2) = 1$.

Solution: Write the equation as

$$y\frac{dy}{dt} = t \;\Rightarrow\; y\, dy = t\, dt.$$

Integrating both sides, we get

$$\int y\, dy = \int t\, dt$$

$$\frac{1}{2}y^2 = \frac{1}{2}t^2 + C$$

$$\frac{1}{2}(1)^2 = \frac{1}{2}(2)^2 + C.$$

We obtain $C = \frac{3}{2}$. Then we solve our equation below for y in terms of x

$$\frac{1}{2}y^2 = \frac{1}{2}t^2 - \frac{3}{2},$$

to obtain

$$y = \sqrt{t^2 - 3}.$$

Note that when we took the square root of both sides in solving for y, we might have written

$$y = \pm\sqrt{t^2 - 3},$$

but we choose the positive square root to due the boundary condition, giving a positive y-value at $t = 2$.

Recall that the integration rule

$$\int \frac{1}{x}\, dx = \ln x + c$$

holds for $x > 0$. The more general formula is

$$\int \frac{1}{x}\, dx = \ln |x| + c \ \text{ for } x \neq 0.$$

Note that unless the independent variable x is known to be positive, we must use this new formula.

Recall that we have seen that the differential equation $y' = ky$ has the solution $y(t) = y_0 e^{kt}$, where $y_0 = y(0)$. We will now see that we can derive that solution via separation of variables.

Example 26.4 (Exponential Growth)

Use separation of variables to solve

$$\frac{dy}{dt} = ky.$$

Solution: We write

$$\frac{1}{y}\frac{dy}{dt} = k \quad \Rightarrow \quad \frac{1}{y}\,dy = k\,dt.$$

Since we do not have an initial condition, we integrate both sides using indefinite integrals to get

$$\int \frac{1}{y}\,dy = \int k\,dt$$

$$\ln|y| = kt + c$$

$$|y| = e^{kt+c} = c_1 e^{kt},$$

where $c_1 = e^c$. Notice that we assumed that $y \neq 0$ in order to rearrange the equation and use separation of variables. However, if $y \equiv 0$, then

$$\frac{dy}{dt} = k(0) = 0,$$

which implies that $y(t) = c_2$, where c_2 is another constant. Thus, we can write the general solution to the differential equation as

$$y(t) = ce^{kt},$$

where c can take on any real value, including 0.

Example 26.5 (Restricted Growth)

Suppose that there is an upper bound b on the size of a population y. The growth rate tends to 0 as y approaches b and is proportional to $b - y$. For this situation, find *all* solutions to

$$\frac{dy}{dt} = k(b - y),$$

under the assumption that $y(0) < b$.

(Continued)

Solution: Since $y(t) < b$, we have

$$\frac{1}{b-y}dy = kdt.$$

Since we do not have a specific initial condition, we integrate both sides using indefinite integrals to get

$$\int \frac{1}{b-y}\,dy = \int k\,dt$$

$$-\ln(b-y) = kt + c$$

$$\ln(b-y) = -kt - c$$

$$b - y = e^{-kt-c} = c_1 e^{-kt},$$

where $c_1 = e^{-c}$. Then we obtain as a general solution

$$y = b - c_1 e^{-kt}.$$

Example 26.6 (Allometric Growth [16])

If x and y are measurements of parts of an organism, then $\frac{dx}{dt}$ and $\frac{dy}{dt}$ are their respective rates of growth. However, much more biological information is conveyed by the specific or relative growth rates

$$\frac{1}{x}\frac{dx}{dt} \quad \text{and} \quad \frac{1}{y}\frac{dy}{dt}.$$

For example, the significance of a growth rate of $\frac{dx}{dt} = 5$ pounds per month would be much different for a human than for an elephant. By forming the specific growth rates, we can compare the growth of the present status of the organism. There is a growth hypothesis that states that over an interval of time (a, b), the specific growth rates $\frac{1}{x}\frac{dx}{dt}$ and $\frac{1}{y}\frac{dy}{dt}$ are directly proportional. Use this information to show that the equation we used for allometric growth in Chapter 4,

$$y = \alpha x^k,$$

is valid.

Solution: If the specific growth rates $\frac{1}{x}\frac{dx}{dt}$ and $\frac{1}{y}\frac{dy}{dt}$ are directly proportional, then there exists some constant $k > 0$ such that

$$\frac{1}{y}\frac{dy}{dt} = k\left(\frac{1}{x}\frac{dx}{dt}\right).$$

Note that

$$\frac{dy}{dx} = \frac{\frac{dy}{dt}}{\frac{dx}{dt}};$$

(Continued)

thus, we can write

$$\frac{dy}{dx} = k\frac{y}{x} \quad \Rightarrow \quad \frac{1}{y}dy = \frac{k}{x}dx.$$

Note that since x and y are measurements of the body parts of an organism, we can assume that $x, y > 0$. Using separation of variables, we have

$$\int \frac{1}{y}\,dy = \int \frac{k}{x}\,dx$$

$$\ln(y) = k\ln(x) + c$$

$$= \ln\left(x^k\right) + c$$

$$y = e^{\ln\left(x^k\right)+c}$$

$$= \alpha x^k,$$

where $\alpha = e^c$.

26.2 Matlab Skills

The solutions to differential equations are continuous functions. However, there are methods to approximate values of the solution function over a series of values of the independent variable (usually time). For example, we could approximate the solution to the differential equation $\frac{dy}{dx} = xy$ by approximating the solution $y(x)$ at x-values $0, 1, 2, 3, 4, \ldots$.

In this text, we do not introduce these methods for numerically approximating the solutions to differential equations. Some of these methods would be discussed in detail in a differential equations course. See Brannan and Boyce's *Differential Equation* textbook [9] or Burden and Faires's *Numerical Analysis* textbook [10] for an explanation of the Euler Method (the simplest numerical method for approximating the solution to a differential equation).

For numerical approximations of solutions to differential equations in this text, we turn to Matlab. Matlab provides built-in functions to numerically "solve" differential equations, and we discuss how to use these ordinary differential equation "solvers" here.

To numerically solve a differentia equation in Matlab, we must first define the differential equation; that is, we must tell Matlab what differential equation we are using. We do this by creating an m-file that we will call `odef.m`. If we wish to numerically solve the differential equation $\frac{dy}{dx} = xy$, then the function `odef.m` would look like the following:

```
─────────────────────── odef.m ───────────────────────
1   % Creates the function odef(x,y)
2   % Input: x (a real number), y (a real number)
3   % Output: dy
4
5   function dy = odef(x,y)
6   dy = x*y;
```

Notice that the function `odef.m` calculates the value of the derivative of y; that is, it finds $\frac{dy}{dx}$ at the point (x, y).

Now, when we are numerically solving differential equations, we need to tell Matlab which specific solution (i.e., which of the infinitely many solutions in the family of solutions) we are interested in finding. We do this by providing Matlab with the initial or boundary condition. To solve the differential equation that we defined in `odef.m`, we use the following command:

```
[X,Y] = ode45(@odef, xarray, y0)
```

where `@odef` tells Matlab to use the differential equation defined by `odef.m`, `xarray` is a vector that specifies the x-values over which we want to approximate the numerical solution to the differential equation, and `y0` is the initial or boundary condition. The outputs of this built-in function are `X` and `Y`. Both of these outputs are arrays (or vectors) of values. The vector `X` contains x-values, and `Y` contains the corresponding y-values at the given x-values. Thus, if we want to display the solution to the differential equation visually, we can plot the vectors `X` and `Y`:

```
plot(X,Y)
```

If we create the function `odef.m` to define the differential equation $\frac{dy}{dx} = xy$, we can then solve the differential equation over the interval $[0,1]$ using various different boundary conditions:

```
———————————————— Command Window ————————————————
>> [X,Y] = ode45(@odef, [0 1], 1);
>> plot(X,Y)
```

Note that we place the semicolon at the end of the first line because we do not want to see the printout of the arrays `X` and `Y`. Instead, we plot the arrays against each other using the `plot(X,Y)` command in the next line, which produces the graph shown in Figure 26.1.

If we want to explore a variety of different initial conditions, we can construct an m-file with a for-loop, where each pass through the loop numerically approximates the solution using a different initial condition:

```
———————————————— ODESolutions.m ————————————————
1   % Vector of initial conditions
2   y0 = 0.2: 0.2 : 2;
3
4   % For loop
5   for i = 1:length(y0)
6       [X,Y] = ode45(@odef, [0 1], y0(i));
7       plot(X,Y)
8       hold on;  % keep same plot window open
9   end
10
11  hold off; % next plot will be in a new window
```

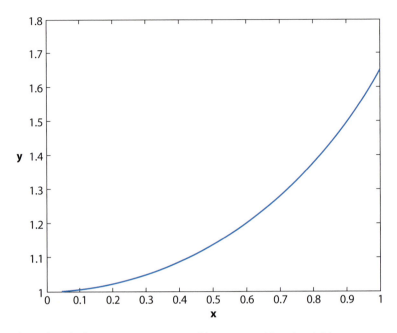

Figure 26.1 Graph produced using `ODESolutions.m` with `odef.m` set to solve $dy/dx = xy$.

This m-file, however, has the same color of each of the different solutions plotted. We would like to be able to differentiate each of the solutions plotted. To do this, we make the following modifications:

ODESolutions.m

```
1    % Vector of initial conditions
2    y0 = 0.2: 0.2 : 2;
3
4    % Create an array of different colors
5    % We need as many colors as we have initial conditions
6    c=hsv(length(y0));
7
8    %Create an empty array for the legend text
9    lt=[];
10
11   % For loop
12   for i = 1:length(y0)
13       [X,Y] = ode45(@odef, [0 1], y0(i));
14       plot(X,Y,'color',c(i,:))
15       lt = strvcat(lt,num2str(y0(i))); % legend text for this loop
16       hold on;  % keep same plot window open
17   end
18
19   legend(lt,'Location','EastOutside');  % create legend
20   hold off; % next plot will be in a new window
```

This file uses some new built-in Matlab functions. The `hsv(n)` function takes an integer number *n* as an input and then produces an $n \times 3$ matrix, where each row has three values that, together represent a distinct color. The `num2str(n)` function takes any real number *n* and stores it as

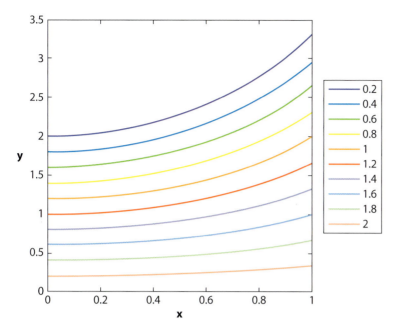

Figure 26.2 Graph produced using `ODESolutions.m` with `odef.m` set to solve $dy/dx = xy$.

text for printing (in this case, to print in the legend of the graph). The function `strvcat(S1, S2, S3, ...)` creates a matrix of text where each row of the matrix is one word of text. In the m-file `ODESolutions.m` we use `strvcat` to add a new line of text to the matrix `lt` with each pass through the loop.

Using `ODESolutions.m` and `odef.m` where $\frac{dy}{dx} = xy$, we produce the graph shown in Figure 26.2.

Now, suppose that we are told that the rate of change of a population is given by the equation

$$\frac{dy}{dt} = 0.25y(t)\ln\left(\frac{1000}{y(t)}\right)$$

and we want to know how the population changes over 20 years if the population starts with 50, 100, 500, 1000, and 1500 individuals. We can use the m-files we have already created to approximate the solution curves $y(t)$ for each of the initial conditions.

First, we modify the m-file `odef.m` appropriately:

──────────────── odef.m ────────────────

```
1   % Creates the function odef(x,y)
2   % Input: x (a real number), y (a real number)
3   % Output: dy
4
5   function dy = odef(x,y)
6   dy = 0.25*y*log(1000/y);
```

In this file, x would represent the independent variable t; however, t does not show up explicitly in the differential equation we are using.

Next, we modify ODESolutions.m so that we have initial conditions 50, 100, 500, 1000, and 1500:

ODESolutions.m

```
1   % Vector of initial conditions
2   y0 = [50 100 500 1000 1500];
3
4   % Create an array of different colors
5   % We need as many colors as we have initial conditions
6   c=hsv(length(y0));
7
8   %Create an empty array for the legend text
9   lt=[];
10
11  % For loop
12  for i = 1:length(y0)
13      [X,Y] = ode45(@odef, [0 20], y0(i));
14      plot(X,Y,'color',c(i,:))
15      lt = strvcat(lt,num2str(y0(i))); % legend text for this loop
16      hold on;  % keep same plot window open
17  end
18
19  legend(lt,'Location','EastOutside');  % create legend
20  hold off; % next plot will be in a new window
```

Notice that there were only two changes to the file. First, we changed line 2, where we defined the parameter y0. Next, we changed one value in line 13 so that the solutions shown are for time values from 0 to 20. Now, if we run the m-file ODESolutions.m we produce the graph shown in Figure 26.3.

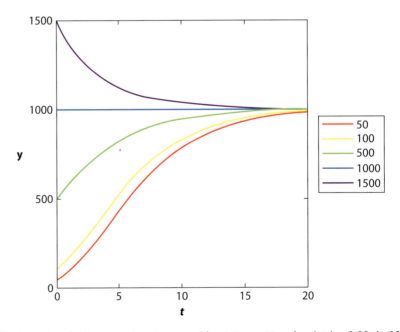

Figure 26.3 Graph produced using ODESolutions.m with odef.m set to solve $dy/dx = 0.25y \ln(1000/y)$.

26.3 Exercises

26.1   Find all solutions to each of the following differential equations.

(a) $\dfrac{dy}{dt} = 3t^4 + 5$

(b) $\dfrac{dy}{dt} = e^{3t} - 2$

(c) $\dfrac{dy}{dt} = \sin t$

26.2   Find the solution to each of the following differential equations using the given condition.

(a) $\dfrac{dy}{dt} = 3t^4 + 5$, with $y(1) = 10$

(b) $\dfrac{dy}{dt} = e^{3t} - 2$, with $y(0) = 5$

26.3   Use separation of variables to find all solutions to each of the following differential equations.

(a) $\dfrac{dy}{dt} = 10 - y$, with $y(0) < 10$

(b) $\dfrac{dy}{dt} = 10 - y$, with $y(0) > 10$

(c) $\dfrac{dy}{dt} = y \sin t$

(d) $\dfrac{dy}{dt} = t^2 y$

(e) $\dfrac{dy}{dt} = e^{t-y}$

26.4   Find the solution to each of the following boundary value problems.

(a) $\dfrac{dy}{dt} = -t/y$, with $y(3) = 4$

(b) $\dfrac{dy}{dt} = 2ty^2$, with $y(0) = 2$

26.5   Find all the solutions to the differential equation

$$y^4 e^{2x} + \frac{dy}{dx} = 0.$$

26.6   Find all the solutions to the differential equation

$$2x + (xy + 3x)\frac{dy}{dx} = 0,$$

with $x \neq 0$.

26.7   Find the solution to the differential equation

$$xy' - (2x + 1)e^{-y} = 0,$$

with $y(1) = 2$. Think of y' as $\frac{dy}{dx}$.

26.8   Suppose that a fish stock (with units in tons) with exponential growth is harvested at a constant rate (0.1 tons/week); then the differential equation for this fish stock would be

$$\frac{dy}{dt} = 1.08y - 0.1.$$

Suppose that $y(0) = 20$. Find the solution to the differential equation.

26.9   Find all the solutions $y(x)$ to the differential equation

$$(\sec x)y' - 2y = 0.$$

26.10 Solve for $y(x)$, such that

$$y' = \frac{2xy}{1 + x^2}$$

and $y(2) = 5$.

26.11 In 1980, the population of the kingdom of Edenia was 1 million. The 2000 census then found a population of 1.44 million. Assuming Malthusian growth, that is, that the growth rate is proportional to the population size, how many people would have lived in Edenia in 1990?

26.12 A ball is thrust up vertically from the ground into the air and hits the ground 2.5 seconds later. What is the maximum height of the ball in feet? Assume that air resistance is negligible. Use the fact that acceleration due to gravity is -32 ft/sec^2. Note that acceleration is the second derivative of the position function. You must take antiderivatives twice to get the position function.

26.13 Given the initial value problem

$$\frac{y'}{2 + \cos x} - e^{-y} = 0,$$

with $y(0) = 0$, find the following.

(a) Determine the solution $y(x)$
(b) Find $y(\pi/2)$

26.14 At time $t = 0$, a tank contain 10 pounds of salt dissolved in 100 gallons of water. Assume that water containing 0.25 pounds of salt per gallon is entering the tank at a rate of 3 gallons per minute and that a well-stirred solution is leaving the tank at the same rate. Find an expression for the amount of salt in the tank at time t.

26.15 Starting at 9:00 a.m., oil is pumped into a storage tank at a rate of $150t^{\frac{1}{2}} + 25$ (in gal/hr), for time t in hours after 9:00 a.m. How many gallons will have been pumped into the tank by 1:00 p.m.?

Equilibria and Limited Population Growth

Throughout this text, we have noted the ubiquity of change in biological systems of all types, including genetic regulatory mechanisms, which modify chemical reactions inside a cell; metabolic processes, which depend on temperature and the availability of oxygen; the photosynthetic uptake of CO_2, which changes depending on the level of light; and population sizes, which change because of varying birth rates and death rates. We illustrated how changes can be modeled though the use of difference equations, derivatives, differential equations, and matrix models that allow us to project forward in time for systems that can be broken down into compartments, such as age-groups in demography.

While change is ubiquitous, another major characteristic of biological systems is that *lack of change* is also ubiquitous. Many biological processes appear, at least on some time scales of measurement, to be relatively unchanging. We say that this process is then *at equilibrium*. Body temperatures for homeotherms are an example, for which any significant deviation from normal body temperature is indicative of an illness. The term *homeostasis* refers to biological processes that not only have relatively small changes across some relevant time scale but also tend to return to an appropriate state when perturbed. So human body temperature would normally return to near 98.6°F when a person recovers from an influenza infection. The processes that allow biological systems to return to a particular state, such as this 98.6°F body temperature, are called **homeostatic** and arise from diverse genetic, physiological, behavioral, and population processes. In the case of body temperature regulation in humans, sweating, which causes evaporative heat loss from the skin, is a homeostatic mechanism to reduce body temperature. In dogs, panting is a similar mechanism to reduce body temperature. Much of the study of illness is an attempt to determine what is causing a certain physiological process to be "unstable," meaning that it is not returning to its normal value. When your heart rate returns to its normal "resting" rate (typically about 60 beats per minute) after exercise, this happens because of a complex series of physiological responses. These homeostatic mechanisms are a type of feedback in biological systems that was discussed in the introduction to this unit.

Inherent in the above discussion of homeostasis is the time scale on which the process is considered. There is strong evidence that body temperature varies throughout the day in any given individual but also varies between individuals. So you should think of 98.6°F as being an average body temperature across many healthy humans measured at many times of the day. If you have never done so, you might find it instructive to measure your body temperature throughout the course of a day. Note also that body temperatures can vary on longer time scales, such as in recovering from an infection, and in women there are changes in body temperature across the menstrual cycle. For essentially all biological processes that appear to be homeostatic, this really arises from taking an average (e.g., an integral over time divided by the length of the time period) over an appropriate time period. When your blood pressure is taken, this is a measure of the pressure exerted by your blood, typically in mm Hg, against the walls of the artery (in your arm if you are being measured through a standard arm cuff, the sphygmomanometer). You are typically given two numbers: systolic pressure, which is the maximum pressure exerted when the heart contracts, and diastolic, which is the minimum pressure when the heart muscle relaxes, across a single cardiac cycle. The actual pressure exerted varies within each cardiac cycle, but medical practice focuses on using the systolic and diastolic pressures to diagnose potential disease.

We have already discussed a method to mathematically analyze a system in equilibrium when we discussed matrix models. Remember that in the case of landscape changes over time, which is modeled as a Markov chain, the eigenvector corresponding to an eigenvalue of 1 gave the long-term fraction of the landscape in each of the classes (e.g., fractions in trees, shrubs, grass, etc.). This is an equilibrium state for the system, and in all the cases we considered, it was also homeostasis because if the landscape was perturbed, such as because of a fire, the long-term fraction of each landscape type eventually returns to that specified by the eigenvector. Similarly, in the Leslie model for age-structured populations, the long-term fraction of individuals in each age class, given by the eigenvector for the largest eigenvalue (which gave the long-term population growth rate), was an equilibrium. In the Leslie model, the total population size may be growing or declining, but the fraction in each age class remains the same eventually (we called it the stable age distribution). The homeostatic mechanisms that allow this to happen are the feedbacks between the fecundities and survivorships in the different age classes on the population fraction in each class.

Our objective in this chapter is to develop the ideas of equilibria and stability for systems modeled by differential equations, and we will focus mostly on models for populations. Unlike the Leslie model, we are therefore dealing with populations changing continuously in time, not just at discrete time intervals or generations. Intuitively, since equilibrium means "not changing," what we do first is decide how we find such an equilibrium by realizing that if the population size is not changing, then the derivative of population size must be 0. As in the Leslie model case for which the age distribution changes got smaller and smaller over time, we'll show how to determine for a differential equation model whether an equilibrium population size is approached over time so that the changes in population size get smaller and smaller. This can happen because the population size feeds back on the population growth rate. This idea of feedbacks is critical to understanding the impacts of harvesting on populations and whether we can sustainably manage a renewable resource, such as a forest or a fishery. Although this inherently involves economic ideas, the field of bioeconomics, dealing with how decisions about natural resource management interact with economics, is beyond the scope of this text. The notions of equilibria and stability that we present are central to economics, not only in a biological context, and extensions of the mathematical ideas we cover have been the basis the work of many who have won Nobel prizes in economics.

27.1 Models of Limited Population Growth

Suppose that we want to model some population and that, in studying the population, we find that the rate at which the population is growing or declining at some time t depends on how large the population is at time t. Let $N(t)$ be the population size at time t and $\frac{dN}{dt}$ be the instantaneous growth rate of the population. In earlier chapters, we considered the case in which the population was growing (or declining) exponentially. Since no real-world population can continue to grow forever without constraints, we will consider in this chapter how we can model a population in which there are limits to growth.

One method for limiting growth rate is to have the instantaneous growth rate become negative when the population N is too large. For example, consider the population whose rate of change is modeled by the differential equation

$$\frac{dN}{dt} = k(A - N).$$

Here the rate of growth of the population is directly proportional to how close the population is to some threshold, A. The proportionality constant k represents the growth rate of the population. Thus, if the growth rate of the population is 50% of the difference between the threshold and the current population size, then $k = 0.50$. Notice that when $N < A$, then $\frac{dN}{dt} > 0$, and the population is increasing. However, when $N > A$, then $\frac{dN}{dt} < 0$, and the population is decreasing.

Example 27.1 (Mountain Goats [29])

A certain nature reserve can support no more than 4000 mountain goats. Assume that the rate of growth of the population is directly proportional to the difference between 4000 and the current population level with a proportionality rate of 20% per year. There are currently 1000 goats in the nature reserve.

(a) Write a differential equation for the growth of this mountain goat population and solve it to obtain a function describing the population at time t years.
(b) How many mountain goats will there be in the nature reserve in 5 years?

Solution:

(a) Since the nature reserve can support no more than 4000 mountain goats, the instantaneous growth rate of the population should become negative when $N > 4000$. Thus, we will model the rate of growth of the population by

$$\frac{dN}{dt} = 0.20(4000 - N).$$

(Continued)

Using separation of variables and assuming that $N(t) < 4000$, we find that

$$\int \frac{1}{4000 - N} \, dN = \int 0.2 \, ds$$
$$-\ln(4000 - N) = 0.2t + C$$
$$4000 - N = e^{-C}e^{-0.2t}$$
$$N(t) = 4000 - e^{-C}e^{-0.2t}.$$

Thus, using $N(0) = 1000$, we solve for the remaining constant e^{-C} and obtain

$$N(t) = 4000 - 3000e^{-0.2t}.$$

(b) In 5 years, there will be

$$N(5) = 4000 - 3000e^{-0.2(5)} \approx 2896$$

mountain goats in the nature reserve.

Recall that if $N(t)$ is the population size at time t, then $\frac{1}{N}\frac{dN}{dt}$ is the specific growth rate of the population N. The assumption that the specific rate of change is equal to a function of the population size, or $\frac{1}{N}\frac{dN}{dt} = f(N)$ or $\frac{dN}{dt} = Nf(N)$ for some function f, is called the **density-dependent hypothesis**. The function f is sometimes referred to as the **growth function**.

Logistic Growth

One classic form of density dependent growth in which the population size is limited by some carrying capacity is called **logistic growth**, which is usually defined by the differential equation

$$\frac{dN}{dt} = rN\left(1 - \frac{N}{K}\right),$$

where $K > 0$ is the carrying capacity of the population. Here we see that $f(N) = r\frac{K-N}{K}$. We see that when the population size is 0, the growth function reduces to $f(0) = r$. Also, when the population is at carrying capacity K, $f(K) = 0$.

We can use separation of variables to show that

$$\frac{dN}{dt} = \frac{r}{K}N(K - N)$$
$$\frac{1}{N(K - N)} \, dN = \frac{r}{K} \, dt.$$

Notice that

$$\frac{1}{N(K-N)} = \frac{1}{KN} + \frac{1}{K(K-N)}.$$

Integrating both sides, we get

$$\int \frac{1}{N(K-N)}\, dN = \int \frac{r}{K}\, dt$$

$$\int \left[\frac{1}{KN} + \frac{1}{K(K-N)}\right] dN = \int \frac{r}{K}\, dt$$

$$\frac{1}{K}\ln|N| + \left(-\frac{1}{K}\ln|K-N|\right) = \frac{r}{K}t + c$$

$$\ln\left|\frac{K-N}{N}\right| = -rt - c.$$

For the case of $n(0) < K$, we have

$$\left|\frac{K-N}{N}\right| = e^{-rt-c}$$

$$\frac{K}{N} - 1 = e^{-rt-c}$$

$$\frac{K}{N} = 1 + e^{-rt-c}$$

$$\frac{N}{K} = \frac{1}{1+e^{-rt-c}}$$

$$N(t) = \frac{K}{1+e^{-rt-c}}.$$

To solve for c, let $N(0) = N_0 > 0$. Then

$$N_0 = \frac{K}{1+e^{-r(0)-c}}$$

$$\frac{K}{N_0} = 1 + e^{-c}$$

$$e^{-c} = \frac{K}{N_0} - 1 = \frac{K-N_0}{N_0}.$$

Thus,

$$N(t) = \frac{K}{1+e^{-rt}\frac{K-N_0}{N_0}} = \frac{KN_0}{N_0 + (K-N_0)e^{-rt}}.$$

This formula is also valid for $N_0 > K$.

Now, taking limits, we see

$$\lim_{t \to \infty} N(t) = \frac{KN_0}{N_0 + (K - N_0)(0)} = K.$$

Thus, for $N_0 > 0$, in the limit, the population approaches the carrying capacity K.

Gompertz Growth

Let $V(t)$ be the volume of a tumor at time t and $\frac{dV}{dt}$ be the rate of growth of the tumor. Suppose that the rate of growth of the tumor is directly proportional to the size of the tumor. So we can write the model

$$\frac{dV}{dt} = \beta V.$$

Thus, we know that V must have a form like

$$V(t) = V_0 e^{\beta t}.$$

The *specific growth rate* (or percent growth rate) is $\frac{1}{V} \frac{dV}{dt}$.

It is unreasonable for a tumor to grow to infinite size. Indeed, there is strong concordance among cancer specialists that tumor growth rate is reduced because of lack of nutrient availability in the core. Thus, it is necessary to modify the model to limit the growth. One possible assumption is that $\frac{1}{V} \frac{dV}{dt}$ is an exponential decay function:

$$\frac{1}{V} \frac{dV}{dt} = ke^{-\alpha t}. \tag{27.1}$$

Let us use the separation of variables method on this new model to obtain

$$\int_{V(0)}^{V(t)} \frac{1}{V} \, dV = \int_0^t ke^{-\alpha s} \, ds$$

$$V(t) = V(0) \exp\left[\frac{k}{\alpha}\left(1 - e^{-\alpha t}\right)\right].$$

This type of curve is called a *Gompertz growth curve*.

Note that

$$\lim_{t \to \infty} V(t) = V(0) \exp\left(\frac{k}{\alpha}\right).$$

Furthermore, since $V(t)$ is an increasing function, this limit is the maximum of V. Let

$$V_M = V(0) \exp\left(\frac{k}{\alpha}\right)$$

denote the maximum value of V.

Example 27.2 (Tumor Growth)

A researcher is studying the growth of tumors in mice. For a certain mouse, when the tumor is first detected, it is measured to have a volume of 1.2 cm^3. Over the next 2 weeks, the researcher monitors the growth of the tumor and determines that the growth rate of the volume of the tumor is given by the equation

$$\frac{dV}{dt} = 1.45 V e^{-0.356t},$$

where t is measured in days.

(a) Estimate the volume of the tumor after the 2 weeks of observation.
(b) What is the maximum expected volume of the tumor given this model?

Solution:

(a) Since that the tumor was first observed to have a volume of 1.2 cm^3, $V(0) = 1.2$. The equation $\frac{dV}{dt} = 1.45 V e^{-0.356t}$ has the same form as Equation (27.1); thus, the solution is

$$1.2 \exp\left[\frac{1.45}{0.356}\left(1 - e^{-0.356t}\right)\right].$$

We can now calculate

$$V(14) \approx 68.5 \text{ cm}^3.$$

(b) Since V is an increasing function, we take the limit as $t \to \infty$ to find

$$V_M = 1.2 \exp\left(\frac{1.45}{0.356}\right) \approx 70.5 \text{ cm}^3.$$

Thus, we expect the tumor to grow no larger than 70.5 cm^3.

27.2 Equilibria and Stability

When we were studying the dynamics of populations using matrix models, we were able to determine when the population was in equilibrium by checking that the dominant eigenvalue was equal to 1. Recall that when the dominant eigenvalue of a population modeled by matrix model is 1, the population size remains constant. We now wish to determine when populations modeled by differential equations are at *equilibrium*. Since a population or system will be at equilibrium when its size is not changing with respect to time, this means that, the first derivative of the population size with respect to time will be 0. Thus, if we wish to find the equilibrium of a differential equation, we set

$$\frac{dN}{dt} = 0$$

and solve for N.

Let us do this for the logistic growth differential equation. When we set the derivative of the population size with respect to time equal to 0, we obtain

$$0 = \frac{dN}{dt}$$
$$= \frac{r}{K}N(K - N)$$
$$= N(K - N).$$

We can see that $\frac{dN}{dt}$ will be 0 when either $N = 0$ or $N = K$. Thus, when $N = 0$ (the population is extinct) or $N = K$ (the population is at carrying capacity), the size of the population remains constant through time.

Graphing Possible Solutions

We would like to know what the graph of the solution looks like given different initial conditions, $N(0) = N_0$. We will assume that $N_0 \geq 0$ since $N(t)$ represents the size of a population at time t, and a negative population size does not make sense.

Case 1: $N_0 = 0$

Since $N = 0$ is an equilibrium point, if the population starts at 0, it will remain at 0. Thus, in this case if there are no individuals to start with, there will continue to be no individuals.

Case 2: $0 < N_0 < K$

Looking at the differential equation, we see that this implies that

$$\frac{dN}{dt} = \frac{r}{K}N(K - N) > 0,$$

so the population is increasing on the time interval $(0, \infty)$. But what shape does it have? If we look at the second derivative of the population size with respect to time, we can determine the concavity of the curve:

$$N''(t) = \frac{r}{K}(K - 2N)N'(t) = \left(\frac{r}{K}\right)^2 (K - 2N)N(K - N).$$

Thus, the second derivative is 0 when $N = 0, \frac{K}{2}, K$. Using a sign chart (see Figure 27.1), we can show that over the interval $(0, K/2)$, $N(t)$ is concave up, while over the interval $(K/2, K)$, $N(t)$ is concave down. Thus, $N = K/2$ is an inflection point. Therefore, if the population initially starts somewhere between 0 and $K/2$, the population will increase with an S-shaped curve (with a point of inflection when $N = K/2$) and approach the carrying capacity as $t \to \infty$. If the population initially starts somewhere between $K/2$ and K, then the population will increase toward the carrying capacity with a concave-down curve and approach K as $t \to \infty$. Note that using the equation we derived for logistic growth, we can show that

$$t = \frac{1}{r} \ln \left(\frac{K - N_0}{N_0} \right)$$

when $N = K/2$.

Case 3: $N_0 = K$

Since $N = K$ is an equilibrium point, if the population starts at the carrying capacity K, it will remain at the carrying capacity.

Case 4: $N_0 > K$

Looking at the differential equation, we see that this implies that

$$\frac{dN}{dt} = \frac{r}{K} N(K - N) < 0,$$

so the population decreases. Recall that, the second derivative $N''(t) = 0$ when $N = 0, \frac{K}{2}, K$. Using the sign chart above, we see that over the interval (K, ∞), $N(t)$ is concave up. Therefore, if the population initially starts somewhere above carrying capacity, the population will decrease with a concave-up curve and approach K as $t \to \infty$.

This information can be quickly determined using sign charts for $N'(t)$ and $N''(t)$ (see Figure 27.1). Additionally, we can use this information to construct some sample logistic growth curves given different initial conditions (see Figure 27.2).

Notice that all the solutions tend toward the equilibrium $N = K$ as time grows larger, while none of the solutions tend toward the equilibrium $N = 0$. We can construct an equilibria chart to represent this information (see Figure 27.1(c)). In constructing such a chart, if arrows on both sides of an equilibrium point toward the equilibrium, then we call the equilibrium *stable*. If both arrows point away from the equilibrium, then we call the equilibrium *unstable*. If one arrow points toward the equilibrium and one points away, then we refer to the equilibrium as *neutral* (however, technically, this equilibrium is unstable). Thus, for the logistic growth equation, the equilibrium $N = 0$ is unstable, while the equilibrium $N = K$ is stable.

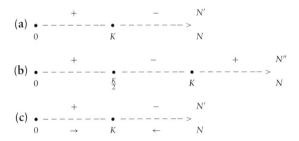

Figure 27.1 Sign charts for the (a) first derivative and (b) second derivative of the population with respect to time where the points on the sign chart are population sizes, N. The equilibria chart in (c) shows that solutions for the case $0 < N_0 < K$ and N increases toward K (as represented by the arrow), whereas solutions for which $N_0 > K$ decrease toward K (as represented by the arrow).

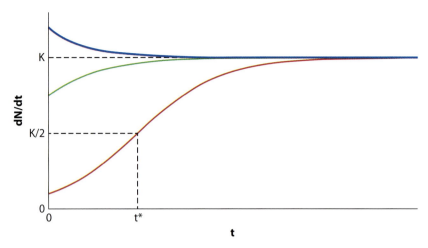

Figure 27.2 Logistic growth. The red curve shows logistic growth when $0 < N_0 < \frac{K}{2}$, the green when $\frac{K}{2} < N_0 < K$, and the blue when $N_0 > K$, where K is the carrying capacity of the population. Note that $t^* = \frac{1}{r} \ln \left(\frac{K - N_0}{N_0} \right)$.

Example 27.3 (Equilibria for Gompertz Growth)

Given the differential equation in Example 27.2 that defined the growth of a tumor, find all the equilibria and classify their stability.

Solution: Recall that the differential equation defining the growth of a tumor in Example 27.2 was

$$\frac{dV}{dt} = 1.45 V e^{-0.356t}.$$

When $\frac{dV}{dt} = 0$, then $V = 0$ (recall that $e^{\alpha} > 0$ for all real α). Since V represents the volume of a tumor, we can ignore negative volume. Thus, we construct the following equilibrium chart:

$$\bullet\ \underset{0\ \ \ \ \ \ \ \ \ \ \rightarrow}{\text{- - - - - - - - ->}}\ \underset{N}{\overset{N'}{}}$$

Since the only arrow is pointing away from the equilibrium $V = 0$, we classify this equilibrium as unstable.

Example 27.4 (Equilibria for a Mountain Goat Population)

Given the differential equation in Example 27.1 that defined the growth of a mountain goat population, find all the equilibria and classify their stability.

(Continued)

Solution: Recall that the differential equation defining the growth of the mountain goat population in Example 27.1 was

$$\frac{dN}{dt} = 0.20(4000 - N).$$

When $\frac{dN}{dt} = 0$, then $N = 4000$. Since N represents population size, we can ignore negative population sizes. Thus, we construct the following equilibrium sign chart:

Notice here that when $N = 0$, we use ∘ instead of • to draw attention to the fact that $N = 0$ is *not* an equilibrium point. Here the only equilibrium is $N = 4000$, which our equilibrium chart indicates is stable.

27.3 Homeostasis

As noted at the beginning of this chapter, homeostasis is a key property of living systems in which some characteristics of organisms, such as body temperature in mammals, are maintained close to a "set point." If for some reason the characteristic is perturbed away from this set point, then a variety of homeostatic mechanisms allow the organism to return to this set point. Thus, the set point is an equilibrium and it is stable because, under normal circumstances, the organism's characteristic moves back to this set point.

Thermoregulation is the method used by homeotherms to regulate their body temperature so that, when perturbed away from a set point, the body temperature returns to the set point. In humans, the typically assumed set point for body temperature is 98.6°F as measured orally. Mechanisms of thermoregulation in mammals include behavioral ones, such as moving to places with different temperatures, and numerous physiological ones. For example, sweating causes evaporative heat loss, which can reduce body temperature, and shivering, which arises from rapid contraction and relaxing of muscles, produces heat through enhanced respiration and can increase body temperature. Numerous metabolic processes are involved in thermoregulation, as are certain complexities associated with individual differences in set points and diurnal variation of set points. Figure 27.3 illustrates one aspect of this, showing a return to daytime set point after a period of reduced body temperature during sleep.

Let's consider the key components of homeostasis that we would want to include in a mathematical model if our objective were to provide an overall method to mimic the dynamics of a return to set point following a perturbation. While we will emphasize a return to normal daytime body temperature in our model, the same underlying methods apply to other processes under homeostatic control, such as regulation of blood pH and blood glucose, as well as those at completely different levels of biological organization, such as that of populations discussed in the previous section. The detailed mechanisms may be quite different in each of these cases of homeostasis, but the underlying mathematics used to describe the situation is often quite similar.

First, our mathematical description needs to include a set point, or equilibrium. This means that if $T(t)$ is the variable describing body temperature at time t, then an equilibrium temperature

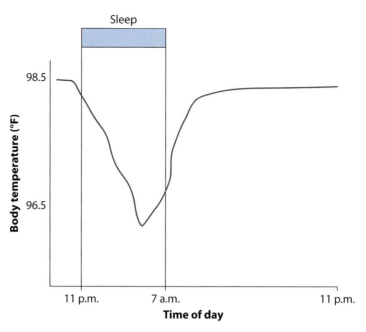

Figure 27.3 Illustration of human body temperature change following a period of sleep (from [91]).

occurs at a temperature for which $T'(t) = 0$. Suppose that we let S be such a temperature. Second, in order for the equilibrium S to be stable, temperature must increase toward S if the current temperature is below S and must decrease toward S if the current temperature is above S. Remembering that the sign of the derivative informs us as to whether a function is increasing or decreasing, this means that we need to have $T'(t) > 0$ if $T(t) < S$ and that we need $T'(t) < 0$ if $T(t) > S$. These properties arise in systems in which there is "negative feedback," in which the dynamics of the process cause it to move back toward an equilibrium regardless of whether the current value is above or below the equilibrium.

We have already seen an example of a model that behaves in the manner needed to describe temperature dynamics under thermoregulation in the discussion of limited population growth above. So an appropriate model is

$$\frac{dT}{dt} = k(S - T).$$

Here the variable T gives body temperature in °F, S is the daytime set point in °F, and k is a constant that has units 1/hours if time t is measured in hours. As we saw in the mountain goat situation (Example 27.1) and before it in Example 26.5, we can now easily solve to find how body temperature changes through time. If, for example, body temperature at the end of the period of sleep is $T(0) = T_0$, then

$$T(t) = S - (S - T_0)e^{-kt}.$$

The only remaining item to deal with is determining k, which specifies how rapidly body temperature returns to the set point. The larger the value of k, the more rapidly body temperature will approach the set point. Using Figure 27.3, we see that the body temperature reaches approximately 97.5°F at about 8:00 a.m., or 3 hours after the return to daytime temperature begins to occur at about 5:00 a.m., and that the set point for this individual is about 98.5°F.

The body temperature at 5:00 a.m. is about 96.2°F. So we can find k from this using $S = 98.5$, $T_0 = 96.2$, and $T(3) = 97.5$. From this, $k = 0.2776$. Note that in this model, body temperature never quite reaches the set point. For example, at 11:00 p.m., body temperature would be given by $T(18) = 98.484$. So this is not a perfect model in that it is only an approximation of the response of body temperature.

27.4 Exercises

27.1   (From [22]) The weights of Guernsey cows are approximated by the equation

$$\frac{dW}{dt} = 0.015(486 - W),$$

where W is the weight in kg after t weeks.

(a) Find the solution to the initial value problem with $W(0) = 28\,\text{kg}$.
(b) What limit does the weight approach as t goes to infinity? How does this limit compare to the weight at $t = 200$ weeks?

27.2   (From [29, 66]) The evaporation index I is a measure of soil moisture. An article on 10- to 14-year-old heath vegetation describes the rate of change of I with respect to W, the amount of water available, by the equation

$$\frac{dI}{dW} = 0.088(2.4 - I).$$

(a) According to the article, I has a value of 1 when $W = 0$. Solve the initial value problem.
(b) What happens to I as W becomes larger and larger?

27.3   (From [29]) An isolated fish population is limited to 5000 by the amount of food available. If there are now 150 fish and the population is growing with a growth constant of 1% per year, find the expected population at the end of 5 years.

27.4   In logistic growth, the rate of change of the population becomes negative when the population is too big (above carrying capacity) and positive when the population is smaller than carrying capacity. However, there are populations whose rate of change over time will be negative if the population is too small. This is known as the Allee effect and can be modeled using the differential equation

$$\frac{dN}{dt} = rN\left(1 - \frac{N}{K}\right)\left(\frac{N}{L} - 1\right),$$

where $L < K$. Find the population sizes at which the population N will be at equilibrium. Determine the stability of each equilibria. Then describe what will happen to the population over time when the following are true.

(a) $0 < N_0 < L$
(b) $L < N_0 < K$
(c) $N_0 > K$

27.5   Find the solution to this boundary value problem.

$$\frac{dy}{dt} = y(1 - y), \text{ with } y(0) = 3.$$

27.6   (From [41]) Red blood cells are formed from stem cells in the bone marrow. The red blood cell density x satisfies the differential equation

$$\frac{dx}{dt} = \frac{2x}{1 + x^2} - x.$$

Find all the equilibria and determine their stability.

27.7   Let

$$\frac{dx}{dt} = x - x^3.$$

Find all equilibria and classify their stability.

27.8   The Gompertz growth model can has also been written in the form

$$\frac{dx}{dt} = x \left(k - \alpha \ln \left(\frac{x}{x_0} \right) \right),$$

where $x_0 = x(0)$ and both α and k are positive constants. Assuming that $x_0 > 0$, find all equilibria for this model and classify them as stable or unstable.

27.9   (From [2]) A mathematical model for the growth of a population is

$$\frac{dx}{dt} = \frac{2x^2}{1 + x^4} - x,$$

where x is the population density and $x(0) = x_0 > 0$. Find all the equilibria of this model and determine their stability.

27.10   Using a similar approach to that used for body temperature following the end of sleep, develop a model for the time to return to a body temperature set point of 98.6°F following a fever arising from an infection. Suppose that an individual has a fever of 101.8°F and that the fever "breaks" at 11:00 a.m., meaning that at that time the person's body temperature starts to decline. Suppose that the body temperature at 5:00 p.m. is 100.3°F. Ignoring the fact that there is a circadian rhythym for body temperature, from your model estimate the first time the body temperature will return to within .1°F of the set point.

Implicit Differentiation and Related Rates

Biology is replete with examples of situations that rely on interactions between numerous factors or components in order to describe the situation. Sometimes this occurs because different biological processes co-occur; for example, photosynthesis and respiration occur concurrently in plants that utilize the C3 photosynthetic pathway. Thus, determining the rate of production of oxygen by the plant (or the rate of utilization of carbon dioxide) requires understanding the inter-actions of the two processes. However, this also depends on the scale of observation since if you considered the entire area around the plant, it would be necessary to take account of respiration by soil microorganisms as well to obtain a reasonable understanding of measurements of CO_2 near the plant.

Biological processes rarely happen in isolation. Interactions such as those between predators and prey, competitors for limited resources, and many biochemical processes at the cellular level are major drivers of biological complexity. These interactions are also rarely linear; that is, we typically cannot derive how one process affects another by assuming that one is proportional to the other. Photosynthetic rate depends nonlinearly on factors such as light level, temperature, and humidity.

Determining how rapidly the uptake of carbon dioxide by a plant varies throughout a day requires translating how the rates of change in, light, temperature and humidity affect the rate of change of carbon dioxide uptake. There is a way to do this, using the idea of related rates discussed in this chapter. To carry it out, we first need to have some way of relating the processes to each other. For example, we saw in Example 15.1 and again in Section 17.4 that the photo-synthetic rate for leaves on a soybean plant increases with respect to light level by a function of the form

$$P(I) = \frac{abI}{aI + b} - d,$$

where P is the photosynthetic rate measured in μmol CO_2 m^{-2} s^{-1}, I is the light level measured in μmol photons m^{-2} s^{-1}, and a, b, and d are constants. Since light levels change with the time of day, the photosynthetic rate will also change with the time of day. Thus, there is also a relationship between dP/dt (the rate of change of the photosynthetic rate with respect to time) and dI/dt (the rate of change of light level with respect to time).

In this example for the relationship between photosynthetic rate and light level, we can express the relationship between P and I such that P can be described explicitly as a function of I. However, it may not be feasible to simply solve for one variable in terms of the others. In these cases, we would still like to determine the relationship between the rates of change of each variable. In this chapter, we develop the notion of related rates (the relationship of multiple rates of change) and define the method of implicit differentiation as a means for determining the relationship between rates of change.

28.1 Explicitly and Implicitly Defined Functions

There are two ways to define functions: explicitly and implicitly. An *explicit function* provides a formula for determining the dependent variable y in terms of the independent variable x. Thus, for the explicitly defined function,

$$y = 2t + 3,$$

given the value $t = 2$, we can determine that $y = 7$. An *implicit function*, on the other hand, is a function in which the dependent variable y has not been defined "explicitly" in terms of the independent variable t. For example, the function

$$y^3 + y^2 t - t^3 y = 3$$

is an implicit function because the exact formula for how y depends on t is not explicitly shown in the equation. For many implicit functions, it may be difficult or even impossible to solve for one variable in terms of the other variables.

Consider this example of an equation involving gypsy moths to illustrate the idea of implicitly defined functions. The gypsy moth, *Lymantria dispar*, is perhaps the most destructive forest defoliator in North America (see [19] for background and a model for this species). *Lymantria dispar nucleopolyhedrosis virus* is a virus that infects only gypsy moths. At a time t, the (uninfected) gypsy moth density N (number of individuals per hectare) is related to the density of virus V and the density of infected gypsy moths I:

$$1 - I = (1 + \alpha c[NI + \rho V])^{\frac{1}{c}}.$$

From this equation, one can view N as an implicit function of V and I, with the other elements of the equation being parameters that do not vary.

28.2 Implicit Differentiation

In Unit 5, we learned how to find the derivative of an explicitly defined function. In this chapter, we learn how to compute the derivative of a dependent variable whose relationship to the independent variable is shown only in an implicit function. This *implicit differentiation* allows us to find $\frac{dy}{dt}$ without first solving an implicitly defined function for y in terms of t.

Implicit Differentiation

In implicit differentiation, we take the derivative of both sides of the equation with respect to t (the independent variable) and use the chain rule,

$$\frac{d}{dt}\left(f(y)\right) = \frac{df}{dy}\frac{dy}{dt},$$

for the dependent variable y. We think of y as the "inside function" here.

First, let us consider an example for which we can apply both explicit and implicit differentiation.

Example 28.1 (Explicit versus Implicit Differentiation)

Find $\dfrac{dy}{dt}$ if $y^3 + t^2 = 1$.

Solution: In the explicit method of differentiation, we would first find

$$y(t) = (1 - t^2)^{1/3}.$$

Then

$$\frac{dy}{dt} = \frac{1}{3}(1 - t^2)^{-2/3}(-2t) = -\frac{2t}{3(1 - t^2)^{2/3}} = -\frac{2t}{3y^2}.$$

Alternatively, using implicit differentiation, we take the derivative of both sides with respect to t to get

$$3y^2\frac{dy}{dt} + 2t = 0.$$

Now, solving for $\frac{dy}{dt}$, we obtain

$$\frac{dy}{dt} = -\frac{2t}{3y^2}.$$

Using implicit differentiation, we obtain the same solution as with explicit differentiation. Notice that using implicit differentiation leads to the solution being expressed in terms of both t and y. In most applied problems, however, this is not a serious drawback.

Now, let us consider some examples for which we cannot solve for y explicitly in terms of t.

Example 28.2 (Implicit Differentiation)

Find $\frac{dy}{dt}$ if $\sin y + 2t^2 = y$.

Solution: Differentiating both sides with respect to t gives

$$(\cos y)\frac{dy}{dt} + 4t = \frac{dy}{dt}.$$

Notice that $\frac{dy}{dt}$ appears on both the left- and the right-hand side of the equation. Solving for $\frac{dy}{dt}$, we obtain

$$\frac{dy}{dt} = \frac{4t}{1 - \cos y}.$$

Example 28.3 (Implicit Differentiation)

Find $\frac{dy}{dt}$ if $1 = ye^{ty}$.

Solution: Here we will need to utilize the product rule and the chain rule. If we take the derivative with respect to the independent variable t on both sides of the equation, we have

$$\frac{d}{dt}(1) = \frac{d}{dt}\left(ye^{ty}\right)$$

$$0 = \frac{dy}{dt}e^{ty} + y\frac{d}{dt}\left(e^{ty}\right) \qquad \text{(Application of product rule)}$$

$$= \frac{dy}{dt}e^{ty} + ye^{ty}\frac{d}{dt}(ty) \qquad \text{(Application of chain rule)}$$

$$= e^{ty}\frac{dy}{dt} + ye^{ty}\left((1)(y) + t\frac{dy}{dt}\right) \qquad \text{(Application of product rule)}$$

$$= e^{ty}\frac{dy}{dt} + y^2 e^{ty} + tye^{ty}\frac{dy}{dt} \qquad \text{(Distribute } ye^{ty})$$

$$-\frac{dy}{dt}\left[e^{ty}(1 + ty)\right] = y^2 e^{ty} \qquad \text{(Factor out } \frac{dy}{dt})$$

$$\frac{dy}{dt} = -\frac{y^2 e^{ty}}{e^{ty}(1 + ty)} = -\frac{y^2}{1 + ty} \qquad \text{(Solve for } \frac{dy}{dt} \text{ and simplify).}$$

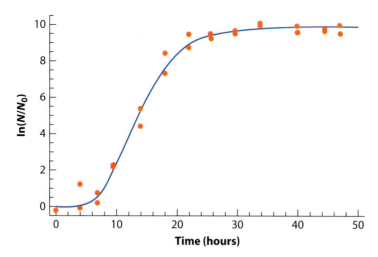

Figure 28.1 Data collected (in red) by Zanoni et al. [80] of *Enterococcus faecium* bacteria in bologna sausage over 50 hours at a constant 32°C. Note that the vertical axis is scaled as $\ln(N(t)/N_0)$, where $N(t)$ is the number of cells per gram and $N_0 = N(0)$. The blue curve is a Gompertz growth model described by Equation (28.2), which approximates the collected data.

Now, let us consider an application for which researchers have developed a model of bacteria growth with the number of bacteria at a particular time being defined by an implicit equation.

Example 28.4 (Bacteria in Sausage)

Enterococcus faecium is a bacteria that can be pathogenic, causing abdominal infections, surgical wound infections, urinary tract infections, and infections in the bloodstream. This bacteria is resistant to multiple forms of antibiotics, making it particularly challenging to treat. Understanding how quickly *Enterococcus faecium* grows under various conditions is useful in treating infections.

   Zanoni et al. [80] studied how populations of *Enterococcus faecium* bacteria grow in bologna sausage at a range of different temperatures. Figure 28.1 shows data collected by Zanoni, et al. of *Enterococcus faecium* bacteria in bologna sausage over 50 hours at a constant 32°C. Note that the vertical axis is scaled as $\ln(N(t)/N_0)$, where $N(t)$ is the number of cells per gram and $N_0 = N(0)$. Zanoni et al. proposed using the model

$$\ln\left(\frac{N(t)}{N_0}\right) = A \exp\left(-\exp\left(\frac{2.718\mu}{A}(\lambda - t) + 1\right)\right), \tag{28.1}$$

where t is measured in hours, A is the asymptote found by taking the limit of $\ln(N(t)/N_0)$ as $t \to \infty$, μ is the maximum growth rate of the bacteria, and λ is a parameter measuring lag time. Note that we use the notation $\exp(x)$ to mean e^x. Zanoni et al. estimated the values of A, μ, and λ (the model parameters) by experimentally growing the bacteria in bologna sausage over 50 hours under a range of different temperatures.

(Continued)

When the temperature was held at $32°C$, the estimated parameter values were $A = 9.8901$, $\mu = 0.7886$, and $\lambda = 7.1157$. When we substitute these values into Equation 28.1 and simplify, we obtain the equation

$$\ln\left(\frac{N(t)}{N_0}\right) = 9.8901 \exp\left(-12.7068e^{-0.2167t}\right). \tag{28.2}$$

Use implicit differentiation to determine the rate at which the bacteria are growing after 5 hours; that is, find $N'(5)$. Zanoni et al. started their experiments with bacteria cultures containing between 100 and 1000 bacteria per gram. Assume that $N_0 = 500$.

Solution: Using implicit differentiation and a careful application of the chain rule, we obtain

$$\frac{\frac{N'(t)}{N_0}}{\frac{N(t)}{N_0}} = 9.8901 \exp\left(-12.68e^{-0.2167t}\right)(-12.7068)\exp\left(-0.2167t\right)(-0.2167).$$

When simplified, we obtain

$$\frac{N'(t)}{N(t)} = 27.23 \exp\left(-12.7068e^{-0.2167t} - 0.2167t\right)$$

$$N'(t) = 27.23N(t)\exp\left(-12.7068e^{-0.2167t} - 0.2167t\right).$$

Notice that the equation for the growth rate of the bacteria population depends on both t and $N(t)$. Thus, to calculate the growth rate at time t, we also need to know how many bacteria per gram are present at time t. We can use Equation 28.2 to determine this quantity. At $t = 5$ (5 hours into the experiment),

$$\ln\left(\frac{N(5)}{500}\right) = 9.8901\exp\left(-12.7068e^{-0.2167(5)}\right) \approx 0.134248$$

$$N(5) \approx (500)(0.134248) = 67.124 \text{ bacteria/gram.}$$

Thus, we calculate

$$N'(5) = 27.23N(5)\exp\left(-12.7068e^{-0.2167(5)} - 0.2167t\right)$$

$$\approx (27.23)(67.124)(0.0109) \approx 19.92 \text{ bacteria/gram/hour.}$$

Thus, the number of bacteria in each gram of bologna sausage is increasing by 19.92 bacteria per hour, 5 hours into the experiment.

28.3 Related Rates

Recall the equation relating the (uninfected) gypsy moth density N (number of individuals per hectare) to the density of virus V and the density of infected gypsy moths I:

$$1 - I = (1 + \alpha c [NI + \rho V])^{\frac{1}{c}} .$$

Time is really the underlying variable in this equation since $N, I,$ and V all depend on time t [19]. The derivatives of $N, I,$ and V with respect to time would be related because of the equation relating these three quantities as time varies.

We will investigate some simpler examples with equations with two dependent variables (y and x), both of which depend on one dependent variable (t). If these equations are defined implicitly, we can proceed with implicit differentiation, taking the derivative of both sides of the equation with respect to the independent variable, t. Then we will have a new equation relating the derivatives $\frac{dy}{dt}$ and $\frac{dx}{dt}$.

Example 28.5 (Implicit Differentiation with Two Dependent Variables)

Find the relationship between $\frac{dy}{dt}$ and $\frac{dx}{dt}$ if

(a) $y = 4x^3$ and
(b) $x^2 + y^2 = 4 + t^2$.

Solution:

(a) Differentiating both sides with respect to t gives

$$\frac{dy}{dt} = 12x^2 \frac{dx}{dt}.$$

(b) Differentiating both sides with respect to t, we get

$$2x \frac{dx}{dt} + 2y \frac{dy}{dt} = 2t.$$

Solving for $\frac{dy}{dt}$, we get

$$\frac{dy}{dt} = \frac{2t - 2x \frac{dx}{dt}}{2y}.$$

Notice in the example above that we get a relation between the rate $\frac{dy}{dt}$ and the rate $\frac{dx}{dt}$. Let us consider a few biological examples involving cases such as this that relate the rates of two different processes.

Example 28.6 (Blood Flow [29])

In Chapter 24, we saw that blood flows faster the closer it is to the center of a blood vessel because of the reduced friction with vessel walls. According to Poiseuille's Law, the velocity V is given by

$$V = k(R^2 - r^2),$$

where R is the radius of the blood vessel, r is the distance of a layer of blood flow from the center of the vessel, and k is a constant (see Figure 24.5). We assume here that $k = 375$. Suppose that a skier's blood vessel has radius $R = 0.08$ millimeters and that cold weather is causing the vessel to contract at a rate of $\frac{dR}{dt} = -0.01$ millimeters per minute. How fast is the velocity of blood changing?

Solution: We want to find $\frac{dV}{dt}$ when $R = 0.08$ and $\frac{dR}{dt} = -0.01$. We will assume that we are looking at the velocity of blood at a given r; thus, we can treat r as a constant. Using implicit differentiation, we find

$$V = 375 \left(R^2 - r^2 \right)$$

$$\frac{dV}{dt} = 375 \left(2R\frac{dR}{dt} - 0 \right)$$

$$= 750R\frac{dR}{dt}.$$

Notice that the rate of change of velocity with respect to time, $\frac{dV}{dt}$, depends on both the radius of the blood vessel at time t and the rate of change of the radius of the blood vessel at time t, $\frac{dR}{dt}$.

Since $R = 0.08$ and $\frac{dR}{dt} = -0.01$, the velocity of blood at constant radius r is changing at rate

$$\frac{dV}{dt} = 750(0.08)(-0.01) = -0.6.$$

That is, the velocity of the blood is decreasing at a rate of -0.6 mm min^{-2}. This is a deceleration (negative acceleration) since it gives the rate of change of velocity.

Example 28.7 (Ichthyosaurs [49])

Ichthyosaurs make up a group of marine reptiles that were fish shaped and were comparable in size to dolphins. They became extinct during the Cretaceous period, which began about 144 million years ago and ended about 65 million years ago. A study of 20 fossil skeletons

(Continued)

found that the skull length (in cm) and backbone length (in cm) of an individual were related through the allometric equation

$$S = 1.162B^{0.933},$$

where S is the skull length and B is the backbone length [4]. How is the growth rate of the backbone related to the growth rate of the skull?

Solution: Let x denote the age of the ichthyosaur; then

$$S(x) = \text{skull length at age } x, \text{ and}$$
$$B(x) = \text{backbone length at age } x.$$

We are interested in the relationship between $\frac{dS}{dx}$ and $\frac{dB}{dx}$. Using implicit differentiation on the equation

$$S(x) = 1.162\,[B(x)]^{0.933}.$$

we find that

$$\frac{dS}{dx} = (1.162)(0.933)\,[B(x)]^{0.933-1}\,\frac{dB}{dx}.$$

Rearranging the terms on the right-hand side, we have

$$\frac{dS}{dx} = \underbrace{(1.162)\,[B(x)]^{0.933}}_{S(x)}\,(0.933)\,\frac{1}{B(x)}\,\frac{dB}{dx}.$$

If we divide both sides by $S(x)$, then

$$\frac{1}{S(x)}\,\frac{dS}{dx} = 0.933\,\frac{1}{B(x)}\,\frac{dB}{dx}.$$

A growth rate $\frac{dN}{dt}$ that has been divided by the quantity $N(t)$ is called a specific growth rate. Thus, $\frac{1}{S(x)}\,\frac{dS}{dx}$ and $\frac{1}{B(x)}\,\frac{dB}{dx}$ are specific growth rates. The factor 0.933 is less than 1, which indicates that skulls grow less quickly than backbones.

28.4 Exercises

28.1   Using implicit differentiation, find $\frac{dy}{dt}$.

(a) $4t^2 + 3y^2 = 6$

(b) $6t^2 + 8ty + y^2 = 6$

(c) $t^3 = y^2 + 4$

(d) $3t^2 = \dfrac{y-2}{y+2}$

(e) $\sqrt{y} = 4 - \sqrt{t}$

(f) $t^2 e^y + y = t^3$

(g) $t + \ln y = t^2 y^3$

(h) $\sin(ty) = t$

(i) $\tan(y) + t = 4$

28.2   Assume that x and y are functions of t. Find the relation between $\frac{dx}{dt}$ and $\frac{dy}{dt}$.

(a) $y^2 - 5x^2 = -1$

(b) $xy - 5x + 2y^3 = 70$

(c) $\dfrac{x^2 + y}{x - y} = 9$

(d) $xe^y = 1 + \ln x$

28.3   Assume that x and y are functions of t. Find $\frac{dy}{dt}$ using the implicit function and the additional information about $\frac{dx}{dt}$, x, and y.

(a) $8y^3 + x^2 = 1; \quad \dfrac{dx}{dt} = 2, \ x = 3, \ y = -1$

(b) $4x^3 - 9xy^2 + y = -80; \quad \dfrac{dx}{dt} = 4, \ x = -3, \ y = 1$

(c) $y \ln(x) + xe^y = 6; \quad \dfrac{dx}{dt} = 5, \ x = 1, \ y = 0$

28.4   (From [29]) Researchers have found a correlation between respiratory rate and body mass in the first 3 years of life. This correlation can be expressed by the function

$$\log R(w) = 1.83 - 0.43 \log w,$$

where w is body weight (in kg) and $R(w)$ is the respiratory rate (in breaths per minute) [24].

(a) Find $R'(w)$ using implicit differentiation.
(b) Find $R'(w)$ by first solving the equation for $R(w)$.
(c) Note that $R'(w)$ is the rate of change in the respiratory rate with respect to body weight at weight w. How does this rate of change depend on $R(w)$? To answer this, did you use information from the differentiation in (a) or (b)?

28.5   (From [29]) The relationship between the number of species in a genus (x) and the number of genera (y) comprising x species is given by

$$xy^a = k,$$

where a and k are constants [42]. Find $\frac{dy}{dx}$ and interpret the biological meaning of $\frac{dy}{dx}$.

28.6   In Example 28.4, we noted that Zanoni et al. proposed a modified Gompertz growth model (see Equation 28.1) to approximate data on the number of *Entercoccus faecium* bacteria per gram over time at a constant temperature. In Example 28.4, we considered the data collected at 32°C. Here we consider the data collected at 50°C.

   The graph below shows the data collected by Zanoni et al. of *Entercoccus faecium* bacteria in bologna sausage at a constant temperature of 32°C (red data) and 50°C (green data). Note that the vertical axis is scaled as $\ln(N(t)/N_0)$, where $N(t)$ is the number of bacteria cells per gram, and $N_0 = N(0)$. The red and green curves are described by the function $N(t)$ given in Equation (28.1), where

$$\text{at } 32°C, A = 9.8901, \ \mu = 0.7886, \ \text{and } \lambda = 7.1157 \ \text{(red curve) and}$$

$$\text{at } 50°C, A = 5.2667, \ \mu = 0.6231, \ \text{and } \lambda = 2.2819 \ \text{(green curve)}.$$

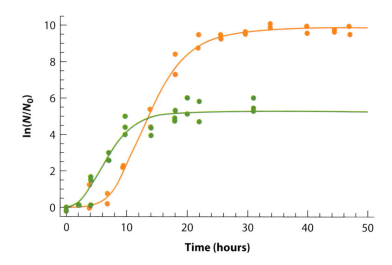

(a) Using only the graph above, estimate which data set has a larger growth rate 5 hours into the experiment.

(b) Using Equation (28.1), the parameter value for 50°C, and implicit differentiation, find the growth rate of the bacteria at $t = 5$ when the temperature is held at a constant 50°C. Is the growth rate larger or smaller than when the temperature is held at a constant 32°C? Does this agree with your answer in (a)?

(c) Using only the graph above, estimate which data set has a larger growth rate 10 hours into the experiment.

(d) Using Equation (28.1), find an expression for $N'(t)$ in terms of $N(t)$, N_0, t, A, μ, and λ.

(e) Use your expression for $N'(t)$ from (d) to estimate the growth rate at $t = 10$ when the temperature is held at a constant 32°C and when the temperature is held at a constant 50°C. Do your answers agree with the hypothesis you made in (c)?

28.7   The radius of a spherical bacteria cell is increasing at a rate of 0.01 μm per second when its radius is 20 μm. At what rate is its volume changing? At what rate is its surface area changing? Many cell processes involve the movement of materials in and out of the cell through the cell surface, so the surface-to-volume ratio is important. How is that ratio changing as the radius changes?

28.8   The surface area of an African elephant can be estimated in terms of the length of its trunk,

$$S = 53x^{0.7},$$

with x being the length of the trunk in feet and S being the surface area in square feet. If a some point in an elephant's life its trunk is changing at a rate of .01 feet per day at which time its trunk length is 3 feet, what is the rate of change of its surface area at that time?

28.9   (From [29]) A cross-country skier has a history of heart problems. She takes nitroglycerin to dilate her blood vessels, thus avoiding angina (chest pain) due to blood vessel contraction. Use Poiseuille's Law with $k = 555.6$ to find the rate of change of the blood velocity when $R = 0.02$ mm and R is changing by 0.003 mm per minute due to the effect of this drug. Assume that r is constant.

28.10   (From [29]) The brain weight of a fetus can be estimated using the total weight of the fetus by the function

$$b = 0.22w^{0.87},$$

where w is the weight of the fetus (in g) and b is the brain weight (in g) [76]. Suppose that the brain weight of a 25-g fetus is changing at a rate of 0.25 g/day. Use this to estimate the rate of change of the total weight of the fetus, $\frac{dw}{dt}$.

28.11   (From [29]) The energy of bird flight as a function of body weight is given by

$$E = 429w^{-0.35},$$

where w is the weight of the bird (in g) and E is the energy expenditure (in cal/g/hr) [57]. Suppose that the weight of a bird weighing 10 g is increasing at a rate of 0.001 g/hr. Find the rate at which the energy expenditure is changing with respect to time.

Unit 7 Student Projects

Budworm Differential Equation

(Adapted from 44) Suppose that the spruce budworm, in the absence of predation by birds, will grow according to this differential equation with logistic growth:

$$\frac{dB}{dt} = rB\left(1 - \frac{B}{K}\right).$$

Budworms feed on the foliage of trees. The size of the carrying capacity K would depend on the foliage of the trees. At first, we take K to be constant. Note that we have an explicit formula for the solution of this differential equation.

1. Draw graphs using Matlab for how this population would change if $r = .48$ and $K = 150$. Use several different initial conditions.
2. Change r to 1.2 and plot graphs using the same K and initial conditions as for the previous part. Discuss the changes that you observe with this different r.
3. Introduce the effect of predation by birds into this model. Assume that for small levels of budworms, there is very little predation but that for larger levels, birds are attracted to this food source. The effect of predation is limited since birds can consume only a limited number of budworms. The predation function giving the rate at which birds consume budworms is

$$P(B) = a\frac{B^2}{b^2 + B^2},$$

    with a and b being positive constants. Plot the graph of $P(B)$ using several values for a and b. Describe how the graph changes as you change a and b.
4. Our new model for the budworm population would be

$$\frac{dB}{dt} = rB\left(1 - \frac{B}{K}\right) - a\frac{B^2}{b^2 + B^2}.$$

Use **ode45** from Matlab to solve this differential equation using $r = .48$, $a = b = 2$, and $K = 15$. Plot your solution graph. Then solve the differential equation again, changing it so that $K = 17$. Note that you need to decide on the length of your time interval and the initial conditions. Discuss the change in the graphs of the solutions when K was changed.

5. Solve the differential equations for a longer time. See if you can deduce any information about possible equilibrium points. You may wish to vary the initial conditions.

Write a one-page report, describing your procedures and results. Attach copies of your m-files on separate pages. ■

Allee Growth

We will investigate the effects of different growth terms in a differential equation that is used to model populations. Consider the differential equation with logistic growth:

$$\frac{dN}{dt} = rN\left(1 - \frac{N}{K}\right).$$

1. Choose the carrying capacity K to be 10 and $r = 0.4$. Using various initial conditions, what do you observe about the limit of the population as t gets large? For an initial condition bigger than K, what happens? What happens for initial conditions less than K? Note that we have previously found an explicit formula for the solution of this differential equation.

2. Next, we investigate the behavior of Allee growth in a differential equation,

$$\frac{dN}{dt} = rN\left(1 - \frac{N}{K}\right)(a - N),$$

with $a < K$.

Compare the graph of the two growth functions:

$$f(N) = rN\left(1 - \frac{N}{K}\right).$$

$$g(N) = rN\left(1 - \frac{N}{K}\right)(a - N).$$

For a few values of r, N, and a, plot the graphs of these two growth functions and explain the differences.

3. Now use **ode45** in Matlab to solve the differential equation with Allee growth for $K = 10$, $r = .4$, and $a = 2$. Plot your resulting solutions using various initial conditions. Tell how the graphs change as you vary the initial conditions. What happens when you use an initial condition below $a = 2$? The parameter a is a threshold for growth, meaning that a population starting below this level will decay. How is this effect different from the solutions from the differential equation with logistic growth?

Write a two-page report, describing your procedures and results. Attach copies of your m-files on separate pages. ■

Bibliography

Note: *Numbered weblink references can be found at the end of the bibliography.*

[1] D. Agrawal, R. Hawk, R. L. Avila, H. Inouye, and D. A. Kirschner. Internodal myelination during development quantitated using x-ray diffraction. *Journal of Structural Biology*, 168:521–526, 2009.

[2] L. J. S. Allen. *An Introduction to Mathematical Biology*. Pearson/Prentice Hall, 2007.

[3] A. Arifah and P. Lees. Pharmacodynamics and pharmacokinetics of phenylbutazone in calves. *Journal of Veterinary Pharmocology and Therapeutics*, 25:299–309, 2002.

[4] M. Benton and D. Harper. *Basic Paleontology*. Addison Wesley and Longman, 1997.

[5] P. Bergsø, D. W. Denman, H. J. Hoffman, and O. Meirik. Duration of human singleton pregnancy: A population-based study. *Acta Obstetricia et Gynecologica Scandinavica*, 69:197–207, 1990.

[6] M. L. Bittinger, N. Brand, and J. Quintanilla. *Calculus for the Life Sciences*. Pearson/Addison Welsey, 2006.

[7] M. H. Bornstein and H. G. Bornstein. The pace of life. *Nature*, 259(5544):557–559, 1976.

[8] C. A. Bosch. Redwoods: A population model. *Science*, 172:345–349, 1971.

[9] J. R. Brannan and W. E. Boyce. *Differential Equations: An Introduction to Modern Methods and Applications*. Wiley, 2010.

[10] R. L. Burden and J. D. Faires. *Numerical Analysis*. PWS-KENT Publishing, 2004.

[11] C. A. Busso and B. L. Perryman. Seed weight variation of Wyoming sagebrush in northern Nevada. *BioCell*, 29(3): 279–285, 2005.

[12] H. Caswell. *Matrix Population Models: Construction, Analysis and Interpretation*. Sinauer Associates, 2nd ed., 2001.

[13] J. Q. Chambers, N. Higuhi, and J. Schimel. Ancient trees in Amazonia. *Nature*, 391: 135–136, 1998.

[14] A. J. Clark. *Comparative Physiology of the Heart*. Macmillan, 1927.

[15] T. H. Clutton-Brock, A. F. Russell, L. L. Sharpe, A. J. Young, Z. Balmforth, and G. M. McIlrath. Evolution and development of sex differences in cooperative behavior in meerkats. *Science*, 297:253–256, 2002.

[16] M. R. Cullen. *Mathematics for the Biosciences*. PWS Publishers, 1983.

[17] M. Denny and S. Gaines. *Chance in Biology: Using Probability to Explore Nature*. Princeton University Press, 2000.

[18] S. Duncan, K. Sturner, and S. Lenhart. Using probability to understand biodiversity. *High School Math Journal, NCTM*, 2:20–24, 2014.

[19] G. Dwyer, J. Dushoff, J. S. Elkinton, and S. A. Levin. Pathogen-driven outbreaks in forest defoliators revisited: Building models from experimental data. *American Naturalist*, 430:105–120, 2000.

[20] B. J. Enquist and K. J. Niklas. Global Allocation Rules for Patterns of Biomass Partitioning in Seed Plants. *Science* 295:1517–1520, 2002.

[21] S. L. Fassino, K. D. Gwinn, S. Lenhart, A. M. Jack, and H. P. Denton. Modeling the effect of abiotic factors on tobacco-specific nitrosamines. *Tobacco Science Journal*, 49:41–46, 2012.

[22] P. A. Fay and A. K. Knapp. Responses to short-term reductions in light in soybean leaves: Effects of leaf position and drought stress. *International Journal of Plant Sciences*, 159:805–811, 1998.

[23] J. France, J. Kijkstra, and M. S. Dhanoa. Growth functions and their applications in animal science. *Annales de Zootechnie*, 45 (Supplement):165–174, 1996.

[24] L. Gagliardi and F. Rusconi Respiratory rate and body mass in the first three years of life. *Archives of Disease in Children*, 76:151–154, 1997.

[25] M. A. Gilchrist. Combining Models of Protein Translation and Population Genetics to Predict Protein Production Rates from Codon Usage Patterns. *Mol Biol Evol*, 24(11):2362–2372, 2007. http://mbe.oxfordjournals.org/cgi/content/full/24/11/2362

[26] F. R. Giordano, M. D. Weir, and W. P. Fox. *A First Course in Mathematical Modeling*. Brooks/Cole, 2nd ed., 1997.

[27] L. Glass and M. C. Mackey. *From Clocks to Chaos: The Rhythms of Life*. Princeton University Press, 1988.

[28] B. Gompertz. On the nature of the function expressive of the law of human mortality. *Philosophical Transactions of the Royal Society of London*, 1825.

[29] R. N. Greenwell, N. P. Ritchey, and M. L. Lial. *Calculus for the Life Sciences*. Addison Wesley, 2002.

[30] L. J. Gross. Some lessons from fifteen years of education initiatives at the interface between mathematics and biology: The entry-level course. In *Undergraduate Mathematics for the Life Sciences: Models, Processes, and Directions*, pp. 121–126, G. Ledder, J. P. Carpenter and T. D. Comar (eds.). Mathematical Association of America, 2013.

[31] T. L. Hennessey, A. L. Freeden, and C. B. Field. Environmental effects on circadian rhythms in photosynthesis and stomatal opening. *Planta*, 189:369–376, 1993.

[32] M. Heron, D. L. Hoyert, S. L. Murphy, J. Xu, K. D. Kochanek, and B. Tejada-Vera. Deaths: Final data for 2006. *National Vital Statistics Reports*, 57(14), 2009. http://www.cdc.gov/nchs/data/nvsr/nvsr57/nvsr57_14.pdf.

[33] B. Horelick, S. Koont, and S. F. Gottlieb. Modules in undergraduate mathematics and its applications. In *Modeling the Nervous System: Reaction Time and the Central Nervous System*, volume 67 of *Applications of Biological Calculus*. COMAP, 1980.

[34] D. Hughes-Hallett, W. G. McCallum, A. M. Gleason, D. Mumford, D. E. Flath, B. G. Osgood, P. Frazer Lock, D. Quinney, D. O. Lomen, K. Rhea, D. Lovelock, J. Tecosky-Feldman. *Calculus: Single and Multivariable*. Wiley, 4th ed., 2005.

[35] J. T. Jorgenson, M. Festa-Bianchet, and W. D. Wishart. Effects of population desnity on horn development in bighorn rams. *Journal of Wildlife Management*, 62:1011–1020, 1998.

[36] P. D. Keightley and A. Eyre-Walker. Deleterious mutations, and the evolution of sex. *Science*, 290:331–333, 2000.

[37] P. Kulesa, G. Cruywagen, S. R. Lubkin, P. K. Maini, J. Sneyd, M. W. J. Ferguson, and J. D. Murray. On a model mechanism for the spatial patterning of teeth primordia in the alligator. *Journal of Theoretical Biology*, 180:287–296, 1996.

[38] J. Ledolter and R. V. Hogg. *Applied Statistics for Engineers and Physical Scientists*. Prentice Hall, 2010.

[39] K. Lim and J. Yin. Computational fitness landscape for all gene-order permutations of an RNA virus. *PLoS Computational Biology*, 5(2):e1000283, 2009.

[40] N. Linthorne. Optimum release angle in shot put. *Journal of Sports Sciences*, 19:359–372, 2001.

[41] J. D. Logan and W. R. Wolesensky. *Mathematical Methods in Biology*. Wiley, 2009.

[42] A. J. Lotka. *Elements of Mathematical Biology*. Dover Publications, 1956.

[43] D. Ludwig, D. D. Jones, and C. S. Holling. Qualitative analysis of insect outbreak systems: The spruce budworm and forests. *Journal of Animal Ecology*, 47:315–332, 1978.

[44] N. Maganaris and J. P. Paul. Tensile properties of the *in vivo* human gastrocnemius tendon. *Journal of Biomechanics*, 35:1639–1646, 2002.

[45] J. M. Mahaffy and A. Chávez-Ross. *Calculus: A Modeling Approach for the Life Sciences.* Pearson Custom Publishing, 2005.

[46] C. Mezzadra, R. Paciaroni, S. Vulich, E. Villarreal, and L. Melucci. Estimation of milk consumption curve parameters for different genetic groups of bovine calves. *Animal Production*, 49:83–87, 1989.

[47] D. D. Mooney and R. J. Swift. *A Course in Mathematical Modeling.* Mathematical Association of America, 1999.

[48] N. C. Myrianthopoulos and S. M. Aronson. Population dynamics of Tay-Sachs disease. *American Journal of Human Genetics*, 18(4):313–327, 1966.

[49] C. Neuhauser. *Calculus for Biology and Medicine.* Pearson/Prentice Hall, 2nd ed., 2004.

[50] G. Nord and J. Nord. Sediment in Lake Coeur d'Alene, Idaho. *The Mathematics Teacher*, 91:292–295, 1998.

[51] K. M. Palaniswamy and K. A. Gomez. Length-width method for estimating leaf area of rice. *Agronomy Journal*, 66(3):430–433, 1974. http://agron.scijournals.org/cgi/reprint/66/3/430.

[52] A. S. Perelson. Receptor clustering on a cell surface. *Mathematical Biosciences*, 53:1–39, 1981.

[53] G. W. Pierce. *The Songs of Insects.* Harvard University Press, 1948.

[54] P. Presrud and K. Nilssen. Growth, size, and sexual dimorphism in artic foxes. *Journal of Mammalogy*, 76:522–530, 1995.

[55] G. Reed and J. Hill. Measuring the thermic effect of food. *American Journal of Clinical Nutrition*, 63:164–169, 1996.

[56] P. M. Reis, S. Jung, J. M. Aristoff, and R. Stocker. How cats lap: Water uptake by *Felis catus*. *Science*, 330:1231–1234, 2010.

[57] C. T. Robbins. *Wildlife Feeding and Nutrition.* Academic Press, 2nd ed., 1993.

[58] A. Rogers, D. J. Allen, P. A. Davey, P. B. Morgan, E. A. Ainsworth, C. J. Bernacchi, G. Cornic, O. Dermody, F. G. Dohleman, E. A. Heaton, J. Mahoney, X. G. Zhu, E. H. Delucia, D. R. Ort, and S. P. Long. Leaf photosynthesis and carbohydrate dynamics of soybeans grown throughout their life-cycle under free-air carbon dioxide enrichment. *Plant, Cell, and Environment*, 27:449–458, 2004.

[59] R. R. Rykaczewski and D. M. Checkley. Influence of ocean winds on the pelagic ecosystem in upwelling regions. *Proceedings of the National Academy of Sciences*, 105(6):1965–1970, 2008.

[60] M. L. Samuels and J. A. Witmer. *Statistics for the Life Sciences.* Prentice Hall, 2003.

[61] J. A. Schneider, K. Bradley, and J. E. Seegmiller. Increased cystine in leukocytes from individuals homozygous and heterozygous for cystinosis. *Science*, 157:1321–1322, 1967.

[62] C. Schwartz and K. Hundertmark. Reproductive characteristics of Alaskan moose. *Journal of Wildlife Management*, 57(3):454–468, 1993.

[63] R. Skjærven, A. J. Wilcox, N. Øyen, and M. Per. Mothers' birth weight and survival of their offspring: Population based study. *British Medical Journal*, 314(7091): 1376, 1997. http://www.bmj.com/cgi/content/full/314/7091/1376.

[64] C. D. Skow and E. M. Jakob. Effects of maternal body size on clutch size and egg weight in a pholcid spider (*Holocnemus pluchei*). *Journal of Arachnology*, 31(2):305–308, 2003. http://www.bioone.org/doi/full/10.1636/01-85.

[65] C. C. Smith and S. D. Fretwell. The optimal balance between size and number of offspring. *American Naturalist*, 108(962):499–506, 1974.

[66] R. L. Specht. Dark island heath (Ninety-Mile Plain, South Australia) V: The water relationships in heath vegetation and patures on the Makin Sand. *Australian Journal of Botany*, 5(2):151–172, 1957.

[67] J. F. Speer, V. E. Petrosky, M. W. Retsky, and R. H. Wardwell. Stochastic numerical model of breast cancer growth that simulates clinical data. *Cancer Research*, 44:4124–4130, 1984.

[68] K. P. Sreekumar and G. Nirmalan. Estimation of the total surface area in indian elephants (*Elephas maximus indicus*). *Veterinary Research Communications*, 14:5–17, 1990.

[69] D. W. Stephens and J. R. Krebs. *Foraging Theory*. Princeton University Press, 1987.

[70] M. Sternstein. *Barron's How to Prepare for the AP Advance Placement Exam Statistics*. Barron's Press, 2004.

[71] T. D. Steury and D. L. Murray. Modeling the reintroduction of lynx to the southern portion of its range. *Biological Conservation*, 117:127–141, 2004.

[72] J. Stewart. *Calculus*. Brooks/Cole, 1999.

[73] J. H. M. Thornley. *Mathematical Models in Plant Physiology: A Quantitative Approach to Problems in Plant and Crop Physiology*, volume 8 of *Experimental Botany: An International Series of Monographs*. Academic Press, 1976.

[74] J. E. Vermaat and M. K. Hanif. Performance of common duckweed species *Lemnaceae* and the waterfern *Azoll filiculoides* on different types of waste water. *Water Research*, 32(9):2569–2576, 1998.

[75] S. Vogel. *Life's Devices: The Physical World of Animals and Plants*. Princeton University Press, 1988.

[76] S. Wanderley, M. Costa-Neves, and R. Rega. Relative growth of the brain in human fetuses: First gestational trimester. *Archives d'anatomie, d'histologie et d'embryologie*, 73: 43–46, 1990.

[77] J. B. West. *Respiratory Physiology*. Williams and Wilkins, 4th ed., 1990.

[78] H. Xin, I. Berry, T. Barton, and G. Tabler. Feed and water consumption, growth, and mortality of male broilers. *Poultry Science*, 73(5): 610–616, 1994.

[79] E. K. Yeargers, R. W. Shonkwiler, and J. V. Herod. *An Introduction to the Mathematics of Biology with Computer Algebra Models*. Birkhauser, 1996.

[80] B. Zanoni, C. Garzaroli, S. Anselmi, and G. Rondinini. Modeling the growth of *Enterococcus faecium* in bologna sausage. *Applied and Environmental Microbiology*, 59(10):3411–3417, 1993.

[81] M. S. Zeilik, S. Gregory, and E. Smith. *Introductory Astronomy and Astrophysics*. Saunders College Publishing, 1992.

Weblinks

[82] Sickle cell disease. http://science.education.nih.gov/supplements/nih1/genetic/activities/activity2_database.htm.

[83] Estimates of Funding for Various Research, Condition, and Disease Categories (RCDC). Technical report, National Institutes of Health, January 2009. http://report.nih.gov/rcdc/categories/.

[84] Estimates of Funding for Various Research, Condition, and Disease Categories (RCDC). Technical report, National Institutes of Health, April 2013. http://report.nih.gov/rcdc/categories/.

[85] http://www.fieldtripearth.org/article.xml?id=776&ordinal=5

[86] http://www.tiem.utk.edu/ gross/bioed/webmodules/aminoacid.htm

[87] http://en.wikipedia.org/wiki/Correlation.

[88] http://www.vendian.org/mncharity/dir3/paper_rulers/

[89] http://www.wikihow.com/Measure-the-Height-of-a-Tree

[90] http://daac.ornl.gov/OTTER/guides/Timber_Data.html

[91] http://www.cyber-dyne.com/~tom/03/natural_amazon.html

[92] http://maps.google.com

[93] http://www.physicalgeography.net/fundamentals/9i.html

[94] http://birds.audubon.org/historical-results

[95] http://www.epa.qld.gov.au/wetlandinfo/site/factsfigures/SummaryInformation/FloraAndFauna/Flora.html

[96] http://schools-wikipedia.org/wp/m/Matrix_multiplication.htm

[97] http://science.education.nih.gov/supplements/nih3/sleep/guide/info-sleep.htm

[98] http://www.mathworks.com/matlabcentral/

[99] http://www.math.utah.edu/lab/ms/matlab/matlab.html/

[100] http://en.wikipedia.org/wiki/Limit_(mathematics)

Getting Started with Matlab

This is a very basic introduction to the elements of Matlab that will be used in the first part of this text. There is also extensive documentation on Matlab available at [98] or see the tutorial located at [99].

Matlab is a mathematics package that allows you to easily solve many of the quantitative problems that arise in the life sciences. This document briefly describes some of the key elements in using Matlab to

1. do descriptive statistics,
2. carry out matrix algebra,
3. incorporate probability, and
4. analyze discrete difference equations.

All these topics are covered in detail in the text, and this appendix is designed to get you started using Matlab and to describe some of the commonly used functions you will find in each of the "Matlab Skills" sections near the end of each chapter of this text.

A.1 Starting Matlab

To get started, open Matlab. When the program opens, it should like Figure A.1.

There are a couple of things you should notice. First, look at the top of the window where it says "Current Directory." This is the directory on your computer that you are currently working from. This will become more important later when you are working with files. Make sure that the Current Directory is pointed to the directory of your choice. If you need to change the Current Directory, click on the button with the "..." on it. This will allow you to change directories. To view the files in the Current Directory, click on the Current Directory tab.

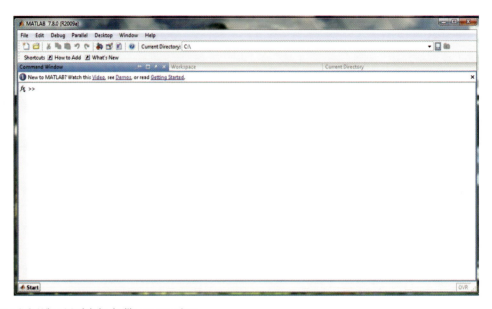

Figure A.1 What Matlab looks like on opening.

Next, click back over to the Command Window tab. This is the main location from which Matlab is run. Commands are typed into the command window and run by pressing the Enter key. Notice the » in the command window. This is where you will type in commands.

A.2 Working from the Command Window

Basic Arithmetic in Matlab

Let us get started by trying out some basic arithmetic commands. We can do this right in the command window. Try typing in the following commands. After typing each line, press the Enter key.

```
>> 5 + 7
>> 6 - 9
>> 2 * 9
>> 25 / 2
>> 3^5
```

Matlab also has special commands for other common operations. Use the `log(x)` command to find the natural logarithm of the number x. For example, try typing into the command window

```
>> log(5)
```

If you need to find the logarithm with a base other than $e \approx 2.71828183$, use the log property

$$\log_a x = \frac{\ln x}{\ln a}.$$

For example, if you wanted to find $\log_2 5$, you would type

```
>> log(5)/log(2)
```

into the command window.

Matlab uses the command $\exp(x)$ to find e^x. For example, if you wanted to find $e^{5.46}$, you would type

```
>> exp(5.46)
```

into the command window.

Some other common commands are the following:

- $\sin(x)$, finds the sine of the number x
- $\cos(x)$, finds the cosine of the number x
- $\tan(x)$, finds the tangent of the number x
- $\text{factorial}(x)$, finds $x! = x \cdot (x-1) \cdot (x-2) \cdots 3 \cdot 2 \cdot 1$
- $\text{sqrt}(x)$, finds the square root of the number x, that is, $\sqrt{x}$
- $\text{nthroot}(x,n)$, finds the nth root of the number x, that is, $\sqrt[n]{x}$

Note that the commands $\text{nthroot}(x,n)$ and $x^\wedge(1/n)$ will not produce the same results for negative numbers. Thus, if you want to compute the cube root of -2 (which is a real number), $(-2)^\wedge(1/3)$ will return a complex number. However, $\text{nthroot}(-2,3)$ will correctly return -1.2599.

Some Useful Matlab Functions

ROUNDING

Several different functions are used for rounding, depending on the rule you want to use for rounding. The function $\text{round}(x)$ will round the number x to the nearest integer. It will round up if the decimal portion of x is greater than or equal to 0.5 and down otherwise. If you always want to round up, use the function $\text{ceil}(x)$ to round up to the nearest integer. If you always want to round down, use the function $\text{floor}(x)$ to round down to the nearest integer. If you want to round to, say, the hundredths place, you multiply x by 100, use the appropriate rounding function (round, ceil, or floor), and then divide by 100:

```
──────────────────── Command Window ────────────────────
>> pi
ans =
     3.1416

>> round(pi)
ans =
     3

>> ceil(pi)
ans =
     4
```

```
>> floor(pi)
ans =
      3

>> round(pi*100)/100
ans =
      3.1400

>> ceil(pi*10)/10
ans =
      3.2000

>> floor(pi*1000)/1000
ans =
      3.1410
```

Note that in Matlab, `pi` represents $\pi \approx 3.14159265\ldots$. When these commands were run in the command window, Matlab was set to display five digits for any answer. Thus, the initial value for `pi` shown was rounded to the nearest ten-thousandth place.

FINDING ROOTS OF POLYNOMIALS

A polynomial of degree n, written generally as

$$P(x) = a_n x^n + a_{n-1} x^{n-1} + \cdots + a_2 x^2 + a_1 x + a_0,$$

will have n roots (values of x where $P(x) = 0$), some of which may be imaginary numbers. To calculate the value of the roots of a polynomial in Matlab, we use the function `roots(c)`, where $c = [a_n \ a_{n-1} \ \cdots \ a_2 \ a_1 \ a_0]$ is the array of coefficients of the polynomial in descending order.

For example, suppose that we want to find the roots of the polynomials

$$f(x) = x^2 - 4x - 5 \quad \text{and} \quad g(x) = \frac{1}{2} - 24x^2 + 64x^3 - 48x^4.$$

To find the roots of $f(x)$, we would use the command `roots([1 -4 -5])`. Notice that the coefficient in front of the x^2 term is 1. Additionally, we must make sure to include the negative signs in font of the coefficients of the x term and the constant term. To find the roots of $g(x)$, we would use the command `roots ([-48 64 -24 0 0.5])`. Notice that we list the coefficients in descending order (starting with the highest-order term, x^4) even though the terms are not written in that order. Additionally, note that there is no x term in $g(x)$, so it has a coefficient of 0:

```
─────────────── Command Window ───────────────
>> roots([1 -4 -5])

ans =

      5
     -1

>> roots([-48 64 -24 0 0.5])
```

```
ans =

   0.6321 + 0.1922i
   0.6321 - 0.1922i
   0.1929
  -0.1237
```

Thus, we find that for the function $f(x)$, the roots are $x = -1$ and $x = 5$. Indeed, if you evaluate $f(x)$ at $x = -1$ and $x = 5$, you will calculate a function value of 0. For the function $g(x)$, we find that there are two real roots and two imaginary roots (recall that imaginary roots always come in pairs).

A.3 Working with Arrays (Vectors and Matrices)

The name Matlab is actually short for Matrix Laboratory. One of Matlab's main features is that it quickly and easily handles and manipulates matrices. In this section, we learn how to enter matrices into Matlab and do some basic matrix computation.

Entering Vectors and Matrices into Matlab

Suppose that you have the following data.

x	2	5	2	4	6
y	4	7	5	8	11

Data can be entered in two ways. The first method is to enter a vector for the x data and a vector for the y data. In the command window, this would look like the following:

```
———————————————— Command Window ————————————————
>> x = [2 5 2 4 6]
x =
     2     5     2     4     6

>> y = [4 7 5 8 11]
y =
     4     7     5     8    11
```

Notice that we use square brackets, [ ], to denote the start and end of the vectors. To separate vector entries, we use spaces. Also, notice that when you press Enter after typing in the vector, Matlab prints the vector back out for you. If you wish to suppress this, place a semicolon at the end of the line, like this:

```
———————————————— Command Window ————————————————
>> x = [2 5 2 4 6];

>> y = [4 7 5 8 11];
```

You can place a semicolon at the end of any command line to suppress the output of that line in the command window.

If you want to represent the x and y data as column vectors instead of row vectors, place a ' after the closing bracket] of the vector to take the transpose:

```
─────────────────────────  Command Window  ─────────────────────────
>> x = [2 5 2 4 6]'
x =
      2
      5
      2
      4
      6
```

The other way to enter data into Matlab is to enter both the x and the y data into the same matrix. We can let the first column of the matrix represent the x data and the second column of the matrix represent the y data. In that case, we would type the following into the command window:

```
─────────────────────────  Command Window  ─────────────────────────
>> data = [ 2 4; 5 7; 2 5; 4 8; 6 11 ]
data =
      2      4
      5      7
      2      5
      4      8
      6     11
```

Notice that we still use the square brackets to define the beginning and end of the matrix; however, now we use a semicolon (within the brackets) to denote when to start the next row in the matrix. An alternative way to enter the matrix is to put a row of data on a new line:

```
─────────────────────────  Command Window  ─────────────────────────
>> data = [ 2 4
      5 7
      2 5
      4 8
      6 11 ]
data =
      2      4
      5      7
      2      5
      4      8
      6     11
```

You may use whichever method you like for entering matrices. You might want to check the length of a vector or the size of a matrix once you have entered it. Use the length(x) command to find the size of a vector labeled x. Use the size(A) command to find the size of a matrix labeled A. This function will output two numbers, the first being the number of rows in A the second being the number of columns in A. Also, if you need the sum of all the elements in your vector or the sum of the columns in your matrix, use sum(x) or sum(A), respectively:

```
─────────────── Command Window ───────────────
>> length(x)
ans =
     5

>> size(data)
ans =
     5     2

>> sum(x)
ans =
          19

>> sum(data)
ans =
          19    35
```

Accessing Entries within a Matrix

Once we have our matrix of data, we would like to be able to access just the *x* data and just the *y* data if needed. Matlab makes this easy. If we have a matrix A entered into the Matlab, then we can use the structure A(i,j) to access portions of the matrix, where i indicates the row or rows that we want and j indicates the column or columns we want. Use A(1,1) to get the entry in the first row, first column. The following sequence of commands entered into the command window will (1) check the entry in the second row, first column of the matrix data; (2) name the first column x; (3) name the second column y; (4) name the third row foo; and (5) create a new matrix called newdata that contains only rows 2 through 4:

```
─────────────── Command Window ───────────────
>> data(2,1)
ans =
     5

>> x = data(:,1)
x =
     2
     5
     2
     4
     6

>> y = data(:,2)
y =
     4
     7
     5
     8
    11
```

```
>> foo = data(3,:)
foo =
     2     5

>> newdata = data(2:4,:)
newdata =
     5     7
     2     5
     4     8
```

Constructing Special Matrices in Matlab

If you have an array of values that increase by regular intervals, say, a list of years, there is a shortcut to entering the data. Suppose that you wanted to make a vector of each year from 1980 to 2010. It would be tedious to enter each year by hand, so we use the shortcut [a:b:c], where a is the smallest value, b is the size of the increment, and c is the largest value:

```
━━━━━━━━━━━━━━━━ Command Window ━━━━━━━━━━━━━━━━
>> years = [1980:1:2010]
years =
  Columns 1 through 7
      1980     1981    1982    1983    1984    1985    1986
  Columns 8 through 14
      1987     1988    1989    1990    1991    1992    1993
  Columns 15 through 21
      1994     1995    1996    1997    1998    1999    2000
  Columns 22 through 28
      2001     2002    2003    2004    2005    2006    2007
  Columns 29 through 31
      2008     2009    2010

>> every5years = [1980:5:2010]
every5years =
      1980     1985    1990    1995    2000    2005    2010
```

Notice that when there are too many columns for Matlab to print on one line, it will label which columns are being printed on each line.

If you want to construct a matrix of all 0s or all 1s, you can use the functions zeros(m,n) and ones(m,n), respectively, where m is the number of rows you would like the matrix to have and n is the number of columns you would like the matrix to have.

A.4 M-files

So far, every command we have run we have typed into the command window. This is convenient for quick calculations, but often we have a series of commands we would like to run several times, perhaps with only a small modification each time. For this, Matlab provides the capability of writing and running m-files. An m-file is a Matlab file that contains a series of commands and can be run from the command window by typing the name of the m-file.

Steps to Creating and Running an M-file

1. Check the location of the Current Directory. Make sure the Current Directory is pointed to the folder where you want to save your m-file. If necessary, change the location of the Current Directory.
2. Open a new m-file (`File -> New -> M-File` or click on the 🗋 button in the top left corner of the Matlab window).
3. Type a few commands into the m-file. See an example of an m-file below.
4. Save the m-file in the Current Directory (`File -> Save As...`). Save the m-file as *filename*.m, where *filename* is the name of your file. M-file names must start with an alphabetic character, may contain any alphanumeric characters or underscores, and must be no longer than 63 characters.
5. To run your m-file, type the filename (without the .m) into the command window and press Enter.

An example m-file follows:

```
─────────────────────── test.m ───────────────────────
1  | % This is a comment in an m-file. Anything typed after a percent sign
2  | % in an m-file will not be read as a command. This allows you to put
3  | % comments and notes in your m-files.
4  |
5  | x = [0:2.5:10];      % Enter some vector
6  |
7  | m = length(x)            % Find the length of x
8  |
9  | % Note the lack of ; after the previous command will cause the output
10 | % of this line to be displayed when this file is run.
```

If we save this file as test.m, then, when this file is run in Matlab, it looks like the following:

```
─────────────────── Command Window ───────────────────
>> test
m =
     5
```

Notice that we can add lines into the file that are not interpreted as commands by putting a % at the front of the line or after a command.

A.5 Nicely Formatted Output: `fprintf`

After you get the hang of writing m-files, you will be able to write m-files that generate a lot of different output. It is often useful to display this output in a nicely formatted way. For this, we use the `fprintf` function. This function is best explained through some examples.

Suppose that we have an array of numbers and we want to compute the minimum and maximum values in that array. We could use the following m-file to do this:

```
                              ─── MinMax.m ───
1   % Filename: MinMax.m
2   % M-file to
3   %    - calculate the minimum & maximum of an array
4   %    - display that min and max
5
6   % Create array
7   x = [20 45 81 6 -3 -23 99];
8
9   % Find minimum and maximum of array
10  minx = min(x);
11  maxx = max(x);
12
13  % Display the min & max using fprinft
14  fprintf('The max is %d\n',maxx)
15  fprintf('The min is %d\n',minx)
16  fprintf('The average of the min and max is %4.1f\n',(maxx+minx)/2);
```

Look at the first `fprintf` line. The text we want displayed is in single quotes, ' ', where the max value we want displayed is replaced with `%d`. The `%` indicates that you want to insert a calculated value here, in this case, `maxx`. The `d` indicates that it is a decimal value. The `\n` just before the close of the single quotes incidates that you want a new line. This way, the next bit of information we print out will start on a new line. After the single quotes there is a comma, followed by the name of the computed value that we want inserted for `%d`. The second `fprintf` line is similar to the first, only we display the minimum value.

Now, look at the last `fprintf` line. Notice that we now use `%4.1f` instead of `%d`. The `f` indicates that this is real number and we will not use scientific notation. If we did want to use scientific notation, we would use an `e`. The `4.1` indicates that we require four spaces to display our number and that there will be one digit after the decimal. Note that the decimal point counts as a space. So `4.1` indicates that there will be two digits before the decimal, the decimal, and then one digit after the decimal, for a total of four spaces. The other difference to note in this line is that we can do calculations within the `fprintf` command. The value that we wish to display in the `%4.1f` slot is computed using `maxx` and `minx`.

B.1 Mathematical Notation

Sigma and Pi Notation

Suppose that we have a sequence of n numbers:

$$x_1, x_2, x_3, \ldots, x_n.$$

There are times when we would like to take the sum or product of all n numbers, but writing out this sum or product can be tedious. In mathematics, we use sigma and pi notation as shorthand to write out sums and products of n numbers.

$$\sum_{k=1}^{n} x_k = x_1 + x_2 + x_3 + \cdots + x_n$$

$$\prod_{k=1}^{n} x_k = x_1 \cdot x_2 \cdot x_3 \cdots x_n$$

Note we use the Σ, or sigma, to denote the sum since Σ is the Greek letter for uppercase S and the Π, or pi, to denote the product since Π is the Greek letter for uppercase P. In each case, underneath the Σ or Π is an index (usually i, j, or k) and the value at which that index starts. In the equations above, we start at $k = 1$ because we start with element 1 in the sequence. If we wanted to add up all the numbers in the sequence starting with element 5, we would use $k = 5$ below the Σ. Notice that the index also appears as a subscript on the x after the Σ or Π. This indicates that we are adding or multiplying the elements of the x sequence. If we had another sequence of y values,

$$y_1, y_2, y_3, \ldots, y_n,$$

we could replace the x_k with a y_k after the Σ or Π to add or multiply the elements in the y sequence. Finally, notice that above the Σ and Π is n. The value above the Σ or Π indicates the last element of the sequence that is added to the sum or multiplied by the product, respectively. Thus, the last element we add is the nth element of the sequence. Thus,

$$\sum_{k=1}^{5} x_k$$

indicates that we sum elements 1 through 5 of the x sequence, while

$$\prod_{k=3}^{10} y_k$$

indicates that we take the product of elements 3 through 10 of the y sequence.

Precise Definition of a Limit

For $p \in \mathbb{R}$ (a real number), we write $\lim_{x \to p} f(x) = L$ if for every number $\varepsilon > 0$ there is a $\delta > 0$ such that $|f(x) - L| < \varepsilon$ whenever $|x - p| < \delta$.

For $p = \infty$, we write $\lim_{x \to \infty} f(x) = L$ if for every number $\varepsilon > 0$ there is an $N > 0$ such that $|f(x) - L| < \varepsilon$ for all $x > N$.

Likewise, for $p = -\infty$, we write $\lim_{x \to -\infty} f(x) = L$ if for every number $\varepsilon > 0$ there is an $N < 0$ such that $|f(x) - L| < \varepsilon$ for all $x < N$.

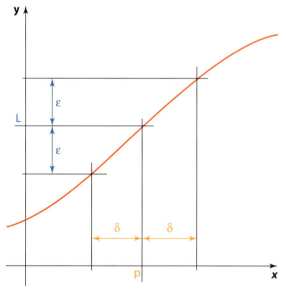

Figure B.1 The formal definition of a limit. Here L is the limit of $f(x)$ as $x \to p$ if the distance between x and p is less than δ; then the distance between $f(x)$ and L is less than ε. Source [100].

B.2 Proofs

For those interested in seeing the mathematical rigorous proofs of various properties and theorems not included in the main text, we provide them here.

Power Rule

If $f(x) = x^n$, where n is a constant, then

$$\lim_{x \to a} \frac{f(x) - f(a)}{x - a} = \lim_{x \to a} \frac{x^n - a^n}{x - a}.$$

Given the formula

$$x^n - a^n = (x - a)(x^{n-1} + x^{n-2}a + \cdots + xa^{n-2} + a^{n-1}),$$

we have

$$\lim_{x \to a} \frac{x^n - a^n}{x - a} = \lim_{x \to a} x^{n-1} + x^{n-2}a + \cdots + xa^{n-2} + a^{n-1}$$

$$= a^{n-1} + a^{n-2}a + \cdots + aa^{n-2} + a^{n-1}$$

$$= na^{n-1}.$$

Thus, $f'(x) = nx^{n-1}$. This is known as the **power rule**.

Chain Rule

Given $y = f(g(x))$, let $u = g(x)$ so that we can express y as $y = f(u)$.

Let $v = \dfrac{g(x+h) - g(x)}{h} - g'(x)$. Note that $v \to 0$ as $h \to 0$ (by the limit definition of a derivative) and that

$$g(x + h) = (v + g'(x))h + g(x). \tag{B.1}$$

Let $w = \dfrac{f(u+k) - f(u)}{k} - f'(u)$. Note that $w \to 0$ as $k \to 0$ (by the limit definition of a derivative) and that

$$f(u + k) = (w + f'(u))k + f(u) = \big(w + f'(g(x))\big)k + f(g(x)). \tag{B.2}$$

Note that to find the derivative of $f(g(x))$, we need to evaluate the limit

$$\lim_{h \to 0} \frac{f(g(x+h)) - f(g(x))}{h}.$$

We start by rewriting just the numerator of the difference quotient using Equation (B.1):

$$f(g(x+h)) - f(g(x)) = f\left(\underbrace{(v + g'(x))h}_{k} + \underbrace{g(x)}_{u}\right) - f(g(x)) \tag{B.3}$$

Note that the first term on the right-hand side of Equation (B.3) is $f(u+k)$, where $u = g(x)$ and $k = (v + g'(x))h$ (note that $k \to 0$ as $h \to 0$). Using Equation (B.2), we can rewrite the first term on the right-hand side of Equation (B.3):

$$\begin{aligned} f(g(x+h)) - f(g(x)) &= f\left((v + g'(x))h + g(x)\right) - f(g(x)) \\ &= \left(w + f'(g(x))\right)\left((v + g'(x))h\right) + f(g(x)) - f(g(x)) \\ &= \left(w + f'(g(x))\right)\left(v + g'(x)\right)h. \end{aligned} \tag{B.4}$$

Now, we divide both sides of Equation (B.4) by h and take the limit as $h \to 0$. Thus, we obtain

$$\lim_{h \to 0} \frac{f(g(x+h)) - f(g(x))}{h} = \lim_{h \to 0} \left(w + f'(g(x))\right)\left(v + g'(x)\right) = f'(g(x))g'(x) \tag{B.5}$$

since $w \to 0$ and $v \to 0$ as $h \to 0$. Thus, we have derived the chain rule.

Derivative of Exponential Functions

Special Case: Derivation of derivative of $f(x) = e^x$

We will need the fact that the constant e is defined as

$$e = \lim_{h \to 0} (1 + h)^{1/h}.$$

Since the limit of the function $(1 + h)^{1/h}$ is e, for small h we can approximate the value of e by $(1 + h)^{1/h}$, which we denote by $e \approx (1 + h)^{1/h}$ (i.e., the symbol $\approx$ means "approximately equals"). Using this approximation, we find that we can approximate e^h by

$$e^h \approx \left[(1 + h)^{1/h}\right]^h = (1 + h),$$

that is, $e^h \approx (1 + h)$. Using this approximation, we find

$$\lim_{h \to 0} \frac{e^h - 1}{h} \approx \lim_{h \to 0} \frac{1 + h - 1}{h} = \lim_{h \to 0} \frac{h}{h} = 1,$$

and thus we have

$$f'(x) = \lim_{h \to 0} \frac{f(x+h) - f(x)}{h}$$

$$= \lim_{h \to 0} \frac{e^{x+h} - e^x}{h}$$

$$= \lim_{h \to 0} \frac{e^x e^h - e^x}{h}$$

$$= \lim_{h \to 0} e^x \frac{e^h - 1}{h}$$

$$= e^x \lim_{h \to 0} \frac{1 + h - 1}{h}$$

$$= e^x.$$

Note that e^x is the *only* function that is equal to its own derivative. This is one of the properties of the function $f(x) = e^x$ that makes it so intriguing.

General Case: Derivation of derivative of $f(x) = a^x$
We can rewrite

$$f(x) = a^x = e^{(\ln a)x}.$$

Now, using the chain rule, we get

$$f'(x) = (\ln a)e^{(\ln a)x} = (\ln a)a^x.$$

Sine

Let $f(x) = \sin x$; then

$$f'(x) = \lim_{h \to 0} \frac{\sin(x+h) - \sin(x)}{h}$$

$$= \lim_{h \to 0} \frac{\sin x \cos h + \sin h \cos x - \sin x}{h}$$

$$= \lim_{h \to 0} \sin x \frac{\cos h - 1}{h} + \lim_{h \to 0} \cos x \frac{\sin h}{h}$$

$$= \lim_{h \to 0} \sin x \cdot \lim_{h \to 0} \frac{\cos h - 1}{h} + \lim_{h \to 0} \cos x \cdot \lim_{h \to 0} \frac{\sin h}{h}.$$

Notice that $\lim_{h \to 0} \sin x = \sin x$ and $\lim_{h \to 0} \cos x = \cos x$ since those functions have no dependence on h. Thus, we have

$$f'(x) = \sin x \cdot \lim_{h \to 0} \frac{\cos h - 1}{h} + \cos x \cdot \lim_{h \to 0} \frac{\sin h}{h}.$$

Now, we must find the limits of $\frac{\sin h}{h}$ and $\frac{\cos h - 1}{h}$. To find the limit of $\frac{\sin h}{h}$, we will use the sandwich theorem. One can verify that for $x > 0$, $0 < \sin x < x$, and thus

$$0 < \frac{\sin x}{x} < 1.$$

Additionally, it can be shown that

$$\frac{\sin x}{\cos x} > x \implies \frac{\sin x}{x} > \cos x.$$

Thus, we have

$$\cos x < \frac{\sin x}{x} < 1.$$

We know that

$$\lim_{h \to 0} \cos x = 1 \text{ and } \lim_{h \to 0} 1 = 1.$$

Thus, by the sandwich theorem,

$$\lim_{h \to 0} \frac{\sin h}{h} = 1.$$

Next, to find the limit of $\frac{\cos h - 1}{h}$, notice that

$$
\begin{aligned}
\lim_{h \to 0} \frac{\cos h - 1}{h} &= \lim_{h \to 0} \frac{\cos h - 1}{h} \times \frac{\cos h + 1}{\cos h + 1} \\
&= \lim_{h \to 0} \frac{\cos^2 h - 1}{h(\cos h + 1)} \\
&= \lim_{h \to 0} \frac{-\sin^2 h}{h(\cos h + 1)} \\
&= -\lim_{h \to 0} \frac{\sin h}{h} \times \lim_{h \to 0} \frac{\sin h}{\cos h + 1} \\
&= -(1) \times \frac{0}{1 + 1} = 0.
\end{aligned}
$$

Thus, we find that

$$\frac{d}{dx}(\sin x) = \sin x \times \lim_{h \to 0} \frac{\cos h - 1}{h} + \cos x \times \lim_{h \to 0} \frac{\sin h}{h} = (\sin x)(0) + (\cos x)(1) = \cos x.$$

Cosine

Let $g(x) = \cos x$; then

$$g'(x) = \lim_{h \to 0} \frac{\cos(x+h) - \cos(x)}{h}$$

$$= \lim_{h \to 0} \frac{\cos x \cos h - \sin x \sin h - \cos x}{h}$$

$$= \lim_{h \to 0} \cos x \cdot \lim_{h \to 0} \frac{\cos h - 1}{h} - \lim_{h \to 0} \sin x \cdot \lim_{h \to 0} \frac{\sin h}{h}$$

$$= (\cos x)(0) - (\sin x)(1)$$

$$= -\sin x.$$

Alternatively, we could use the fact that $\cos x = \sin\left(\frac{\pi}{2} - x\right)$. Then, using the chain rule, the derivative we derived for $\sin x$, and the fact that $\cos\left(\frac{\pi}{2} - x\right) = \sin x$, we find that

$$\frac{d}{dx} \cos x = \frac{d}{dx}\left[\sin\left(\frac{\pi}{2} - x\right)\right] = \cos\left(\frac{\pi}{2} - x\right) \cdot (-1) = -\sin(x).$$

Answers to Selected Problems

Chapter 1

1.1 (a) 970.09, 906   (b) 951.02,932.19

1.2 8.47, 8.6, 8.7

1.3 Multiply each statistic by the positive constant

1.4 (a) 20, 22.7   (b) 59, 329.79, 18.16

1.5 96, 663.82, 25.77

Chapter 2

2.3

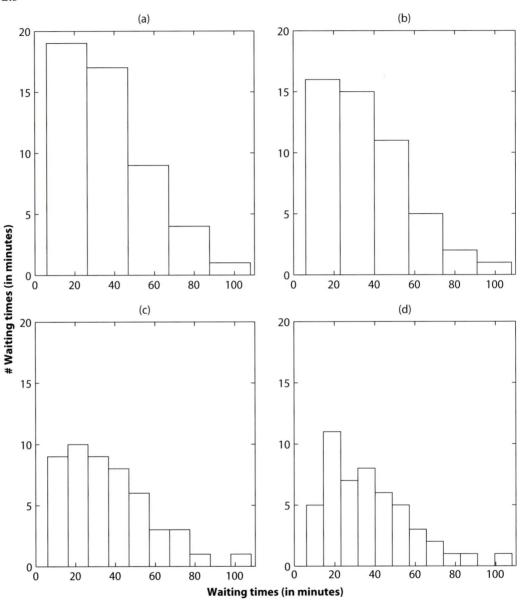

(e) Unimodal   (f) No

2.4 (a) 0.22

2.5

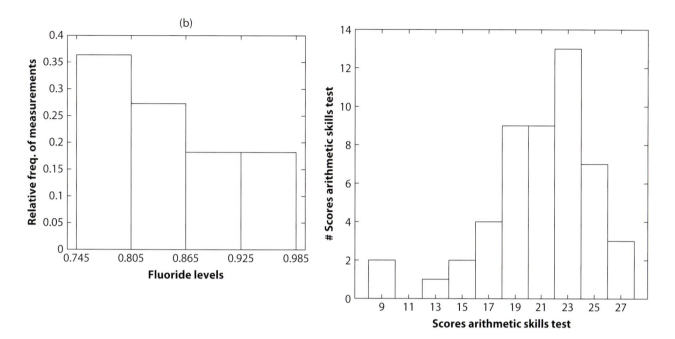

(b)

2.6

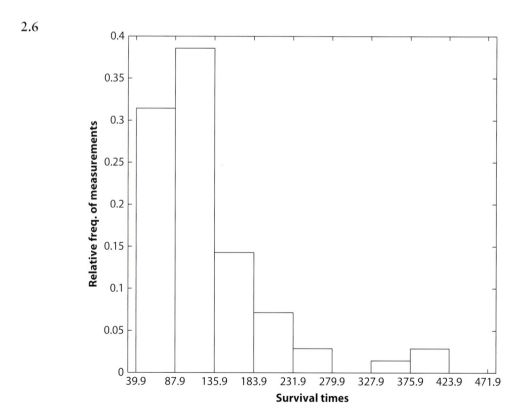

Chapter 3

3.4 (a)

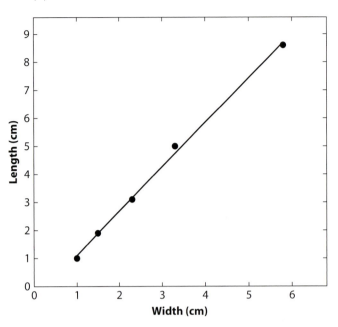

(d) $\hat{y} = 1.585\hat{x} - 0.487$ (e) $0.999, 99.75\%$

3.5 (a)

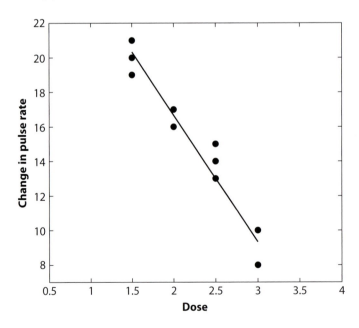

(d) $\hat{y} = -7.333\hat{x} + 31.333$ (e) $-0.971, 94.38\%$

3.7 **(a)**

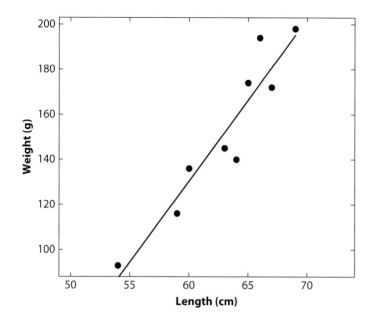

(d) $\hat{y} = 7.191\hat{x} - 301.087$ **(e)** 0.944, 89.05%

3.8 **(a)**

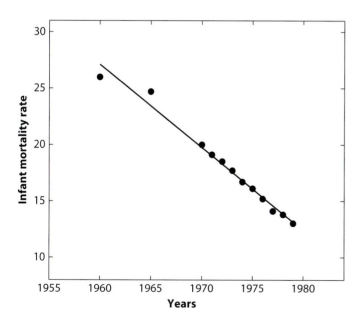

(d) $\hat{y} = -0.737\hat{x} + 1471.965$ **(e)** −0.991, 98.23%

3.9 (a)

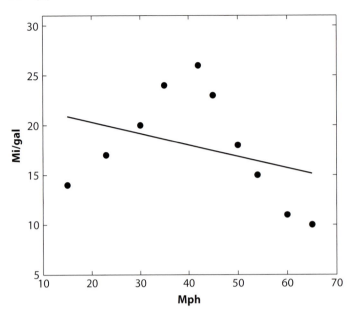

(d) $\hat{y} = -0.115\hat{x} + 22.612$ (e) $-0.341, 11.60\%$

Chapter 4

4.1 (a) 3 (c) $-\ln 0.4$ (e) $\frac{1}{3}$ (g) $-5\ln 4$

4.2 (a) 10^{1000} (c) 60 (d) $\frac{3^{16}}{2}$

4.3 (a) 1800 (b) 300 (c) $\frac{\ln 2}{\ln 6}$

4.5 (a) $f(x) = 100(\sqrt{1.6})^x$ (b) 5.899

4.6 (a) Allometric (b) $\alpha = 4.13, \beta \approx 0.293$ (c) $A \approx 406$

4.7 (a) $L = 10^{-\frac{2}{3}} R^{\frac{2}{3}}$ (b) Plant A has $2^{\frac{2}{3}}$ more leaf biomass than plant B

4.10 (a) 66.19, 222.39 (b) $x = g(z) = 0.454z$ (c) $f(z) = 19.7(0.454z)^{0.753}$

Chapter 5

5.2 (a) $\left\{0, \frac{1}{4}, \frac{2}{7}, \frac{3}{10}, \frac{4}{13}\right\}; \frac{1}{3}$ (c) $\left\{1, -\frac{1}{4}, \frac{1}{16}, -\frac{1}{64}, \frac{1}{256}\right\}; 0$

5.3 (a) $x_n = 0.3^n \times 10$ (c) $x_n = 10 + 6n$ (e) $x_n = \left(10 + \frac{5}{2}\right) 3^n - \frac{5}{2}$ (f)
$x_n = (10 + 2) 2^n - 2$ (g) $x_n = 10 - 4n$ (h) $x_n = (10 - 1) (-3)^n + 1$ (i)
$x_n = (10 + 1) \left(\frac{2}{3}\right)^n - 1$

5.5 (a) $x_{n+1} = 1.1 \times x_n$ (b) $x_n = 1.1^n \times 50$

5.6 (a) $x_{n+1} = 0.9 \times x_n$ (b) $x_n = 0.9^n \times 50$ (c) 33

5.7 (a) $x_{n+1} = 0.9 \times x_n$ (b) $x_n = 0.9^n \times 180$ (c) 162 mg

5.8 (a) $x_n = (800)1.1^n + 200$ (b) No   (c) 68

5.9 (a) n = 5   (b) Extinction

5.10 (a) 443.274 mg   (b) 443.274 mg   (c) 2.23 hours

5.11 1.82 hours

Chapter 6

6.2 (a) $\begin{bmatrix} 5 \\ 0 \\ 3 \end{bmatrix}$ (b) $\begin{bmatrix} 1 \\ 2 \\ -3 \end{bmatrix}$ (c) Undefined   (d) $\begin{bmatrix} 4 \\ -4 \\ -6 \\ 8 \end{bmatrix}$ (e) Undefined

(f) $\begin{bmatrix} -4 \\ -4 \\ 14 \end{bmatrix}$ (g) $\begin{bmatrix} -1 \\ 8 \\ 12 \\ 12 \end{bmatrix}$

6.3 $\begin{bmatrix} S \\ I \\ Q \end{bmatrix} = \begin{bmatrix} 0.65 \\ 0.20 \\ 0.15 \end{bmatrix}$

6.4 (a) $\begin{bmatrix} W \\ M \\ Q \end{bmatrix} = \begin{bmatrix} 250 \\ 180 \\ 2 \end{bmatrix}$ (b) $\begin{bmatrix} M \\ F \end{bmatrix} = \begin{bmatrix} 0.417 \\ 0.583 \end{bmatrix}$

6.5 (a) $\begin{bmatrix} 0.65 & 0.82 \\ 0.35 & 0.18 \end{bmatrix} \begin{bmatrix} S \\ I \end{bmatrix}$ (b) $\begin{bmatrix} 0.8 & 0.25 & 0.08 \\ 0 & 0.75 & 0.01 \\ 0.2 & 0 & 0.91 \end{bmatrix} \begin{bmatrix} S \\ G \\ T \end{bmatrix}$

(c) $\begin{bmatrix} 0.85 & 0.01 & 0 & 0 \\ 0.05 & 0.98 & 0 & 0.2 \\ 0 & 0.01 & 1 & 0 \\ 0.1 & 0 & 0 & 0.8 \end{bmatrix} \begin{bmatrix} G \\ S \\ W \\ O \end{bmatrix}$

6.10 $\begin{bmatrix} 0.8 & 0.8 \\ 0.2 & 0.2 \end{bmatrix}$

Chapter 7

7.1 (a) $x = \frac{9}{2}, y = 2$ (b) No solution   (c) $x = 4, y = -1$

7.2 $\frac{5}{2}, \frac{7}{2}$

7.4 (a) $\begin{bmatrix} 2.4 & -1.2 \\ 6 & 3.6 \end{bmatrix}$ (c) $\begin{bmatrix} 2 & 0 & 4 \\ 3 & 7 & -1 \\ 0 & 0 & 0 \end{bmatrix}$ (e) $\begin{bmatrix} 10 & -5 & 8 \\ 14 & 4 & 9 \end{bmatrix}$ (f) $\begin{bmatrix} 18 & 2 \\ 13 & 10 \\ 0 & 0 \end{bmatrix}$

(g) $\begin{bmatrix} 11 & -1 & 6 \\ 11 & 3 & 4 \end{bmatrix}$ (i) Undefined (j) $\begin{bmatrix} -10 & 1 & -18 \\ -3 & -3 & 21 \end{bmatrix}$ (k) $\begin{bmatrix} -12 & -10 \\ 23 & 9 \end{bmatrix}$

7.6

Week	S	I
1	0.6840	0.3160
2	0.7037	0.2963
3	0.7004	0.2996
5	0.7008	0.2992
10	0.7009	0.2991
25	0.7009	0.2991
50	0.7009	0.2991

7.7

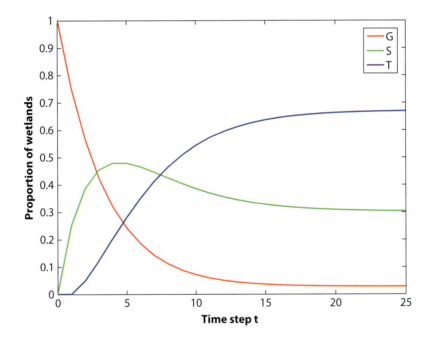

Chapter 8

8.3 $\begin{bmatrix} 2/3 \\ 1/3 \end{bmatrix}$

8.6 $\begin{bmatrix} .75 \\ .25 \end{bmatrix}$

8.8 (a) $\begin{bmatrix} S \\ I \end{bmatrix} = \begin{bmatrix} 0.7009 \\ 0.2991 \end{bmatrix}$ (b) $\begin{bmatrix} S \\ I \end{bmatrix} = \begin{bmatrix} 0.6212 \\ 0.3788 \end{bmatrix}$ (c) $\begin{bmatrix} S \\ I \end{bmatrix} = \begin{bmatrix} 0.7664 \\ 0.2336 \end{bmatrix}$

8.9 (a) $\begin{bmatrix} G \\ S \\ W \\ O \end{bmatrix} = \begin{bmatrix} 0 \\ 0 \\ 1 \\ 0 \end{bmatrix}$ (b) 1102 (c) 1092 (d) 1097

8.10 (a) $\begin{bmatrix} 0.88 & 0.02 & 0 \\ 0.06 & 0.97 & 0.05 \\ 0.06 & 0.01 & 0.95 \end{bmatrix} \begin{bmatrix} P \\ W \\ Z \end{bmatrix}$ (b) $\begin{bmatrix} P \\ W \\ Z \end{bmatrix} = \begin{bmatrix} 8 \\ 86 \\ 6 \end{bmatrix}$

(c) $\begin{bmatrix} P \\ W \\ Z \end{bmatrix} = \begin{bmatrix} 0.1064 \\ 0.6383 \\ 0.2553 \end{bmatrix}$

Chapter 9

9.1 (a) 1.1639

(b) $\begin{bmatrix} 0.6599 \\ 0.3401 \end{bmatrix}$

(c) The long-term growth rate is 1.1639, 66% of the population in the first class and 34% of the population in the second class

9.3 (a) 2.2

(b) $\begin{bmatrix} 0.9524 \\ 0.0476 \end{bmatrix}$

(c) The long-term growth rate 2.2; 37.5% of the population in the first class and 62.5% of the population in the second class

9.4 (a) 0.9

(b) $\begin{bmatrix} 0.5 \\ 0.5 \end{bmatrix}$

(c) The long-term decay rate is 0.9, 50% of the population in the first class and 50% of the population in the second class

9.5 (a) 0.8

(b) $\begin{bmatrix} 0.6667 \\ 0.3333 \end{bmatrix}$

(c) The long-term decay rate is 0.8, 67% of the population in the first class and 33% of the population in the second class

9.6 (a) 0.9

(b) $\begin{bmatrix} 0.6 \\ 0.4 \end{bmatrix}$

(c) The long-term decay rate is 0.9, 80% of the population in the first class and 20% of the population in the second class

9.7 (a) $\begin{bmatrix} 0 \\ 400 \\ 0 \end{bmatrix}$ (b) $A^3 = 2A = \begin{bmatrix} 0 & 40 & 0 \\ 0.2 & 0 & 0 \\ 0 & 0.4 & 0 \end{bmatrix}$ and $A^4 = 2A^2 = \begin{bmatrix} 4 & 0 & 0 \\ 0 & 4 & 0 \\ 0.04 & 0 & 0 \end{bmatrix}$

(c) 1.4142, growing

9.8 $A^4 = 2A^2 = \begin{bmatrix} 0 & 0 & 1 \\ 0.7 & 0 & 0 \\ 0 & 0.8 & 0.9 \end{bmatrix} \begin{bmatrix} C \\ Y \\ A \end{bmatrix}$

9.12 (a) $\begin{bmatrix} 0 & 0.8 & 3.5 \\ 0.8 & 0 & 0 \\ 0 & 0.9 & 0 \end{bmatrix} \begin{bmatrix} Immature \\ Adolescent \\ Adult \end{bmatrix}$ (b) 1.5170; growing

(c)

Week	I	Adolescent	Adult
1	0	80	0
2	64	0	72
3	252	51	0
4	41	202	46
5	323	33	181
6	661	258	29
7	310	529	232
8	1236	248	476
9	1864	989	223
10	1571	1492	890

(d) $\begin{bmatrix} 0.5434 \\ 0.2866 \\ 0.1700 \end{bmatrix}$

Chapter 10

10.1 (a) {HHH, HHT, HTH, HTT, THH, THT, TTH, TTT} (d) {AA, Aa, aa}
(e) {Aa, aa} (f) {Aa} (g) {AA, AB, AC, BA, BB, BC, CA, CB, CC}
(h) {A+, B+, AB+, O+, A−, B−, AB−, O−}

10.2 (a) $\frac{1}{2}$ (c) $\frac{1}{9}$

10.3 (a) $\frac{1}{2}$ (b) $\frac{1}{4}$

10.4 $\frac{5}{32}$

10.10 $\frac{1}{2}$

10.11 (b) 1 (c) $\frac{1}{2}$

10.12 (a) AO and BO (c) $\frac{1}{4}$

10.13 (a) Bb (b) BB, Bb (c) $\frac{1}{2}$

Chapter 11

11.1 (a) {1, 2, 3, 4, 5}, {1, 3} (b) {2, 4, 6} (c) {6} (d) No

11.2 a = "right-handed females," b = "left-handed males," c = "right-handed males"

11.3 (a) e = "male and developed emphysema and not a heavy smoker" (b) g (c) h

11.4 (a) Sum of 10 or doubles, {(1, 1), (2, 2), (3, 3), (4, 4), (5, 5), (6, 6), (4, 6), (6, 4)}
(c) doubles and no 6, {(1, 1), (2, 2), (3 ,3), (4, 4), (5, 5)}

11.5 (a) {(H, H, H)} (b) {(T, T, H), (H, T, H), (T, H, H)} (c) {(H, H, H)}

11.6 0.5

11.7 0.3

11.8 (a) Mutually exclusive (b) Not mutually exclusive (c) Mutually exclusive

11.9 (a) a = "cowboy and beer drinker," b = "cowboy and not a beer drinker,"
c = "not a cowboy and beer drinker," d = "not a cowboy and not a beer drinker"
(b) 0.7

11.10 (a) a = "resistant to Rifampin and Ethambul," 0.15 (b) 0.55, $b \cup c$

11.11 (a) 22%, g

Chapter 12

12.1 (a) Not independent (b) Independent (c) Not independent

12.2 $\frac{350}{470}, \frac{350}{580}, \frac{120}{470}, \frac{300}{420}$

12.3 0.625

12.4 (a) 0.64 (c) 0.32

12.5 0.25

12.6 (a) $\frac{1}{18}$

12.7 (a) $\frac{1}{2}$ (b) $\frac{1}{4}$

12.8 $\frac{1}{4}$

12.9 (a) aa/Bb, AA/bb (b) green $\frac{1}{2}$, white $\frac{1}{2}$ (c) green $\frac{9}{16}$, yellow $\frac{3}{16}$, blue $\frac{1}{16}$, white $\frac{3}{16}$

12.10 (a) $\frac{1}{8}$ (b) $\frac{7}{8}$ (c) $\frac{4}{8}$ (d) $\frac{7}{8}$

12.11 0.8926

12.12 0.6513

Chapter 13

13.4 (a) 0.0488 (c) 0.512

13.5 (a) $\frac{3}{26}$ (b) 1 (c) 0

13.8 0.1007

13.9   0.3

13.10 0.5

13.11 0.05

13.12 0.24

13.13 (a) $\frac{3}{8}$,   (b) $\frac{1}{16}$

13.14 (a) $\frac{1}{4}$,   (b) $\frac{49}{64}$

Chapter 14

14.1 0.04, 0.32, 0.64

14.2 0.4, 0.48

14.5 (a) 0.494   (b) 0.306

14.7 (a) 0.04875   (c) 0.16

14.8 (b) 0.2719, 0.064, 0.6641   (c) 0.3611

14.9 (b) $q_1 = 0.0045$, $q_2 = 0.0034$, $q_3 = 0.0025$, $q_4 = 0.0019$, $q_5 = 0.0014$

Chapter 15

15.1 (a) 2   (b) DNE   (d) 1   (f) $-\frac{1}{2\sqrt{2}}$ (h) $-\frac{1}{2}$ (i) 3   (j) 3   (k) -174 (m) $\frac{1}{2\sqrt{3}}$

15.2 (a) 3   (c) 3 (d) 0

15.3 (a) -2 (c) 1   (d) -3

15.4 12, yes

15.5 0

15.6 (a) 65   (b) 71.8

15.7 (a) 36.20 cm   (b) Yes, at 155 cm

Chapter 16

16.1 (a) $(-\infty, \infty)$ (c) $(-\infty, -2) \cup (-2, \infty)$ (d) $(-\infty, 3) \cup (3, \infty)$ (e) $(-\infty, \infty)$
(f) $(-\infty, 0) \cup (0, \infty)$ (g) $(-\infty, -1) \cup (-1, 1) \cup (1, \infty)$

16.2 (a) Infinite   (b) Jump

16.3 (a)   **(c)**

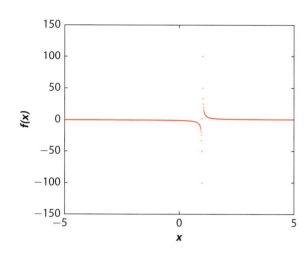

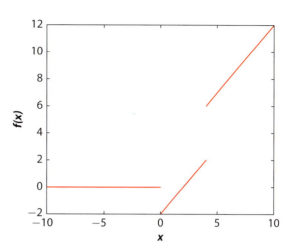

16.5 (a) 1.0995×10^{12} **(b)** $2^{30.034}$ or $1,099,511,628$

16.6 (a) $y = \begin{cases} 0.675x + 120 & 0 \le x \le 40 \\ -x + 180 & 40 \le x \le 60 \end{cases}$

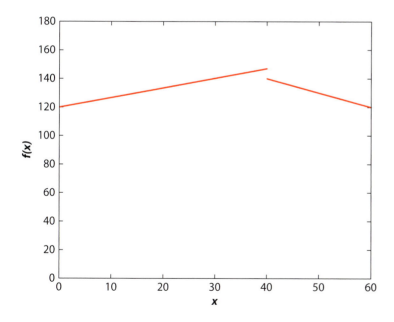

　　　　(b) No, 40

16.7 (a) No   **(b)** No

16.8 (a) 687.16   **(b)** No

(c)

Chapter 17

17.1 28, 7, -15 m/sec

17.2 Using $(-110, 3)$ and $(-30, 0)$, we obtain -26.667; using $(-30, 0)$ and $(30, 3)$, we obtain 20

17.4 (a) 11.1 m/s   (b) 6.2 m/s

17.7 (a) If an adult consumes an average of 1800 calories per day, his or her weight should be 155 pounds. If an adult consumes an average of 20,000 calories per day, his or her weight should remain constant
(b) Pounds calories per day

17.8 (a) Liters cm of water   (c) At 10 cm of water   (d) Because when the lung is nearly full, the relative change in the volume of the lung is small compared to the change in pressure reduction.

Chapter 18

18.1 (a) 2   (c) 4   (d) -1 (e) $\frac{1}{2}$

18.2 (a) $2x$ (c) 3

18.3 (a) 12   (c) 3

18.4 $y = 30x - 75$

18.5 $y = 48x - 64$

18.6 $y = 5x$

18.7 (a) $(-\infty, \infty)$ (c) $(0, \infty)$ (e) $(-\infty, -1) \cup (-1, \infty)$ (f) $(-\infty, 0) \cup (0, \infty)$
(g) $(-\infty, 3) \cup (3, \infty)$

Chapter 19

19.1 (a) $30x^9 + 10x^4 + 1 - \cos(x) + e^x - \frac{1}{x}$ (c) $2x\cos(x) - (x^2 - 1)\sin(x)$
(f) $5(x^2 + 4x + 6)^4(2x + 4)$ (h) $-\sin(tanx)\sec^2 x$ (j)
$\sin(x^{-1}) + (-x)\cos(x^{-1})x^{-2}$
(l) $\frac{\cos(x)}{\sin(x)}$ (n) $\frac{1}{\sin^2(x^2-1)} 2\sin(x^2 - 1)\cos(x^2 - 1)2x$

19.2 (a) 43.72 (c) $-\frac{3}{2}$

19.3 (a) $y = 41x - 35$ (c) $y = -\frac{3}{2}x + \frac{5}{2}$

19.4 (a) $f'(0)$ DNE   (b) $y = 320x + 32$

19.5 (a) $6x, 6$ (d) $\frac{1}{2}t^{-\frac{1}{2}}, \frac{-1}{4}t^{-\frac{3}{2}}$

19.6 (b) 2720

19.7 $f(6) \approx 1.25$ billion, $f'(6) \approx 0.01738$ billion per year, $f(6)$ estimates the population in 1999, and the derivative shows that the population is increasing

19.10 $m'(t) = k(-r)e^{-rt}, m''(t) = kr^2 e^{-rt}$

19.12 (a) A dosage of 120 mg is required for a person who weighs 140 pounds
(b) A dosage change of 3 mg more is required for an increase of 1 pound in weight in a 140-pound person

Chapter 20

20.1 (a) Critical points: $x = 2$(global max). Concave down: $(-\infty, \infty)$
(c) Critical points: $x = 0$, $x = 1$(local min). Concave up: $(-\infty, 0) \cup \left(\frac{2}{3}, \infty\right)$,
concave down: $\left(0, \frac{2}{3}\right)$. Inflection point: $x = 0$ and $x = \frac{2}{3}$
(d) Critical points: $x = -2$(global min), $x = 0$(local max), $x = 2$(global min).
Concave up: $\left(-\infty, -\sqrt{\frac{4}{3}}\right) \cup \left(\sqrt{\frac{4}{3}}, \infty\right)$, concave down: $\left(-\sqrt{\frac{4}{3}}, \sqrt{\frac{4}{3}}\right)$.
Inflection point: $x = \pm\sqrt{\frac{4}{3}}$
(e) Critical points: $x = 2 - \frac{1}{\sqrt{3}}$(local max), $x = 2 + \frac{1}{\sqrt{3}}$(local min). Concave down:
$(-\infty, 2)$, concave up: $(2, \infty)$. Inflection point: $x = 2$
(f) Critical points: $x = -2$(local max), $x = 2$(local min), $x = 0$(vertical asymptote). Concave up: $(0, \infty)$, concave down: $(-\infty, 0)$

20.3 (b) $x = -3$(local max), $x = 0$(vertical asymptote), $x = 3$(local min)
(c) $x = 0$(local min), $x = -3$(inflection point)

20.4 (a) $x = 0$ (min), $x = 4$ (max)   (b) $x = -1$ (min), $x = 4$ (max)

20.14 Area$= 2\sqrt{\frac{4}{3}}\left(4 - \frac{4}{3}\right) \approx 6.15$ square units

20.15 $w = 10.5$ pounds

20.16 $t = \frac{300}{720} \approx 0.42$ years

20.17 Radius $= \frac{7}{12}$ miles

20.18 Distance traveled by highway $= 53.4912$ miles, by water $= 11.9320$ miles

20.19 (a) $t = 15$ days   (b) $P(t) = 16.875\%$

20.20 $x \approx 13.1341$ miles

Chapter 21

21.1 (a) 5 units^2 (c) 3.25 units^2

21.2 (a) 3.25 units^2 (b) 0.95 units^2

21.3 (a) 6 units^2 (b) 1.6833 units^2 (c) 3.25 units^2

21.4 0.492

21.8 (a) $y = b_0 \left(\frac{t}{7}\right)^{b_1} e^{b_2 \frac{t}{7}}$ (b) 167.56, 170.13, 171.26, 171.32. Yes, to 171.3
(c) 229.95, 234.46, 236.65, 236.8 Yes, to 236.8   (d) Nelore

21.9 3.59

Chapter 22

22.1 (a) $F' = f$ (b) $F(x) = G(x) + C$

22.2 e^{x^2}

22.3 (a) $6x + C$ (c) $\frac{1}{3}x^3 - 2x^2 + 5x + C$ (d) $10x^{\frac{3}{2}} + \frac{2}{3}x^3 + C$ (e)
$5\tan(x) + 8\cos(x) + C$ (f) $3e^{2x} + C$ (h) $\frac{2^x}{\ln(2)} + \frac{3^x}{\ln(3)} + \frac{4^x}{\ln(4)} + \frac{5^x}{\ln(5)} + C$

22.4 (a) -42 (c) 44.9004   (d) 3.1945   (e) 0.6931

22.5 (a) 6   (c) $\frac{10}{3}$ (e) $\frac{32}{3}$ (f) 0.6321

22.7 (a) $f(t) = -e^{-0.01t} + C$ (b) 0.0952

22.8 33.815 cm

Chapter 23

23.1 (a) $\frac{2}{3}(x^2 + 1)^{\frac{3}{2}} + C$ (c) $-\ln|\cos(x)| + C$

23.2 (a) $-xe^{-x} - e^{-x} + C$ (c) $\frac{1}{3}x\sin(3x) + \frac{1}{9}\cos(3x) + C$ (d) $x\sin(x) + \cos(x) + C$

23.4 0.876

23.6 644

23.8 (a) 9.5981   (c) 0.7337   (d) $\frac{1}{2}$ (e) $\frac{1}{3}$ (f) $\frac{14}{3}$ (h) $\frac{24}{5}$

23.9 (a) 7.01   (b) 1249.88

Chapter 24

24.2 $16.2809 + 14.5464 = 30.8273$

24.3 $\frac{56}{15}\pi$

24.5 8π

24.6 $\pi^2 - 2\pi$

24.7 (a) $42.03\ \mu$g/ml   (b) $22.4404\ \mu$g/ml

24.8 (a) $14.9667°$C   (b) Yes   (c) $15.0417°$C

24.9 726.242 millijoules

24.10 (a) m_2 (b) 2 words

24.11 7.716

24.12 Using $v'(0) = 0$, show that $b = 0$, and using, $v(R) = 0$, show that $c = -aR^2$

24.15 (a) $\rho(x) = 5$ (b) $\rho(x) = 2x + 1$ (c) $\rho(x) = 150x - 3x^2$ (d) $\rho(x) = 100e^{-2x}$
(e) $\rho(x) = xe^{-0.1x}$

24.16 (a) (1) 500 cod, (2) constant (3) 100 cod, 20%
(c) (1) 625 cod, (2) 0 m, (3) 125 cod, 20%
(d) (1) 100 cod, (2) 30 m, (3) 24 cod, 24.48%
(e) (1) 569 cod, (2) 40 m, (3) 133 cod, 23.4%
(f) (1) 159 cod, (2) 25 m, (3) 49 cod, 30.9%

24.17 (a) 309,522 krill (b) 1000 m (c) 635,854 krill

Chapter 25

25.1 0.86

25.2 $2\sqrt{2} - 2$

25.3 $\frac{11}{8}$

25.5 (a) variance $\frac{(b-a)^2}{12}$, standard deviation $\frac{b-a}{2\sqrt{2}}$,   (b) $\frac{1}{\sqrt{3}}$

25.6 $\frac{2}{3}$

Chapter 26

26.1 (a) $\frac{3}{5}t^5 + 5t + C$ (c) $-\cos(t) + C$

26.2 (a) $\frac{3}{5}t^5 + 5t + \frac{22}{5}$ (b) $\frac{1}{3}e^{3t} - 2t + \frac{14}{3}$

26.4 (a) $\pm\sqrt{-t^2 + 25}$ (b) $\frac{1}{\frac{1}{2} - t^2}$

26.5 $\left(\frac{3}{2}e^{2x} - 3C\right)^{-\frac{1}{3}}$

26.6 $-3 \pm \sqrt{9 - 4x + 2C}$

26.11 1.2 million

26.12 25 feet

26.13 (a) $\ln|2x + \sin(x) + 1|$ (b) $\ln(2 + \pi)$

26.14 $\frac{dS}{dt} = 0.75 - \frac{3}{100}S(t)$

26.15 900 gallons

Chapter 27

27.1 (a) $486 - 458e^{-0.0152t}$,   (b) 486, $w(200) = 463$

27.2 (a) $2.4 - 1.4e^{-0.088W}$ (b) Tends to 2.4

27.3 387 fish

27.4 0 (stable), L (unstable), and K (stable)   (a) Extinction   (b) Increase and approach to K   (c) Decrease and approach to K

27.5 $\frac{3}{3 - 2e^{-t}}$

27.6 -1 (stable), 0 (unstable), and 1 (stable)

27.7 -1 (stable), 0 (unstable), and 1 (stable)

27.8 0 (unstable) and $e^{\frac{K}{\alpha}}x_0$

Chapter 28

28.1 (a) $-\frac{4t}{3y}$ (c) $\frac{3t^2}{2y}$ (d) $\frac{3t(y+2)^2}{3}$ (e) $-\sqrt{\frac{y}{t}}$ (f) $\frac{3t^2 - 2te^y}{t^2 e^y + 1}$ (h) $\frac{1 - \cos(ty)y}{t\cos(ty)}$ (i) $-\frac{1}{\sec^2(y)}$

28.2 (a) $\frac{dy}{dt} = \frac{10x}{2y}\frac{dx}{dt}$ (c) $\frac{dy}{dt} = \frac{9 - 2x}{10}\frac{dx}{dt}$ (d) $\frac{dy}{dt} = \frac{\frac{1}{x} - e^y}{xe^y}\frac{dx}{dt}$

28.3 (a) $-\frac{1}{2}$ (c) -5

28.4 (a) $-0.43\frac{R}{w}$ (b) $-0.43\frac{R}{w}$ (c) Proportional, Yes

28.5 $-\frac{y}{\alpha x}$, rate of change in number of genera with respect to number of species

28.8 0.27 ft^2/day

28.10 1.98 g/day

Index

absolute maximum, 377, 393–394
absolute minimum, 377, 393–394
albinism, 185, 195–198, 235–236
Allee growth, 556
alleles. *See also* population genetics
   defined, 183
   Hardy-Weinberg equilibrium, 247–250
   Matlab Skills, probability, 211–212
allometry
   defined, 54–55
   growth rates, separation of variables, 521–522
   rates of change and, 315–316
American bison project, 172–174
antiderivatives
   average values, 450–453
   defined, 442–444
   Exercises, 456–458
   families of, 446
   fundamental theorem of calculus, 444–445
   integrals and, 446–449
   Matlab Skills, 453–455
   overview, 441
   separation of variables method, 518
   substitution method of integration, 459–465
antidifferentiation. *See* integration
area
   rectangle, 415
   trapezoid, 415
   between two curves, 472–476
area under curve. *See also* calculus, fundamental
     theorem of
   area below horizontal axis, 430–433
   estimating with rectangles, 417–424

estimating with trapezoids, 424–426
estimating, overview, 414–417
estimation, accuracy of, 426–430
Exercises, 436–439
Matlab Skills, 433–436
arithmetic mean, 4, 5, 6–7
arithmetic sequences, 93
arrays, Matlab Skills, 566–568
arteries, rate of flow, 481–482
asymptotes, graphs of, 380–382
Audubon Society, 85
average, *See* mean
average rate of change
   derivatives of exponential functions, 330–334
   overview, 306–308
   populations, 315
average value of a function, 450–453
average velocity, 309–311
axes
   linear analysis of data, 32
   rescaling data, log-log and semilog graphs, 55–62
   scatter plots, 23–24

bacteria, growth in sausage, 547–548
bar charts
   construction of, 16–18
   frequency distributions, 15–16, 17
   vector representations, 109–110
Bayes' theorem, 238–242
Beer-Lambert Law, 52–53
Bernoulli experiment, 188–189
bimodal data, 22
binomial experiments, 188–189

binomial probability, 188–189
binomial probability distribution, 188–189
biodiversity, Student Projects, 299–301
bioinformatics, 176
biological data types, 3–4
biomass rate of change, 315, 330–334
bird watching, waiting times, 500–507
birth rates. *See* Leslie matrix models
bivariate data
    correlation, 37–40
    defined, 31
    exercises, 43–45
    least-squares fit, 35–37
    Matlab Skills, 41–43
blood flow rates, 481–482, 550
blood type
    compound events, overview of, 201–204
    genotype frequency, 249
    independence, 223
    probability, computing, 185–186, 208
body temperature, homeostasis, 539–541
boundary condition, 363, 517, 519
budworm differential equation, 555–556

calculus. *See also* limits
    definition, 259–260
    history of, 326
calculus, fundamental theorem of
    antiderivatives and calculus, 446–449
    average values, 450–453
    Exercises, 456–458
    integrals, defined, 441–442
    Matlab Skills, 453–455
    overview, 440–441, 444–445
carbon dating, 365–366
cardiovascular system, rate of flow, 481–482, 550
carriers, recessive alleles, 184
carrying capacity, populations
    density dependent growth, 532–534
    graphing of, 536–539
    logistic growth equation, 269–270
    logistic growth curve, 388–391
categorical data, 4
causation, 37
central limit theorem, 497
central tendency, measures of, 4–6
chain rule, derivatives
    fundamental theorem of calculus, 445
    overview, 354–359
    proof of, 573–574
    substitution method of integration, 460–461
change. *See* rate of change
characteristic equation, 158–159
Christmas Bird Count, 85
chromosomes, 183–186
circadian rhythms, 324–325, 328–330, 334–336
circulatory system, rate of flow, 481–482, 550
coefficient of determination, 39

coefficient of variability (variance), 7
coin toss, probability and, 177–179, 190–195
column vectors, 131
combinations, probability and, 186–188
complement, event probability, 203–204, 205
complex roots, 162
composition of functions, chain rule and, 354–359
compound events
    computing probability of, 204–209
    Exercises, 213–215
    Matlab Skills, 210–212
    overview, 201–204
concave down, defined, 386
concave up, defined, 386
conditional probability
    Bayes' theorem, 238–242
    Exercises, 230–232
    independence, 220–225
    Matlab Skills, 225–230
    overview, 216–220
constant coefficients, linear difference equation
    and, 93–96
constant functions, derivatives and, 329, 336, 353
constant multiple, derivatives, 353
consumption, Holling's functional responses, 359–362,
    370–371
continuity. *See also* continuous random variables;
    derivatives
    derivatives and, 336–341
    overview of, 259–260
    properties of, 284–289
    Student Projects, 299–301
continuous data, 4, 15–16, 17
continuous functions, limits of
    continuity, properties of, 284–289
    Exercises, 295–298
    intermediate value theorem, 290–292
    Matlab Skills, 292–294
    overview, 282–283
    right and left limits, 283
continuous random variables
    defined, 490
    Exercise, probability, 507–509
    expected and mean values, 493–495
    Matlab Skills, waiting times, 500–507
    normal distributions, 495–498
    probability and, 490–493
    waiting times, 498–500
correlation, 37–40
correlation coefficient (ρ)
    Matlab Skills, 41–43
    overview, 37–40
cosecant, reciprocal rule, 362
cosine function
    chain rule and, 359, 361–362
    higher derivatives, 372
    proof of derivative, 577
cotangent, chain and quotient rules, 361–362

cricket chirp rates, 305–306, 308, 317–320
critical numbers, 379
critical points
    concavity, 386–387
    defined, 379
    second derivative test, 391–394
curves, area between two, 472–476
curves, area under. *See also* calculus, fundamental
        theorem of; integrals
    area below horizontal axis, 430–433
    estimating with rectangles, 417–424
    estimating with trapezoids, 424–426
    estimating, overview, 414–417
    estimation, accuracy of, 426–430
    Exercises, 436–439
    Matlab Skills, 433–436

data displays
    bar charts, 16–18, 109–110
    exercises, 27–29
    frequency distributions, 15–16, 17
    histograms, 18–23, 24–25
    overview, 14–15
    rescaling data, log-log and semilog graphs, 55–62
    scatter plots, 23–24
data types
    continuous *vs.* discrete, 4
    interval scales, 3
    nominal scales, 4
    ordinal scales, 3–4
    ratio scale, 3
death rates, 445. *See also* Leslie matrix models
death, time of, 366–367
decay (half-life), 49–51, 365–366
decay constant, 50
decay rate, 365–366
decreasing function, defined, 377
definite integral
    approximation of, 451–453
    defined, 446
    examples of, 448–449
    integration by parts, 466–469
    substitution method of integration, 463–465
demography, 80, 82–83, 157–163
density functions
    distribution function and, 492–493
    overview, 482–485
    probability density function, 490–492
density-dependent hypothesis, 532–535
dependent variables, defined, 514
derivatives. *See also* antiderivatives; calculus,
        fundamental theorem of
    allometric properties, 315–316
    average rate of change, 306–308
    biomass rate of change, 315
    chain rule, 354–359
    concavity, 385–394
    constant function, 329, 336

constant multiple rule, 353
continuity and, 336–341
    derivative at a point, defining, 316
    derivative product rule, 330, 336
    derivative sum rule, 328–329, 336
    drug absorption rates, 315
    Exercises, 320–323, 349–351, 372–375, 404–409
    exponential functions, 330–334, 336, 362–369
    exponential growth model, 363–364
    first derivative test, 377–382
    function rules, summary chart, 353
    higher derivatives, 369–372
    integrals and, 441–442
    limit definition of functions, 326–330
    linear function, 329, 336
    logarithmic functions, 341–345
    Matlab Skills, 316–320, 345–349, 402–404
    maxima and minima, overview, 376–377
    mean value theorem, 382–385
    Newton's Law of Cooling, 366–369
    optimization problems, 394–401
    overview of, 303–304, 324–326
    photosynthesis, rates of, 311–314
    population growth rates, 315
    power rule, 353–354
    quotient and reciprocal rules, 359–362
    radioactive decay, 365–366
    rate of change estimations, 308–309
    rate of change, overview, 305–306
    second derivative test, 391–394
    trigonometric functions, 334–336
    velocity, 309–311
descriptive statistics. *See also* bivariate data; data
        displays; exponential functions; linear regression;
        logarithmic functions
    central tendency, measures of, 4–6
    data types, 3–4
    dispersion, measures of, 6–9
    Exercises, 11–13
    Matlab skills, 9–11
    overview of, 1–2
    Student Projects, 71–77
design of experiment, 1–2
determinant, 159–162
dice, probability and, 177–180, 182–183, 206, 207,
        221–222
difference equations
    boundary conditions, solving with, 519
    defined, 91, 514
    Exercises, 102–106
    first-order difference equations, 91
    logistic difference equations, 102
    overview, 513–514
difference quotient, 316, 326–330
differentiable at a point, 337–341
differentiable on an open interval, 337–341
differentiable, defined, 337

differential equations
   defined, 362–363
   Exercises, equilibrium and stability, 541–542
   Exercises, implicit differentiation and related rates, 551–554
   Exercises, separation of variables, 527–528
   Gompertz growth, 534–535
   homeostasis, 539–541
   implicit differentiation, 543–548
   limited population growth models, 531–535
   logistic growth, 532–534
   Matlab Skills, separation of variables, 522–526
   population growth, overview, 529–530
   related rates, 549–551
   separation of variables, 515–522
   Student Projects, 555–556
discontinuous functions, 284–289, 336–341
discrete data
   bar charts, 16–18
   defined, 4
   frequency distributions, 15–16, 17
discrete difference equations
   Exercises, 102–106
   linear difference equation with constant coefficient, 93–96
   logistic difference equation, 102
   Matlab Skills, 100–103
   overview, 90–91
   pharmacokinetics, 97–100
discrete time modeling, 79–83. *See also* discrete difference equations; matrices; sequences; vectors
disease. *See also* Leslie matrix models, epidemiology
   AIDS, spread of, 516–517
   genetic diseases, 184
   spread of, 176, 514
dispersion, measures of, 6–9
displays. *See* data displays
distributions
   expected value, 493–495
   frequency, 15–16, 17
   median value, 493–495
   normal distribution, 495–498
   probability function, 492–493
   uniform distribution, 493
diversity, measures of, *See also* Simpson's Index of Diversity, 9
dominant alleles, 183
dominant eigenvalue, 159
doubling time, 49–51, 364
drugs
   absorption rates, 315, 336–338
   blood concentration, changes in, 308–309
   dose determination, 97–100
   drug testing, 218–219, 227–230, 240–241
dynamics, See also differential equations, discrete difference equations. vectors, 112–120

earthquakes, Richter scale, 53–54
ecology, Holling's functional responses, 359–362, 370–371
eigenvalues
   dominant eigenvalue, 159
   Exercises, 168–170
   long-term population structures, 163–164
   Matlab Skills, 165–168
   oscillating populations, 162–163
   overview, 156–163
eigenvectors
   Exercises, 149–151, 168–170
   long-term population structures, 163–164
   Matlab Skills, 149, 165–168
   overview of, 142–146
   stability and, 147–149
elementary events, 178–181
empty set, event probability, 204
epidemiology
   AIDS, spread of, 516–517
   infections, spread of, 514
   probability and, 176
equilibrium
   concavity and, 385–394
   defined, 143
   eigenvectors, 142–146
   Exercises, 149–151, 253–254, 404–409, 541–542
   feedback, overview, 529–530
   first derivative test, 377–382
   graphing of, 535–539
   Hardy-Weinberg equilibrium, 247–254
   homeostasis, 303–304, 529–530, 539–541
   limited population growth models, 531–535
   Matlab Skills, 149, 402–404
   maxima and minima, overview, 376–377
   mean value theorem, 382–385
   optimization problems, 394–401
   overview of, 141–142
   stability, 147–149
   Student Projects, foraging, 410–412
equilibrium population, 80–81
equiprobable sample space, 181
errors in measurement, 495
estimation, 2
evenness, diversity measures, *See also* Simpson's Index of Diversity, 9
events
   combinations and permutations, 186–188
   compound events, overview, 201–204
   compound events, probability computation, 204–209
   Exercises, 213–215
   Matlab Skills, compound events, 210–212
   Matlab Skills, conditional probability, 225–227
   Matlab Skills, probability, 193–195
   probability of an event, 181–186
   probability, conditional, 216–220
   probability, independence and, 220–225
   probability, overview of, 178–181

sequential, Bayes' theorem, 238–242
sequential, partition theorem, 233–238
evolution. *See* population genetics
Exercises
  area under the curve, 436–439
  bivariate data, 43–45
  derivatives of functions, 349–351, 372–375
  descriptive statistics, 11–13
  differential equations, separation of variables, 527–528
  discrete difference equations, 102–106
  equilibrium and eigenvectors, 149–151
  equilibrium and stability, 541–542
  exponential functions, 67–70
  implicit differentiation, 551–554
  integrals and antiderivatives, 456–458
  integrals, area and volume, 485–488
  integration, methods of, 469–470
  Leslie matrix model, 168–170
  limits of continuous functions, 295–298
  linear regression, 43–45
  logarithmic functions, 67–70
  matrices, 138–140
  maxima and minima values, 404–409
  probability, 198–200
  probability in continuous context, 507–509
  probability, compound events, 213–215
  probability, conditional, 230–232
  probability, population genetics, 253–254
  probability, sequential events, 242–245
  rates of change, 320–323
  related rates, 551–554
  sequences, 102–106
  vectors, 120–122
  visual data displays, 27–29
expansion by minors, 160
expected value, 493–495
experiment, defined, 1
experimental design, 1–2
explicit function, 544
exponential functions
  allometry, 54–55
  Beer-Lambert Law, 52–53
  continuous functions, 286
  derivatives of, 330–334, 336, 353, 362–369
  derivatives, proof of, 574–575
  Exercises, 67–70
  exponential decay function, 47
  exponential density function, 499–500
  exponential growth function, 47, 363–364, 520
  Matlab Skills, 62–63, 500–507
  overview, 46–47
  oxygen consumption, 51–52
  rescaling data, log-log and semilog graphs, 56–62
extrapolate, 37, 41
eye color, probability, 205, 255–257

factoring limits, 267
family of antiderivatives, 443–444

family of solutions, 516–517
family planning, 224–225
fecundity rate, 155. *See also* Leslie matrix models
feedback mechanisms, 513, 529–530. *See also* difference equations
Fibonacci sequence, 91
first derivative test, 377–382
first derivatives, 370–372
first-order difference equation, 91
first-order differential equation, 514, 515–522. *See also* differential equations
fish, ocean density and, 482–484
fishery populations, 95–96
flow rates, Poiseuille's law, 477, 480–482, 485–488
for loops, Matlab, 101–103
foraging, optimization of, 410–412
founder effect, 176
fprintf, Matlab output, 570
freehand linear fit, 33–34
frequency distributions, 15–16, 17, 22–23
functions, average values, 450–453
functions, increasing and decreasing, 377
functions, limits of. *See* limits
fundamental theorem of calculus
  antiderivatives and integrals, 446–449
  average values, 450–453
  Exercises, 456–458
  Matlab Skills, 453–455
  overview, 440–441, 444–445

gender, probability and, 206, 224–225
genetics and genotypes. *See also* blood type; population genetics
  albinism, partition theorem, 235–236
  albinism, probability estimates, 195–198
  combinations and permutations, 187–188
  conditional probability, 216–220
  genes, defined, 91
  genetic testing, 241–242
  genetics terminology, 183–186
  Hardy-Weinberg selection model, 250–253
  human pedigree, defined, 233, 234
  Matlab Skills, 211–212, 225–227
  Mendel's pea plants, 180–181, 182
  population genetics, probability and, 176
  probability examples, 183–186
  Tay-Sachs disease, 237–238
genomics, probability and, 176
Geographical Information Systems (GIS), 81–83
geometric growth, 80
geometric mean, 5
geometric sequences, 91–93
global maximum, 377
global minimum, 377
Gompertz growth curve, 534–535, 538
Gompertz survival function, 445
Gompertz, Benjamin, 445
growth function, 532–535

growth rates
  intrinsic growth rate, 334
  per capita growth rate, 334

half-life, 49–51, 365–366
Hardy-Weinberg equilibrium, 247–250, 253–254
Hardy-Weinberg selection model, 250–254
harmonic mean, 5
heterozygous genes, 183
histograms, 15–23, 24–25
Holling's functional responses, 359–362, 370–371
homeostasis. *See also* equilibrium
  differential equations and, 539–541
  feedback, overview, 529–530
  rates of change, measuring, 303–304
homogeneous difference equation, 93–96
homozygous dominant genes, 183
homozygous recessive genes, 183
horizontal asymptote, 269
human pedigrees, 233, 234, 237
hypertension, rate of flow, 481–482
hypotheses
  inferential statistics and, 2
  linear analysis and, 30

implicit differentiation, 543–548, 551–554
implicit function, 544
increasing function, defined, 377
indefinite integral, 446, 447–448, 459–463
independence, probability and, 220–225
indeterminate form, 270
index, Matlab *for loops*, 101–103
infection. *See also* Leslie matrix models
  AIDS, 516–517
  probability and, 176
  spread of, 514
inferential statistics, overview, 2
infinite discontinuity, 284–285
infinity, limits and, 272–273
inflection point, 386–394
initial condition, 363, 517
inner function, 354
instantaneous growth rate, 315
instantaneous per capita growth rate, 315
instantaneous rate of change. *See also* derivatives
  defined, 308–309
  derivative at a point, defining, 316
  overview, 303–304, 325–326
  photosynthesis, 312–314
integrals. *See also* integration
  antiderivatives and, 446–449
  applications, overview, 471–472
  approximation of, 451–453
  area between two curves, 472–476
  average values, 450–453
  defined, 441–442
  density functions, 482–485
  Exercises, 456–458, 469–470, 485–488

expected and median values, 493–495
  fundamental theorem of calculus, 444–445
  integration by parts, 465–469
  Matlab Skills, 453–455, 500–507
  normal distributions, 495–498
  probability density function, 490–492
  probability distribution function, 492–493
  properties of, 446–447
  separation of variables method, 518
  Student Projects, 510–511
  substitution method of integration, 459–465
  volume of a solid of revolution, 477–482
  waiting times, 498–500
integrand, 441
integration. *See also* integrals
  area below horizontal axis, 430–433
  area estimation, accuracy of, 426–430
  area under curve, overview, 414–417
  area, rectangle estimates, 417–424
  area, trapezoid estimates, 424–426
  Exercises, 436–439, 469–470
  limits of, 441
  Matlab Skills, 433–436
  overview, 413
  by parts, 465–469
  substitution method, 459–465
intermediate value theorem, 290–292
interpolate, 37, 41
intersection, event probability, 202–203, 207
interval scales, 3
intrinsic growth rate, 334, 364

jump discontinuity, 284–285

Krebs, John, 410

lake pollution, measuring, 472–476, 511
laminar flow, 477, 479, 550
landscapes, descriptions of, 81–83
Law for Rate of Flow, Poiseuille's, 477–482, 550
Law of Cooling, Newton, 366–369
LD 50, 4–5
least-squares fit, 35–37
least-squares regression, 317–318
left-hand limits, 283
Leibniz, Gottfried Wilhelm, 326
Leslie matrix models
  American bison project, 172–174
  eigenvalues, 156–163
  Exercises, 168–170
  long-term population structure, 163–164
  Matlab Skills, 165–168
  oscillating populations, 162–163
  overview, 152–156
light. *See also* photosynthesis, rates of
  Beer-Lambert Law, 52–53
  Michaelis-Menten curve, 263
limit, sequences, 87–90

limits
    carrying capacity, 269–270
    derivative functions and, 326–330
    Exercises, 277–281
    of functions, 262–266
    of integration, 441
    mathematical notation of, 571–572
    Matlab Skills, 274–277
    overview of, 259–260
    properties of, 266–274
    simplifying, 267–268
    Student Projects, 299–301
limits, continuous functions
    continuity, properties of, 284–289
    Exercises, 295–298
    intermediate value theorem, 290–292
    Matlab Skills, 292–294
    overview, 282–283
    right and left limits, 283
linear difference equation
    with constant coefficients, 93–96
    pharmacokinetics, 97–100
linear functions, derivatives and, 329, 336, 353
linear regression
    correlation, 37–40
    defined, 32
    Exercises, 43–45
    least-squares fit, 35–37
    Matlab skills, 41, 63–67
    overview, 30–31
    rescaling data, 58–67
linear relationships, 31–32
local maximum, 377
local minimum, 377, 392–394
loci, chromosomes, 183
log-log graphs, 55–62
logarithmic functions
    continuous functions, 286
    derivatives of, 341–345, 353
    Exercises, 67–70
    half-life, 49–51
    integration by parts, 466
    Matlab Skills, 62–63
    overview, 46–49
    Richter Scale, 53–54
logistic difference equation, 102
logistic growth equation, 269, 388–391,
    532–534
long-term dynamics. *See* equilibrium;
    Leslie matrix models
long-term growth rate, 157–163
loops, Matlab Skills, 101–103

m-file. *See* Matlab Skills
Malthus, Thomas, 80, 157
Malthusian growth rate, 80, 157–163
mathematical notation
    limit, definition of, 571–572

    pi, 571–572
    sigma, 571–572
Matlab Skills
    albinism, probability of, 195–198
    area under the curve, 433–436
    arrays (vectors and matrices), 566–568
    basic arithmetic, 562–563
    command window, working from, 562–565
    correlation coefficients, 41–43
    dataset entry, 9–10
    derivatives of functions, 345–349
    descriptive statistics, 10–11
    differential equations, separation of variables, 522–526
    eigenvectors, 149
    exponential functions, 62–63
    extrapolation, 41
    histograms, 24–25
    integrals, 453–455
    interpolation, 41
    least-squares regressions, 317–318
    Leslie matrix models, 165–168
    limits of continuous functions, 292–294
    limits of functions, 274–277
    linear regression, 41
    logarithmic functions, 62–63
    loops, 101–103
    m-files, creating and running, 568–569
    m-files, writing as a function, 191–193
    matrix operations, 133–138
    maxima and minima values, 402–404
    output, formatting of, 570
    polynomials, roots of, 564–565
    probability, 189–198
    probability, compound events, 210–212
    probability, conditional, 225–230
    random number generation, 189–191
    rates of change, 316–320
    rescaling data and linear regression, 63–67
    rounding, 563–564
    scatter plots, 25–27
    starting Matlab, 561–562
    Student Project: dendrology measurement, 74–76
    Student Project: tobacco measurements, 76–77
    waiting times, 500–507
matrices. *See also* Leslie matrix models
    applications of, 129–133
    column vectors, 131
    defined, 124
    determinant, 159–162
    dominant eigenvalue, 159
    Exercises, 120–122, 138–140
    Matlab Skills, entering and using, 133–138, 566–568
    Matlab Skills, Leslie matrix models, 165–168
    matrix addition, 124–125
    matrix multiplication, 116, 123–128
    square matrices, 131
    Student Projects, 171–174
    transfer matrix, 116

matrices (*continued*)
  vector algebra, 110–112
  vector dynamics, 112–120
  vectors, overview, 107–110
maxima
  absolute maximum, 377, 393–394
  concavity, 385–394
  Exercises, 404–409
  first derivative test, 377–382
  Matlab Skills, 402–404
  mean value theorem, 382–385
  optimization problems, 394–401
  overview, 376–377
  second derivative test, 391–394
  Student Projects, foraging, 410–412
mean
  arithmetic mean, 4, 5, 6, 16
  geometric mean, 5
  harmonic mean, 5
  mean value theorem, 382–385
  mean waiting time, 499
measurement errors, 495
measurement scales, 3–4
median, 4–5, 493–495
memoryless property, 500
Mendel, Gregor, 180–181, 211–212
Mendel's pea plants, 180–181, 182, 187–188, 189
Michaelis-Menten curve, 89–90, 263, 313
midrange, 5
minima
  absolute minimum, 377, 393–394
  concavity, 385–394
  Exercises, 404–409
  first derivative test, 377–382
  Matlab Skills, 402–404
  mean value theorem, 382–385
  optimization problems, 394–401
  overview, 376–377
  second derivative test, 391–394
  Student Projects, foraging, 410–412
minors, expansion by, 160
mode, 5
modeling. *See also* Leslie matrix models,
    differential equations
  discrete time modeling, 79–83
  exponential growth model, 363–364
  Hardy-Weinberg equilibrium, 247–250
  Hardy-Weinberg selection model, 250–253
  limited population growth models, 531–535
  predator-prey models, 171–172
modern synthesis, population genetics, 246
multimodal data, 22
multiplication principle, 180
multivariate data, overview of, 14–15
mutation rates, 60–62
mutually exclusive events, 204, 220–225
myelin sheath, 341–345

natural selection. *See* population genetics
negative infinity, 272–273
nervous system growth, 341–345
neurons, 341–345
neutral equilibrium, 537
Newton, Sir Isaac, 326
Newton's Law of Cooling, 366–369
nitrogen, soil concentration of, 490–492, 494–495
nodes of Ranvier, 341–345
noise, 495
nominal data, 15–17
nominal scales, 4, 9
normal distributions, 15–16, 17, 495–498
normal probability density, 495–498
null set, event probability, 204

observation, defined, 1
ocean, density functions, 482–485
ocean, salinity of, 485
optimal foraging theory, 410–412
optimization problems, 394–401
ordinal data, 15–16, 17
ordinal scale, 3–4
ordinary differential equations, 514. *See also* differential
    equations
oscillating populations, 162–163
outer function, 354
oxygen consumption, 51–52

parabola, solid of revolution, 479–480
parameter estimation, 2
particular solution, 94–96
partition theorem, 233–238
  Bayes' theorem and, 238–242
partition, defined, 233–234
pea plants, probability and, 180–181, 182, 187–188,
    189, 211–212
pedigrees, human, 233, 234, 237
per capita growth rate, 334
perfectly negatively correlated data, 39–40
perfectly positively correlated data, 39
permutations, probability and, 186–188
pesticides, timing of application, 429
pharmacokinetics
  absorption rates, 315, 336–338
  blood, drug concentration in, 308–309
  dose determination, 97–100
  drug testing, 218–219, 227–230, 240–241
phenotype, 183–184
photosynthesis, rates of
  chain rule derivatives for, 354–357
  circadian rhythms and, 324–325, 328–330, 334–336
  estimating by time interval (integration), 422–424,
    428–433
  integration methods, 459–465
  intermediate value theorem and, 290–292
  limits of functions, 261–266, 270–272
  mean rate of change, 383–385

Michaelis-Menten curve, 263
    rate of change and, 311–314
    sequences, 86–90
    tools for measuring, 282, 285
pi, 564, 571–572
placebos, probability of, 218–219, 227–230
point-intercept form, 32
point-slope form, 32
points of inflection, 386–394
Poiseuille, Jean, 477
Poiseuille's Law, 477–482, 550
pollution in a lake, 472–476, 511
polynomial functions
    continuous functions, 286
    graph, sketching of, 379–380
    roots of, Matlab Skills, 564–565
population biodiversity, Student Project, 299–301
population genetics
    Exercises, 253–254
    Hardy-Weinberg equilibrium, 247–250
    Hardy-Weinberg selection model, 250–253
    overview, 246
    probability and, 176, 183–186
    Student Projects, bison extinction, 257–258
population growth. *See also* equilibrium
    American bison matrix model, 172–174
    average value of a function, 450
    bacteria in sausage, 547–548
    biomass measures, 315
    carrying capacity, 269–270
    chain rule, population cycles, 359
    derivatives of exponential functions, 330–334
    differential equations, separation of variables, 520–522
    discrete time modeling, overview, 79–83
    doubling time, 49–51
    eigenvalues, long-term growth, 156–163
    equilibria, graphing of, 536–539
    Exercises, Leslie matrix model, 168–170
    exponential growth, 46, 47, 363–364
    geometric and arithmetic sequences, 92–93
    Gompertz growth curve, 534–535
    Gompertz survival function, 445
    intrinsic growth rate, 334
    Leslie matrix models, overview, 152–156
    linear difference equation, 93–96
    logarithmic growth, 46
    logistic growth equation, 269, 388–391, 532–534
    long-term population structure, 163–164
    Malthusian growth rate, 157
    models of limited growth, 531–535
    oscillating populations, 162–163
    per capita growth rate, 334
    predator-prey models, 171–172
    rates of change, 315
    rescaling data graphs, 57–58
    specific growth rate, 534–535
    Student Projects, 556
    world population estimates, 430

power function
    continuous functions, 286
    defined, 47
    power rule, derivatives, 286, 353
    substitution method of integration, 463
power rule, derivatives, 353–354, 573
predator-prey models, 171–172
pregnancy, length of, 497–498
probability. *See also* population genetics
    of an event, 181–186
    Bayes' theorem, 238–242
    binomial experiments, 188–189
    combinations and permutations, 186–188
    compound events, computing probability,
        204–209
    compound events, overview, 201–204
    conditional probability, 216–220
    in continuous context, 489–493
    dart board analogy, 209–210
    Exercises, compound events, 213–215
    Exercises, conditional probability, 230–232
    Exercises, events, 198–200
    Exercises, in continuous context, 507–509
    Exercises, population genetics, 253–254
    Exercises, sequential events, 242–245
    expected value and median value, 493–495
    genetics examples, 183–186
    human pedigrees, 233, 234
    independence, 220–225
    Matlab Skills, compound events, 210–212
    Matlab Skills, conditional probability, 225–230
    Matlab Skills, events, 189–198
    Matlab Skills, waiting times, 500–507
    normal distribution, 495–498
    overview of, 175–178
    partition theorem, 233–238
    sample spaces and events, 178–186
    Student Projects, 255–258, 299–301
    waiting times, 498–500
probability density function
    Exercises, 507–509
    expected and median values, 493–495
    overview, 490–492
    waiting times, 499
probability distribution function, 492–493
probability function, 181–186
product of functions, derivatives and, 329–330, 353
product rule, derivatives, 330
proofs
    chain rule, 573–574
    cosine derivative, 577
    exponential functions, derivatives of, 574–575
    power rule, 573
    sine derivative, 575–576
proportional relationships, 31
Punnett square, 184
Punnett, Reginald C., 184

quartiles, 5
queuing theory, 499–500
quotient rule, derivatives, 359–362

radiant flux density, 52–53
radii, Poiseuille's law, 477–482
radioactive decay rate, 365–366
random number generation, 189–191
random samples, probability and, 490–493
random variables
    defined, 210
    Exercises, probability, 507–509
    expected and median values, 493–495
    Matlab Skills, waiting times, 500–507
    normal distributions, 495–498
    waiting times, probability and, 498–500
range, 6, 18–23
ranking data, 5
rate of change. See also derivatives; integration
    allometric properties, 315–316
    average rate of change, 306–308
    average value of a function, 450–453
    biomass, 315
    chain rule, 354–359
    concavity, 385–394
    derivative at a point, defining, 316
    derivatives of exponential functions, 330–334
    drug absorption, 315
    estimation of, 308–309
    Exercises, 320–323
    Holling's functional responses, 359–362
    instantaneous rate of change, 303–304, 308–309
    Matlab Skills, 316–320
    mean value theorem, 382–385
    optimization problems, 394–401
    overview, 305–306
    photosynthesis example, 311–314
    population growth, 315
    velocity, 309–311
rate of cooling, 366–369
rate of flow, Poiseuille's law for, 477, 480–482, 485–488
rate-of-decay, pharmacokinetics, 97–100
rates, as measurements, 261–262
ratio scale, 3
rational functions, 463
real sequence, 86
recessive alleles, 183
reciprocal rule, derivatives, 359–362
rectangles, area estimates and, 417–424, 426–430
rectangles, area of, 415
regression, defined, 31. See also linear regression
related rates, 549–551
relative frequency, histograms, 22–23
rescaling data, log-log and semilog graphs, 55–62
    Matlab Skills, 63–67
revolution, solid of, 477–482
Rh classification. See blood type
rho (ρ), 37–39

richness, diversity measures, 9
Richter scale, 53–54
right-hand limits, 283
rounding, Matlab, 563–564

salinity, ocean water, 485
sample
    defined, 490
    probability and, 490–493
    sample variance, 6–7
sample spaces
    combinations and permutations, 186–188
    overview, 178–181
    partition theorem, 233–238
    probability of an event, 181–186
scalars, 108, 125
scales. See also logarithmic functions
    dispersion, nominal scale data and, 9
    importance of, 14–15
    interval scale, 3
    nominal scale, 4
    ordinal scale, 3–4
    ratio scale, 3
    rescaling, log-log and semilog graphs, 55–62
    Richter scale, 53–54
scatter plots, 23–24, 25–27, 32–33
secant, chain and quotient rules, 362
second derivative test, 391–394
second derivatives, 370–372
semilog graphs, 55–62
sensitivity, probability and, 239–242
separation of variables
    Exercises, 527–528
    Matlab Skills, 522–526
    overview of, 515–522
sequences
    discrete difference equations and, 90–91
    Exercises, 102–106
    Fibonacci sequence, 91
    geometric and arithmetic, 92–93
    limits, 87–89
    linear difference equation, constant coefficient, 93–96
    Matlab Skills, 100–103
    overview, 84–87
sequential events
    Bayes' theorem, 238–242
    Exercises, 242–245
    partition theorem, 233–238
sickle cell anemia, 247–248, 250
sigma, notation of, 571–572
sign chart, 378
Simpson's Index of Diversity (SID), 9, 299–301
sinusoidal (sine) functions, derivatives of
    chain and quotient rules, 361–362
    maxima and minima, 394
    overview, 334–336, 353
    proof of, 575–576
    substitution method of integration, 461

slant asymptote, 380–381
slope, 32, 325–326
smiley test, 392
soil, nitrogen concentration, 490–492, 494–495
solid of revolution, 477–482
solutions, family of, 516–517
species evenness, 9
species richness, 9
specific growth rate, 534–535
specificity, 240–242
speed, 310
square matrices, 131
stability
    Exercises, 541–542
    feedback and, 529–530
    graphing of, 535–539
    limited population growth models, 531–535
    long-term growth rates, 153–163
    overview, 147–149
stable equilibrium, 537
standard deviation, 7, 16, 495–498
Stephens, David, 410
Student Projects
    American bison matrix model, 172–174
    American bison population extinction, 257–258
    biodiversity, limits and continuity, 299–301
    descriptive statistics, 71–77
    differential equations, 555–556
    eye color, human, 255–257
    optimal foraging theory, 410–412
    predator-prey models, 171–172
    probability, 255–258
    trapezoid rule, 510
substitution method of integration, 459–465, 468, 469–470
succession, ecological. *See also* equilibrium
    discrete time modeling, 81–83
    Matlab Skills, matrices, 133–138
    matrix applications, 129–133
    vector dynamics, 112–120
sum rule, derivatives, 328–329, 353
survival rate, 155. *See also* Leslie matrix models
system of discrete difference equations, 172

tangent
    chain and quotient rules, 361–362
    derivative functions and, 325–326
    derivatives of logarithmic functions, 341–345
    equation of tangent line, 327
Tay-Sachs disease
    Hardy-Weinberg selection model, 250–253
    partition theorem and, 237–238
    probability of, 216–218, 223–227
temperature, homeostasis, 539–541
temperature, rate of cooling, 366–369
tendon strain, measures of, 451
theory
    defined, 1

thermoregulation, homeostasis, 539–541
third derivatives, 370–372
time. *See also* integration
    average value of a function, 450–453
    homeostasis, overview of, 530
    photosynthesis rate of change, 311–314
    time of death, 366–367
    velocity and, 309–311
    waiting times, probability and, 498–500
time modeling. *See also* discrete time modeling; equilibrium; Leslie matrix models; matrices; population genetics; vectors
    Matlab Skills, plotting time dynamics, 137–138
    vector dynamics, 112–120
transfer matrix
    defined, 116
    eigenvalue of, 160
    long-term population structures, 163–164
trapezoid, area estimates and, 424–430, 510
trapezoid, area of, 415
trigonometric functions
    continuous functions, 286
    derivatives of, 334–336
tumor growth rate, 534–535, 538

uncorrelated data, 39
uniform distribution
    defined, 493
    Exercises, probability, 507–509
    waiting times, 503
uniform sample space, 181
unimodal data, 22
union, event probability, 202, 207
unstable equilibrium, 537
upwelling, chain rule and, 357–358

variables
    allometry, 54–55
    bivariate data, 31
    dependent variables, defined, 514
    differential equations, variable separation, 515–522
    Exercises, separation of variables, 527–528
    Matlab Skills, separation of variables, 522–526
    random variable, defined, 210
variance
    measures of, 6–7
    normal distributions, 496–498
variation. *See* probability
vectors
    column vectors, 131
    dynamics, 112–120
    eigenvectors, 142–146, 163–164
    equilibrium, 142
    Exercises, 120–122, 138–140
    Matlab Skills, entering and using, 566–568
    Matlab Skills, matrices, 133–138
    matrix addition, 124–125
    matrix applications, 129–133

vectors (*continued*)
   matrix multiplication, 123–128
   overview, 107–110
   vector algebra, 110–112
velocity, 309–311, 326, 477
Venn diagrams, 203
vision, probability and, 176
visual data displays
   bar charts, 16–18
   exercises, 27–29
   frequency distributions, 15–16, 17
   histograms, 18–23, 24–25
   overview, 14–15

rescaling data, log-log and semilog graphs, 55–62
   scatter plots, 23–24
Vogel, Steven, 400
volume of a solid of revolution, 477–482

waiting times, 498–500
   Matlab Skills, 500–507
water, density functions, 482–485
wound healing rate, 367–369

XTick, 25

y-intercept, 32